16th International Symposium on Boron, Borides and Related Materials 2008

Journal of Physics: Conference Series Volume 176

Matsue, Shimane, Japan
7-12 September 2008

ISBN: 978-1-61738-307-6

Printed from e-media with permission by:

Curran Associates, Inc.
57 Morehouse Lane
Red Hook, NY 12571
www.proceedings.com

Some format issues inherent in the e-media version may also appear in this print version.

Printed by Curran Associates, Inc. (2010)

Institute of Physics
Dirac House, Temple Back
Bristol BS1 6BE UK

Phone: 44 1 17 929 7481
Fax: 44 1 17 920 0979

techtracking@iop.org

TABLE OF CONTENTS

This volume of *Journal of Physics: Conference Series* contains invited and contributed peer-reviewed papers that were presented at the 16th International Symposium on Boron, Borides and Related Materials (ISBB 2008), which was held on 7–12 September 2008, at Kunibiki Messe, Matsue, Japan. This triennial symposium has a half-century long history starting from the 1st meeting in 1959 at Asbury Park, New Jersey. We were very pleased to organize ISBB 2008, which gathered chemists, physicists, materials scientists as well as diamond and high-pressure researchers. This meeting had a strong background in the boron-related Japanese research history, which includes the discovery of superconductivity in MgB2 and development of Nd-Fe-B hard magnets and of YB66 soft X-ray monochromator.

The scope of ISBB 2008 spans both basic and applied interdisciplinary research that is centered on boron, borides and related materials, and the collection of articles defines the state of the art in research on these materials. The topics are centered on:

1. Preparation of new materials (single crystals, thin films, nanostructures, ceramics, etc) under normal or extreme conditions.

2. Crystal structure and chemical bonding (new crystal structures, nonstoichiometry, defects, clusters, quantum-chemical calculations).

3. Physical and chemical properties (band structure, phonon spectra, superconductivity; optical, electrical, magnetic, emissive, mechanical properties; phase diagrams, thermodynamics, catalytic activity, etc) in a wide range of temperatures and pressures.

4. Applications and prospects (thermoelectric converters, composites, ceramics, coatings, etc)

There were a few discoveries of new materials, such as nanomaterials, and developments in applications. Many contributions were related to 4*f* heavy Fermion systems of rare-earth borides. Exotic mechanisms of magnetism and Kondo effects have been discussed, which may indicate another direction of development of boride. Two special sessions, 'Boron chemistry' and 'Superconductivity', were also held at the symposium. The session on Boron chemistry was planned to honor the scientific work in boron chemistry of Professor J Bauer on the occasion of his retirement. Many recent results were discussed in the session, and Professor Bauer himself introduced novel rare-earth–boron–carbon compounds $RE_{10}B_7C_{10}$ (RE = Gd - Er) in his lecture. In the latter session, on the basis of recent discoveries of superconductivity in MgB_2 and in β-boron under high pressure, the superconductivity of boron and related materials was discussed and the superconductivity of boron-doped diamond was also addressed.

More than 120 participants from 16 countries attended the ISBB 2008, and active presentations (22 invited, 33 oral and 68 posters) and discussions suggest that research on boron and borides is entering a new phase of development. This volume contains 46 articles from 52 submitted manuscripts. The reviewers were invited not only from symposium participants but also from specialists worldwide, and they did a great job of evaluating and commenting on the submitted manuscripts to maintain the highest quality standard of this volume.

Recent discoveries of superconductivity in boron under high pressure, synthesis of a new allotrope of boron and of various boron and boride nanostructures will lead this highly interdisciplinary field of science, which will further grow and gain attention in terms of both basic and applied research. In this context, we are very much looking forward to the next symposium, which will be held in Istanbul, Turkey, in 2011, organized by Professor Onuralp Yucel, Istanbul Technical University. Turkey currently has the world highest share of borate production and is expected to be involved more in boron-related research.

Acknowledgements
We gratefully acknowledge the style improvement by Dr K Iakoubovskii, and sincerely thank Shimane Prefecture and Matsue City for their financial support. The symposium was also supported by Tokyo University of Science, Suwa and foundations including, the Kajima Foundation, Foundation for Promotion of Material Science and Technology of Japan and Nippon Sheet Glass Foundation for Materials Science and Engineering, as well as companies including JFE Steel Corporation, Shincron Co, Ltd, Toyo Kohan Co, Ltd, Fukuda Metal Foil and Powder Co, Ltd, Japan New Metals Co, Ltd, H C Starck Ltd and Fritsch Japan Co, Ltd.

Editors

Chair
Takaho Tanaka (National Institute for Materials Science, Japan)

Vice chairs
Koun Shirai (Osaka University, Japan)
Kaoru Kimura (The University of Tokyo, Japan)
Ken-ichi Takagi (Tokyo City University, Japan)
Touetsu Shishido (Tohoku University, Japan)
Shigeru Okada (Kokushikan University)
Hideaki Itoh (Nagoya University,Japan)
Katsumitsu Nakamura (Nihon University, Japan)

Organizing committee of ISBB 2008

K Takagi *Chairman* (Tokyo City University)

T Tanaka *Program Committee Chairman* (National Institute for Materials Science)

K Kimura *Secretary* (The University of Tokyo)

J Akimitsu (Aoyama University)

H Itoh (Nagoya University)

K Nakamura (Nihon University)

K Nishiyama (Tokyo University of Science, Suwa)

S Okada (Kokushikan University)

K Shirai (Osaka University)

T Shishido (Tohoku University)

K Soga (Tokyo University of Science)

M Takeda (Nagaoka University of Technolo

Y Yamazaki (Toyo Kohan Co, Ltd)

International Scientific Committee 0f ISBB (2008-2011)

K Takagi *Chairman* (Japan)

B Albert (Germany)

M Antadze (Georgia)

J Bauer (France)

I Boustani (Germany)

D Emin (USA)

M Engler (Germany)

N Frage (Israel)

Yu Grin (Germany)

V N Gurin (Russia)

J-F Halet (France)

H Hillebrecht (Germany)

W Jung (Germany)

K Kimura (Japan)

T Mori (Japan)

P D Ownby (USA)

P Rogl (Austria)

S Shalamberidze (Georgia)

N Shitsevalova (Ukraine)

M Takeda (Japan)

T Tanaka (Japan)

R Telle (Germany)

M Trenary (USA)

O Tsagareishvili (Georgia)

H Werheit (Germany)

G Will (Germany)

O Yucel (Turkey)

G Zhang (China)

All papers published in this volume of *Journal of Physics: Conference Series* have been peer reviewed through processes administered by the proceedings Editors. Reviews were conducted by expert referees to the professional and scientific standards expected of a proceedings journal published by IOP Publishing.

Superconductivity research on boron solids and an efficient doping method

Koun Shirai, Haruhiko Dekura and Akira Masago[†]

Nanoscience and Nanotechnology Center, ISIR, Osaka University, 8-1 Mihogaoka, Ibaraki, Osaka, Japan

[†] Department of Materials Engineering, University of Tokyo, Hongo 7-3-1, Bunkyo-ku, Tokyo, 113-8656, Japan

E-mail: koun@sanken.osaka-u.ac.jp

Abstract. Boron solids are, similar to diamond, strong electron-phonon coupling systems and therefore potentially have high-T_c superconductivity. For realizing high transition temperature T_c, however, careful control of the valence electrons is necessary, which requires detailed understanding of the electronic structure. By critically reviewing recent development in this field, along with the situation in heavily doped diamond and clathrates, we show a promising route for high T_c in boron system. Although both α- and β-boron become metallic and exhibit superconductivity at high pressure, the strong covalent character still remains, which significantly limits its T_c achieved by applying high pressure. Heavy doping of related compounds is desirable, as long as the basic electronic structure is retained, but it is extremely difficult at normal conditions. A promising method to achieve this is high pressure synthesis. Similar technique is useful for heavy doping of diamond. To design high-pressure experiment, the role of theory is important through predicting the phase diagram, which was not available for boron previously.

1. Introduction

It was believed that superconductivity of phonon mediated mechanisms could not exceed 30 K in its critical temperature T_c; it was only possible in cuprates. Now this old belief must be completely discarded. Many high-T_c materials such as MgB_2 and K_3C_{60} have been discovered, although their superconductivity is most likely described by the phonon-mediated mechanisms [1, 2]. In this respect, boron and diamond systems are particularly interesting, because they exhibit strong electron-phonon coupling. For boron case[1], the superconductivity was once discussed more than twenty years ago in terms of small bi-polaron mechanism [3]. Unfortunately, the superconductivity of boron was not discovered in this way.

Big breakthroughs for superconductivity have occurred after 2000 in different context. The first and probably most influential one is discovery of MgB_2 by the group of Akimitsu [4]. The superconductivity mechanism of MgB_2 is now understood within the context of classical Bardeen-Cooper-Schrieffer (BCS) theory in terms of phonon-mediated mechanism. However, the case of MgB_2 is in many respects beyond our expectations based on empirical rules which were useful for searching candidates before [5]. The second breakthrough was discovery of the superconductivity in β-rhombohedral boron at high pressure by Eremets *et al* [6]. Hereafter,

[1] By using a word boron, the authors intend to include boron-rich solid such as boron carbides.

we abbreviate β-rhombohedral boron to β-boron, unless there is a risk to confuse. It is known that most semiconductors undergo phase transitions at high pressures and many of them exhibit superconductivity. Eremets *et al* showed that boron is not an exception, although the transition pressure is extremely high ~160 GPa. The third breakthrough is discovery of superconductivity in heavily doped semiconductors. It was first found for B-doped diamond by Ekimov *et al* [7] and then for B-doped silicon [8] and B-doped SiC [9]. Further discoveries followed in doped InN [10] and AlN [11]. The significance of these findings is that superconductivity appears even at normal pressure, which is quite appealing for applications.

Regarding the structure of the superconducting phase of β-boron, there are many controversies; some experimentalists found a structural transition to amorphous phase [12], some found a transition to a tetragonal phase [13]. Theoretically, α-Ga type was suggested [14, 15]. When the present authors started the structural study, it was surprising that there was no phase diagram available for boron; therefore, we decided to calculate it. According to our phase diagram, α-boron is the stable phase at low temperatures and high pressures [16, 17].

Seeing the stability of α-boron at high pressure, along with other reasons, the present authors and Japanese experimentalists started a collaboration on superconductivity of α-boron. Very recently, the superconductivity has been found for α-boron at high pressure [18]. Although the transition pressure and the critical temperature of α-boron are almost the same as those of β-boron, this discovery is not just a duplication of the results for β-boron. The crystal structure including the icosahedral unit is proven to survive even after the superconducting transition; the rhombohedral structure is retained up to 200 GPa [19]. This stability is consistent with our prediction of the phase diagram. To our knowledge, it is the first time that a semiconductor exhibits superconductivity at high pressure without any structural transformation.

By analyzing the electronic structure, we convinced ourselves that the present T_c (below 10 K) is by no means the upper bound which boron can have. In fact, for boron and diamond systems, there are many predictions of high T_c around 40 K. If the electronic structure is carefully controlled, in particular the Fermi level, it is possible to obtain higher T_c. A promising method of controlling the Fermi level is doping. Doping idea was proposed more than ten years ago [20, 21, 22]. However, doping α-boron is difficult in experiment. In order to overcome this difficulty, the present authors have proposed an efficient doping method using high pressure [23, 24].

Looking at the case of heavily B-doped diamond, we notice that a similar technique of high pressure was used. Actually, superconductivity in B-doped diamond was demonstrated on samples produced by high-pressure and high-temperature (HPHT) method. Unfortunately, this technique is unavailable to many experimentalists. Once the superconductivity was found for films synthesized by chemical vapor deposition (CVD) [25], most researchers started using CVD method, and the T_c value was increased. However, the increase in the concentration of B and in T_c has almost saturated around several at.% of B and $T_c \sim 10$ K. Apparently, a breakthrough over this technical difficulty is needed. Recently, the present authors have demonstrated the effectiveness of high-pressure synthesis for B-doped diamond [26, 27].

The above development shows that no exotic mechanism is needed to obtain high T_c, but instead, new ideas of compounds or new technologies such as high pressure and heavily doping are important. In this time of International Symposium on Boron, Borides and Related Materials, it was timely to organize a special session focusing on the superconductivity of boron solids. It is a good opportunity to share the establishments of this field in the last decade. However, retrospective reviewing is not the most important purpose of this session. Much effort should be focused on further development of material research. Questions of how to control these electronic structures and how synthesize such promising materials are of central importance. This view point is emphasized throughout this article. We apologize in advance for frequent referring to unpublished works or informal sources of discussion, because the research is ongoing.

The present study contains our proposed answers to the above question. In the first part of this paper, we review recent development of the structural study and the electronic structure of boron at high pressures. High-pressure research has provided many new insights into physics of boron in the last decade, since an interesting study on the pressure dependence of Raman spectra of α-boron [28]. This part gives the background ideas upon which we have conceived the use of high-pressure process for heavy doping. Since many features are common in boron and in B-doped diamond, the basic strategy for high-T_c semiconductors is described by treating both cases in parallel, in Sec. 3. The second part (Sec. 4) is the most important part of this paper, which describes our proposal for heavy doping of boron and diamond. By the nature of material research, the subjects treated here are encompassed from the basic properties of materials to synthesis, and hence summary is provided at the end of each section.

2. High-pressure structural study

2.1. Crystal structures

We now know that the superconducting behaviors of α- and β-boron are almost the same with respect to T_c and the transition pressure [6, 18]. The superconductivity of boron seems insensitive to the details of crystal structure. Does it mean that the role of structural study of boron is not important for the superconductivity research? The answer is definitely no. In the following, we will see that theories based on incorrect or hypothetical structures eventually lead to totally wrong understanding, even though there is a fortuitous agreement with experiment, when we focus on a specific property only. Structural study plays a central role in superconductivity research on boron solids, because boron has many complicated polymorphic modifications. However, many theoretical studies treated only simple structures such as fcc or bcc [29]. A simple reason for this is that computation is easy. Another reason might be a belief that materials favor close-packed structures, which is simple, at high p. Now, we must admit that the latter believe is wrong for boron at least up to 200 GPa. A recent experiment shows stability of α-boron up to 200 GPa [19]. Although a phase transition of α-boron to α-Ga type structure was predicted [30], there is experimentally no indication of such a transition. Therefore, a prediction of T_c based on this structure does not reflect the actual situation, although it is theoretically interesting.

For β-boron, the situation is more complicated. Since Eremets first demonstrated superconductivity [6], experiments were mainly carried out on this phase. Sanz *et al* found a structural transition to amorphous phase at $p > 100$ GPa at room temperature [12]. Irreversible transformation to amorphous phase was observed in clathrate compounds: Ba_8Si_{46} (at 40 GPa) [31] and K_8Si_{46} (32 GPa), etc. Sanz *et al* also reported that another rhombohedral structure appeared at $p > 20$ GPa under nonhydrostatic conditions. Ma *et al* found a transition to a tetragonal phase [13] at $p > 10$ GPa and $T > 1500$ K. There is no consensus about the structure when superconductivity takes place.

Step features were observed in the pressure dependence of the resistivity of β-boron [6] at 30 and 130 GPa. These steps were interpreted as a sign of phase transitions. Sanz *et al* suggested that the step at 30 GPa corresponds to the transition to another rhombohedral structure [12]. Haussermann *et al* suggested that the step at 130 GPa is due to the band closure of β-boron, while the step at 30 GPa is due to a transition to α-boron [30]. Shimizu group found similar step structures in the resistivity of α-boron [18]. On interpreting the step structure for α-boron, the present authors attribute the step at 50 GPa to a subtle structural change, even though the crystal structure is maintained. This issue is described later.

When the authors started research on boron solids, they were surprised by absence of boron-related phase diagrams. Although α-boron is obtained commonly at relatively low temperatures, it was widely believed that β-boron is the most stable phase over the whole temperature range [32]. After completing first-principles calculations on the total energy, along with other thermodynamic functions, we found that α-boron is the ground state of boron, while β-boron is stable only at high temperatures [16]. The transition temperature $T_{\alpha\beta}$ is calculated as about 1000 K at $p = 0$. The mechanism underlying this difference is simple - stiff materials are stable at low T, while soft materials can be stable at high T because of contribution of entropy. The situation is very much like that of diamond-graphite system. It is a common observation that the dense and stiff phase (diamond) is stable at low T and at high p, while the dilute and soft phase (graphite) is stable at high T and low p.

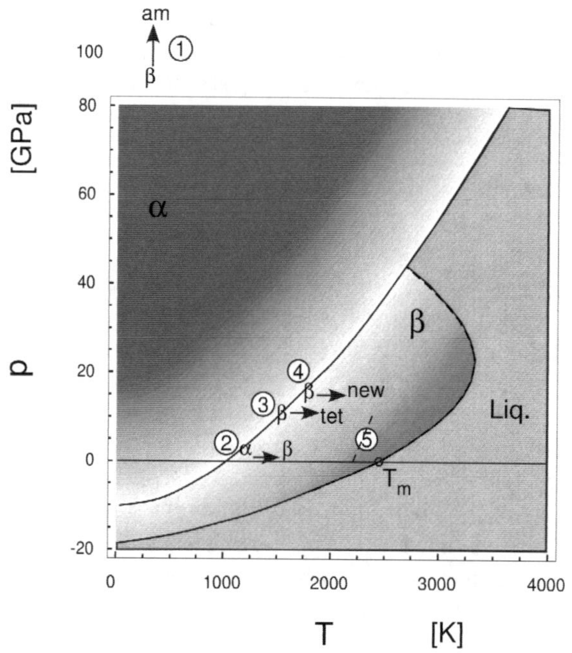

Figure 1. Phase diagram of boron. The calculated phase diagram is taken from Ref. [17]. Experimental data are compared; 1: Sanz *et al* [12], 2: Runow [33], 3: Ma *et al* [13], 4: Wentorf [34], and 5: melting curve by Brazhkin *et al* [35].

After that work, similar conclusion followed by others [36]. On the other hand, there are still debates on the stability of α- and β-boron. Widom and Mihalkovič [37] showed that β phase becomes stable even at T =0 K, if all the defects which were reported by Slack *et al* [38] are taken into account. Further studied are in progress [39, 40]. In our study, even one kind of defect causes the stability of β phase, which significantly reduces the transition temperature $T_{\alpha\beta}$ down to about 200 K [16]. The effect of defects is important to this extent. In this regard, examining defects in α-boron is interesting, because α-boron is so far considered as perfect occupancy system. The first study on this subject is made by us [41], with a tentative conclusion that defects do not occur in α-boron at such concentrations that thermodynamical stability is significantly influenced.

Although still uncertain, the conclusion that α phase is the ground state seems reasonable, considering the experimental facts. Crystals of α phase are obtained in molten metal, where the eutectic temperature is as low as 830 °C for B-Pt system [42] and similar for Pd-B system[43]. Molten metal synthesis is close to thermodynamic equilibrium. Kimura's group showed that, by starting from amorphous boron, α phase is obtained around 1200°C [44]. Below that temperature, no crystallization occurred. If β phase is stable, β phase should be grown in these conditions, because there is no special conditions which favor the α phase.

Based on the ground-state energy calculations along with phonon calculations, a phase diagram of boron was calculated over full $p - T$ range [17]. Anharmonic effects are taken into account within quasiharmonic approximation. The calculated phase diagram is shown in figure 1. This is the first prediction of the phase diagram of boron. Only α- and β-boron phases are considered. Although there are other polymorphs, the authors think that these two are representative of boron in many respects. For example, atomic density is the highest for the α and lowest for the β phase, see Table I of Ref. [45]. The reader should understand this diagram as providing the limiting cases; other phases, if present, could appear in an intermediated region.

Let us check experimental situation on the phase diagram. So far, no systematic experiment has been performed. Only small pieces of data are available in scattered $p - T$ points. Runow observed a transition from α to β phase at an estimated temperature of 1370°C [33]. Although the lattice spacing was perfect, there was a deficiency of 9 atoms for β-boron. Starting from a mixture of β phase and amorphous phase, Wentorf was able to synthesize a new structure of pure boron [34]. The density (2.46-2.52 g/cm^3) was recorded as the highest among known boron polymorphs. Very recently, Dubrovinsky has rediscovered and characterized this phase [46].

The equations of states have been examined for various polymorphs; in a range 0-10 GPa [47], 0-100 GPa [12] and 0-30 GPa at elevated T [13] for β-boron, 0-100 GPa [48, 49] for α-boron, and 0-11 GPa [50] for boron carbide. Murnaghan's equation-of-states [51]

$$\frac{V(p)}{V_0} = \left(1 + \frac{B_1}{B_0} p\right)^{-1/B_1} \tag{1}$$

fits well to all cases. The thus obtained bulk modulus B_0 at $p = 0$ and its pressure derivative B_1 are in good agreement with calculation. Here, we give typical values for later discussion; $B_0 = 210 - 240$ GPa and $B_1 = 4$.

Recently, a systematic examination of the transition between α and β phases up to 50 GPa has been conducted by Mori *et al* [52]. A difficulty experienced by them is that, as usual in high-temperature experiments, reactions with unwanted parts of the apparatus sometimes take place, which appear as unknown peaks in X-ray diffraction study. Historically, such unknown peaks have been sometimes recognized since Decker and Kasper reported the first the crystal structure of α-boron [53].

This prediction of phase diagram is also useful for obtaining good-quality crystals of α-boron, which is very important for the superconductivity application. One way is to use high-temperature and high-pressure method. Starting with $p > 50$ GPa and $T > 2000$ K, crystal of α-boron can be grown directly from the liquid phase. After calculating this phase diagram, the authors came to know an experiment on the melting of boron [35]. As shown in the figure, there is a large difference in the melting data. In Ref. [17], the melting curve was calculated from existing data on the free volume and the latent heat of fusion, on assumption that Eq. (1) is valid for both solid and liquid phases. After publishing paper [17], we came to know an experimental data for the melting curve of boron at high pressure [35], which is shown in this figure. There is a large discrepancy, in particular, for the initial gradient dp/dT. The data of the free volume and the latent heat of fusion in our calculation are largely different from data of [35].

In summarizing this section, we are able to delineate the phase diagram of boron. At low T and high p, α-boron is the most stable phase. Although there are still uncertainties in density functional theory (DFT) calculations, this conclusion seems reasonable. Indeed, it is otherwise difficult to interpret the crystal growth of α-boron at low temperatures. For β-boron, phase transitions certainly occurs at high pressures, because β-boron is thermodynamically unstable there. However, what structure appears around the superconducting transition is still an open question. A good news for superconductivity research is that α-boron exhibits superconductivity, and that the original structure is retained even after the superconducting transition, as indicated by the calculated phase diagram.

2.2. Electronic properties

Numerous theoretical studies were published immediately after the discovery of superconductivity of β-boron, even though there had already been many published papers on the electronic structures of boron solids [54, 55, 56, 57, 29, 58]. At early stage, calculations were performed only on simple structures, such as α-boron [59, 60] or even simpler hypothetical structures [61, 15, 30, 62, 63]. Later, calculations on β-boron appeared [64, 16, 37]. All calculations after 1990 are modern DFT calculations. Here, we focus mainly on the pressure dependence of the electronic structure. Studies on the superconductivity are described in the next section. Discussion on hypothetical structure of α-Ga type is omitted here, for the reason described earlier.

For α-boron, the change of the electronic structure at high pressure has been studied by Zhao and Lu [59] using the generalized gradient approximation (GGA) as the electron-correlation part. Evidently, the intention of this article is to explain the metallization of boron which was experimentally shown for β-boron [6]. However, the main results they show are those on α-boron, because of computational costs. They show that α-boron is stable up to 270 GPa and that the band gap continuously decreases to zero around 160 GPa. Owing to the similarities of α- and β-boron, it is inferred that the band-gap closure occurs also for β-boron at similar pressure. We now know that α-boron indeed retains its rhombohedral structure up to 200 GPa [19]. Then, the work of Zhao and Lu [59] is really a good reference for α-boron. Even though the gap value is somewhat ambiguous due to a problem inherent to DFT, overall behavior is consistent with the experiment. In α-boron, at zero pressure, an icosahedral unit is bonded to six nearest neighboring icosahedral units; these strong bonds of the length 1.71 Å bring covalent character to this crystal. In the second-nearest neighbors, six icosahedra are connected in the *ab* plane. These bonds are the so-called three-center bonds; they are very weak with the bond length of 2.03 Å. These three-center bonds are significantly enhanced at high pressures and eventually form a high coordination number around an icosahedron. High coordination numbers are a typical feature of metals.

We have also confirmed this band closure and the structural change by our own calculations. The three-center bond has its chief component in the conduction bands, and when this bond is shortened, the bottom of the conduction band eventually merges with the top of the valence band. It should be noted that even though the electronic structure turns to be metallic, there is yet a remnant of semiconducting origin at a very small density-of-state (DOS) at the Fermi level, which qualitatively separates the parts of valence and conduction bands. The metallic phase of α-boron (probably β-boron too) has both characters of metal and semiconductor. Metals possessing such a mixed character are known, for example, in WC [65, 66] and in α-Ga [67].

Although we are satisfied with this explanation, our calculations resolve further interesting features. Actually, the way of deforming the lattice frame and the icosahedral units are contrasted between α-boron and β-boron [19]. This fact itself was found in model calculation on the elastic response of boron solids [68]; see a review [45]. Among α-, β-boron and boron carbides, the apex angle $\alpha_{\rm rh}$ of the rhombohedral lattice is slightly deviated from 60°; just 60° transforms the rhombohedral Bravais lattice to fcc. The deviation $\Delta\alpha_{\rm rh}$ from 60° is negative for α-boron, while it is positive for β-boron and boron carbides. In all these cases, the pressure dependence of $\Delta\alpha_{\rm rh}$ is such that the deviation increases with increasing pressure. Increase in $\Delta\alpha_{\rm rh}$ means crystal anisotropy, which implies more covalent character for boron solids. In this sense, the covalent character of boron solids is initially increased on applying pressure. But, there must be some point at which this change ceases. After this point, metallic character is expected to increase. For α-boron, this turning point is about 50 GPa, which was found in a collaboration of theory and experiment [19]. This pressure is well correlated to a small change in the Raman spectrum. The librational mode of α-boron has many exceptional features. One of them is almost no pressure dependence of the vibration frequency [28]. However, for $p > 50$

GPa, slight lowering of frequency with increasing pressure is observed [19, 48]. The frequency of the librational mode ω_l is governed by angle forces, which arise from strongly oriented character of the covalent bonds [69]. Forming high coordination number reduces the covalent character, and hence angle force and ω_l decrease.

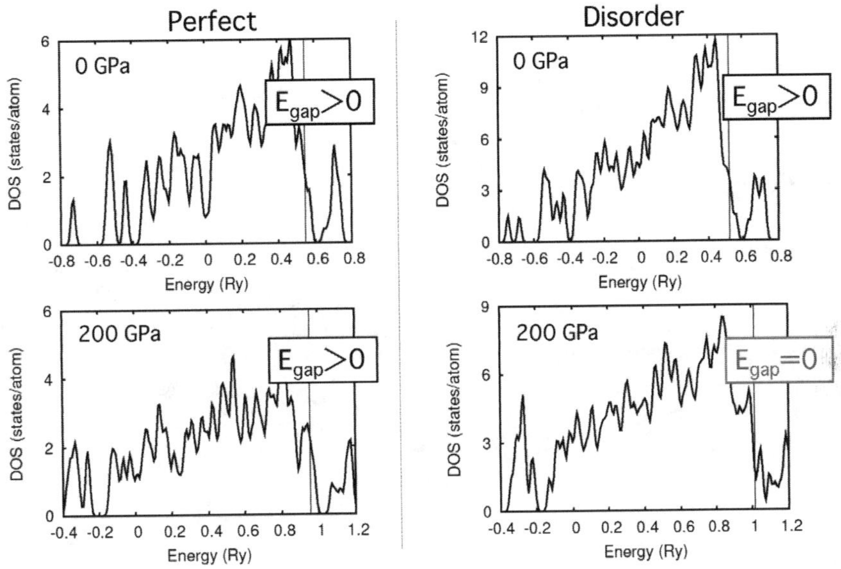

Figure 2. Energy gap of β-boron with and without defects.

In the electrical resistivity of α-boron, Kaneshige *et al* [18] found step structures in the pressure dependence, in a way similar to β-boron. A step appears around 50 GPa. As discussed above, this feature is not due to phase transition but it should be ascribed to some change within the rhombohedral structure of α-boron. Accurate band calculations show that turning from the covalent to metallic character, although still semiconductor, makes a change in the pressure dependence of the energy gap. The change in the energy gap alters the pressure dependence of resistivity, which is observed in experiment. The details will be given in a forthcoming paper [70].

For β-boron, effect of defects is problematic. One time of band calculation is manageable in the present status of computational technology, partly shown in [59]. However, we now know that β-boron has a variety of defects. The presence of defects significantly alters the stability of crystal [16, 37]. There is a report demonstrating softening of phonons at high T [36]. Of course, it modifies the gap states too. Here, in figure 2, we show our preliminary results of DOS structure of β-boron, which indicate how strongly the defects influence the gap state. Because there are so many configurations of defects, no conclusive results have been drawn concerning β-boron with defects.

We summarize the metallization process of α-boron at high pressure. If we judge the metallization by closure of the band gap, the transition is obviously abrupt around 150 GPa. However, the change in the electronic structure is continuous as indicated by band calculations, and hence it is difficult (and maybe meaningless) to say when metallization begins. In spite

Table 1. Prediction of T_c and the related parameters, $N(E_F)$, $\langle\omega\rangle$, λ, and μ^*. Crystals are conveniently grouped into three. The first and second groups are boron crystals including boron carbide; the first group gathers hypothetical structures, while the second real systems of icosahedron-based polymorphs. The last group gathers other related materials. Pressure p is given in GPa.

structure (ref.)	$N(E_F)$ 1/eV·atom	$\langle\omega\rangle$ meV	λ	μ^*	T_c K	notes
boron (Papaconstantopoulos and Mehl [61])						
fcc	0.15	107	0.39-0.53	0.09-0.13	5-15	$p=307$
(Häussermann *et al* [30])						
α-Ga	0.02					$p=74$
(Ma *et al* [62])						
α-Ga	0.22	124	0.38	0.12	6	$p=160$
(Bose *et al* [63])						
bct		97	1.02	0.17	>50	high p
fcc		136	0.84	0.17	>50	high p
boron carbide (Calandra *et al* [72])						
B_{12}(CBC)	0.24	65.8	0.81	0.1	36.7	
α-boron (present work)						
B_{12}	0.03					$p=200$
B-doped diamond (Xiang *et al* [73])						
diamond	0.03	105	0.39	0.10	4.4	B: 2.8 %
BC_5 (Calandra *et al* [74])						
diamond ?	0.12	67.4	0.89	0.1	45	B: 16.7 %
MgB_2 (An and Pickett [75] and Kortus *et al* [76])						
hex	0.19	62	1.0	0.1-0.15	32-46	$N(E_F)$ per B
Li_xBC (Rosner *et al* [71])						
hex	0.23	58	1.5	0.09	>40	$x = 0.5$

of this, we may regard 50 GPa as an estimate for the beginning of metallization. The signs of transition are: (i) the change in pressure dependence of the apex angle $\Delta\alpha_{\rm rh}$, which indicates a transition from anisotropic to isotropic deformations; (ii) slight softening in the librational mode, which suggests weakening of covalent character; (iii) the step structure in the resistivity, which is a consequence of the change in the energy gap.

3. Idea for high T_c materials

After the discovery of superconductivity of β-boron, many theoretical studies on T_c have been published on boron and related solids [61, 71, 72, 62, 63]. Indeed, the predicted T_c is scattered from 5 [61] to more than 50 K [63]. Obviously, listing of such a widely spread numbers alone does not make sense for prediction.

Combination of DFT calculation and Eliashberg theory makes theoretical prediction of T_c more reliable for simple metals and semiconductors; this is particularly true for the electron-phonon coupling parameter λ. A difficult problem is evaluation of the pseudopotential parameter μ^*, which presents the retarded Coulomb repulsion. A common practice is fitting to the experimental T_c. This practice is particularly effective when the pressure dependence of T_c is known [77, 78]. In this sense, this approach is not regarded as first-principles. Recently, there was an attempt to calculate T_c without explicitly using these parameters λ and μ^*. Gross *et al* have succeeded to formulate the ground state of superconductivity within the framework of DFT by introducing an additional variable; the order parameter is taken as a fundamental variable; see a review [79]. The agreement with experiment is profound, 34.1 K for MgB_2 [80]. However, the accuracy depends on the electron correlation functional, which is formally present but its concrete form is not known.

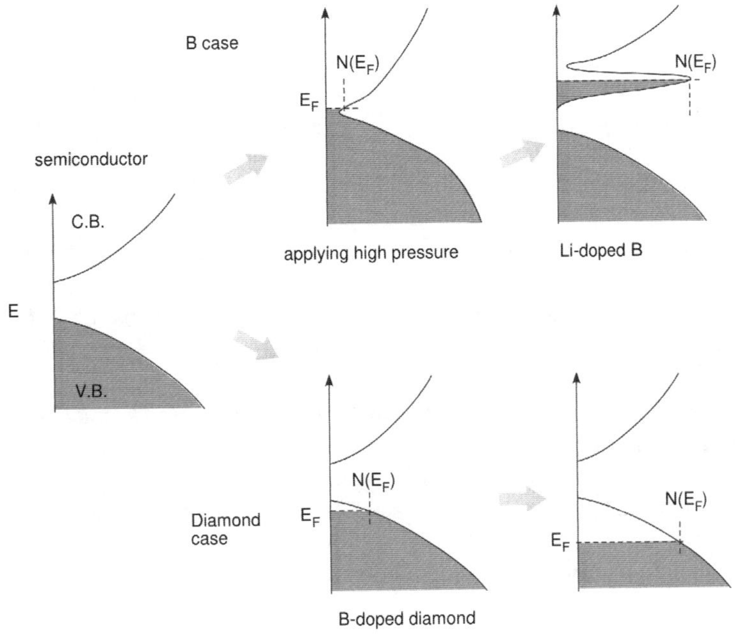

Figure 3. Two ways of valence control of semiconductors. Starting from the normal state of a semiconductor (left figure), applying pressure causes band closure (middle upper). For B-doped diamond, heavy doping yields Fermi surface at the top of the valence band (middle lower). Further doping increases $N(E_{\mathrm{F}})$ (right).

In this way, accurate prediction of T_c is still difficult. Nevertheless, the authors believe that if we take enough care for the structure, we can find a coherent understanding from those scattered predictions on T_c. In Table 1, we compare those scattered T_c for various boron crystals and related materials. The data are grouped according to the structures. Some rows are added for comparison, even though they supply the density of state $N(E_F)$ only. Readers should be aware that only a selected set of data is picked out from the original reference, and that calculation approaches are different from author to author. For example, Ma *et al* deduced μ^* in order to match T_c [62], so by construction, the T_c agrees with the experiment. Since the present authors are not specialists in T_c calculation, assessment of these results is not our intention. The list is of course incomplete; there are references missing in this table, for example, the superconductivity of dodecaboride such as ZrB_{12} [81]. Our aim here is an insight into design of high-T_c materials among boron-rich solids. By classifying the predicted T_c according to the structures, we see some trends in T_c.

First, we note that the conditions of the prediction by Bose *et al* [63] are unrealistic, although challenging. Their calculations assume simple structures. The listed values in Table 1 are those for the atom volume $\Omega_{\mathrm{atom}} = 20$ Bohr^3. Although they do not show the corresponding pressure, we can estimate it using Murnaghan's equation of state (1): $p = 730$ GPa for bct and 660 GPa for fcc structure. Here, we used data of Ref. [29] for B_0 and V_0, while B_1 =4 is assumed. Both pressures are out of our currently accessible region.

Next, we see that for simple metal structures $N(E_F)$ is large, more than 0.1 /eV·atom, in all cases investigated. On the other hand, for semiconductors, which include α-boron and diamond, $N(E_F)$ is smaller by one order of magnitude, even though these crystals become metallic by applying pressure or doping. The case of α-Ga type boron is intermediate between metal and semiconductor. As stated before, α-boron becomes metallic above 150 GPa, but it still has the semiconducting character. This character is reflected in the smallness of $N(E_F)$.

Boron carbide case is different. Viewing boron carbide as C-doped boron may be wrong, as the phonon spectrum suggested [82]. Rather, strong C bonds seem to exhibit their own bonding environment. Based on the standard band calculation, it is concluded that boron carbide $B_{13}C_2$ is a metal from the beginning, for an obvious reason of the odd number of valence electrons. Hence, large $N(E_F)$ is obtained as expected. In spite of this, experiment shows the semiconducting character. There are arguments concerning this problem [83, 84, 85, 86]. Probably the disorder which is commonly observed in this system is a cause of the insulating character [87, 88, 89, 90]. A related reason may be Mott-Hubbard insulator caused by electron correlation [91, 92]. So far, no consensus has been established for the electronic nature of boron carbides.

Anyway, the semiconducting property of boron carbide is an experimental fact. Hence, it is no surprise to see that superconductivity has not been discovered for boron carbide at normal pressure, although numerous efforts were dedicated to search superconductivity in this material [93]. High-pressure technology may provide useful insights into boron carbides too. An unusual positive pressure dependence of the resistivity was found up to 2.5 GPa [85]. Further examination of high-pressure properties of boron carbides is interesting, which is now in progress by Japanese collaborators.

From the above consideration, we can summarize what happens around boron crystals at high pressure and in heavily doped diamond. Figure 3 illustrates the situations of both semiconductors. For α-boron case, by applying pressure, the band gap is closed. But, its strong covalent character still remains in the shape of DOS. Overall separation of the valence and conduction bands is still present, and thereby the Fermi level is located close to the minimum point in DOS. Accordingly, T_c does not increase as expected. For obtaining higher T_c, moving the Fermi level E_F deeply inside the conduction or valence band is required. This can be done by doping.

The situation is similar for heavily B-doped diamond. Upon heavy doping, the Fermi level is shifted to the top of the valence band [94]. Even though now T_c reaches 10 K [95], the Fermi level E_F is still just under the top of valence band [94]. Hence, if we push down E_F deeply into the valence band, we could obtain high T_c. In actual situation of diamond experiment, of course, not all the B atoms can enter at the substitutional site. Some part of B atoms form dimers or may go to interstitial site, which are electrically inactive [96, 97]. On the other hand, there are arguments that in some cases the disorder enhances T_c [98, 99]. We will not be concerned too much with this problem here. Our standpoint is that disorder is not good for superconductivity. Very recently, Calandra and Mauri have shown an interesting calculation that cubic BC_5 (actually slightly distorted) has a deep E_F in the valence band [74]. The predicted T_c is above 40 K. This shows a good solution to avoid the effect of disorder in impurity systems.

Lastly, we mention a technical issue in calculating E_F, which is important to predict T_c. In many cases, doping level significantly modifies E_F, so that accurate determination of E_F is crucial. In calculating Fermi surface, fine mesh of Brillouin zone and interpolation method are used. However, interpolation sometimes causes spurious components, in particular for those materials possessing nesting Fermi surface. Sophisticated algorithms utilizing symmetry are needed. A utility program TSPACE developed by Yanase is such a powerful tool for doing this

[100]. The present authors and Yanase have developed an improved scheme of band interpolation specialized to obtaining Fermi surfaces [101]. An improved algorithm shows a beautiful Fermi surface for Li-doped α-boron [101].

We have seen that, for both boron and diamond, the basic strategy for obtaining high T_c is doping as heavily as possible. It is recommended to use stoichiometric compounds like BC_5, as long as such a compound retains the basic electronic structure of the parent crystal. Then, any problem caused by disorder would be avoided. Li-doped α-boron (LiB_{12}), which we propose here, actually belongs to this category, even though we often refer this as impurity system. Similar usage of terminology is made in doped fullerides [102], in which the highest T_c is now recorded as 38 K for Cs_3C_{60} [103]. Clathrate compounds such as Ba_8Si_{46} are called doped-clathrates, and they exhibit superconductivity [104]. It is unclear whether or not boron carbide belongs to this category. Examination of superconductivity at high pressure is very important.

4. High pressure doping method

Doping idea itself is very common, and anyone can conceive this. A practical problem is how to achieve heavy doping of a material under consideration. In usual semiconductor processes, the doping level is far less than 1 at.%. Contrary to this situation, in superconductor research, heavy doping above 1 at.% is required. Efficient doping method is indispensable.

In boron case, this problem is crucial. Doping of various metallic elements to β-boron has been investigated over the years [105]. However, doping of impurities yields gap states, which hinder effective doping. In this respect, α-boron is a good candidate for doping, as stated earlier. Although it was predicted that Li doping to α-boron is easy from energetics consideration [21, 22, 106], this doping has not been succeeded experimentally [44]. Something must be wrong with these predictions.

We have revisited this issue of Li doping to α-boron in a more systematic way, and have found that it is difficult to dope Li at normal conditions [23, 24], contrary to previous studies. Further investigation led us to find an effective method of doping Li. The method is utilizing high pressure [23, 24, 107]. By analyzing our result, we have established the principal mechanism of high-pressure method for doping [26].

Before proceeding to discussion on the efficient doping method, we would like to appreciate a pioneering work of Gunji and Kamimura more than a decade ago [20, 21] to find Li as a good dopant among various impurities. We have confirmed the suitability of Li doping which they found, except for the problem of formation energy. Among various impurities, Li is ideal in that the change in the electronic structure is best described by rigid-band shift with a quite high $N(E_F) = 0.28$ /eV·atom [24]. The rigid-band shift due to doping is sometimes referred to, but actually is scarce in real materials. Few cases are K_3C_{60} [108] and Li-doped carbon clathrates [109].

4.1. Theory of solubility limit

The solubility c of impurity can be controlled through the formation enthalpy h_f as

$$c = \exp\left(-\frac{h_f}{kT}\right), \tag{2}$$

where k is Boltzmann's constant [110]. The formation enthalpy h_f is a material constant if the host crystal A and impurity atom B are fixed. Nonetheless, h_f can be controlled by changing external conditions. The formation enthalpy h_f is defined by the difference in energy E between the reactants and the product as

$$h_f = E_{AB_m} - (mE_B + E_A), \tag{3}$$

where $m = 1/c$. In writing this equation, it is understood that the starting materials are of their most stable phases in the ambient conditions. At finite T and p, all the energies appearing in Eq. (3) should be replaced by the corresponding chemical potentials μ. Chemical potentials depend on concentration as well as on T and p. In semiconductor processes, their dependence on the charge state is also important for controlling the impurity concentration. By changing the starting states of reactants, we can control h_f and thereby the concentration of the impurity.

The solubility c is determined by the chemical equilibrium condition

$$\mu_A^2 = \mu_A^1, \tag{4}$$

where the superscripts 1 and 2 indicate the reactant and the product systems, respectively. The impurity system 2 has an entropy contribution due to random distribution, $kT \ln c$.

We will show in the following the usefulness of the high-pressure process for Li doping to α-boron, based on the foregoing theory. The theory is quite general. Recently, we have analyzed an interesting method of Ekimov's experiment on heavily B-doped diamond and succeeded to describe the experimental conditions [26, 27], based on the same theory. An application to B-doped diamond is also given below.

4.2. Li doping to α-boron

Let us first describe the situation of the Li doping to α-boron [23]. This example corresponds to a situation at $T = 0$ K, so that the analysis is easier. The involved reaction is expressed as

$$B_{12} + Li \rightarrow Li@B_{12}. \tag{5}$$

and accordingly the formation enthalpy $h_{f,\mathrm{Li}}$ of Li in α-boron can be calculated by Eq. (3). Full calculation by the first-principles pseudopotential method was done [23]. At $p = 0$, $h^0_{f,\mathrm{Li}}$ was obtained as +0.55 eV. This positive formation energy is very common in semiconductors, so that Li doping is realized only with very small concentrations, well below at.%. Our previous paper [23] illustrates that this situation can be changed by applying high pressure. The pressure dependence of $h_f(p)$ is shown in figure 4. The sign of $h_{f,\mathrm{Li}}(p)$ is changed at $p = 8$ GPa, so that above 8 GPa heavy doping appears possible.

The main cause of this pressure dependence comes from the starting state of Li, *i.e.*, metallic Li. Metallic Li is a soft material compared with boron, so that the most part of the pressure dependence is carried by the bulk modulus of the starting state of Li. From this observation, we can deduce a simple rule that *softer materials can be easily doped in stiffer materials under high pressure.*

This rule is simple but has wide applicability. The rule is founded by a well-known principle of thermodynamics, Le Chatelier's principle: a perturbation induces a process that attenuates the initial perturbation. In the present case, the perturbation is volume change. In fact, it is a common observation by high-pressure experimentalists the inclusion of gaseous atoms such as He in the specimen at high pressures. An affirmative example is seen in doped clathrates. Yamanaka *et al* use high-pressure process for synthesizing Ba_8Si_{46} [104]. The reaction they used is

$$8BaSi_2 + 30Si \rightarrow Ba_8Si_{46}. \tag{6}$$

They point out that the free volume of Ba_8Si_{46} is smaller by 15 % than that of the reactant mixture, from which they found it efficient to use high-pressure synthesis.

4.3. B-doped diamond

Ekimov *et al.* synthesized heavily doped diamond with boron concentrations above 2 at.% in the boron carbide-graphite system under high pressure and high temperature, at $p = 8 - 9$ GPa

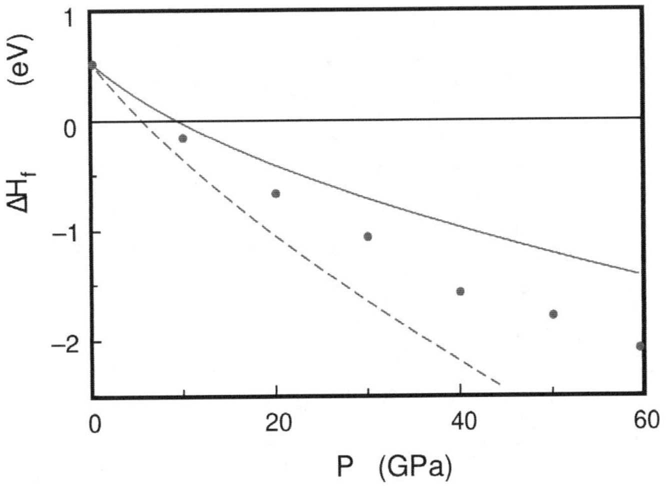

Figure 4. Pressure dependence of the formation enthalpy of Li for α-boron. The dots are the results of full calculation of $h_{f,\mathrm{Li}}$. The solid (red) line indicates the model calculation. The dashed (blue) line indicates a part of the solid Li. The data are from Ref. [26].

and $T = 2600 - 2800$ K [111]. In this case, the analysis is more involved, because of finite temperature and multi-component systems. The details should be referred to Ref. [26], and here we describe the main results only.

The reaction of Ekimov's method is symbolically written as

$$\mathrm{B_4C} + \mathrm{G} \rightarrow \mathrm{B@D}. \tag{7}$$

Here, G and D denote graphite and diamond, respectively. Actually, this reaction is composed of the following three reactions,

$$4\mathrm{B} + \mathrm{C} \rightarrow \mathrm{B_4C} + g_{f,\mathrm{B_4C}}, \tag{8a}$$

$$\mathrm{D} \rightarrow \mathrm{G} + g_{f,\mathrm{DG}}, \tag{8b}$$

$$\mathrm{B} + m\mathrm{D} + g_{f,\mathrm{B@D}} \rightarrow \mathrm{B@D}_m, \tag{8c}$$

The impurity concentration at arbitrary p and T can be obtained by calculating

$$\mu_\mathrm{B}^2 - \mu_\mathrm{B}^1 = \frac{1}{4} g_{f,\mathrm{B_4C}} + m g_{f,\mathrm{DG}} + g_{f,\mathrm{B@D}} + kT \ln c_\mathrm{B} = 0. \tag{9}$$

Calculations of the T and p dependence of the free energies were performed by using various available data. Appreciable dependence appears only for the free energy difference between graphite and diamond $g_{f,\mathrm{DG}}$. The T dependence has been calculated by taking only the phonon contribution within the harmonic approximation. A problem for determining the solubility limit is the formation energy of boron in diamond $g_{f,\mathrm{B}}$. In particular, the concentration-dependent formation energy is required for such a high concentration regime. For diamond, there is no reliable data on the formation energy $g_{f,\mathrm{B@D}}$. Hence, in a previous paper, we calculated $g_{f,\mathrm{B@D}}$ by *ab initio* pseudopotential method [27]. Even though there are still some difficulties to determine $g_{f,\mathrm{B@D}}$, we have estimated $g_{f,\mathrm{B@D}} = 1.0$ eV, if the situation of a supercell $\mathrm{BC_{63}}$ is regarded as the representative of B doping.

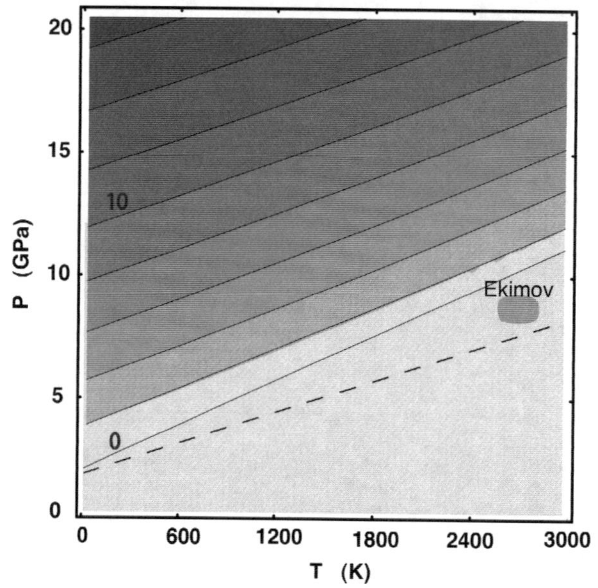

Figure 5. Calculated solubility of boron in diamond synthesized by Ekimov's method. A contour map for c_B is plotted with an interval 2 at.%. This figure is taken from Ref. [27].

In this way, we are able to calculate the B concentration in diamond at arbitrary p and T, based on the reaction (7). A contour map of the B concentration in the $p-T$ plane is shown in figure 5. In the figure, the experimental line for the phase boundary between graphite and diamond [112] is indicated by a dashed line. By considering that many thermodynamic quantities are involved in this calculation, the agreement seems good. The experimental conditions, where Ekimov *et al* [7] reached boron concentration of 2.8 at.%, are indicated by a red region. As clearly shown in this map, as increasing pressure, we can obtain higher boron concentrations in diamond.

A problem of the phase diagrams obtained by such calculations is that these are obtained by thermodynamic consideration alone. Any idea how fast or how slow phase transformations are completed is missing in the phase diagram. In many cases, occurrence of phase transformations depends on the process in which the experiment has been performed. This complicates direct comparison of the prediction and experiment. Practical problems regarding this phase diagram are discussed in Ref. [27].

In spite of the complications of the real situation, the authors think that heavily B-doped diamond (above 10 at.%) could be obtained by HPHT method, if the experimental conditions are chosen wisely. In particular, the starting materials including the crucible material are important. Recently, a diamond-structure compound BC_5 was synthesized by Solozhenko [113]. This compound is predicted to have T_c above 40 K [74]. This compound should not be stable, based on our preliminary calculation. However, Solozhenko's result shows that even such an unstable compound could be synthesized at high pressure and high temperature with using appropriate starting materials.

5. Conclusion

We have reviewed a recent development of superconductivity research on boron solids. Throughout this paper, the authors tried to discuss the situation of boron in connection with other related materials, such as B-doped diamond and clathrates, in order to make discussion as general as possible.

Crystal of α-boron is ideal for superconductivity research in many respects: structural simplicity, stability over a wide range of pressure, relatively defect-free structure and controllability of the valence electrons. Hence, the discovery of superconductivity in α-boron is significant.

Although superconducting transition has been discovered with relatively low T_c in both α- and β-boron, our investigation shows that its potential for high-T_c has not been fully disclosed yet. There is still potential to raise T_c, if we control appropriately the valence electrons. Heavy doping or use of compound as the alternation could improve T_c significantly. Although the doping idea itself can be easily found in the literature, the present idea is based on our deep understanding of the electronic structure of boron. The basic feature of solid boron is that the covalent character is so strong that it survives even after metallic transition, which makes $N(E_F)$ rather small around the superconducting transition pressure.

A practical problem for doping is how to synthesize such heavily doped boron or another compound. A promising method is high-pressure synthesis, which has been proven to be useful for doped clathrates. By using this method, we have showed in calculation that heavy doping of Li to α-boron can be achieved. For B-doped diamond case, further application of high pressure could increase the B concentration, which results in higher T_c.

In both cases of boron and diamond, accurate construction of phase diagrams is very important, which was partly achieved in our study. This is also true for a new compound BC_5, because this compound can be obtained only when special starting materials are chosen.

Added note After submitting this paper, another interesting paper was published concerning a high-pressure phase of boron. Solozhenko's group found a new phase of boron at high pressure [114]. It seems to us that this "new" phase is the same as the orthorhombic phase which has recently re-discovered by Dubrovinsky's group [46]. This phase was originally discovered by Wentorf in 1965 [34]. However, at that time, he showed only the diffraction pattern and the structure was left as undetermined.

Acknowledgments

As the organizer of the special session on the superconductivity of boron solids, the author (K.S.) himself is undoubtedly the person who benefits most from the organizing activity, especially discussions with those guest speakers: Prof. J. Akimitsu, D. C. Larbalestier, Y. Takano, M. Eremets, and N. Vast. Also thanks should be addressed to our collaborators on superconductivity of α-boron, Prof. K. Shimizu, Y. Mori, Y. Mita, and K. Kimura. We hope that writing this lengthy review brings benefits back to all of them. This work was partly supported by a Grant-in-Aid for Scientific Research in Priority Areas "Areas of New Materials Science Using Regulated Nano Spaces" (No. 20045009) of MEXT, Japan.

References

[1] Leggett A J 2006 *Quantum Liquids: Bose Condensation and Cooper Pairing in Condensed-Matter Systems* (New York: Oxford)

[2] 2008 *International workshop on "the Possibility of Room Temperature Superconductivity and Related Topics"* vol 468 (Physica C) especially, articles of A. A. Abrikosov, M. Tachiki, N. W. Ashcroft, and W. E. Pickett are valuable.

[3] Emin D 1991 *AIP Conf. Proc.* vol 231 pp 65–76 this paper itself does not discuss the superconductivity, but just describes a small bi-poloron mechanism in the transport phenomena. People used to discuss possibility of superconductivity for boron carbides in connection with this mechanism.

[4] Nagamatsu J, Nakagawa N, Muranaka T, Zenitani Y and Akimitsu J 2001 *Nature (London)* **410** 63

[5] Pickett W E 2008 *Physica C* **468** 126

[6] Eremets M I, Struzhkin V V, Mao H and Hemley R J 2001 *Science* **293** 272
[7] Ekimov E A, Sidorov V A, Bauer E D, Mel'nik N, Curro N J, Thompson J D and Stishov S M 2004 *Nature* **428** 542
[8] Bustarret E, Marcenat C, Achatz P, Kacmarcik J, Lévy F, Huxley A, Ortéga L, Bourgeois E, Blase X, Débarre D and Boulmer J 2006 *Nature* **444** 465
[9] Ren Z A, Kato J, Muranaka T, Akimitsu J, Kriener M and Maeno Y 2007 *J. Phys. Soc. Jpn.* **76** 103710
[10] Inushima T 2006 *Sci. Tech. Advanced Materials* **7** S112
[11] Akimitsu J 2008 *16th Int. Symp. Boron, Borides and Related Materials* (Matsue, Japan)
[12] Sanz D N, Loubeyre P and Mezouar M 2002 *Phys. Rev. Lett.* **89** 245501
[13] Ma Y, Prewitt C T, Zou G, Mao H and Hemley R J 2003 *Phys. Rev. B* **67** 174116
[14] Zhu W J and Henley C L 2000 *Europhys. Lett.* **51** 133
[15] Segall D E and Arias T A 2003 *Phys. Rev. B* **67** 064105
[16] Masago A, Shirai K and Katayama-Yoshida H 2006 *Phys. Rev. B* **73** 104102
[17] Shirai K, Masago A and Katayama-Yoshida H 2007 *Phys. Status Solidi B* **244** 303
[18] Kaneshige M, Shimizu K, Hyodo H and Kimura K 2007 *48th High-Pressure Conference of Japan* (Kurayoshi, Japan: Jpn. Soc. High Pressure Sci. Technol.) p 1C06
[19] Mori Y, Fujii Y, Dekura H, Shirai K, Hyodo H and Kimura K Structural study of α-boron at high pressure in preparation
[20] Gunji S, Kamimura H and Nakayama T 1993 *J. Phys. Soc. Jpn.* **62** 2408
[21] Gunji S and Kamimura H 1995 *Phys. Rev. B* **54** 13665
[22] Hayami W 1999 *Phys. Rev. B* **60** 1523
[23] Dekura H, Shirai K and Katayama-Yoshida H 2007 *J. Phys.: Condens. Matter* **19** 365241
[24] Dekura H, Shirai K and Katayama-Yoshida H 2007 *Physica B* **401-402** 702
[25] Takano Y, Nagao M, Sakaguchi I, Tachiki M, Hatano T, Kobayashi K, Umezawa H and Kawarada H 2004 *Appl. Phys. Lett.* **85** 2851
[26] Shirai K, Dekura H and Yanase A 2009 *Phys. Status Solidi B* **246** 673
[27] Shirai K and Dekura H 2008 In submission
[28] Vast V, Baroni S, Zerah G, Besson J M, Polian A, Grimsditch M and Chevin J C 1997 *Phys. Rev. Lett.* **78** 693
[29] Mailhiot C, Grant J B and McMahan A K 1990 *Phys. Rev. B* **42** 9033
[30] Häussermann U, Simak S I, Ahuja R and Johansson B 2003 *Phys. Rev. Lett.* **90** 065701
[31] Miguel A S, Merlen A, Toulemonde P, Kume T, Floch S L, Aouizerat A, Pascarelli S, Aquilanti G, Mathon O, Bihan T L, It'e J P and Yamanaka S 2005 *Europhys. Lett.* **69** 556
[32] Hoard J H and Hughes R E 1965 *The Chemistry of Boron and Its Compounds* ed Mueterties E L (New York: Wiley) chap 2
[33] Runow P 1972 *J. Mater. Sci.* **7** 499
[34] R H Wentorf J 1965 *Science* **147** 49
[35] Brazhkin V V, Taniguchi T, Akaishi M and Popova S V 2004 *J. Mater. Sci.* **19** 1643
[36] Shang S, Wang Y, Arroyave R and Liu Z K 2007 *Phys. Rev. B* **75** 092101
[37] Widom M and Mihalkovič M 2008 *Phys. Rev. B* **77** 064113
[38] Slack G A, Hejna C I and Kasper J S 1988 *J. Solid State Chem.* **76** 52
[39] van Setten M J, Uijttewaal M A, de Wijs G A and de Groot A 2007 *J. Am. Chem. Soc.* **129** 2458
[40] Ogitsu T, Gygi F, Reed J, Motome Y, Sxhwegler E and Galli G 2009 *J. Am. Chem. Soc.*
[41] Dekura H, Shirai K and Yanase A 2008 (Matsue, Japan) in this volume
[42] Horn F H 1959 *J. Electrochem. Soc.* **106** 905
[43] Tallant D R, Aselage T L, Campbell A N and Emin D 1989 *Phys. Rev. B* **40** 5649
[44] Soga K, Oguri A, Araake S, Terauchi M, Fujiwara A and Kimura K 2004 *J. Solid State Chem.* **177** 498
[45] Shirai K, Masago A and Katayama-Yoshida H 2004 *Phys. Status Solidi B* **241** 3161
[46] Zarechnaya E Y, Dubrovinsky L, Dubrovinskaia N, Miyajima N, Filinchuk Y, Chernyshov D and Dmitriev V 2008 *Sci. Technol. Adv. Mater.* **9** 044209
[47] Nelmes R J, Loveday J S, Allan D R, Besson J M, Hamel G, Grima P and Hull S 1993 *Phys. Rev. B* **47** 7668
[48] Polian A, Chervin J C, Munsch P and Gauthier M 2008 *Physica C* **468** 126
[49] Mori Y, Fujii Y, Hyodo H, Takarabe K and Kimura K 2008 *13th Int. Conf. High Pressure Semiconductor Physics* (Fortaleza, Brazil)
[50] Nelmes R J, Loveday J S, Wilson R M, Marshall W G, Besson J M, Klotz S, Hamel G, Aselage T L, and Hull S 1995 *Phys. Rev. Lett.* **74** 2268
[51] Murnaghan F D 1944 *Proc. Natl. Acad. Sci. U.S.A.* **30** 244
[52] Fujii T, Fujii Y, Fujii A, Nakano H, Mori Y and Sata N 2008 *Fall Meeting of the Physical Society of Japan*

(Iwate) pp 22pWF–11
[53] Decker B F and Kasper J S 1959 *Acta Cryst.* **12** 503
[54] Longuet-Higgins H C and de V Roberts M 1955 *Proc. Roy. Soc.* **A224** 336
[55] Perrot F 1981 *Phys. Rev. B* **23** 2004
[56] Bullett D B 1982 *J. Phys. C: Solid State Phys.* **15** 415
[57] Lee S, Bylander D M and Kleinman L 1990 *Phys. Rev. B* **42** 1316
[58] Morrison I, Bylander D M and Kleinman L 1992 *Phys. Rev. B* **45** 10872
[59] Zhao J and Lu J P 2002 *Phys. Rev. B* **66** 092101
[60] He J, Wu E, Wang H, Liu R and Tian Y 2005 *Phys. Rev. Lett.* **94** 015504
[61] Papaconstantopoulos D A and Mehl M J 2002 *Phys. Rev. B* **65** 172510
[62] Ma Y, Tse J S, Klug D D and Ahuja R 2004 *Phys. Rev. B* **70** 214107
[63] Bose S K, Kato T and Jepsen O 2005 *Phys. Rev. B* **72** 184509
[64] Prasad D L V K, Balakrishnarajan M M and Jemmis E D 2005 *Phys. Rev. B* **72** 195102
[65] Mattheiss L F and Hamann D R 1984 *Phys. Rev. B* **30** 1731
[66] Liu A Y, Wentzcovitch R M and Cohen M L 1988 *Phys. Rev. B* **38** 9483
[67] Bernasconi M, Chiarotti G L and Tosatti E 1995 *Phys. Rev. B* **52** 9988
[68] Shirai K 1997 *Phys. Rev. B* **55** 12235
[69] Shirai K and Katayama-Yoshida H 1998 *J. Phys. Soc. Jpn.* **67** 3801
[70] Shirai K, Dekura H and Yanase A Electronic structure and electrical resistivity of α-boron under high pressure submitted to J. Phys. Soc. Jpn.
[71] Rosner H, Kitaigorodsky A and Pickett W E 2002 *Phys. Rev. Lett.* **88** 127001
[72] Calandra M, Vast N and Mauri F 2004 *Phys. Rev. B* **69** 224505
[73] Xiang H J, Li Z, Yang J, Hou J G and Zhu Q 2004 *Phys. Rev. B* **70** 212504
[74] Calandra M and Mauri F 2008 *Phys. Rev. Lett.* **101** 016401
[75] An J M and Pickett W E 2001 *Phys. Rev. Lett.* **86** 4366
[76] Kortus J, Mazin I I, Belashchenko K D, Antropov V P and Boyer L L 2001 *Phys. Rev. Lett.* **86** 4656
[77] Chang K J, Dacorogna M M, Cohen M L, Mignot J M, Chouteau G and Martinez G 1985 *Phys. Rev. Lett.* **54** 2375
[78] Dacorogna M M, Cohen M L and Lam P K 1986 *Phys. Rev. B* **34** 4865
[79] Lüders M, Marques M A L, Floris A, Profeta G, Lathiotakis N N, Franchini C, Sanna A, Continenza A, , Massidda S and Gross E K 2006 *Ψ-k News Letter* **76** 54
[80] Floris A, Profeta G, Lathiotakis N N, Lüders M, Marques M A L, Franchini C, Gross E K, Continenza A, and Massidda S 2005 *Phys. Rev. Lett.* **94** 037004
[81] Teyssier J, Kuzmenko A B, van der Marel D, Marsiglio F, Liashchenko A B, Shitsevalova N and Filippov V 2007 *Phys. Rev. B* **75** 134503
[82] Shirai K and Emura S 1996 *J. Phys.: Condens. Matter* **50** 10919
[83] Wood C and Emin D 1984 *Phys. Rev. B* **29** 4582
[84] Emin D 1988 *Phys. Rev. B* **38** 6041
[85] Samara G A, Tardy H L, Venturini E L, Aselage T L and Emin D 1993 *Phys. Rev. B* **48** 1468
[86] Werheit H 2008 *16th Int. Symp. on Boron, Borides and Related Materials* (Matsue, Japan) in this volume
[87] Kwei G H and Morosin B 1996 *J. Phys. Chem.* **100** 8031
[88] Schmechel R and Werheit H 1999 *J. Phys.: Condens. Matter* **11** 6803
[89] Lazzari R, Vast V, Besson J M, Baroni S and Corso A D 1999 *Phys. Rev. Lett.* **83** 3230
[90] Mauri F, Vast N and Pickard C H 2001 *Phys. Rev. Lett.* **87** 085506
[91] Bylander D M, Kleinman L and Lee S 1990 *Phys. Rev. B* **42** 1394
[92] Bylander D M and Kleinman L 1991 *Phys. Rev. B* **43** 1487
[93] Akimitsu J Private communication
[94] Yokoya T, Nakamura T, Matsushita T, Muro T, Takano Y, Nagao M, Takenouchi T, Kawarada H and Oguchi T 2005 *Nature (London)* **438** 647
[95] Takano Y 2006 *Sci. Technol. Adv. Mater.* **7** S1
[96] Goss J P and Briddon P R 2006 *Phys. Rev. B* **73** 085204
[97] Bourgeois E, Bustarret E, Achatz P, Omnès F and Blase X 2006 *Phys. Rev. B* **74** 094509
[98] Baskaran G 2006 *Sci. Tech. of Advanced Materials* **7** S49
[99] Shirakawa T, Horiuchi S, Ohta Y and Fukuyama H 2007 *J. Phys. Soc. Jpn.* **76** 014711
[100] http://www.cmp.sanken.osaka-u.ac.jp/~yanase/
[101] Dekura H, Shirai K and Yanase A *J. Comput. Theor. Nanosci.* In press
[102] Gunnarsson O 1997 *Rev. Mod. Phys.* **69** 575
[103] Ganin A Y, Takabayashi Y, Khimyak Y Z, Margadonna S, Tamai A, Rosseinsky M J and Prassides K 2008 *Nature Materials* **7** 367

[104] Yamanaka S, Enishi E, Fukuoka H and Yasukawa M 2000 *Inorg. Chem.* **39** 56
[105] Matsuda H, Nakayama T, Kimura K, Murakami Y, Suematsu H, Kobayashi M and Higashi I 1995 *Phys. Rev. B* **52** 6102
[106] Hayami W, Tanaka T and Otani S 2005 *J. Phys. Chem. A* **109** 11975
[107] Dekura H, Shirai K and Yanase A Li-doped α-boron and new doping method in preparation
[108] Saito S, Umemoto K and Miyake T 2004 *"Fullerene-Based Materials - Structural and Properties", Structure and Bonding* vol 109 ed Prassides K (Springer) pp 41–57
[109] Rey N, noz A M, Rodríguez-Hernández P and Miguel A S 2008 *J. Phys.: Condens. Matter* **20** 215218
[110] Lannoo M and Bourgoin J 1981 *Point Defects in Semiconductors I* (Berlin: Springer)
[111] Ekimov E A, Sidorov V A, Mel'nik N, Gierlotka S and Presz A 2004 *J. Mat. Sci.* **39** 4957
[112] Bundy F P, Bassett W A, Weathers M S, Hemley R J, Mao H K and Goncharov A F 1996 *Carbon* **34** 141
[113] Solozhenko V L, Kurakevych O O, Andrault D, Godec Y L and Mezouar M 2009 *Phys. Rev. Lett.* **102** 015506
[114] Oganov A R, Chen J, Gatti C, Ma Y, Ma Y, Glass C W, Liu Z, Yu T, Kurakevych O O and Solozhenko V L 2009 *Nature* **457** 863

Journal of Physics: Conference Series **176** (2009) 012002 doi:10.1088/1742-6596/176/1/012002

Boron carbides from first principles

Nathalie Vast, Jelena Sjakste and Emmanuel Betranhandy

Laboratoire des Solides Irradiés,
CEA Direction des Sciences de la Matière - CNRS UMR 7642 - École Polytechnique, 91120 Palaiseau, France

Abstract. In this work, we focus on the understanding gained from the investigation of the physical properties of boron carbides with theoretical methods based on density functional theory (DFT). Together with the examination of the DFT total energies of various atomic configurations in the unit cell, comparison with the experiments of the theoretical vibrational or NMR spectra has led to the determination of the atomic structure of B_4C as C-B-C chains linking mostly $B_{11}C$ icosahedra, and a few percents of $B_{10}C_2$ icosahedra. In the icosahedron, the carbon atom is found to be in the polar site (B_4C^p). When there are two carbon atoms, they are found to be in antipodal polar positions. At carbon concentrations other than 20%, we find that only four structural models have a negative formation energy with respect to a formation from α-boron + diamond. Moreover, they all have a positive formation energy with respect to B_4C^p, showing a tendency to decompose into B_4C^p + α-boron or B_4C^p + diamond. This metastability explains actual difficulties in the synthesis of clean samples, in particular for $B_{13}C_2$. Finally, the idea of combining high hardness and superconductivity in the same material by doping boron-rich solids has emerged. We show results on the strength of the electron-phonon coupling constant obtained with DFT-based methods in $B_{13}C_2$.

E-mail: `nathalie.vast@polytechnique.edu`

1. Introduction

Density functional theory (DFT) has become a crucial *from-first-principles* tool to investigate ground state properties of materials [1, 2, 3]. Theoretical results on boron carbides presented in this paper rely on DFT, and therefore we briefly recall in this introduction the main purpose of DFT and its limitations. DFT maps a system of interacting electrons on a system of noninteracting electrons in an effective potential, which contains the so-called exchange and correlation potential. In fact, DFT relies on assumptions for the exchange and correlation energy, which aim at modeling terms difficult to take into account in the mapping : *(i)* the difference in kinetic energy between a system of interacting electrons and noninteracting ones; *(ii)* all many-body electron-electron quantum interactions - such as Pauli's principle - beyond the Hartree energy (the classical electrostatic energy for interacting charge densities).

In the simplest local density approximation (LDA), each small volume of the system is expected to contribute to the exchange and correlation energy as would do an equal volume of a homogeneous electron gas at the same density [4]. Equilibrium properties calculated within DFT-LDA such as lattice parameters and internal atomic positions usually agree extremely well -within 1%- with experiment. There are however a number of applications in which this accuracy is not high enough. It is worth pointing out, for instance, that most of today's computational electronic spectroscopy is performed at the experimental lattice constant to avoid the lattice misfit [5].

Exchange and correlation functionals beyond LDA thus become required in order to evaluate the equilibrium volume with a greater accuracy. The generalized gradient approximation (GGA) has been developed with such an objective [6]. It is important for properties which explicitly depend on the equilibrium volume V_{eq}, such as the tensor of elastic constants C_{ijkl}, defined as the second order derivative of the (DFT) total energy E with respect to the strain tensor ε_{ij}, where i, j, k and l are cartesian indexes

$$C_{ijkl} = \frac{1}{V_{eq}} \left(\frac{\partial^2 E}{\partial \varepsilon_{ij} \partial \varepsilon_{kl}} \right)_{\varepsilon=0} . \tag{1}$$

In above equation, the explicit dependence on the equilibrium volume V_{eq} makes the error bar on elastic constants much smaller when calculated within the generalized gradient approximation than within LDA [7]. Nuclear magnetic resonance (NMR) is another type of property which directly probes the local atomic structure through the electronic response to an external magnetic field. For the same reasons, calculations of the chemical shift tensor performed with the GGA do show a better agreement with experiment in comparison with LDA ones [8, 9, 10]. We therefore used GGA for our NMR results spectra presented in subsection 2.4.

Among known limitations of DFT, forbidden energy gaps in insulators and semiconductors are underestimated. For energy gap calculations, theoretical frameworks beyond DFT are required, which take into account the electron-electron interaction beyond the mean-field approximation. Energy gaps are currently computed with a Green's function approach which allows to calculate the quasiparticle corrections to DFT energies to first-order in perturbation [11, 12, 13].

Excited states also are out of reach of DFT. They are treated either by time-dependent density functional theory (TDDFT) [14, 15], or by the solution of the Bethe-Salpeter equation, when, like in optical absorption, it is necessary to take into account the electron-hole interaction [5, 16]. These methods nonetheless rely on DFT, for the ground state wavefunctions and energies are often (if not always) used in, either as a starting point, or to calculate key quantities like the dielectric tensor.

Finally, it is not yet clear whether the above-mentioned many-body methods based on Green's functions are sufficient in materials where electronic correlations are of primary importance, like in oxides with multiply occupied $3d$ orbitals. The dynamical mean field theory presents an alternative method to model the correlations [17]. It should be noted however that both classes of methods have been successfully applied to the electronic spectroscopy of vanadium dioxide [18, 19].

Being *sp* bonded materials, boron-rich solids (BRS) are expected to have ground state properties well described by DFT based methods. The crystal symmetry of BRS is rhombohedral with $R\bar{3}m$ space group. The atomic structure consists in one distorted icosahedron per primitive unit cell (figure 1), eventually plus one multi-atomic chain. Each atom in the icosahedron is the apex of a pentagonal-based pyramid. Two crystallographic sites in the icosahedron are defined by the intericosahedral bonding. In polar site p, one atom is linked by a covalent two-center p-p bond to one atom in the polar site of a neighboring icosahedron. In the second site e, the equatorial atom is bonded to the chain-end atom of a multiatomic chain that links neighboring icosahedra. The intericosahedral chain can be absent like in α-boron. In most of icosahedral solids, there is one chain per unit cell. The chain can be diatomic - like in $B_{12}P_2$ and $B_{12}As_2$ - or triatomic like in boron carbides (figure 1).

The nature of the bonding makes boron-rich solids widely different from fullerenes, a class of materials with which they are nonetheless often compared. Due to electron deficiency [21, 22, 23], a "standing alone" B_{12}, $B_{11}C$ or $B_{10}C_2$ neutral cluster is unstable. Thus BRS can not be compared to fullerene crystals [24, 25]. The comparison of the respective compressibilities of the intra and interatomic bonds by *ab initio* methods has proved that neither α-boron nor

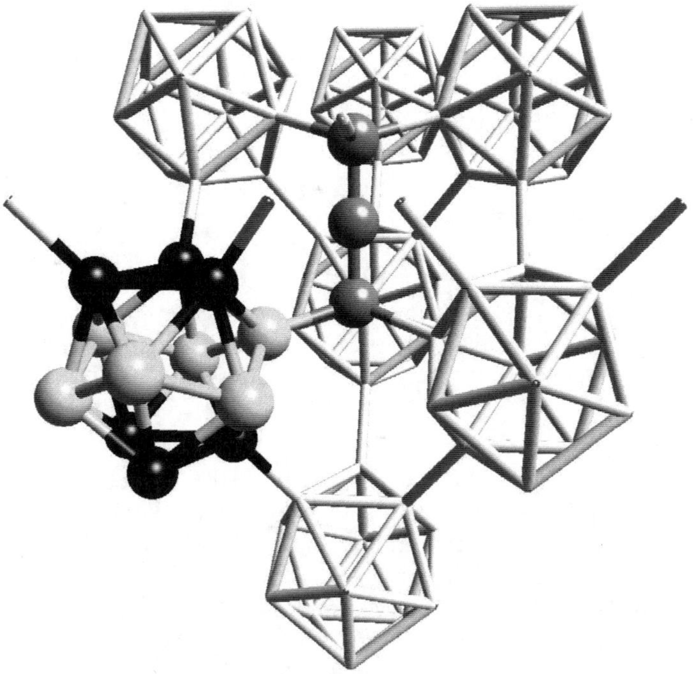

Figure 1. Atomic structure of icosahedral boron carbide from Ref. [20]. Seven distorted icosahedra are shown. The rhombohedral Bravais lattice vectors (not shown) link icosahedra centers. Black balls : atoms at the *p* polar site. Light grey balls : atoms at the *e* equatorial site. Dark grey balls : atoms at chain center and chain ends. Polar atoms are shown to be bonded to neighboring icosahedra *via* the covalent two-center *p-p* bonds. Equatorial atoms are linked to one atom at the end of the triatomic chain. (Reproduced with permission from [20])

icosahedral boron carbide with 20% of carbon concentration (B_4C) are molecular crystals. They do not form either an inverted molecular crystal, in which the intericosahedral bonds would be "stronger" than the intraicosahedral ones [25, 26, 27]. Icosahedral boron-rich solids form a special case of covalent materials adopting an unusual structure because of the electronic deficiency that is intrinsic to the boron atom.

The icosahedral *sp* bonding is by far more complex than *sp* bonding in diamond or other covalent materials with a tetrahedral coordination. Density functional theory is expected to be well suited to explain physical properties related to this complex *sp* bonding.

2. The atomic structure of B_4C from first principles

In this section, our aim is to review what has been established about the atomic structure of B_4C by comparing experimental data to results computed *from first principles*.

The atomic structure of $B_{12}C_3$ (B_4C) boron carbide contains 15 atoms in four occupied crystallographic sites (figure 1). Two sites form a distorted icosahedron, the polar and equatorial sites, with 6 atoms *per* site. The third and fourth sites are respectively the center of the triatomic chain (one atom) and the chain ends (two atoms).

Experimentally, the three carbon atoms are difficult to distinguish from the twelve boron atoms. Indeed, the scattering length of a ^{13}C nucleus is very close to the scattering length of a ^{11}B nucleus, and only cell parameters have been measured by neutron diffraction [28, 29]. Atomic positions deduced from X-ray diffraction data indicate that the chain is C-B-C and that one carbon atom is substituted in the icosahedron [30, 31]. As the X-ray form factor scales like Z^2, where Z is the atomic number, the form factors for light neighboring elements like B and C are very close, and it is difficult to distinguish between boron and carbon atoms in the icosahedron. The picture provided by X-ray diffraction is the following one : the third carbon atom is statistically substituted on both polar and equatorial sites, with a predominance for the polar site. The formula is $(B_xC_{1-x})_6^p(B_yC_{1-y})_6^e$, in which $0.93 \leq x \leq 1$ for the polar site and $0.74 \leq y \leq 0.86$ for the equatorial site [31, 32].

In order to gain knowledge of the atomic structure of boron carbide with a carbon concentration of 20%, we investigated several structural models with *ab initio* methods based on DFT, with the purpose of comparing various computed physical properties with available experimental data. Calculations were performed within DFT using the plane-wave pseudopotential method. More details can be found in Refs [20, 24, 25].

2.1. Structural models

In this subsection, all our structural models have a 20% carbon concentration and consist in one unit cell of 15 atoms, periodically repeated by translations of the Bravais lattice vectors. One exception is the *bipolar* model, which contains thirty atoms to keep the stoichiometry (table 1). In the *polar* (resp. *bipolar*) model, the chain is C-B-C and the icosahedron has one (resp. two) carbon atom(s) in the polar site. In the *equatorial* model, the chain is C-B-C and one carbon atom is substituted in the equatorial site. We have also investigated a unit cell where the chain is C-C-C and the icosahedron is made of pure boron (*chain* model), or introduced some disorder in the chain, making it C-C-B instead of C-B-C (*disordered chain* model). A last model consists in a B-rich chain C-B-B (B-*rich chain* model) (table 1).

The substitution of a carbon atom in the polar site of the icosahedra offers six possibilities, resulting in substitutional disorder. This makes the observed symmetry of the unit cell *on average* rhombohedral. In our calculations, we do not take explicitly into account the substitutional disorder, as too large supercells would be required. Therefore, the periodic repetition of a carbon atom at a given position in the icosahedron induces a small distortion from the rhombohedral symmetry to a base-centered monoclinic cell. We checked that this approximation does not modify the conclusions of this paper.

2.2. DFT total energy

The total energy obtained within DFT-LDA is reported in table 1. The structure with the lowest energy is the *polar* model. We take this energy as our reference. The *bipolar* model comes second, with an energy higher by 0.25 eV per carbon atom change. This energy difference is very significant and corresponds in terms of thermal energy to a temperature larger than 2500 K. Furthermore, the DFT energetics allows to exclude the *equatorial* model, the *chain* model, the *disordered chain* model and the B-*rich chain* model. It shows that the chain is C-B-C. Exclusion of the *equatorial* and *chain* models will be further confirmed by inspecting the lattice dynamics and the NMR spectra. Moreover, the formation energies of *disordered chain* and B-*rich chain* models show that these models are unstable (table 5, rows 6 and 7, respectively).

The difference in total energy between our various structural models is much more important than actual accuracy of exchange and correlation functionals. The later is indeed limited to 3 kcal/mol (10 meV/atom) [33]. As our models differ only by the location of one carbon atom, the exchange-correlation functional limits the accuracy of our calculations to 10 meV per carbon atom change. Crystal phases differing by less than that amount can hardly be distinguished and this may lead to a wrong prediction of the most stable phase. We see that this is not the case in table 1.

Table 1. Total energy (eV) *per* change of carbon atom location with respect to the *polar* model within DFT-LDA from Ref. [20, 24, 25].

Model	Chain	Icosahedron	Total energy (eV)
Polar	C-B-C	$B_{11}C^p$	(0)
Bipolar	$(C\text{-}B\text{-}C)_2$	$B_{10}C_2{}^p + B_{12}$	0.24
Equatorial	C-B-C	$B_{11}C^e$	0.53
Chain	C-C-C	B_{12}	1.05
Disordered chain	C-C-B	$B_{11}C^p$	2.5
B-*rich chain*	C-B-B	$B_{10}C_2{}^p$	3.9

Furthermore the difference in total energy between our structural models is similar in LDA and in GGA (not shown, Ref. [20]) within 20 meV per change of carbon atom. This gives us confidence in the predictive ability of the DFT energetics for our models. We emphasize that LDA and GGA have the same physical content. Nonetheless, we note that by construction there is a cancellation of errors between the exchange energy and the correlation energy in LDA. This cancellation is not built-in in the GGA functionals [34], which makes the computations both in LDA and in GGA worth performing and comparing. Moreover, by construction, GGA functionals tend to decrease the energy of the less dense phase, and hence to favor their stability [35]. Therefore, it is interesting to compute energy differences both in LDA and GGA, in order to obtain an error bar when the crystal density varies among the polymorphs. These differences in the construction of the functionals do not modify the relative stabilities of our structural models (table 1 and Ref. [20]).

Our results for B_4C are robust against approximations. Contrastingly relative energies of different crystal structures of pure carbon or of pure boron are much less robust. Graphite is found more stable than diamond by 145 meV/atom in DFT-GGA [36], whereas diamond is found more stable than graphite in DFT-LDA by only 15 meV [36]. Zero-point motion might reverse the most stable phase. Clearly, the determination of the stability of pure carbon polymorphs is beyond the capacity of actual exchange and correlation functionals [37]. This is particularly true for graphite, because Van der Waals interactions are not properly taken into account in LDA nor in GGA [38, 39, 40, 41, 42].

Turning to boron, in calculations for perfect crystals, α-B_{12} has been shown to be more stable than perfect 105-atoms-β-boron in DFT-LDA [43]. Recent calculations performed with a denser sampling of the Brillouin zone of α-boron yielded a total energy difference between α- and β-boron of 68 meV per atom in LDA and of 25 meV per atom in GGA, both predicting that α-B_{12} is more stable [44]. The difference between LDA and GGA is very important in this case. The 43 meV per atom energy difference yields an error bar of the exchange and correlation functionals.

On the other hand, calculations including defects predicted that defected β-boron is the most stable phase [43, 45, 46, 47, 48, 49]. The energy difference with respect to perfect α-boron is at most 20 meV/atom, far below the above-mentioned error bar of 43 meV. Also in this case, calculations reach the limit of exchange and correlation accuracy. We believe that the theoretical determination of the stability of pure carbon or pure boron polymorphs deserves further theoretical breakthroughs to overcome the intrinsic limitation of LDA and GGA functionals.

2.3. Lattice dynamics

Turning to lattice dynamics, we have computed [24] the phonon frequencies at the center of the Brillouin zone for the *polar*, *equatorial* and *chain* models (table 1). We note that in our case the electronic structure and the lattice dynamics are performed within the same theoretical framework. This implies that variation of both Hartree potential and of exchange and correlation potential are calculated to first-order in perturbation. This response of the screening potential to the perturbing phonon is computed self-consistently, so that at variance with semi-empirical models, the electronic density does not "rigidly shift" with the atomic motion. Compared to *e.g.* valence-force-field models, our computational scheme for phonon frequencies and displacement patterns does not contain any fitting parameter.

In figure 2, we report the theoretical spectrum [24] for infrared absorption compared with data from Ref [50]. Peak positions are clearly in better agreement for the *polar* model. With respect to the *equatorial* model, this is particularly true for the mode observed near 407 cm^{-1}. This is a vibration of the chain-center atom perpendicular to the chain axis.

The discrepancy between experimental peak positions and *chain* model ones is even more pronounced at high frequencies. The peak observed around 1560 cm^{-1} comes from the stretching of the chain, when the chain-end atoms vibrate antisymmetrically (*ungerade* vibration). The frequency is by far larger when the chain is C-C-C (Panel a) than when the chain is C-B-C (Panels b and c). We therefore conclude that the *chain* model can be disregarded : stretching of C-C-C chains does not occur in measured infrared spectra.

Turning to peak intensities in the *polar* model, we show in panel (d) a fit with one single parameter being the proportion between E_u and A_{2u} polarizations. Indeed, this proportion of polarizations is unknown in experiment. Our fit is in good agreement with experiment, showing that the tensor of effective charges is properly accounted for in the calculations.

If not silent, a vibration in B_4C is observed either in Raman diffusion or in infrared absorption. This is because the centre of the icosahedron is an inversion centre. The parity of a mode with respect to the inversion center is unique and the mode is either infrared active (ungerade E_u and A_{2u} modes) or Raman active (gerade E_g and A_{1g} modes). Therefore it is interesting to compare our computed phonon frequencies also to the peaks observed by Raman scattering. To our knowledge, only one Raman spectrum measured on a single crystal has been reported in the literature, and it is our reference in figure 3. Observed peaks are broadened by disorder, and in the high frequency range, the theoretical frequencies of all our models [24] coincide with the experiment.

Nonetheless, two vibrational modes at 481 cm^{-1} and 534 cm^{-1} can be singled out. Both modes involve bond bending. The mode observed at 481 cm^{-1} corresponds in our calculations to a rotation of the chain-end atoms around the chain-center. The mode observed at 534 cm^{-1} corresponds to the libration of the icosahedron. This mode has been previously observed in α-boron and it was shown to be strongly harmonic [51].

Compared with the disorder-induced broadening of other vibrations, the small widths of the two peaks indicate that the two above mentioned modes are harmonic in boron carbide. The frequency difference between them is therefore a quantity which is expected to be well reproduced in our calculations within the harmonic approximation. Figure 3 shows that the experimental frequency difference is reproduced only by the *polar* model. This difference is smaller by a factor of 2 in the *equatorial* model. Therefore the *equatorial* model can be disregarded as a possible structural model for B_4C.

Last but not least, we note that our calculations show no vibrational modes below 400 cm^{-1} (figure 3). The Raman peaks which are observed below 400 cm^{-1} in most of experiments [52, 53, 54] come from a lift of the selection rules for Raman scattering [24]. If ω is the phonon frequency, the intensity scales like a ω^4 power law at low frequency, which is the signature of a density of states of acoustic phonons. These phonons could be used to check the samples against structural disorder : they are expected to vanish in clean samples.

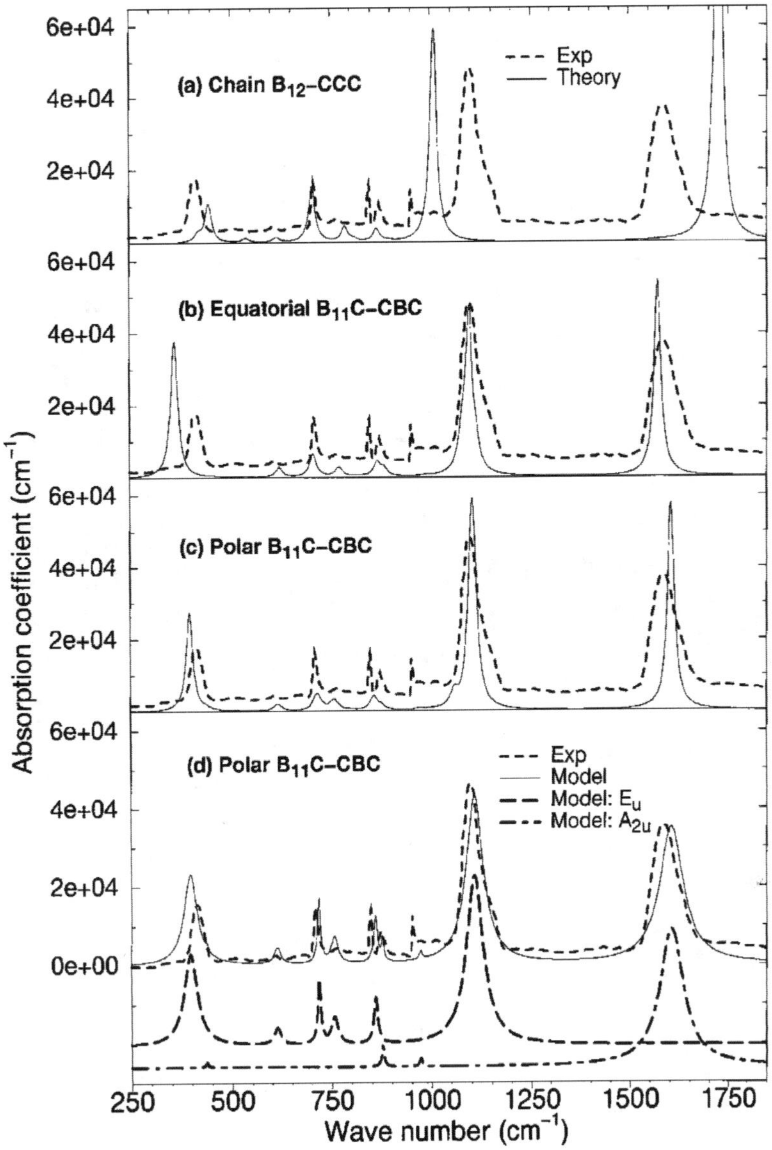

Figure 2. Infrared spectrum of B_4C from Ref. [24]. Dashed line : experiment.[50] Solid line: theoretical spectrum for the *chain* model (panel a); *equatorial* model (panel b); *polar* model (panel c). Panel d : spectrum for the *polar* model where the proportion of E_u and A_{2u} polarizations has been fitted with one single parameter to reproduce the unknown mixing of polarizations in the experiment (see text). (Reproduced with permission from [24])

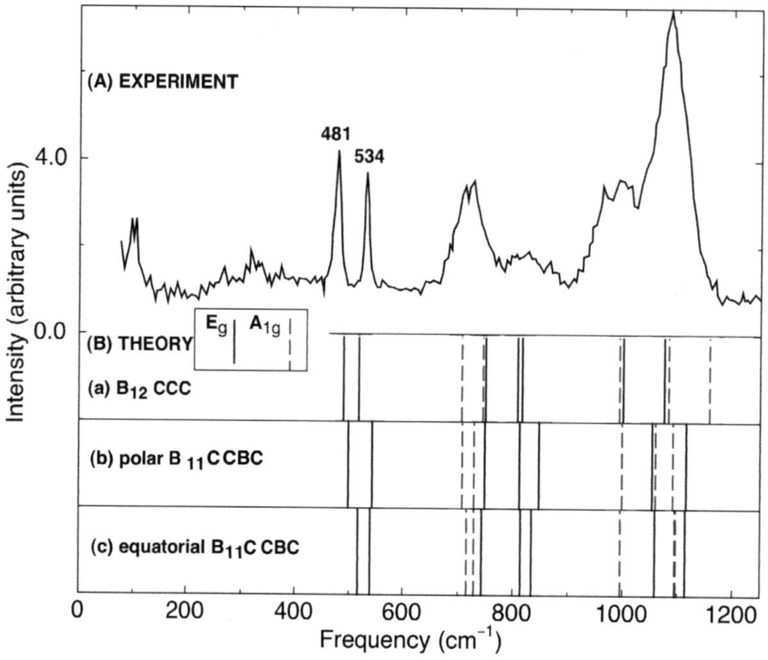

Figure 3. Raman spectrum of B_4C from Ref. [24]. Upper panel : experiment [52]. Lower panel : theory for *chain*, *polar* and *equatorial* models in subpanels a, b, and c respectively. Solid line : E_g mode. Dashed line : A_{1g} mode. (Reproduced with permission from [24])

2.4. Nuclear Magnetic Resonance

In previous subsection, the comparison of the calculated lattice dynamics with experiment allowed to identify the atomic structure of B_4C as a C-B-C chain and a $B_{11}C^p$ icosahedron with the carbon atom substituted in the polar site. But we have also seen that experimental vibrational spectra do show a large broadening caused by substitutional disorder and/or structural defects. Therefore we have decided to compute NMR spectra, with the aim to learn more about these defects [20].

Three values of the chemical shift of ^{13}C have been reported in the experimental literature (table 2). Experiments [55, 56, 57] agree with one another about two signals around 1 and 82 ppm. Only one experimental work reports a signal at 101 ppm.[55] We note that this experiment yielded the best resolved spectrum. This additional peak may also be explained by a different route of synthesis of the sample, yielding more structural disorder.

Our calculations show that the *polar* and *bipolar* models do explain well the experimental signals. Peaks around 1 and 82 ppm come from chain-end carbon atoms and from carbon atoms in the polar sites of the icosahedra, respectively (*polar* model). The *bipolar* model allows to explain the signal at 101 ppm observed in Ref. [55], which comes from a different screening of the magnetic field when two carbon atoms are substituted in the same icosahedron. From the point of view of the total energy, the favorite relative location of two carbon atoms is in antipodal polar positions [20]. Furthermore, we note that a few percents of pure B_{12} icosahedra are necessary to keep the stoichiometry in our calculations.

Turning to the *equatorial* model, the width of the peak predicted by theory that comes from chain-end carbon atoms is greater than 9.1 ppm (table 2). This is significantly larger than in experiment. Moreover, the calculated spectrum of the ^{11}B nucleus has a shoulder on the side of negative chemical shifts for the *equatorial* model (figure 4). At variance with this finding, the observed spectrum shows a shoulder on the side of positive chemical shifts, in agreement with the *polar* model [58]. This allowed us to disregard the substitution of one carbon atom in the equatorial site as a possible structure. Energetics, vibrational and NMR spectra thus agree on this point.

Moreover, a carbon atom at the center of the chain would yield a chemical shift as large as 234 ppm in the ^{13}C spectrum. Such a signal is not observed (table 2). Energetics, vibrational and NMR spectra agree on the fact that no C-C-C chains are present.

Finally, comparison of computed and observed NMR spectra allows to draw the conclusion that the atomic structure of B_4C consists in a C-B-C chain and a $B_{11}C^p$ icosahedron with one carbon atom substituted in the polar site. A few percents of icosahedra have a $B_{10}C_2^p$ or a B_{12} atomic structure. We have estimated that 95% of icosahedra are $B_{11}C^p$, 2.5% are $B_{10}C_2$ and 2.5% are B_{12} ones [20].

Table 2. B_4C : ^{13}C NMR chemical shifts, δ_{TMS} (ppm), and the corresponding peak intensities (in brackets). Experiments from Ref. [55, 56, 57]. Note that the weak resonance at 101.3 ppm is visible only in the experiment of Ref. [55]. Theory : models as in table 1.

Experiments			Theory			
Ref. [56]	Ref. [57]	Ref. [55]	*Bipolar*	*Polar*	*Equatorial*	*Chain*
13±15	-0.6	1.0	-1.7 & -1.1	-3.9 & 1.3	0.8 & 9.9	19.9
(66.7%)	(75.6%)	(66.9%)		(All models : 66.7%)		
85±10	79	81.9		81.6	79.1	233.8
(33.3%)	(24.4%)	(31.4%)		(These three models : 33.3%)		
		101.3	100.8 & 104.			
		(1.7%)	(33.3%)			

2.5. Conclusions

Inspection of both NMR and vibrational spectra yields the same picture for the atomic structure of B_4C : a C-B-C chain and a $B_{11}C^p$ icosahedron with one carbon atom substituted in the polar site.

There are no signatures of C atoms at the chain center. This has important consequences : C-C-C chains can not be responsible for the boron carbide amorphization in shock-wave experiments [59].

Lastly, $B_{10}C_2^p$ icosahedra are present as defects in some B_4C samples. We have estimated that 2.5% of the icosahedra are $B_{10}C_2^p$. These defects have been experimentally observed. The presence of B_{12} icosahedra is inferred by us from the observed stoichiometry. To our knowledge, there are no direct observations of B_{12} icosahedra in B_4C samples.

3. Atomic structure of boron carbides at lower carbon concentrations.

At variance with the case of B_4C, knowledge about boron carbides at lower carbon concentrations must be improved both experimentally and theoretically. The atomic structures are not understood, nor are the phase stabilities [60].

We have computed the equilibrium volume and formation energy of several structural models. Calculations were performed within DFT using the plane-wave pseudopotential method. Exchange and correlation energy was computed in the local density approximation. The

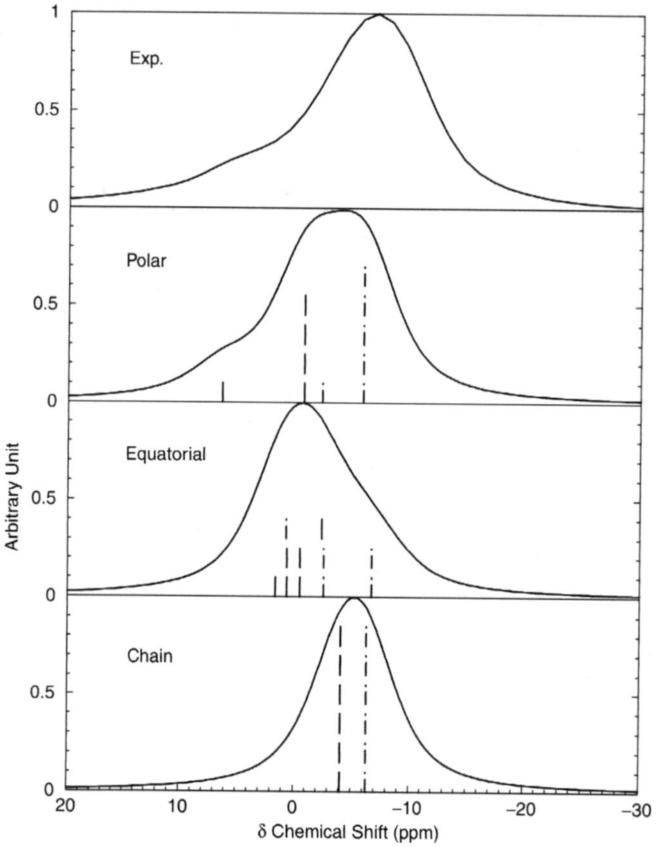

Figure 4. B_4C : ^{11}B NMR chemical shift spectrum. Experimental data are from Ref. [58] and theory for the *polar*, *equatorial* and *chain* models from Ref. [20]. (Reproduced with permission from [20])

pseudopotentials of boron and carbon were developed with the Troullier-Martin scheme [61]. For boron carbides, boron and diamond, plane waves up to a kinetic cutoff of 40 Ry have been included in the basis set. The irreducible wedge of the Brillouin zone has been sampled with 32 points when the unit cell was monoclinic (boron carbides with carbon atoms in the icosahedra), and 10 points when the unit cell was rhombohedral (*e.g.* for α-boron or boron carbides with B_{12} icosahedra) or cubic (diamond). Cell parameters and atomic degrees of freedom have been relaxed to minimize the total energy at theoretical equilibrium.

3.1. Formation energy

Following present knowledge, α-boron (resp. diamond) is the most stable polymorph of pure boron (resp. carbon) within DFT-LDA for defect-free phases (see subsection 2.2). In a first step, formation energies of boron carbides have been computed for a formation from α-boron + diamond.

We first considered a structural model as stable at 0 K when its formation energy E^f is negative with respect to a decomposition into α-boron + diamond, and when E^f is the smallest one among structures with the same carbon concentration. Among many structural models that have been investigated, only the five models of table 3 are found stable [62]. At 20% carbon concentration, the *polar* model B_4C^p has the lowest formation energy. At approximatively 13 % carbon concentration, $B_{13}C_2$ is the most stable phase. It consists in one B_{12} icosahedron and one C-B-C chain. A model combining both B_4C^p and $B_{13}C_2$, doubling the unit cell, is also stable (table 3, 16.7 at. C %).

In a second step, we have tested our structural models against a decomposition into B_4C^p + α-boron (for a carbon concentration lower than 20%) or B_4C^p + diamond (for a carbon concentration greater than 20%). Surprisingly, no phases are found stable against this decomposition in table 4. For $B_{13}C_2$, we found $E^f_{B_4C}$=7 meV per atom. Our result is similar to previous calculations, -21 meV/atom at 0 K and -9 meV/atom at ambient temperature [60]. The model combining B_4C^p and $B_{13}C_2$ also is marginally unstable ($E^f_{B_4C}$ = 7 meV/atom, table 4). These numbers are however within accuracy of actual exchange and correlation functionals (see subsection 2.2).

Replacing the chain-center atom by a vacancy in B_4C^p or in $B_{13}C_2$ yields two models with 21.3 or 14.3% carbon concentration respectively. These crystals with vacancies are found to have a negative formation energy with respect to a formation from α-boron + diamond (table 3). They however have a positive formation energy in table 4 and should decompose into B_4C^p + α-boron or B_4C^p + diamond in thermodynamical conditions.

In conclusion, only B_4C^p is found to be thermodynamically stable in our calculations. Four other phases are found thermodynamically unstable at ambient conditions because : *(i)* at 0 K, they are energetically close to a decomposition into B_4C^p + α-boron or B_4C^p + diamond (table 4); *(ii)* entropy effects are expected to further favor this decomposition at ambient temperature. Having however negative formation energies with respect to α-boron + diamond (table 3), the four phases could nonetheless exist either in a metastable state, like diamond does; or as defects in an otherwise perfect crystal of B_4C^p, as $B_{10}C_2^p$ does (see subsection 2.4).

Turning to unstable structures,[62] our formation energies with respect to α-boron + diamond are gathered in table 5 together with previous studies [46, 60, 63]. Second in formation energy after B_4C^p, which is the most stable phase, come both our model for substitutional disorder (row 2) and the *bipolar* model (row 3). We have inferred from the broadening of the experimental vibrational modes that substitutional disorder is present, and inspection of NMR data tells us that the *bipolar* model is observed as a defect in actual samples. We now compare these two models. Our first approach to account for the effect of substitutional disorder (table 5, row 2) is to double the cell and to substitute the carbon atom of the second icosahedron at different locations in the polar site. The effect of such a substitutional disorder on the formation energy is noticeable (row 2). In fact, starting from two identical icosahedra $B_{11}C^p$ + $B_{11}C^p$, one would expect that the effect on the formation energy of exchanging one carbon atom and one boron atom in the same icosahedron, resulting in non identical $B_{11}C^p$ + $B_{11}C^{p'}$ (row 2) should be much smaller than that of a *bipolar* model, where one carbon atom and one boron atom are exchanged between two neighbouring icosahedra, yielding $B_{10}C_2^p$ + B_{12} (row 3). This is not the case, the two models have a comparable formation energy (table 5, rows 2 and 3). We then conclude that either our study of substitutional disorder lacks convergence with respect to the number of unit cells, and that larger cells should be studied; or that substitutional disorder might be energetically less favoured than a double substitution in the icosahedron.

At a carbon concentration of 20%, we found negative energies for the *equatorial* and *chain* models previously discussed in subsection 2.2 (table 5, rows 4 and 5). Nonetheless, these models have a much higher formation energy than the *bipolar* model, which is consistent with the fact that inspection of spectroscopic data allows to exclude them. Moreover, we found positive formation energy for the *disordered chain* and B-*rich chain* models, which allows to exclude their formation in actual compounds (table 5, rows 6 and 7).

At carbon concentrations other than 20%, all our unstable models show a positive energy of formation, larger than our expected accuracy on exchange and correlation functionals (see subsection 2.2). They can not be produced in a metastable state or as defects. For instance, insertion of carbon atoms in interstitial crystallographic sites is proved to be an inappropriate way of modeling boron-rich boron carbides (table 5, rows 13 and 17). Another model which has been discussed in the literature contains one atom at the centre of the icosahedron [64]. Such a model has a very high positive formation energy (table 5, rows 8 and 14). Therefore we do not expect to observe materials in which boron icosahedra accommodate one atom at the icosahedron centre.

3.2. Discussion

Like B_4C, boron carbides at low carbon concentrations probably are mixings of various building blocks. At carbon concentration other than 20%, the numerous structural models of table 5 can be disregarded as possible candidates. Instead, models of table 3 are useful guides towards a determination of actual atomic structures. The common thought that "intrinsic structural defects (...) originate from the energetically more favorable configuration compared to the ideal structures" [65] lacks a theoretical justification. Indeed, the surprise comes from the small number of possible candidates (table 3). Furthermore, the computation of the formation energy shows that these structures are not stable with respect to a decomposition into B_4C^p boron carbide + α-boron or B_4C^p + diamond (table 4). This metastability explains why physical properties measured on different samples strongly depend on the synthesis route. More precisely, the instability of $B_{13}C_2$ with respect to a decomposition into B_4C^p + α-boron also explains the lack of an actual route to synthesize single crystals of $B_{13}C_2$.[66] Investigation with *ab initio* methods of an eventual stabilization under high pressure should shed some light on this question.

NMR data are crucially lacking for boron carbides at low carbon concentrations. To our knowledge, only the experiment of Ref. [55] yielded data on boron carbides at 13.3 and 10% carbon concentrations. This experiment showed two additional signals beside those of B_4C^p. The intensities of these additional peaks however barely vary between 13.3 and 10% atomic carbon concentrations, so that one can not straightforwardly consider them as fingerprints of boron-rich boron carbides. The precision of NMR technique has drastically increased since 1991, and it would be of utmost importance to perform new experiments on boron carbides. DFT-based methods are mature enough to help understanding these experiments.

Finally, one should note that three models of table 3 contain an odd number of electrons per unit cell, and hence are metals in any band theory. Contrastingly, boron carbides are thought to be semiconductors [66]. We think that actual experimental knowledge might be too much sample dependent. A precise measurement of the resistivity on clean samples is lacking, and would be very helpful in discriminating among the models of table 3. We show in the next section that the strength of the electron-phonon coupling combined with the high average vibrational frequency makes boron carbides very interesting if a metallic state can be achieved.

4. Strength of the electron-phonon coupling

Among possible candidates for the role of the main building block for boron carbides at low carbon concentrations, $B_{13}C_2$ is the favourite one. As described in previous section 3, it consists in one B_{12} icosahedron and one C-B-C chain. The odd number of electrons makes it metallic, while B_4C is semiconducting. $B_{13}C_2$ can be viewed as a B_4C crystal doped with one hole per unit cell, and any band theory in general, and DFT in particular, implies that $B_{13}C_2$ is a metal (figure 5). 90% of the states at the Fermi level are icosahedral states. This was our main motivation to investigate electron-phonon coupling in this material, which is a prototype of metallic icosahedral structures.

We have therefore computed the electron-phonon coupling *ab initio* [8, 69]. We obtained a coupling parameter averaged around the Brillouin zone equal to 0.8. This quantity is 0.9 in MgB_2 [69]. The phonon density of states reported in figure 6, and the Eliashberg function in

Table 3. Formation energy of stable [62] boron carbides at $T = 0$ K, E^f (meV/atom). First column : model name as in table 1. Second column : carbon content. In the third column, excepted for the *polar* model and the elements, the formula of the icosahedron (resp. chain) is given in (resp. out of) brackets. V stands for a vacancy. This work : E^f is computed in LDA for a formation from α-B_{12} + diamond. The formation energies of pure carbon or boron phases are also given.

Model	at. % C	Materials	This work	Theory	Exp.
	100	Graphite		-24[1];0[2]	-35±10[5]
	100	Diamond	0		
	21.4	$(B_{11}C^p)$C-V-C	-15	-11[2]	
Polar	20	B_4C^p	-121	-112[2],-109[3][4]	-88±3[6] (119±24)[7]
	16.7	$(B_{11}C^p)$C-B-C+(B_{12})C-B-C	-94	[-92,-84][4]	
	14.3	(B_{12})C-V-C	-20	-35[2]; -33[4]	
	13.3	(B_{12})C-B-C	-74	-81[2], -79[4]	
	0	α-B_{12}	0	0[2]	

[1] Formation from graphene with Van der Waals forces, Ref. [38].
[2] Formation from graphite and α-B_{12} in GGA, Ref. [63].
[3] LDA, Ref. [60].
[4] GGA, Ref. [46]. Square brackets indicate an energy range for E^f.
[5] Cited in Ref. [38].
[6] Enthalpy of formation at 0 K[67].
[7] Energy of formation at ambient temperature, Ref. [68].

Table 4. Same as in table 3 with $E^f_{B_4C}$ computed in LDA for a formation from B_4C^p + diamond for a carbon concentration larger than 20%, and from B_4C^p+α-boron for a carbon concentration lower than 20%.

Model	at. % C	Materials	$E^f_{B_4C^p+\alpha-B_{12}}$	$\Delta E^f_{B_4C^p+diamond}$
	21.4	$(B_{11}C^p)$C-V-C		104
Polar	20	B_4C^p	0	0
	16.7	$(B_{11}C^p)$C-B-C+(B_{12})C-B-C	7	
	14.3	(B_{12})C-V-C	66	
	13.3	(B_{12})C-B-C	7	

figure 7, allow to understand that the magnitude of the electron-phonon coupling mostly comes from a coupling with the vibrations of the icosahedron.

The pioneering theoretical method to compute the superconductivity now allows to compute the screened Coulomb interaction which is required to evaluate the critical temperature.[70, 71] In our calculation, we have instead used a parameter for the Coulomb repulsion in Mc Millan formula [69]. When this parameter is taken equal to the value of MgB_2, we obtain a critical temperature of 27.6 K, only slightly below the 30.7 K obtained for MgB_2.

To conclude this section, it has been shown that the critical temperature for metallic $B_{13}C_2$ is comparable to the Tc of MgB_2, and that superconductivity is linked to B_{12} building blocks.

Table 5. Unstable [62] boron carbides. Formation energy E^f (meV/atom) from α-B_{12} + diamond. Square brackets indicates an energy range for E_f. Column 1 : model name as in table 1 or row number. Column 2 : carbon concentration. Columns 3 and 4 : description of the chain and of the icosahedron, respectively. V stands for a vacancy. Column 5 indicates if additional atoms are in interstitial positions, and column 6 indicates if the icosahedron contains one atom at its centre.

Model or row number	at. % C	Chain	Icosahedron	Inter-stitial	Ico. center	This work	Theory
	26.7	C-B-C	$B_{10}C_2$			117	
(row 2)	20	$(C\text{-}B\text{-}C)_2$	$B_{11}C^p$+$B_{11}C^{p'}$			[-117,-81]	[-109,-70][1]
Bipolar	20	$(C\text{-}B\text{-}C)_2$	$B_{10}{C_2}^p$+B_{12}			-113	
Equatorial	20	C-B-C	$B_{11}C^e$			-84	-74[1]
Chain	20	C-C-C	B_{12}			-38	
Disordered chain	20	C-C-B	$B_{11}C^p$			55	
B-*rich chain*	20	C-B-B	$B_{10}C_2^p$			127	
(row 8)	18.8	C-B-C	B_{12}		C	648	
	15.4	C-V-V	$B_{11}C^p$				144[2]
	14.3	C-B-V	$B_{11}C^p$				48[2]
	14.3	B-V-C	$B_{11}C^p$				62[2]
	13.3	B-B-C	$B_{11}C^p$				45[2]
(row 13)	12.5	C-B-C	B_{12}	B		49	
(row 14)	12.5	C-B-C	B_{12}		B	504	
	12.5	none	$B_{11}C$+$B_{10}C_2$			579	
	11.1	C-B-C	$B_{11}C$+B_{12}			93	
(row 17)	10.	C-V-C	B_{12}	B_6		355	
	8.3	none	$B_{11}C^p$				127[2]
	7.7	B-V-V	$B_{11}C^p$				212[2]
	7.7	V-B-V	$B_{11}C^p$				212[2]
	7.4	C-B-C	B_{12}+B_{12}			75	
	7.1	C-B-V	B_{12}				65[2]
	7.1	B-V-C	B_{12}				67[2]
	7.1	B-B-V	$B_{11}C^p$				119[2]
	7.1	B-V-B	$B_{11}C^p$				119[2]
	6.7	C-V-V	B_{12}				77[2]
	6.7	B-B-C	B_{12}				14[2]; 16[1]; -58[3];
	6.7	B-B-B	$B_{11}C^p$				75[2]

[1] GGA, Ref. [46].
[2] From graphite and B_{12} in GGA, Ref. [63].
[3] LDA, Ref. [60].

The difficulty comes from the synthesis of single crystals (see subsection 3.2). Extrapolating our results to heavily doped α-boron, its Tc is expected to be greater than the Tc of heavily doped diamond, silicon or silicon carbide, by more than one order of magnitude.

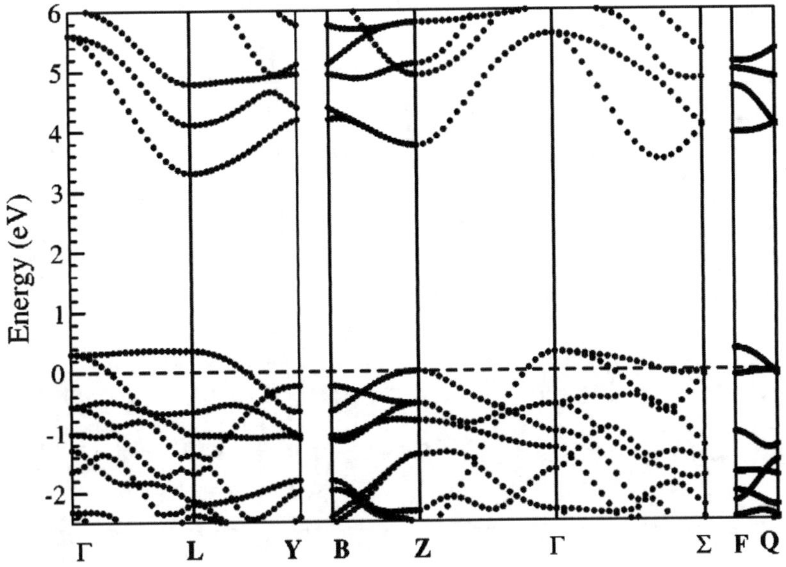

Figure 5. DFT-LDA band structure of $B_{13}C_2$ from Ref. [69]. The energy (eV) is referred to the Fermi level (dashed line). (Reproduced with permission from [69])

5. Conclusion

In this work, we have reviewed actual understanding of the atomic structure of boron carbides gained from *ab initio* calculations. These computations allowed the unambiguous determination of the atomic structure of B_4C. Observed vibrational and NMR spectra are well explained by the *polar* model : one $B_{11}C^p$ icosahedron and one C-B-C chain per unit cell. We have shown that the main structural defect, apart from substitutional disorder of the carbon atom in the polar site, is a $B_{10}C_2^p$ icosahedron. These defects have been observed in one NMR experiment. Moreover stoichiometry implies the existence of B_{12} icosahedra.

This good agreement moreover gives us a great confidence in the ability of DFT to accurately predict the formation energies of boron carbides at concentrations other than 20%. We have shown that very few structures can be regarded as possible candidates to be main structural blocks for boron carbides at carbon concentrations lower than 20%. This goes against the widely spread thought that intrinsic structural defects are energetically favourable with respect to ideal structure. Rather, we find a metastability of any model with respect to B_4C^p + α-boron or B_4C^p + diamond. By shedding light on this metastability, we have explained actual difficulties in the synthesis of clean samples of boron carbides with carbon concentrations lower than 20%.

Last, we have shown that the strength of the electron-phonon coupling in a prototype icosahedral metal, $B_{13}C_2$, is comparable to electron-phonon coupling in MgB_2. From these considerations, we expect electron-phonon coupling in heavily doped α-boron to be more than one order of magnitude greater than that of heavily doped silicon. Combining superconductivity and the known remarkable mechanical properties of boron carbides can lead to materials of high industrial interest.

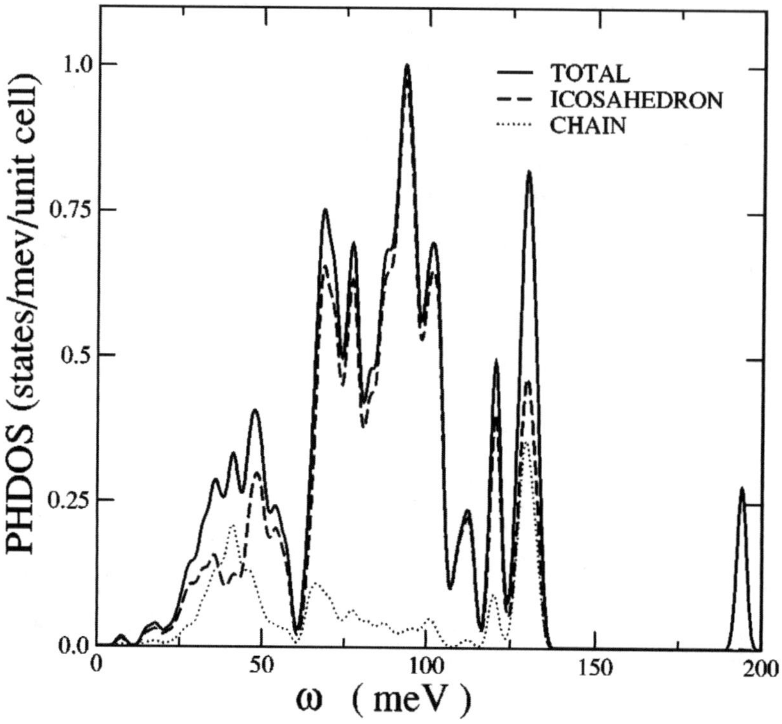

Figure 6. Phonon density of states of $B_{13}C_2$ from Ref. [69]. Solid line : total DOS. Dashed line : contribution from the vibrations of the icosahedron. Dotted line : contribution from the phonon modes of the chain. (Reproduced with permission from [69])

Acknowledgments
The authors thank M. Widom for the communication of results prior to publication [46], and acknowledge useful discussions with M. Widom, I. Boustani, H. Werheit, K. Shirai, H. Dekura and many participants of this ISBB 2008 conference. Support from DGA through a grant is gratefully acknowledged. Results have been obtained with the Quantum Espresso package [4, 72]. Computer time has been granted by CEA/DSM (project p93) and by GENCI (project 2210).

References

[1] Kohn W 2003 *Nobel Lectures, Chemistry 1996-2000* (Singapore: World Scientific Publishing Co.) URL `http://nobelprize.org/nobel_prizes/chemistry/laureates/1998/kohn--lecture.html`
[2] Hohenberg P and Kohn W 1964 *Phys. Rev.* **136** B864
[3] Kohn W and Sham L 1965 *Phys. Rev.* **140** A1133
[4] Baroni S, de Gironcoli S, Corso A D and Giannozzi P 2001 *Rev. Mod. Phys.* **73** 515
[5] Onida G, Reining L and Rubio A 2002 *Rev. Mod. Phys.* **74** 601
[6] Perdew J P and Wang Y 1992 *Phys. Rev.* B **45** 13244
[7] Trinité V 2006 *Etude théorique des phases du titane* Ph.D. thesis Ecole Polytechnique Palaiseau, France URL `http://www.imprimerie.polytechnique.fr/Theses/Files/Trinite.pdf`
[8] Mauri F and Louie S 1996 *Phys. Rev. Lett.* **76** 4246
[9] Cheeseman J, Trucks G, Keith T and Frisch M 1996 *J. Chem. Phys.* **104** 14
[10] Pickard C and Mauri F 2001 *Phys. Rev.* B **63** 245101
[11] Hedin L 1965 *Phys. Rev.* **139** A796

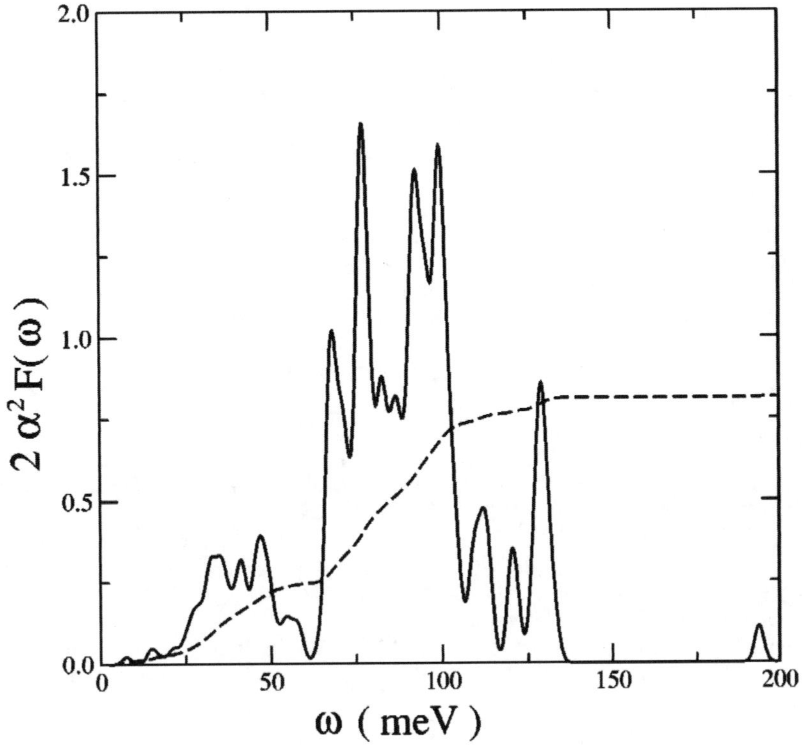

Figure 7. Solid line : Eliashberg function from Ref. [69]. Dashed line : average electron-phonon coupling parameter. (Reproduced with permission from [69])

[12] Hedin L and Lundqvist B 2000 *Solid State Physics* vol 23 ed H Ehrenreich F S and Turnbull D (New York: Academic Press)
[13] Bruneval F, Vast N and Reining L 2006 *Phys. Rev.* B **74** 045102
[14] Runge E and Gross E 1984 *Phys. Rev. Lett.* **52** 997
[15] Gross E K U and Kohn W 1985 *Phys. Rev. Lett.* **55** 2850
[16] Bruneval F, Vast N, Reining L, Izquierdo M, Sirotti F and Barrett N 2006 *Phys. Rev. Lett.* **97** 267601 and references therein.
[17] Georges A, Kotliar G, Krauth W and Rozenberg M J 2001 *Rev. Mod. Phys.* **73** 515
[18] Gatti M, Bruneval F, Olevano V and Reining L 2007 *Phys. Rev. Lett.* **99** 266402
[19] Biermann S, Poteryaev A, Lichenstein A and Georges A 2005 *Phys. Rev. Lett.* **94** 026404
[20] Mauri F, Vast N and Pickard C J 2001 *Phys. Rev. Lett.* **87** 085506
[21] Bullett D W 1982 *J. Phys. C: Solid State Phys.* **15** 415
[22] Jemmis E D, Eluvathingal D, Balakrishnarajan M M and Pancharatna P D 2001 *J. Am. Chem. Soc.* **123** 4313
[23] Jemmis E D, Eluvathingal D and Balakrishnarajan M M 2001 *J. Am. Chem. Soc.* **123** 4324
[24] Lazzari R, Vast N, Besson J, Baroni S and Corso A D 1999 *Phys. Rev. Lett.* **83** 3230
[25] Lazzari R, Vast N, Besson J, Baroni S and Corso A D 2000 *Phys. Rev. Lett.* **85** 4194
[26] Emin D 1987 *Phys. Today* **320** 55
[27] Nelmes R J, Loveday J S, Wilson R M, Marshall W G, Besson J M, Klotz S, Hamel G, Aselage T L and Hull S 1995 *Phys. Rev. Lett.* **74** 2268
[28] Morosin B, Kwei G, Lawson A, Aselage T and Emin D 1995 *J. Alloys Compd.* **226** 121
[29] Kwei G and Morosin B 1996 *J. Phys. Chem.* **100** 8031
[30] Morosin B, Aselage T and Feigelson R 1987 *Mater. Res. Symp. Proc.* **97** 145
[31] Larson A 1986 *Boron Rich Solids Conf. Proc.* vol 140 (New York: AIP)

Journal of Physics: Conference Series **176** (2009) 012002 doi:10.1088/1742-6596/176/1/012002

[32] Morosin B, Aselage T and Emin D 1991 *Boron Rich Solids Conf. Proc.* vol 231 (New York: AIP)
[33] Fernandez-Lima F, Ponciano C R, da Silveira E F and Nascimento M C 2007 *Chem. Phys. Lett.* **445** 147
[34] Marini A, García-Gonzáles P and Rubio A 2006 *Phys. Rev. Lett.* **96** 136404
[35] Lazzeri M, Vittadini A and Selloni A 2001 *Phys. Rev.* B **63** 155409
[36] Janotti A, Wei S H and Singh D 2001 *Phys. Rev.* B **64** 174107
[37] Yin M and Cohen M 1984 *Phys. Rev.* B **29** 6996
[38] Rydberg H, Dion M, Jacobson N, Schröder E, Hyldgaard P, Simak S, Langreth D and Lundqvist B 2003 *Phys. Rev. Lett.* **91** 126402
[39] Fuchs M and Gonze X 2002 *Phys. Rev.* B **65** 235109
[40] Furche F and Voorhis T V 2005 *J. Chem. Phys.* **122** 164106
[41] García-Gonzáles P, Fernandez J J, Marini A and Rubio A 2007 *J. Phys. Chem.* A **111** 12458
[42] Harl J and Kresse G 2008 *Phys. Rev.* B **77** 045136
[43] Masago A, Shirai K and Katayama-Yoshida H 2006 *Phys. Rev.* B **73** 104102
[44] Dekura H and Shirai K 2008 *private communication*
[45] Widom M and Mihalkovič M 2008 *Phys. Rev.* B **77** 064113
[46] Widom M and Mihalkovič M 2009 *in this Proceedings*
[47] Shang S, Wang Y, Arroyave R and Liu Z K 2007 *Phys. Rev.* B **75** 092101
[48] van Setten M, de Wijs M U G and Robert A 2007 *J. Am. Chem. Soc.* **129** 2458
[49] Ogitsu T, Gygi F, Reed J, Motome Y, Schwegler E and Galli G 2009 *J. Am. Chem. Soc.* **131** 1903
[50] Kuhlmann U, Werheit H and Schwetz K 1992 *J. Alloys Compd.* **189** 249
[51] Vast N, Baroni S, Zerah G, Besson J M, Polian A, Grimsditch M and Chervin J C 1997 *Phys. Rev. Lett.* **78** 693
[52] Tallant D R, Aselage T L, Campbell A and Emin D 1989 *Phys. Rev.* B **40** 5649
[53] Kuhlmann U and Werheit H 1993 *Phys. Stat. Sol.* **175** 85
[54] Werheit H, Schmechel R, Kuhlmann U, Kampen T, Mönch W and Rau A 1999 *J. Alloys Compd.* **291** 28
[55] Kirkpatrick R, Aselage T, Phillips B and Montez B 1991 *Boron Rich Solids Conf. Proc.* vol 231 (New York: AIP)
[56] Duncan T 1984 *J. Am. Ceram. Soc.* **106** 2270
[57] Harazono T, Hiroyama Y and Watanabe T 1996 *Bull. Chem. Soc. Jpn* **69** 2419
[58] Simeone D, Mallet C, Dubuisson P, Baldinozzi G, Gervais C and Maquet J 2000 *J. Nucl. Mater.* **277** 1
[59] Fanchini G, McCauley J and Chhowalla M 2006 *Phys. Rev. Lett.* **97** 035502
[60] Bylander D and Kleinman L 1991 *Phys. Rev.* B **43** 1487
[61] Troullier N and Martins J L 1991 *Phys. Rev.* B **43** 1993
[62] We call stable at 0 K a structural model whose formation energy is negative with respect to a decomposition into diamond plus α-boron, and is the lowest one among structures with the same carbon concentration.
[63] Saal J, Shang S and Liu Z K 2007 *Appl. Phys. Lett.* **91** 231915
[64] Hayami W 1999 *Phys. Rev.* B **60** 1523
[65] Werheit H 2000 *Landolt-Börnstein, Boron compounds* vol 41D ed Madelung O (Berlin: Springer-Verlag) p 2
[66] Werheit H 2008 *this Conference*
[67] Chase M 1998 *NIST-JANAF Thermochemical Tables, Part I, Al-Co* 4th ed vol 9 (Nat. Inst. Stand. Technol.)
[68] Smith D, Dworking A and Artsdalen E V 1955 *J. Am. Chem. Soc.* **77** 2654
[69] Calandra M, Vast N and Mauri F 2004 *Phys. Rev.* B **69** 224505
[70] Lüders M, Marques M, Lathiotakis N, Floris A, Profeta G, Fast L, Continenza A, Massida S and Gross E 2005 *Phys. Rev.* B **72** 024545
[71] Marques M, Lüders M, Lathiotakis N, Profeta G, Floris A, Profeta G, Fast L, Continenza A, Gross E and Massida S 2005 *Phys. Rev.* B **72** 024546
[72] Baroni S, de Gironcoli S, dal Corso A, Giannozzi P, Alfé D, Antoniella F, Ballabio G, Bungaro C, Cipriani G, Cococcioni M, Debernardi A, Deinzer G, Fratesi G, Gebauer R, Lazzeri M, Mäder K, Mauri F, Conte A M, Pavone P, Roma G, Stokbro K, Kokalj T and Wentzcovitch R http://www.pwscf.org

Flux pinning property of TiB_2 centers in sintered MgB_2

Yoshihide Kimishima, Syoko Numa, Masatomo Uehara and Tetsuji Kuramoto

Department of Physics, Faculty of Engineering, Graduate School of Yokohama National University, Tokiwadai 79-5, Hodogaya-ku, Yokohama 240-8501, Japan

E-mail: kimi@ynu.ac.jp

Abstract. Bulk $(TiB_2)_x(MgB_2)_{1-x}$ samples were prepared by sintering mixtures of Mg, B and TiO_2 in (10% H_2)/Ar atmosphere at 900 °C for 30 min. Improvement of the critical current density J_c was confirmed at temperature of 20 K and magnetic field of 1T for the x=0.02 sample with no reduction in the critical temperature T_c. TiB_2 phase produces efficient centers of magnetic flux pinning in MgB_2.

1. Introduction

Recently, our group has investigated the effect of metal doping on the magnetic flux pinning in MgB_2. Enhancement of critical current density J_c at temperature of 20 K and magnetic field of 1 T was observed in MgB_2:M (M = Ag, Cu, Zn or Nb) sintered bulk samples containing a few mol% of metal M. Those samples were prepared by sintering mixtures of Mg, B and M_mO_n in Ar or (10% H_2)/Ar atmosphere at 900 °C for 30 min. Then M_mO_n was reduced to M, and M reacted with Mg or B to form extrinsic pinning centers. The impurities of Ag-Mg alloy [1, 2], $MgCu_2$[3], Mg-Zn alloy [4] and NbB_2 [1] acted as strong pinning centers for the penetrating magnetic flux. The largest J_c of M = Ag, Cu, Zn and Nb systems at 20 K and 1 T was 1.33, 1.48, 1.80 and 1.70 times that of pure MgB_2, respectively. The efficiency of small addition of metallic elements was also confirmed for sintered MgB_2 system under high pressure of 3.5 GPa [5], and for MgB_2 with addition of SiC [6].

Here we present enhancement of J_c by TiB_2 pinning centers. Effects of Ti doping into bulk MgB_2 have been reported by several authors [7-9]. Zhao *et al* [7] prepared $Ti_xMg_{1-x}B_2$ by solid-state reaction in flowing Ar at 600 °C for 1 h, at 800 °C for 1 h and at 900 °C for 2 h. They observed extended grain connectivity and high density in $Ti_xMg_{1-x}B_2$ system as well as high J_c value for $x = 0.1$ sample at 20 K and 1 T. Bulk MgB_2, synthesized under high pressure with addition of 10 wt% Ti [8], also showed high J_c. Shen *et al* [9] showed efficient flux pinning by Ti-phase in $Ti_{0.05}Mg_{0.95}B_2$ and $Ti_{0.05}Mg_{0.95}C_{0.05}B_{1.95}$. However, the J_c was unchanged by adding up to 4 wt% of 10-15 nm TiO_2 particles into MgB_2 [10].

As to the TiB_2-doped MgB_2 system, Goto *et al* [11] showed that $Ti_{0.05}(MgB_2)$ had the largest J_c value. Xu *et al* [12] doped TiB_2 into MgB_2 wires and observed high J_c at 4.2 K for $x = 5 - 8$ mol% samples. Thus, J_c enhancement by adding TiB_2 was reported, but systematic study of the effects of TiB_2 pinning centers has not been performed for bulk $(TiB_2)_x(MgB_2)_{1-x}$, $x = 0 - 0.05$ samples. In the following sections, variations of T_c and J_c with TiB_2 content in such samples are presented.

2. Sample preparation

$(TiB_2)_x(MgB_2)_{1-x}$ samples were prepared by sintering mixtures of Mg, B and TiO_2 with mole ratios Mg:B:TiO_2 = $1.6(1-x):2:x$, $x = 0 - 0.05$. The excess amount of Mg was expected to react with the oxygen from TiO_2. Solid phase reaction within the powder mixture was achieved in an electric furnace at 900 °C for 30 min under Ar atmosphere with 10% H_2, followed by cooling in the furnace. Then TiO_2 was reduced to Ti which reacted with B to form extrinsic TiB_2 pinning centers.

3. Experimental results and discussion

Figure 1 shows Cu$K\alpha$ powder X-ray diffractions (XRD) patterns for the $x = 0$, 0.02 and 0.05 samples. The intensities of the (101) major peak of MgB_2 phase are normalized to 100. The intensities of MgO diffraction peaks increased upon a small addition of TiO_2, but the increase saturated with further Ti doping ($0.01 \leq x \leq 0.05$). The (101) peak of TiB_2 was found at 2θ of about 44.4°, and the particle diameter of TiB_2 was estimated at about 8-10 nm from the half width of the (101) peak.

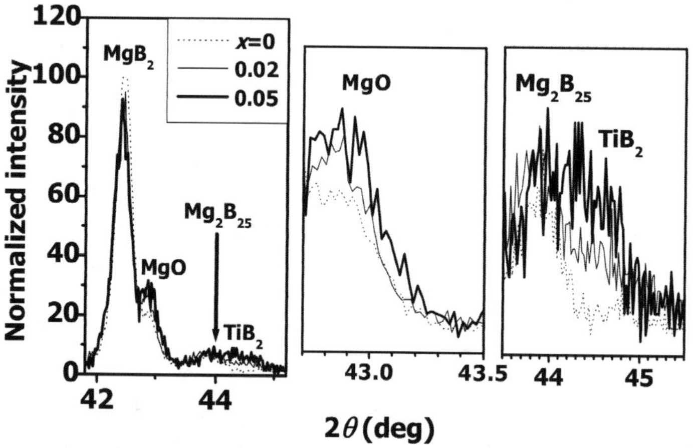

Figure 1. X-ray diffraction patterns of $x = 0$, 0.02 and 0.05 samples.

Electrical resistivity ρ was measured by DC four-terminal method and its temperature dependences $\rho(T)$ are shown in figure 2 for $x = 0 - 0.05$ samples. The x dependence of superconducting transition temperatures T_c (onset, mid and end points of the transition region) are shown in figure 3. The T_c values varied between 38 and 39.5 K for all samples. The absence of significant T_c reduction in our $x = 0.01 - 0.05$ samples means that TiB_2 has not dissolved significantly in the MgB_2 phase. Small change of T_c is preferable for enhanced pinning in this system.

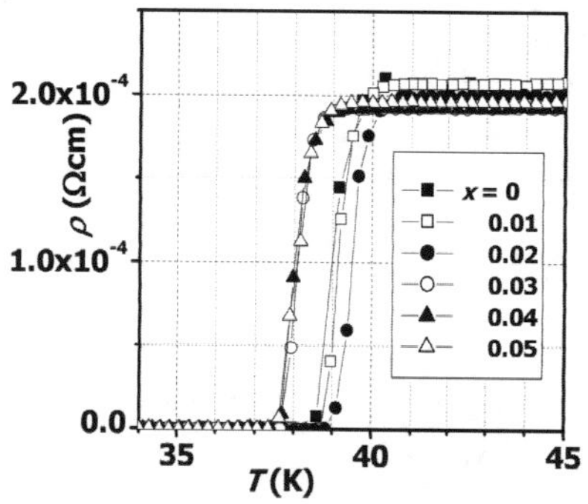

Figure 2. Temperature dependence of resistivity for $x = 0 - 0.05$ samples.

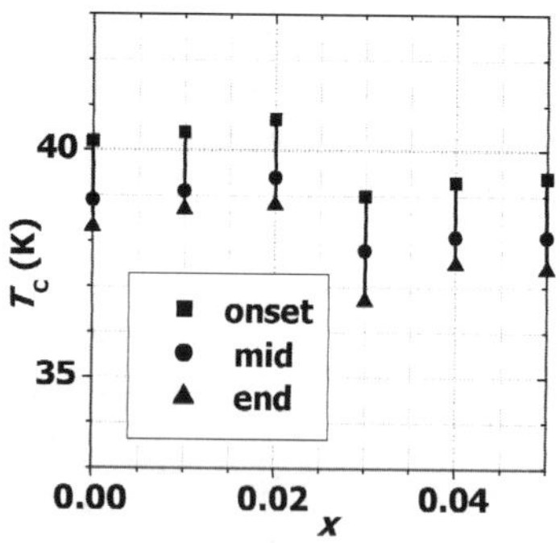

Figure 3. x-dependence of superconducting transition temperature T_c.

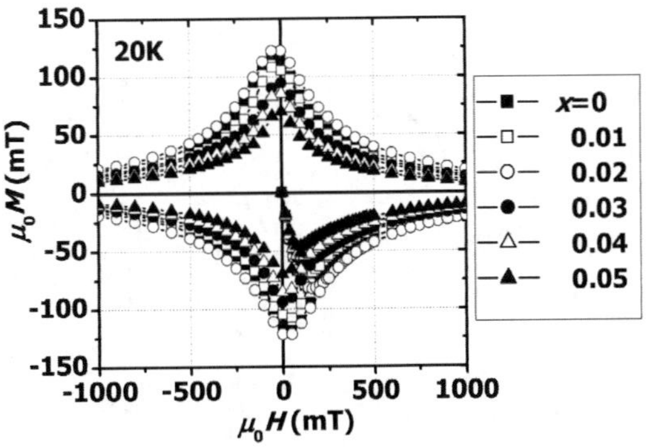

Figure 4. $M(H)$ curves of x = 0–0.05 samples at 20 K.

Magnetization measurements were performed with a MPMS SQUID (Quantum Design) magnetometer at 5 and 20 K varying the magnetic field between −5 and 5 T. The expressions for magnetization by extended critical state model [13-16] were fitted to the experimental data. The $M(H)$ curves and theoretical curves of $\mu_0 a J_c$ at 20 K for x = 0 – 0.05 are depicted in figures 4 and 5, respectively, where $\mu_0 = 4\pi\times10^{-7}$ T·m/A is the vacuum permeability, and $2a$ is the sample thickness. The x dependences of J_c at 20 K at 0 T and 1 T, normalized by those of pure MgB_2, are shown in figure 6. The enhancement of J_c is apparent for nominal $(TiB_2)_{0.02}(MgB_2)_{0.98}$ system. Therefore, in the present case of relatively quick sintering, we can expect that the TiB_2 content $x \sim 0.02$ is optimal for the pinning enhancement. MgB_2 grains of about 0.1 mm in size were observed by scanning electron microscopy in all samples. Substituting 0.1 mm for $2a$, J_c at 20 K and 1 T was estimated as 6.4×10^4 and 7.4×10^4 A/cm^2 for x = 0 and 0.02 samples, respectively. These J_c values are relatively low, but they should improve in samples with higher density and connectivity produced by high-pressure sintering.

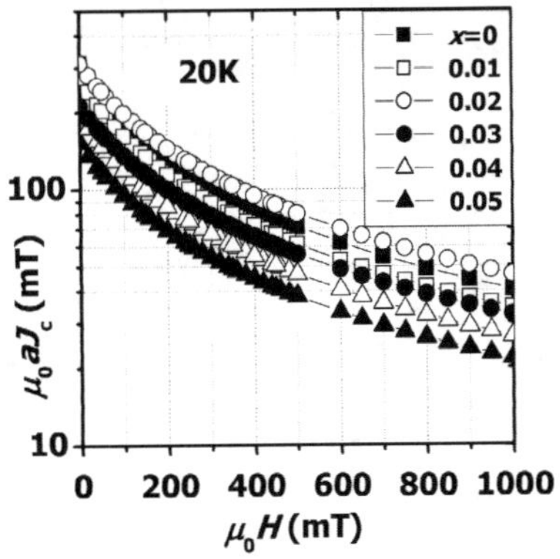

Figure 5. Field dependence of theoretical $\mu_0 a J_c$ for x = 0–0.05 samples at 20 K.

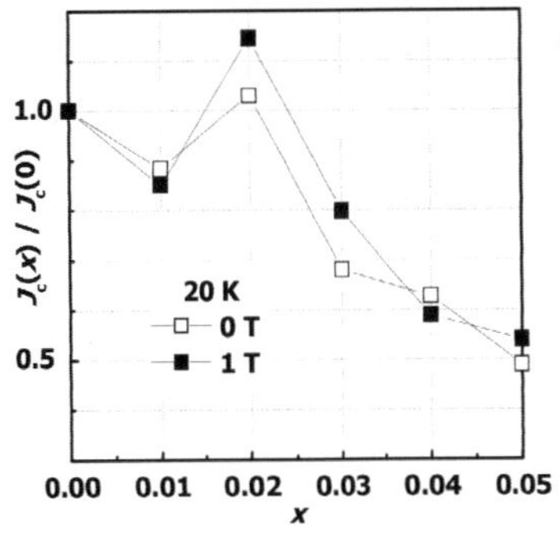

Figure 6. Composition dependence of normalized critical current density at 20 K.

As shown in figure 7, the irregular flux jumps in $M(H)$ curves at 5 K, disappeared in the present TiB_2/MgB_2 system at x above 0.03. Perhaps improved connectivity between MgB_2 grains by TiB_2 impurities increased thermal conductivity and thus suppressed the flux jumps due to the thermal excitation of dendritic flux avalanches in MgB_2 [17]. However, more likely the shielded field and thus the flux jumps were suppressed by the decrease of a or J_c.

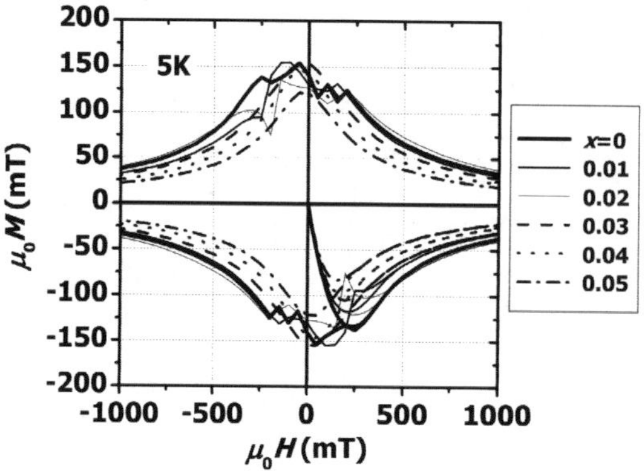

Figure 7. *M*(*H*) curves of $x = 0-0.05$ samples at 5K.

4. Conclusion

We present a method to introduce TiB_2 flux pinning centers into bulk MgB_2 by sintering. Bulk samples of $(TiB_2)_x(MgB_2)_{1-x}$, $x = 0$, 0.01, 0.02, 0.03, 0.04 and 0.05, were prepared from mixtures of Mg, B, and TiO_2 under (10% H_2)/Ar atmosphere at 900 °C for 30 min. Critical current density J_c was estimated from the magnetization data using an extended critical state model. Improvement of J_c with no reduction in T_c was confirmed at 20 K and 1 T for the $x = 0.02$ sample. TiB_2 phase produced efficient pinning centers in this system.

References

[1] Kimishima Y, Takami S, Uehara M, Okuda T and Kuramoto T 2006 *J. Phys.: Conf. Ser.* **43** 488
[2] Kimishima Y, Takami S, Uehara M, Numa S, Sugiyama Y and Kuramoto T 2008 *Physica C* **468** 1181
[3] Kimishima Y, Takami S, Uehara M and Kuramoto T 2006 *Physica C* **445-448** 224
[4] Kimishima Y, Sugiyama Y, Numa S, Uehara M and Kuramoto T 2008 *Physica C* **468** 1185
[5] Kimishima Y, Takami S, Okuda T, Uehara M, Kuramoto T and Sugiyama Y 2007 *Physica C* **463-465** 281
[6] Kimishima Y, Okuda T, Uehara M and Kuramoto T 2007 *Physica C* **463-465** 286
[7] Zhao Y, Feng Y, Cheng C H, Fudamoto Y, Koshizuka N and Murakami M 2001 *Appl. Phys. Lett.* **79** 1154
[8] Hörhager N, Eisterer M, Weber H W, Prikhna T, Tajima T and Nesterenko V F 2006 *J. Phys.: Conf. Ser.* **43** 500
[9] Sheng T M, Li G, Cheng C H and Zhao Y 2006 *Supercond. Sci. Technol.* **19** 1219
[10] Kishan H, Awana V P S, de Oliveira T M, Alam S, Saito M and de Lima O F 2007 *Physica C* **458** 1
[11] Goto D, Machi T, Zhao Y, Koshizuka N, Murakami M and Arai S 2003 *Physica C* **392-396** 272
[12] Xu H L, Feng Y, Xu Z, Yan G, Li C S, Mossang E and Sulpice A 2006 *Physica C* **443** 5
[13] Kimishima Y, Uehara M, Kuramoto T, Ichiyanagi Y, Iriyama Y and Yorimasa K 2002 *Physica C* **377** 196
[14] Yorimasa K, Uehara M, Kuramoto T, Inoue H and Kimishima Y 2003 *Physica C* **392-396** 306
[15] Kimishima Y, Ichikawa H, Takano S and Kuramoto T 2004 *Supercond. Sci. Technol.* **17** S36
[16] Kimishima Y, Uehara M, Kuramoto T, Takano S and Takami S 2004 *Physica C* **412-414** 402
[17] Johansen T H, Baziljevich M, Shantzev D V, Goa P E, Galperin Y M, Kang W N, Kim H J, Choi E M, Kim M–S and Lee S I 2002 *Europhys. Lett.* **59** 599

16th International Symposium on Boron, Borides and Related Materials IOP Publishing
Journal of Physics: Conference Series **176** (2009) 012004 doi:10.1088/1742-6596/176/1/012004

Possibility of imperfections in α-boron

Haruhiko Dekura, Koun Shirai, and Akira Yanase[†]

Nanoscience and Nanotechnology Center, ISIR, Osaka University, 8-1, Mihogaoka, Ibaraki, Osaka, Japan

[†] ISIR, Osaka University, 8-1 Mihogaoka, Ibaraki, Osaka, Japan

E-mail: dekura@cmp.sanken.osaka-u.ac.jp

Abstract. Recently, we have shown by density functional theory calculations that α-boron is the most stable phase at low temperature among icosahedron-based boron crystals. This is consistent with low-temperature crystal growth of α-boron. A problem of the calculation is, however, that the conclusion would be changed if defect states for β-boron, which are observed in experiment, were taken into account. This urges us to suspect the perfectness of α-boron crystal. In this paper, we investigate possibility of imperfections of α-boron from the standpoint of crystal stability. Intrinsic defects of vacancy and self-interstitial have been examined. Based on the total energy calculations, we conclude that intrinsic defects are not present at a level above 0.1 at%.

1. Introduction

The problem of stability of polymorphic modifications has long been an important issue of boron physics and chemistry. There are many polymorphic modifications for boron. Here, we are more specifically concerned with the stability between α- and β-rhombohedral boron, because these two are most often obtained polymorphs showing strong contrast in many respects. Since α-boron (hereafter term rhombohedral will be omitted for brevity) was for the first time isolated from decomposition of boron hydrides [1], it has been recognized that the preparation temperature is relatively low compared with the temperature of β-boron synthesis. McCarty *et al* reported that α-boron was obtained from pyrolytic decomposition of BI_3 on a surface heated to a temperature in the range 800 to 1100°C [1]. At higher temperatures, β phase becomes stable. Later, Horn succeeded to obtain α phase from a molten metal (B-Pt) at the eutectic temperature 830 °C [2] (however 750°C in Ref. [3]). On the other hand, there is no report on crystal growth of β-boron at low temperatures comparable to these data.

In spite of these facts, many researchers believed that β phase is stable over the whole temperature range. Probably, this believe comes from an authoritative reference of Hoard and Hughes [4]. Although they notice the fact of low-temperature synthesis of α-boron, it seems reasonable for them to consider that α phase is obtained only in some specific circumstance, for example, using specific starting materials. However, Kimura's group showed that by starting from amorphous boron, α phase is obtained in a very narrow temperature range around 1200°C with the width $\sim$20 deg [5]. Below this temperature, no crystallization occurred. From this observation, it seems that α phase is the stable phase at low temperatures. Indeed, it is otherwise difficult to understand low-temperature synthesis of α-boron in molten metals (Pt-B) [2] or Pd-B [6] [7].

The stability problem becomes more important in connection with advent of superconductivity of β-boron at high pressure [8]. The stability of β-boron at high pressure was examined by various experimentalists [9, 10]. On theoretical side, relatively simple structural models were examined [11], and α-Ga type had been proposed [12, 13]. However, there was no systematic study on the stability of β-boron before five years ago. In this situation, the authors have studied the stability of α- and β-boron by *ab initio* calculations [14]. Our conclusion is that α phase is the stable phase at low temperature, whereas β phase is stable at high temperature. The transition temperature between α and β is about 700°C. This value is obtained by taking the phonon contribution within the harmonic approximation. Later, by taking anharmonic effects into account within quasiharmonic approximation, we have obtained phase diagram of boron [15]. Considering the anharmonic effect in this way does not significantly change the transition temperature.

Although the calculated phase diagram seems to give reasonable account to experimental facts, there is a problem of defect states affecting the thermodynamic stability. In obtaining the phase diagram mentioned above, we ignored defect states of β-boron. If the defect states in β-boron are taken into account, the transition temperature becomes too low (-100°C). Our calculation shows that even though the energy reduction by defects is only minor, *i.e.*, $\Delta E_{\mathrm{d}} \sim 4$ meV, its thermodynamic consequence becomes appreciable, because the entropy effect is sensitive to the energy difference [14]. Recent study of β-boron by Widom and Mihalkovic extends calculation to include further defect states [16]. It found that β-boron is more stable even at zero temperature that contradicts to the above experimental facts. Further studies were published in this regard [17, 18].

By observing this unfavorable effect of the defects in β-boron, the present authors supposed that even α-boron must have intrinsic defects similar to β-boron [14]. Any calculation ignoring defects in α-boron would lead to a conclusion of relative stability of β-boron at finite temperatures. However, there has been thus far no information about the defect states in α-boron. This is why the problem of defects for α-boron was not pursued in our study [14].

Defects do exist in α-boron, if we are not concerned with the concentration. Since McCarty's first isolation of α-boron, it has been recognized that crystals can be of black or red color [19, 20, 4]. The color can originate from defects and impurities. However, the impurities responsible for the red color of α-boron are not well characterized. These impurities may not be important for the stability problem of α- and β-boron, unless the concentration is too high. In this paper, we are mainly interested in intrinsic defects, *i.e.*, interstitial or vacancy, as in β-boron. Since Decker and Kasper made a precise determination of the crystal structure of α-boron [19], such intrinsic defects have not been considered, because the atomic occupancies of all the lattice sites are 100 %. This clearly contrasts with the situation of β-boron, for which several cites of partial occupancy were found [21, 22, 23, 24]. Among various polymorphs, α-boron is regarded as the most perfect [25]. Of course, X-ray determination of 100 % occupancy does not guarantee perfect crystal with no trace-level defects. Presence of intrinsic defects with a marginal concentration could have a role in the crystal stability.

This is the first study addressing the perfection of α-boron from standpoint of crystal stability. The types of intrinsic defects studied in this paper are simple point defects, i.e., vacancy and interstitial. Complex types are excluded here. Before performing first principles calculations, we give an estimate of the concentration of defects which is relevant to the present issue. Then, the formation energies of defects are calculated by first-principles pseudopotential method.

2. Preliminary

Let us first examine the accuracy of the experimental atomic density of α-boron. For α-boron, the formal value of atoms per cell N_{at} is 12. Decker and Kasper determined this N_{at} from the measurement of the lattice parameters and the mass density [19]. They obtained

$a_0 = 5.057 \pm 0.003$ Å (0.06 % error) and $\alpha = 58.06 \pm 0.05$ deg (0.09 %), from which the cell volume has been calculated as $V = 87.3 \pm 0.16$ Å^3 (0.2 %). The mass density ρ was obtained as $\rho = 2.46$ g/cm^3. Using the nominal value for the mass of boron, $M = 10.811$ amu, we obtain

$$N_{\rm at} = \rho V/M = 11.96 \text{ atom/cell}, \tag{1}$$

with the relative error of 0.3 %. Therefore, concentrations of intrinsic defects below 0.3 % are permissible without causing any contradiction with the existing experimental data. This concentration roughly amounts to one defect in 30 primitive unit cells, with a total number of atoms 360. Although a 400-atom supercell is currently manageable in our code, for calculating the formation energy, usually we do not need such a large supercell. In this study, we examined three primitive unit cells as the supercell. This is conveniently carried out using a hexagonal unit cell.

Concerning the formation energy of defects $\Delta E_{\rm d}$, let us estimate the magnitude of $\Delta E_{\rm d}$ which could have discernible effect on the present problem. In case of β-boron [14], the effect of defects on the free energy ΔF_c has been estimated through the configurational entropy,

$$\Delta F_d = kT \ln\left(\sum_i w_i e^{\Delta E_i/kT}\right), \tag{2}$$

where w_i is the multiplicity of a specific type of defect i. In order for this contribution to yield a sizable effect, temperature T should be higher than the lowest ΔE_i among various defects. The transition temperature between α- and β-boron is around 1000 K, so that in order for the defects of α-boron to affect the thermodynamics through Eq. (2), the $\Delta E_{\rm d}$ should be of the order 1000 K, *i.e.*, 0.1 eV. This means that any defect with $\Delta E_{\rm d} > 1$ eV may be irrelevant with the thermodynamic stability of α- and β-boron.

3. Method

In this study, first-principles pseudopotential method with local density approximation (LDA) is used for the total-energy calculation. The used code is 'Osaka2002' [26]. Troullier-Martins' pseudopotentials [27] are used, with aid of a full separable Kleinman-Bylander form [28].

The cutoff energy for the plane-wave basis is 55 Ry. A Mhonkhorst-Pack mesh of (2×2×2) is used for k-point sampling [29]. With these conditions, convergence of the total energy was below 2 meV/atom, and that of formation energy below 0.5 meV/atom. Structural optimization was performed for atomic positions only.

Structural models of defects in α-boron are constructed on 12-atom primitive unit cell and 36-atom supercell. The 36-atom supercell is constructed as a hexagonal cell as shown in figure 1. For vacancy, there are only two sites: polar (p) site and equatorial (e) site of the icosahedron. In contrast, for interstitial site, there are many possibilities. However, by our experience of impurity study in α-boron [30, 31] and by analogy to other boron-rich solids such as $B_{12}P_2$, the center of the main diagonal (O) site is almost always the most stable site. Hence, here only O site is considered.

The formation energy ΔE_f is obtained by difference of the total energies before and after the reaction. For example, the formation energy ΔE_f of vacancy in the primitive unit cell is obtained

$$\Delta E_f = E(\mathrm{B}_{11}\square) - \frac{11}{12}E(\alpha - \mathrm{B}). \tag{3}$$

Here, vacancy is denoted by $\square$ and interstitial boron is denoted by B_i. The B atom which was missed after creation of a vacancy is supposed to move outside the specimen, but still in the crystal state of α-boron. In this way, ΔE_f can vary by changing the reference state, in other

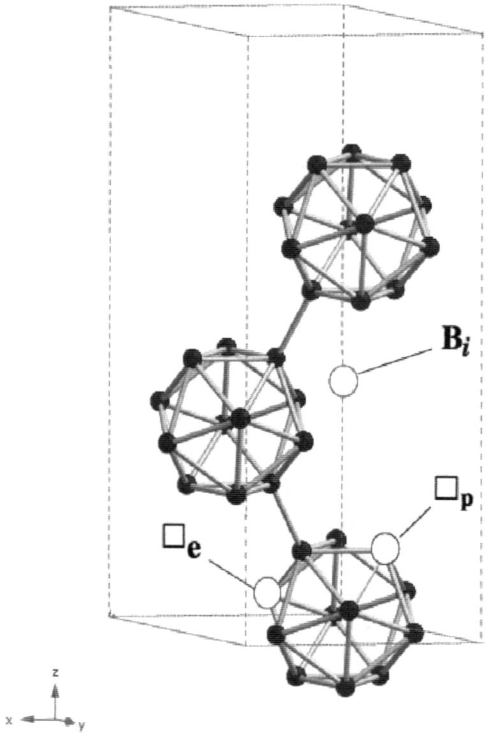

Figure 1. Sites of intrinsic defects in α-boron B_{12}; B_i is self-interstitial, while $\square_p$ and $\square_e$ are vacancy of polar site and equatorial site, respectively. The rhombohedral crystal is described in a hexagonal system. Hence, three icosahedra B_{12} are seen.

words, it is controllable by changing the starting conditions, such as the initial phase (liquid, amorphous, alloy, etc). The sign convention of definition (3) is usual one in impurity physics of semiconductors. Positive ΔE_f means that creation of a defect is difficult.

4. Results and discussion

The calculated ΔE_f of various defects are listed in table 1. The formation energies are all positive, and the values are from 4 to 6 eV. These values are typical for usual semiconductors [32, 33, 34]. For vacancy, there is virtually no difference between the polar and equatorial sites.

The formation energy ΔE_f decreases with decreasing concentration of defect, however the decrease is minor. To see this dependence more clearly, we plot ΔE_f as a function of the concentration n in figure 2. As shown in figure 2, the ΔE_f at low-concentration limit is 4.7 eV for vacancy and 6.3 eV for interstitial. Therefore, judging from the aforementioned criteria on the formation energy, it is concluded that vacancy or interstitial defects are not responsible for the crystal stability of α-boron. The crystal of α-boron must be perfect within about 0.1 at.%.

There are several issues to be considered before establishing this conclusion. One deficiency of our calculations is that the cell parameters were not optimized. Instead, we monitored the residual stress $\Delta\sigma$, which was caused by fixing the lattice parameters. The bulk modulus of

Table 1. Formation energy ΔE_f of defects in α-boron as a function of the defect concentration n. The residual force Δf and residual stress $\Delta\sigma$ are also listed in order to check the accuracy. a_B is the Bohr radius.

Defect	ΔE_f eV/atom	n at.%	Δf $\times 10^{-2}$ Ry/a_B	$\Delta\sigma$ GPa
use of a primitive cell B_{12}				
$B_{11}\square_p$	+4.86	8.3	1.5	29.4
$B_{11}\square_e$	+4.86	8.3	3.1	14.7
$B_{12}B_i$	+6.90	7.7	3.0	19.1
use a conventional cell, $3\times B_{12}$				
$B_{35}\square_p$	+4.77	2.8	0.6	12.2
$B_{35}\square_e$	+4.77	2.8	0.53	6.0
$B_{36}B_i$	+6.54	2.7	1.45	5.3

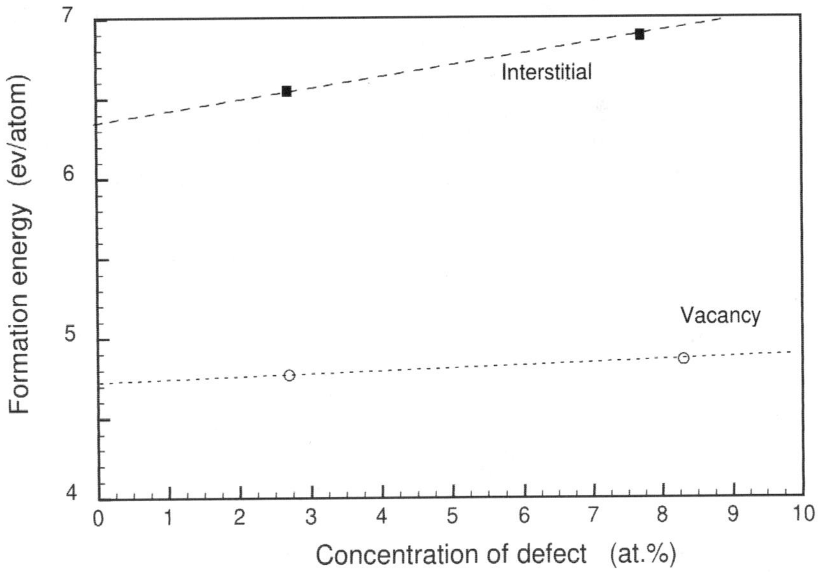

Figure 2. Concentration dependence of ΔE_f. The obtained ΔE_f of interstitial and vacancy at low concentration limit are 6.3 and 4.7 eV/atom, respectively.

boron B is about 200 GPa [14], so that the listed value of the residual stress in table 1, for example $\Delta\sigma = 10$ GPa, amounts to a volume change $\Delta V/V = \Delta\sigma/B = 5\%$. Then, the error due to the cell relaxation is estimated by $\Delta E = (1/2)BV(\Delta V/V)^2$, giving a value of 0.13 eV. This order of magnitude is not significant for the present problem.

Another problem of the calculations is the supercell size. The concentration dependence shown in figure 2 confirms that use of larger supercells makes no significant difference. The effect of zero-point energy is not significant. Use of generalized gradient approximation (GGA) changes the energy difference of perfect and defective α-boron only slightly. Therefore, it is unlikely that further examination of calculation parameters changes the above conclusion.

It is concluded, within the accuracy of density-functional-theory (DFT) calculation, that simple interstitial or vacancy does not occur at a level above 0.1 at.% in α-boron. If we want to ascribe the relative stability of α-boron over β-boron to the presence of defects in α-boron, we have to seek the origin of the discrepancy in the model structure for the defects. Aside from simple interstitial or vacancy, there might be complex forms of native defects, which are not considered here. Experimentally, it is a common recognition that crystal growth of α-boron is possible only when contaminations are carefully removed [1, 19, 20, 4, 6].

5. Conclusion

When defect states of β-boron are taken into account, the existing calculations show that β-boron must be the stable phase even at low temperature, which contradicts to the experiment. One solution of this difficulty is to allow presence of defects in α-boron; then this would cancel out the effect of defects in β-boron. With this motivation, we have studied intrinsic defects of α-boron. Simple model of vacancy and interstitial at O site as a point defect has been examined.

The ΔE_f at low-concentration limit is large - 4.7 eV for of vacancy and 6.3 eV for interstitial. These values are too large to influence the crystal stability of α-boron compared with defective β boron. Therefore, it is concluded within the present modes of structure that intrinsic defects do not exist in α-boron at a level at which thermodynamic stability of α- and β-boron is altered.

Acknowledgments

This work was partly supported by a Grant-in-Aid for Scientific Research in Priority Areas "Areas of New Materials Science Using Regulated Nano Spaces" (No. 20045009) of MEXT, Japan. The computations were performed on the supercomputer facilities of the Cybermedia Center, Osaka University.

References

[1] McCarty L V, Kasper J S, Horn F N, Decker B F and Newkirk A E 1958 *J. Amer. Chem. Soc.* **80** 2592
[2] Horn F H 1959 *J. Electrochem. Soc.* **106** 905
[3] Niemyski T and Zawadzki W 1962 *Phys. Lett.* **2** 30
[4] Hoard J H and Hughes R E 1965 *The Chemistry of Boron and Its Compounds* ed Mueterties E L (New York: Wiley) chap 2
[5] Soga K, Oguri A, Araake S, Terauchi M, Fujiwara A and Kimura K 2004 *J. Solid State Chem.* **177** 498
[6] Tallant D R, Aselage T L, Campbell A N and Emin D 1989 *Phys. Rev. B* **40** 5649
[7] Prof. T. Tanaka of NIMS informed us that boron alloy systems with Pt or Pd have the lowest eutectic temperature in binary systems.
[8] Eremets M I, Struzhkin V V, Mao H and Hemley R J 2001 *Science* **293** 272
[9] Ma Y, Prewitt C T, Zou G, Mao H and Hemley R J 2003 *Phys. Rev. B* **67** 174116
[10] Sanz D N, Loubeyre P and Mezouar M 2002 *Phys. Rev. Lett.* **89** 245501
[11] Mailhiot C, Grant J B and McMahan A K 1990 *Phys. Rev. B* **42** 9033
[12] Häussermann U, Simak S I, Ahuja R and Johansson B 2003 *Phys. Rev. Lett.* **90** 065701
[13] Segall D E and Arias T A 2003 *Phys. Rev. B* **67** 064105
[14] Masago A, Shirai K and Katayama-Yoshida H 2006 *Phys. Rev. B* **73** 104102
[15] Shirai K, Masago A and Katayama-Yoshida H 2007 *Phys. Status Solidi B* **244** 303
[16] Widom M and Mihalkovič M 2008 *Phys. Rev. B* **77** 064113
[17] van Setten M J, Uijttewaal M A, de Wijs G A and de Groot A 2007 *J. Am. Chem. Soc.* **129** 2458
[18] Ogitsu T, Gygi F, Reed J, Motome Y, Sxhwegler E and Galli G 2009 *J. Am. Chem. Soc.*
[19] Decker B F and Kasper J S 1959 *Acta Cryst.* **12** 503
[20] Horn F H 1959 *J. Appl. Phys.* **30** 1611
[21] Hoard J L, Sullenger D B, Kennard G H L and Hughes R E 1970 *J. Solid State Chem.* **1** 268
[22] Callmer B 1977 *Acta Cryst. B* **33** 1951
[23] Slack G A, Oliver D W and Horn F H 1971 *Phys. Rev. B* **4** 1714
[24] Slack G A, Hejna C I and Kasper J S 1988 *J. Solid State Chem.* **76** 52
[25] Schmechel R and Werheit H 1999 *J. Phys.: Condens. Matter* **11** 6803
[26] http://www.cmp.sanken.osaka-u.ac.jp/~koun/osaka.html
[27] Troullier N and Martins L 1991 *Phys. Rev. B* **43** 1993

[28] Kleinman L and Bylander M 1982 *Phys. Rev. Lett.* **48** 1425
[29] Monkhorst H J and Pack J D 1976 *Phys. Rev. B* **13** 5188
[30] Dekura H, Shirai K and Katayama-Yoshida H 2007 *J. Phys.: Condens. Matter* **19** 365241
[31] Dekura H, Shirai K and Katayama-Yoshida H 2007 *Physica B* **401-402** 702
[32] Car R, Kelly P J, Oshiyama A and Pantelides S 1984 *Phys. Rev. Lett.* **52** 1814
[33] Car R, Kelly P J, Oshiyama A and Pantelides S 1985 *Phys. Rev. Lett.* **54** 360
[34] Nichols C S, de Walle C G V and Pantelides S 1989 *Phys. Rev. Lett.* **62** 1049

Journal of Physics: Conference Series **176** (2009) 012005 doi:10.1088/1742-6596/176/1/012005

Metallization of α-boron by hydrogen doping

Haruhiko Dekura, Koun Shirai, and Akira Yanase[†]

Nanoscience and Nanotechnology Center, ISIR, Osaka University, 8-1, Mihogaoka, Ibaraki, Osaka, Japan

[†] ISIR, Osaka University, 8-1 Mihogaoka, Ibaraki, Osaka, Japan

E-mail: `dekura@cmp.sanken.osaka-u.ac.jp`

Abstract. A theoretical assessment of doping α-boron with hydrogen is presented aiming at high T_c superconductor. Dissimilar to other impurities, hydrogen is unique because T site is the most stable site. This position is well suited to the three-center bond. The bottom of conduction band has a large component at this center. Insertion of H atom increases the bonding character of this band through the three-center bond, which is otherwise a weak bond. Enhancement of bonding character of the three-center bond merges the bottom conduction and the top valence bands, yielding a relatively large density of states (DOS) at the Fermi level. This large DOS suggests that H doped α-boron is a promising high T_c material. When the gaseous state H_2 is used as the starting material, the formation energy is above 2 eV at the normal condition, so that heavy doping is not easy. High-pressure process can avoid this difficulty. It is predicted that hydrogen is incorporated easily at $p > 18$ GPa. A thermodynamic investigation has been made for high-pressure doping method when gases are used as the starting materials.

1. Introduction

This paper presents a theoretical assessment of doping α-boron aiming at superconductivity. Boron is a promising high-T_c superconductor because of strong electron-phonon coupling. As boron is however a semiconductor, it cannot exhibit superconductivity at normal conditions. Carrier doping to concentration as high as in metals is necessary, which is practically the most important and difficult problem. In a series of our studies, we have examined a variety of dopants [1, 2]. Among them, Li is the best candidate in terms of small formation energy ΔH_f and desirable band shift [1]. The fact that Li is the most efficient dopant was found more than ten years ago by Gunji and Kamimura [3]. However the formation energy (0.5 eV) of Li is not small enough to achieve heavy doping.

In order to solve this difficulty, we found that high pressure synthesis is an efficient doping method [1, 4]. Indeed, high pressure process is useful in many other cases, such as B-doped diamond [5]. For heavily B-doped diamond, the reason to use high-pressure seems different [6], but actually we found that the same principle of decreasing ΔH_f under pressure is appropriate [7, 8]. The basic principle is simple:

Softer materials can be easily dissolved in stiffer materials under high pressure. (P.1)

[1] In this paper, we take a convention that the positive formation energy means difficult doping.

This is a part of a well-known principle of thermodynamics, Le Chatelier's principle. Actually, this principle was used to realize a heavily doped clathrate Ba_8Si_{46} [9].

In the present paper, we have examined suitability of hydrogen to dope α-boron. In a previous study [2], we have seen that the formation energy ΔH_f of hydrogen is not small, 3.1 eV. But, this value was obtained assuming solid hydrogen as the reference state. Hydrogen is, of course, in the gas state at ambient conditions, and thus the ΔH_f in experiment could be largely different from this value. By considering the above principle, a good reason for using H is that the gas state of hydrogen is very soft, so that doping by hydrogen must be easier at high pressure.

Aside from ease of doping, there is another reason for examining H. Solid H itself has long been thought a candidate for high T_c superconductor [10]. This idea has not been proven yet. However, sometimes hydrogen helps non-superconducting metals to exhibit superconductivity, such as PdH_x [11]. In this case, the role of hydrogen is not as simple as doping. Recently, researchers came to think about more affirmative effect of hydrogen in hydrogenated molecular crystals. In fact, superconductivity has been found in molecular crystals such as SiH_4 [12]. In this case, silane crystal is considered as a high-density hydrogen solid. Similar effect may be expected for hydroboranes. Structural change in decaborane ($B_{10}H_{14}$) is reported under high pressure [13]. The research in this direction has been continuing.

With these reasons, studying H doping is far from a simple extension of previous doping studies. The formation energy is re-examined in a more systematic way. Various impurity configurations are investigated. More importantly, the doping efficiency at finite temperature and finite pressure has been estimated from thermodynamic considerations. Characterization of the electronic structure is a part of the present study from the above interests in a special role of hydrogen.

2. Method

In this study, first-principles pseudopotential method with local density approximation (LDA) based on density functional theory is used for the total-energy calculation. The used code is 'Osaka2002' [14]. Troullier-Martins' pseudopotentials [15] are used, with aid of a fully separable Kleinman-Bylander form [16].

A Monkhorst-Pack k mesh ($2\times2\times2$) is used in both 12-atom (a primitive cell of α-boron) and 36-atom (hexagonal cell) cells [17]. The cutoff energy for the plane-wave basis is 60 Ry. With this cutoff energy, the accuracy for the total energy is within 1 meV/atom.

For calculations of DOS and band structure, a program package 'TSPACE' developed by A. Yanase was used [18]. By virtue of TSPACE, all the symmetry properties of wavefunctions are clarified, and accordingly, accurate band dispersions and its connectivity at zone boundaries can be established. This is important for interpreting the bonding characters. An improved algorithm of calculating Fermi surface enables us to know accurate Fermi surface [19].

Structural parameters were optimized for both cell parameters and atomic positions. We achieved self-consistency with the residual stress less than 1 GPa. The residual forces were 5.0×10^{-3} Ry/Bohr. With this condition, convergence of the formation energy was below 0.15 eV/cell.

In order to compare the formation energies between different concentrations, the formation energy ΔE_f is evaluated by a difference per impurity atom X, as

$$\Delta E_f = \frac{1}{n}\left[E(X_n\mathrm{B}_m) - (nE(X) + mE(\mathrm{B}))\right]. \tag{1}$$

Readers may suspect that the doped α-boron crystals described in this paper are actually compounds with specific stoichiometry. This is true, but, we continue to treat these compounds as impurity systems, in connection to heavily doped semiconductors. Such treatment is common for doped fullerides (*e.g.* K_3C_{60}) and clathrates (*e.g.* Ba_8Si_{46}). In equation (1), the most

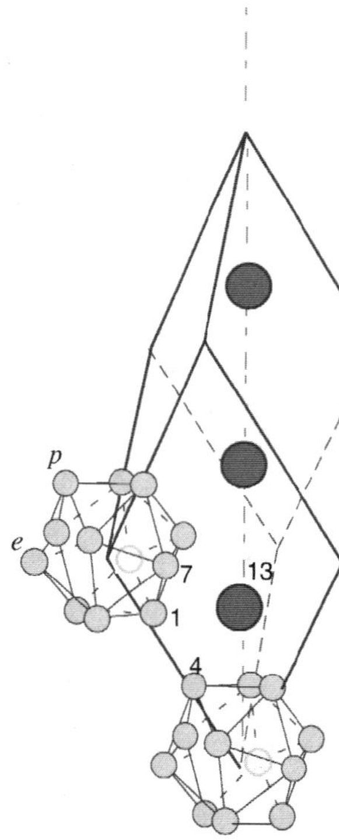

Figure 1. Crystal structure and impurity sites of α-boron. Important atom labels are indicated by numbers. The icosahedron B_{12} has two Wyckoff positions at polar (p) and equatorial (e) sites. Impurity sites are indicated by blue sphere (O site) and red sphere (T site).

stable phase of X is assumed as the reference state. For hydrogen, a proposed model of an orthorhombic structure (space group, $Pca2_1$) was used in our previous paper [2]. However, hydrogen is normally in a gas state H_2. Hence, in the present paper, we use H_2 molecule as the reference state. In this case, we use an experimental value for the binding energy of 0.3483 Ry. Then the total energy of H_2 molecule becomes

$$\frac{1}{2}\left[-2\ (\mathrm{Ry}) - 0.3483\ (\mathrm{Ry})\right] = -15.975\ (\mathrm{eV/atom}). \quad (2)$$

The cohesive energy of H_2 to solid is estimated as 0.011 eV/atom from the facts that the enthalpy of fusion is 117 J/mol (at T=13.96 K and p=0.711 atm) and the enthalpy of vaporization is 904 J/mol (at T=20.39 K and p=1 atm) [20]. This value is very small in the present energy scale, and accordingly either state at low temperatures, the gas state or the solid state, can be used as the reference state within the present accuracy.

Under finite pressure p, enthalpy H plays a central role in an energetic consideration. The formation enthalpy ΔH_f of the impurity X is obtained by adding pV to ΔE_f, where V is the crystal volume. The volume V is determined as a function of p by the structural optimization. At finite temperature, we will use the chemical potential, as described in Sec 3.4.

In our previous study on doping of various impurities [1, 2], it was found that in most cases, impurity atoms enter at the center of the rhombohedron, i.e., O site, which is the most spacious cite. An exception is hydrogen, for which we make further study here. The atomic configuration for H impurity is modeled by supercells of the primitive unit cell of α-boron (B_{12}) and a conventional hexagonal unit cell (B_{36}). The impurity sites to be investigated are O and

Table 1. Formation energy ΔE_f, change in the lattice parameters, and bond length around H atom (NN and 2NN show the nearest neighbor and second neighbor, respectively) of several configurations of H impurity. Changes in the lattice parameter Δa_0, the rhombohedral angle $\Delta\alpha_{\rm rh}$ and the cell volume ΔV are indicated by the relative change to those of non-doped α-boron. Note that all listed bond lengths NN and 2NN are the bonds between H and B atom. In the columns of bond length, entry of 2.07×6 should be read as there are six equivalent bonds whose bond length is 2.07 Å.

Config.	ΔE_f eV/(H atom)	Δa_0 %	$\Delta\alpha_{\rm rh}$ %	NN Å	2NN Å	
HB_{12} (O)	+3.40	0.10	-0.15	0.06	2.07×6	2.57×12
HB_{12} (T)	+3.07	0.82	-0.39	1.8	1.67×3	1.89×3
H_2B_{12} (T×2)	+2.39	1.53	-1.57	2.1	1.70×3	1.84×3
H_3B_{12} (T×2 + O)	+2.77	2.76	-2.4	4.4	1.76×3	1.83×3
HB_{36} (O)	+3.57	0.20	-1.1	-1.1	2.08×6	2.55×12
HB_{36} (T)	+3.43	0.69	-0.84	0.72	1.77×3	1.84×3
H_2B_{36} (T×2)	+2.65	1.43	-2.17	0.81	1.77×3	1.84×3
H_3B_{36} (T×2 + O)	+2.74	1.05	-1.38	0.94	1.82×3	1.84×3

T sites, which are shown in figure 1. The labels for specific atoms which are important in the following discussion are indicated in this figure.

3. Results and discussion

3.1. Impurity site

The obtained hydrogen formation energies are listed in table 1. All energies are positive, so that heavy doping is impossible. However, when comparing ΔE_f between different supercells, we see a definite tendency. For the T or O site, ΔE_f decreases when the smaller supercell is used. This says that the higher the concentration, the smaller the formation energy. Only in the case of a triatomic chain, the formation energy is similar in the supercell and in the primitive cell. Furthermore, when comparing T and $T \times 2$, we see that association of two H atoms in the primitive unit cell is energetically favorable. The binding energy $E_{\rm b}(H_2)$ defined by

$$E_{\rm b}(\mathrm{H_2}) = -2\left[\Delta E_f(\mathrm{H_2B_{12}}) - \Delta E_f(\mathrm{HB_{12}})\right], \tag{3}$$

becomes +1.36 eV.

The smallest formation energy occurs for configuration H_2B_{12} (T×2), for which the original crystal symmetry does not change. Henceforth, the main part of our discussion is focused on this configuration T×2. In this configuration, the hydrogen at T site has a six-fold coordination, unlike suggested by the name "tetrahedral site". The bond length of H-$B_{(e)}$ is 1.70 Å and that of H-$B_{(r)}$ is 1.84 Å. Here, the subscript indicates the Wyckoff positions of boron, as shown in figure 1. Both are substantially longer than typical lengths of hydroborate molecules, for example, in B_2H_6 molecule 1.192 for the terminal B-H and 1.329 Å for three-center B-H bond [21]. In this sense, the H seems not to form covalent bond to B atoms in α-boron. Nevertheless, it participates in the host B-B bonds which has important contribution to metallization, as shown in the next section.

3.2. Charge density

The bonding nature of hydrogen can be understood by seeing the charge density distribution. The charge density of H_2B_{12} configuration is plotted in figures 2 to 4.

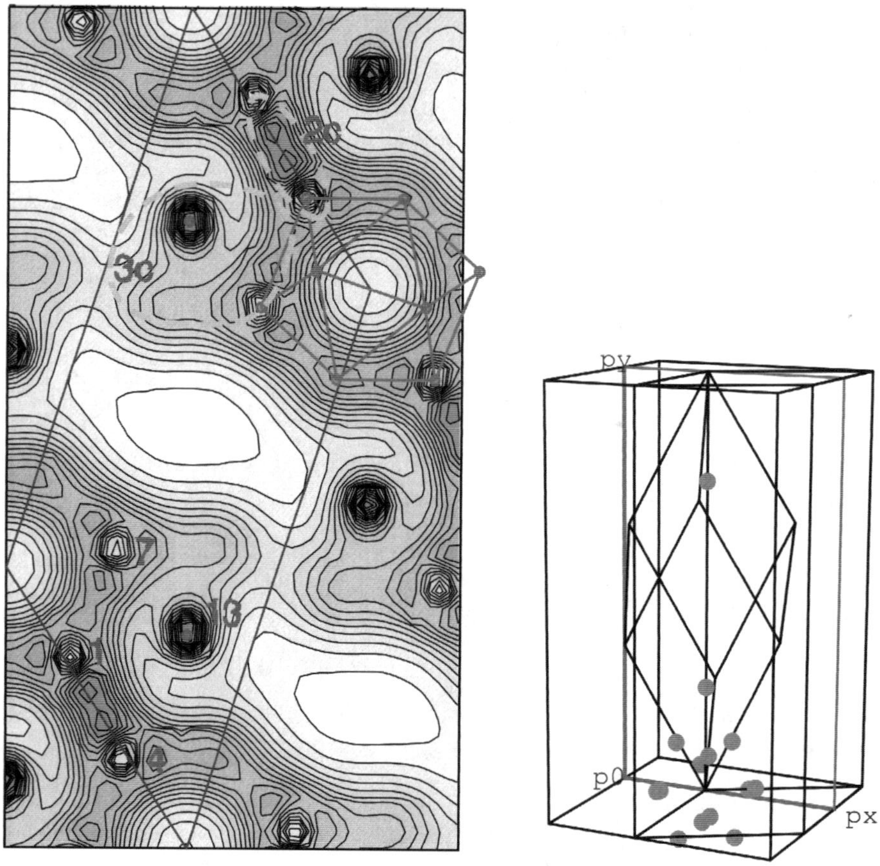

Figure 2. Charge density of H_2B_{12}. The cut plane is one of mirror planes of the rhombohedral system, which is shown on the right side. The blue line indicates the boundary of the primitive unit cell. The charge density is plotted from 0 to 0.2 el/Bohr^3 with steps of 0.01 el/Bohr^3. A part of three center bond is indicated 3c and the *inter*icosahedral bond is indicated by 2c. Important atom sites are indicated according to the scheme in figure 1. One H atom is located at 13, and another one at another symmetry-equivalent site.

Figure 2 shows the charge distribution on a mirror plane of the rhombohedral lattice. The peak of charge density at H site is 0.21 el/Bohr^3, which is almost the same as that of the *inter*icosahedral B-B bond. In this sense, H seems isolated from other B atoms. However, one can see that the H atom has weak bonding to an adjacent icosahedron (13−1−7) in the *ab* plane. On the other hand, there is no bonding to the neighboring icosahedron along the *c* axis (atom 4), which would be expected for the truly tetrahedral coordination.

In order to see this *in-plane* character more closely, we show the charge distribution in an *ab* plane, as shown in figure 3. Note that the cut plane is a plane formed by three equatorial B atoms, and that the H atom is slightly off the plane. For this reason, the maximum charge density in this figure is not the maximum of three-center bond. At the center of a triangle formed by three equatorial B atoms, a peak of charge density is located, which is the characteristic of three-center bond. Hence, the insertion of H atoms in this way enhances the three-center bond.

Around a hydrogen atom, six bonds are formed. This is shown by the charge distribution around the H atom in figure 4. One can see that around the H atom, there are three vertical

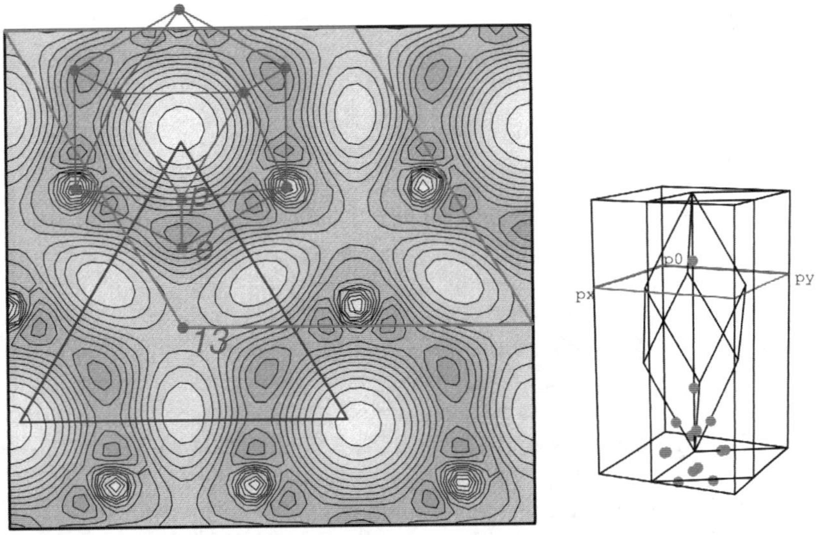

Figure 3. Charge density of H_2B_{12}. The cut plane is shown on the right side, which is an *ab* plane containing the equatorial sites of icosahedra. The green line indicates the boundary of the hexagonal system. The blue line indicates the boundary of the primitive unit cell. The charge density is plotted from 0 to 0.2 el/Bohr^3 with steps of 0.01 el/Bohr^3. Hydrogen atom is located at the center of a blue triangle, but is slightly off the plane.

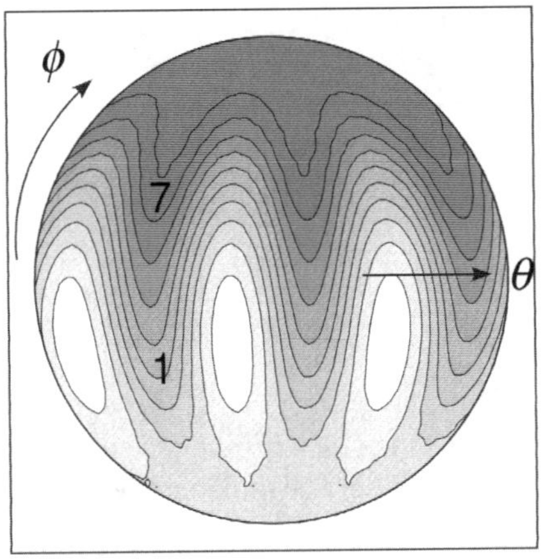

Figure 4. Contour map of the charge density around a hydrogen atom in H_2B_{12}. Shown is the contour map on the sphere surface around the H(13) atom with a radius 0.8 Å. The spherical coordinates ($0 < \theta < 2\pi$, $-\pi/2 < \phi < \pi/2$) are mapped on a circle. Charge density is plotted from 0.026 to 0.073 el/Bohr^3. The upper hemisphere faces three equatorial B atoms, while the lower one faces three polar B atoms.

lines of maximum charge density. These lines are formed by a connection from an equatorial atom 7 to a polar atom 1. In addition, in the upper hemisphere, the charge density is higher than in the lower part. This is a consequence of the fact that the bond length of H to the equatorial B atom H-$B_{(e)}$ is shorter than that of H-$B_{(r)}$.

3.3. Electronic structure

Density of states of H_2B_{12} is calculated and compared with undoped system, as shown in figure 5. In the figure, we see that the whole gap is filled by electrons upon H doping, leading to metallic character. The part of the valence band is not significantly changed. The density of states at Fermi level $N(E_F) = 0.10$ state/(eV·atom). This value is comparable to that of hypothetic metallic fcc phase of boron (see table 1 of Ref. [8]), which is expected to have a moderate value of T_c [22].

Band diagram helps us to understand which band contributes to the metallization. Figure 6 shows a band diagram of doped and undoped α-boron. For labeling of the k-points and symmetry lines, we refer to Ref. [23]. As shown in the figure, lowering of the conduction bottom band is significant toward F point, where the point group of wavevector $\mathbf{k}$ is C_{2h}. On the other hand, the change in the valence band is not substantial, as in DOS.

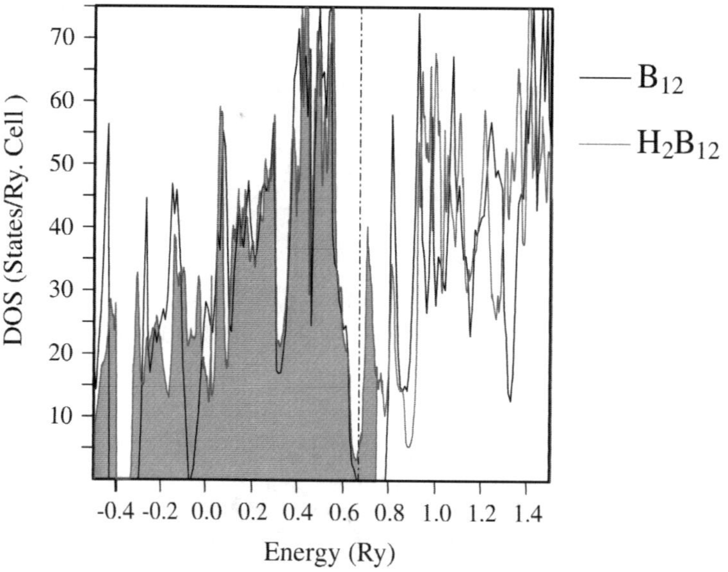

Figure 5. Density of states of H_2B_{12} (red line) and pure α-B (B_{12}) (black line). Occupied states are indicated by a filled region for H_2B_{12}.

The symmetry properties of wavefunctions are very important for interpreting the bonding characters and are given in another paper [24]. Here, only a qualitative description is summarized in table 2.

The conduction bottom band has a component of the inter-icosahedral three-center bond. Therefore, on doping, the three-center bond is enhanced, which results in contraction of the bond length of three-center bond. This makes the rhombohedral lattice more elongated along the c axis. This change is clearly seen in the apex angle of the rhombohedral lattice $\Delta\alpha_{\mathrm{rh}}$ listed in table 1. In a previous paper [2], we showed that decrease in α_{rh} on doping is a common property when inserting various elements apart from carbon. The basic reason of the decrease in α_{rh} is this character of the conduction-bottom band: three-center bonding character.

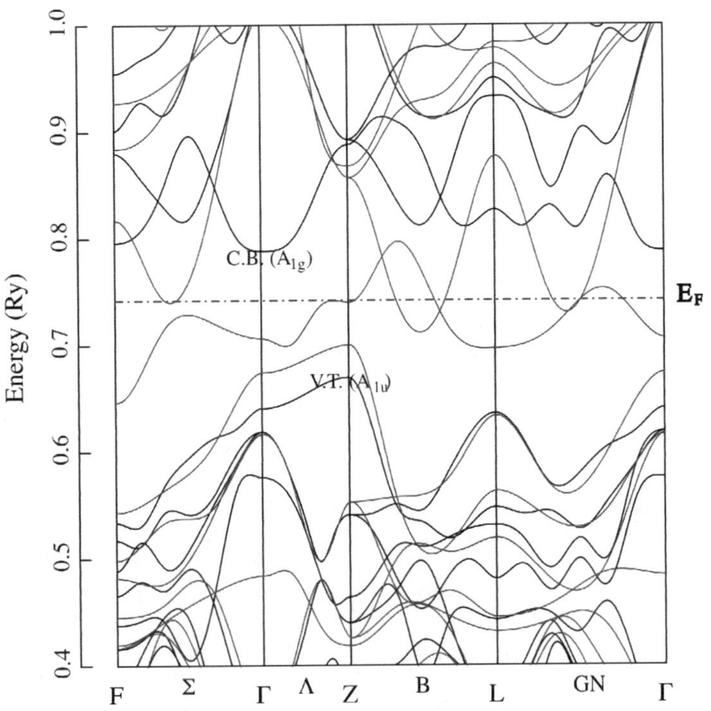

Figure 6. Energy band structures of H_2B_{12} (red line) and B_{12} (black line) near the Fermi energy. The symmetry of wavefunctions are indicated for the valence top (V.T.) and the conduction bottom (C.B.) bands of undoped α-boron.

Table 2. Bonding characters of the valence top (V.T.) and conduction bottom (C.B.) bands; (B) bonding, (AB) antibonding, (N) non-bonding. The *intra*icosahedral bonding is further decomposed to three types of bonds.

component		V.T.	C.B.
symmetry of **k**		A_{1u} (Z)	A_{1g} (Γ)
inter-		AB	B
	$p-p$	N	B
intra-	$p-e$	B	AB
	$e-e$	B	N
three-center		N	B

Another effect of filling the conduction-bottom band is elongation of icosahedron along the c axis. This is because the conduction bottom band has anti-bonding character as to the *intra*icosahedral $p-e$ bond, as shown in table 2. This change will be described in more details in our forthcoming paper [4].

3.4. High pressure doping

We have shown that ΔE_f of hydrogen is not small, when the gas state of H_2 at low temperature is used as a starting material. Now, most experiments are performed at room temperature. At finite temperature, $E(\mathrm{H})$ in equation (1) should be replaced by the chemical potential μ; actually a half of $\mu(H_2)$. Other parts of equation (1) have no significant pressure dependence, so that we keep them constant. Then, we are required to evaluate the chemical potential to the normal state $T = 300$ K and $p = 1$ atm. As to the reference state of hydrogen, whether the state is at $T = 0$ or 20.39 K, or at $p = 0$ or 1 atm makes no difference in our energy scale.

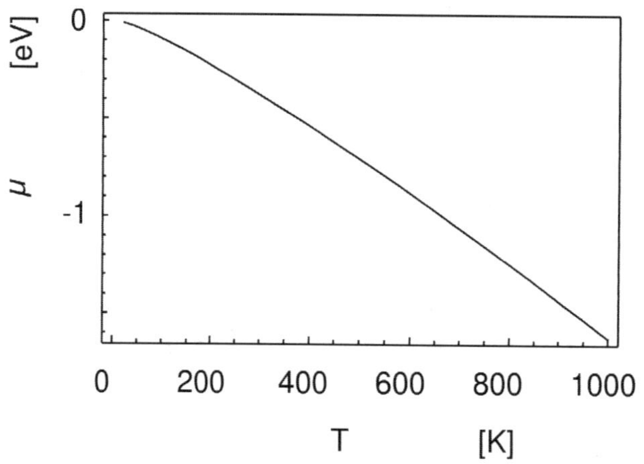

Figure 7. The chemical potential of H_2 as a function of T at the normal pressure.

Liquid hydrogen vaporizes at $T_b = 20.39$ K. Above T_b, we describe the gas state of H_2 by Van der Waals (VdW) equation of state as

$$p = \frac{RT}{v-b} - \frac{a}{v^2}. \tag{4}$$

Validity of the VdW equation is not so wide to cover the present range of pressures [25]. But, at any rate, this form of equation of state is the most available form in the literature. In order to calculate the chemical potential μ, we are required to calculate the internal energy u of gas. A simple form of u compatible with VdW is as follows [26],

$$u = cRT - \frac{a}{v}, \tag{5}$$

where c=5/2 for diatomic molecule. Then, the chemical potential μ is obtained by,

$$\mu = u + pv - RT \ln\left[(v-b)\left(u + \frac{a}{v}\right)^c\right], \tag{6}$$

with a constant term.

For numeric estimation, $a = 0.0248$ Pa/m^6 and $b = 26.6 \times 10^{-6}$ m^3 are used for H_2 [26]. As seen before, the cohesive energy of hydrogen solid compared with a free H_2 molecule is small enough. Thus we can take the energy given by equation (2) as the energy of a free H_2 molecule at $T = 20.39$ K and $p = 1$ atm. Then, variation of μ of H_2 gas is calculated as a function of T at $p = 1$ atm. The result is shown in figure 7. The chemical potential decreases as T increases because of the entropy term. The difference in μ between T = 20 and 300 K is -0.037 eV. Therefore, at the normal condition, the formation enthalpy of hydrogen is

$$\Delta H_f = \Delta E_f - \frac{1}{2}\Delta\mu(\mathrm{H_2}) = 2.39 + 0.037/2 = 2.41 \text{ eV}, \tag{7}$$

where ΔE_f is the formation energy listed in table 1 and $\Delta\mu$(H_2 is change in the chemical potential of H_2 gas from the melting point to room temperature. This says that when the H_2 gas state at the standard condition is referred to, ΔH_f is slightly increased, because the gas state is the most stable state at the standard condition. Doping hydrogen from H_2 gas is again difficult.

In order to overcome the positive ΔH_f, we try to apply high pressure. The basic idea of applying pressure is increase in the energy of soft impurity as stated in the principle (P.1). However, quantitative description becomes complicated for gaseous states. The calculation of the pressure dependence of $\Delta\mu$ has no difficulty. Now that we are treating a gas state, remember that the work is not a thermodynamic property. The result highly depends on the process in which the high pressure is achieved. In the following, we will describe two cases of compression separately, fixing temperature at T = 300 K.

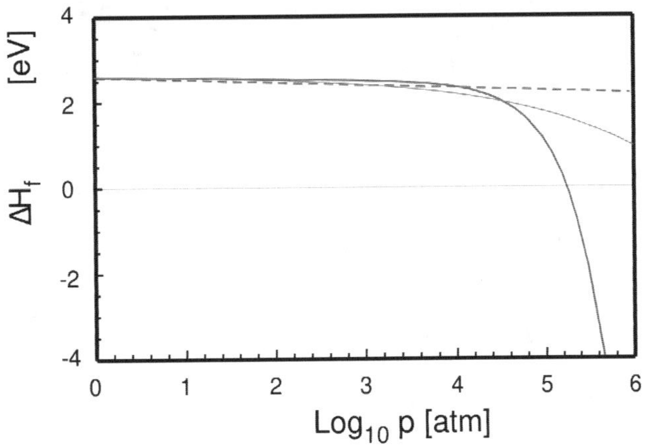

Figure 8. Formation enthalpy of hydrogen in α-boron as a function of p at a fixed T = 300 K. H_2B_{12} is used as the structural model for H-doped α-boron. For isothermal compression, ideal gas (dashed and blue line) and VdW gas (red line) models are compared. For adiabatic compression, a calculation of ideal gas is shown by the blue line.

Isothermal compression First, let us consider an ideal gas. In isothermal compression, enthalpy $u + pv$ does not change for ideal gases. Even though a work was done to compress a gas, the work dissipates to the surroundings leaving no effect on the gas. Only the entropy term $-TS$ has pressure dependence. As shown by the dashed blue line in figure 8, the entropy term $-TS$ has a slight decrease in the formation enthalpy, because on applying pressure, the gas state becomes more dense, which decreases the entropy S.

Next, consider a non-ideal gas. When H_2 gas is described by the VdW model, the formation enthalpy is changed to the red line in figure 8. The situation is almost the same as the ideal gas case until $p = 10^4$ atm (1 GPa). Significant change appears at $p > 10^4$ atm. The primary contribution comes from the first term of equation (4). At $p > 10^4$ atm, the specific volume v becomes comparable to the limiting volume b. The sign of the formation enthalpy is changed at $p = 18$ GPa. Quantitatively, this value is not so accurate, because the validity of VdW description is limited in a wide range of pressure [25]. However, qualitative conclusion seems reasonable by considering the experimental observations. In fact, inclusion of gaseous atoms, such as He, are commonly observed in the specimen at high pressures, when the gas is used as a pressure-transmitted medium.

Adiabatic compression The fact that compressing a gas does not decrease ΔH_f until very high pressure $p > 10^4$ atm, could be changed using adiabatic compression. In this case, the work W which is done by applying pressure is completely transformed to the internal energy of gas, leaving the entropy term unchanged. For ideal gases, the work W can be calculated analytically as

$$W = c\,(p_2 v_2 - p_1 v_1)\,, \tag{8}$$

where the subscripts 1 and 2 indicate the initial and final states, respectively. As shown in figure 8, the effect of using adiabatic compression is not significant. At relatively low pressure, isothermal compression is even slightly better than adiabatic compression because of the entropy term.

In either case, the effect of compression appears only when the pressure exceeds 10 GPa, where the density almost equals that of liquid. Although this is very high pressure, development of high-pressure techniques enables us to reach it easier than before. We see a successful example in doped clathrates. Yamanaka *et al* demonstrated the effectiveness of using high-pressure process for synthesis of Ba_8Si_{46} [9].

4. Summary

The potential of hydrogen doping into α-boron for metallization has been theoretically examined. The role of H is dissimilar to Li case, for which the effect of doping is just filling a lowest part of the conduction band. The case of hydrogen is special. It enters α-boron at T site, where the H has six-fold coordination, and enhances the bonding character of the conduction bottom band through the weak three-center bond. This closes the energy gap yielding the metallic character.

A significant modification of the conduction bottom yields a favorable consequence on the superconductivity. The density of states at Fermi level is relatively large, suggesting a promising candidate for high T_c.

When the gas state H_2 is used as the starting material, the formation energy is above 2 eV at the normal condition, so that heavy doping is difficult. High-pressure process is useful for efficient doping. However, application of the Le Chatelier's principle (P.1) should not be literally taken for high-pressure doping method. Even though gas states are very compressible, isothermal compression has no effect for ideal gases. In either adiabatic compression for any gas or isothermal compression for non-ideal gases, the effect is very small at low pressure. Only at very high pressure $p > 18$ GPa, where the density of gas is comparable to that of liquid, the effect is appreciable. The cutting-edge techniques of high pressure research enable us to utilize high pressure experiment for material processing purpose.

Acknowledgments

We thank Dr. S. Nakano and T. Taniguchi of NIMS for fruitful discussion on high-pressure experiments. This work was partly supported by a Grant-in-Aid for Scientific Research in Priority Areas "Areas of New Materials Science Using Regulated Nano Spaces" (No. 20045009) of MEXT, Japan. The computations were performed on the supercomputer facilities of the Cybermedia Center, Osaka University.

References

[1] Dekura H, Shirai K and Katayama-Yoshida H 2007 *J. Phys.: Condens. Matter* **19** 365241
[2] Dekura H, Shirai K and Katayama-Yoshida H 2007 *Physica B* **401-402** 702
[3] Gunji S and Kamimura H 1995 *Phys. Rev. B* **54** 13665
[4] Dekura H, Shirai K and Katayama-Yoshida H Metallization of α-boron by efficient impurity doping in preparation
[5] Ekimov E A, Sidorov V A, Bauer E D, Mel'nik N, Curro N J, Thompson J D and Stishov S M 2004 *Nature* **428** 542
[6] Dubrovinskaia N, Dubrovinsky L, Miyajima N, Langenhorst F, Crichton W A and Braun H F 2006 *Z. Naturforsch* **61b** 1561
[7] Shirai K, Dekura H and Yanase A 2009 *Phys. Status Solidi B* **246** 673
[8] Shirai K, Dekura H and Masago A 2008 *16th Int. Symp. Boron, Borides and Related Materials* (Matsue, Japan) in this volume
[9] Yamanaka S, Enishi E, Fukuoka H and Yasukawa M 2000 *Inorg. Chem.* **39** 56
[10] Ashcroft N W 1968 *Phys. Rev. Lett.* **21** 1748
[11] Skoskiewics T 1972 *Phys. Status Solidi A* **11** k123
[12] Eremets M I, Trojan I A, Medvedev S A, Tse J S and Yao Y 2008 *Science* **319** 1506
[13] Nakano S, Hemley R J, Gregoryanz E A, Goncharov A F and Mao H K 2002 *J. Phys.: Condens. Matter* **14** 10453
[14] http://www.cmp.sanken.osaka-u.ac.jp/~koun/osaka.html
[15] Troullier N and Martins L 1991 *Phys. Rev. B* **43** 1993
[16] Kleinman L and Bylander M 1982 *Phys. Rev. Lett.* **48** 1425
[17] Monkhorst H J and Pack J D 1976 *Phys. Rev. B* **13** 5188
[18] http://www.cmp.sanken.osaka-u.ac.jp/~yanase/
[19] Dekura H, Shirai K and Yanase A *J. Comput. Theor. Nanosci.* In press
[20] of Japan C S (ed) 2004 *Handbook of Chemistry, 5th ed.* (Tokyo: Maruzen) in Japanese
[21] Greenwood N N 1973 *Comprehensive Inorganic Chemistry* ed Bailar J C, Emeléus H J, Nyholm S R and Trotman-Dickenson A F (Oxford: Pergamon) chap 11
[22] Papaconstantopoulos D A and Mehl M J 2002 *Phys. Rev. B* **65** 172510
[23] Yanase A 1995 *Fortran Program for Space Group (TSPACE)* (Tokyo: Shokabo) in Japanese
[24] Shirai K, Dekura H and Yanase A Electronic structure and electrical resistivity of α-boron under high pressure submitted to J. Phys. Soc. Jpn.
[25] Cengel Y A and Boles M A 2002 *Thermodynamics - an engineering approach, 4th ed.* (New York: McGraw Hill)
[26] Callen H B 1985 *Thermodynamics and an Introduction to Thermostatistics 2nd ed* (New York: John Wiley and Sons)

Novel ternary Y-B-C compound: $Y_{10+x}B_7C_{10-x}$ ($x \approx 0.1$)

Takaho Tanaka, Akira Sato, Katsuaki Watanabe, Masahiro Nagao, Yoshio Matsui, Tetsuya Kawashima and Takashi Aizawa

National Institute for Materials Science, 1-2-1 Sengen, Tsukuba, Ibaraki 305-0047, Japan

TANAKA.Takaho@nims.go.jp

Abstract. A novel ternary Y-B-C compound $Y_{10+x}B_7C_{10-x}$ is established. $Y_{10+x}B_7C_{10-x}$ has a monoclinic crystal structure with space group *C*2/c (No.15) and lattice constants a = 1.1273(14) nm, b = 1.1159(14) nm, c = 2.3566(30) nm and β = 98.15(2)°. $Y_{10+x}B_7C_{10-x}$ crystal rods were grown by floating zone technique and consisted of a predominant crystal phase and impurity phases. A small crystal having sufficient quality for single crystal XRD analysis could be selected. Structure refinement achieved a final $R1$ = 0.044 for 1660 independent reflections with $F_o > 4s(F_o)$. 10 Y sites, 8 B sites, 9 C sites and one C/Y mixed occupancy site with about 0.9/0.1 occupation ratio were assigned. The unit chemical formula is $Y_{10+x}B_7C_{10-x}$ ($x \approx 0.1$), and the unit cell contains 8 formula units. $Y_{10+x}B_7C_{10-x}$ has a layer structure: Y-C double layers, in which Y and C bond alternatively like in the NaCl structure, spread parallel to the (001) plane and stack along the *c*-axis. The Y-C double layer is connected to the neighboring Y-C double layers via C-B-C chains or C-B-B-C chains. This structure was confirmed by electron diffraction and high-resolution electron microscopy. Magnetic susceptibility measurement indicated that the $Y_{10+x}B_7C_{10-x}$ phase has no superconductivity above 2 K. A minor superconducting phase with T_c=7.5 K was observed, but ascribed to an unidentified impurity phase.

1. Introduction

Ternary rare-earth-boron-carbon compounds have attracted continuous scientific interests because of a variety of structures and chemical and physical properties [1]. Discovery of a new intermetallic compound may relate to useful functional properties such as superconductivity and high-efficiency thermoelectric energy conversion. For example, Y_2C_3 exhibits superconductivity bellow a relatively high T_c of 18 K [2,3]. The recent discovery of superconductivity in the Fe-based layered pnictide oxide has explosively raised interest to intermetallic compounds among the superconductivity researchers [4].

We have reported a ternary Sc-B-C compound $Sc_2B_{1.1}C_{3.2}$ that has a hexagonal layer structure, and resembles to graphite-intercalation-compounds [5,6]. In its crystal structure, a Sc_2C layer is sandwiched along the *c*-axis by graphite-like BC_2 layers, in which the unit lattice of the Sc_2C layer and the unit lattice of the BC_2 layer commensurately order with a ratio of 7:9. $Sc_2B_{1.1}C_{3.2}$ shows graphite intercalation compound like behavior [7]. It must be interesting whether or not other rare-earth elements can form an isotypic compound with $Sc_2B_{1.1}C_{3.2}$ by replacing Sc.

Bauer and Nowotony have reported the Y-B-C ternary phase diagram [6], where they have established three ternary phases of YBC, YB_2C_2 and YB_2C. In addition, they have reported the existence of Y_2BC_2 that melts congruently; however, they did not determine the crystal structure. Because of similarity of compositions between $Sc_2B_{1.1}C_{3.2}$ and Y_2BC_2 we may expect that Y_2BC_2 is isotypic with $Sc_2B_{1.1}C_{3.2}$. Meanwhile Rogl has suggested that Y_2BC_2 might rather be $Y_5B_2C_{6+x}$ or Y_2BC_3 and exhibit isotypism with the homologous holmium-containing phases [1].

Here we describe the discovery of $Y_{10+x}B_7C_{10-x}$ ($x \approx 0.1$), which was initially recognized as Y_2BC_2 by Bauer and Nowotony [8]. $Y_{10+x}B_7C_{10-x}$ has a new layer structure that is isotypic with neither $Sc_2B_{1.1}C_{3.2}$ nor $Y_5B_2C_{6+x}$ or Y_2BC_3; however, a structural similarity to the latter could be seen.

2. Experimental

We tried to grow single crystal by floating zone technique using a double halogen-lamp image furnace (Ascal Co., Ltd.). The feed rod for the floating zone crystal growth was prepared by arc-melting method using elemental sources of Y (Furuuchi-Kagaku Inc., flake, nominal purity 99.9%), B (Yamanaka Chem., Inc., granular, nominal purity 98%) and C (Nippon Carbon Co., granular graphite, nominal purity 99.9%). A raw material with composition of Y_2BC_2 was firstly synthesized by the arc-melting method turning the arc-melted button several times to obtain uniform composition. It was then cracked into several pieces, which were put in a groove of a water-cooled Cu plate of the arc-melting furnace. Successive arc-zone-melting through the groove produced a feed rod, 6 mm in diameter and 60 mm long, for the floating zone crystal growth. The crystal growth was carried out under a flowing Ar gas atmosphere using a growth rate of 7 mm/h and a counter-rotation of the growing crystal and the feed rod at ±20 rpm. Similar to rare-earth carbides, the obtained compound was unstable under ambient atmosphere. Therefore, the samples were stored in an airtight sample storage or in kerosene.

A single-crystal specimen for single-crystal X-ray diffraction (XRD) data collection could be obtained by cracking a part of the grown crystallite. The specimen was covered by a nail varnish to shield from atmosphere before the data collection. The single-crystal XRD data were collected by an Enraf-Nonius CAD4 automatic 4-circle diffractometer with graphite-monochromated Mo$K\alpha$ radiation. The crystal data and XRD measurement parameters are given in table 1. The intensity data were corrected for Lorentz and polarization effects. The absorption correction applied to the collected data was empirical, based on Ψ-scans.

Table 1. Crystallographic and data collection parameters

Crystal system	Monoclinic
Space group	*C2/c* (No.15)
Lattice constants	$a = 1.1273(14)$, $b = 1.1159(14)$, $c = 2.3566(30)$ nm $\alpha = 90°$, $\beta = 98.15(2)°$, $\gamma = 90°$
Unit cell volume	2.9346 nm^3
Unit chemical formula	$Y_{10+x}B_7C_{10-x}$ ($x \approx 0.1$)
Z	8
D_x (g/cm^3)	4.946
μ for Mo$K\alpha$ (cm^{-1})	389.0
Crystal dimensions (mm)	0.3×0.15×0.1
$2\theta_{max}$ (deg)	53.80
Reflection measured	$-12 \leqq h \leqq 13$, $-14 \leqq k \leqq 13$, $-29 \leqq l \leqq 26$
Structure refinement program	SHELXL97
R1	0.044 for 1660 ($F_o > 4\sigma(F_o)$), 0.089 for 2601 (all F_o)
w*R2*	0.117(F^2)
Number of variables	166

The initial structure was solved by applying the direct method SIR92 to the data set assuming Y, B and C atoms in the unit cell [9]. The structure was refined using SHELXL97 program [10] and visualized using the crystallographic program CrystalMaker [11].

The high-resolution transmission electron microscopy (HRTEM) and electron diffraction (ED) observations were carried out on a Hitachi H-1500 instrument operated at 800 kV. HRTEM image simulation by multislice method was performed using WinHREM simulation package (HREM Research Inc.) [12].

The magnetic susceptibility measurements were performed using a SQUID magnetometer (Quantum Design, MPMS) at a magnetic field of 10 Oe in the temperature range from 2 to 20 K by zero field cooling (ZFC) and field cooling (FC) processes.

3. Results and discussion

Inspection under optical microscope for the cracked surface of the rod grown by the floating zone method indicated that the rod consisted of mm-size thin crystallites and much smaller inclusions. The thin crystallites had large flat cleavages as compared with its thickness, which suggested a layer characteristic of the crystallites. The inclusions were not identified.

3.1. Structure analysis

An initial solution was obtained according to a monoclinic crystal structure with the space group *C2/c* (No.15) and lattice constants a = 1.1273(14) nm, b = 1.1159(14) nm, c = 2.3566(30) nm, α = 90°, β = 98.15(2)° and γ = 90°. The solution gave 28 atomic positions in which eleven Y sites, eight B sites and nine C sites were assigned. Visualization of the crystal structure suggested that one Y site with partial occupancy should be assigned as a carbon site; however, the electron density was too high for a carbon site, and thus the site was assigned as a C/Y mixed occupancy site. The final refinement of 166 parameters for 1660 independent reflections ($F_o > 4\sigma(F_o)$) resulted in an *R1* value of 4.4% and 8.9% for

Table 2. Atomic coordinates, site occupancy and isotropic displacement parameters

Atom	Site	*x/a*	*y/b*	*z/c*	Occ.	U_{eq}(Å^{2}x10^3)
Y1	8*f*	0.4030(1)	0.2766(1)	0.4247(1)	1	0.0078(3)
Y2	8*f*	0.4665(1)	0.0715(1)	0.3194(1)	1	0.0069(3)
Y3	8*f*	0.3033(1)	-0.0252(1)	0.4281(1)	1	0.0078(3)
Y4	8*f*	0.3667(1)	-0.2206(1)	0.3194(1)	1	0.0063(3)
Y5	8*f*	0.4989(1)	0.4243(1)	0.5752(1)	1	0.0078(3)
Y6	8*f*	0.1750(1)	0.1707(1)	0.3198(1)	1	0.0067(3)
Y7	8*f*	0.7018(1)	0.1736(1)	0.4245(1)	1	0.0082(3)
Y8	8*f*	0.1037(1)	0.3749(1)	0.4245(1)	1	0.0076(3)
Y9	8*f*	0.5751(1)	0.3787(1)	0.3197(1)	1	0.0068(3)
Y10	8*f*	0.2682(1)	0.4752(1)	0.3100(1)	1	0.0088(3)
C/Y	8*f*	0.2703(6)	-0.0245(7)	0.3155(3)	0.911/0.089(5)	0.0260(27)
C1	8*f*	0.6010(9)	0.2251(9)	0.5840(4)	1	0.0057(21)
C2	8*f*	0.1988(9)	0.1750(9)	0.4159(4)	1	0.0078(23)
C3	8*f*	0.6651(9)	0.1756(9)	0.3084(5)	1	0.0091(23)
C4	8*f*	0.5668(9)	-0.1271(7)	0.3069(5)	1	0.0082(23)
C5	8*f*	0.4992(9)	0.0756(7)	0.4165(4)	1	0.0081(22)
C6	8*f*	0.3700(9)	0.2740(10)	0.3074(5)	1	0.0126(25)
C7	8*f*	0.5984(9)	0.3745(9)	0.4165(5)	1	0.0100(23)
C8	8*f*	0.2998(9)	0.4746(10)	0.4110(4)	1	0.0100(21)
C9	8*f*	0.4686(9)	-0.4234(9)	0.3075(5)	1	0.0096(23)
B1	8*f*	0.3154(12)	0.4753(13)	0.4742(6)	1	0.0206(29)
B2	8*f*	0.2806(13)	0.3198(13)	0.5212(7)	1	0.0269(35)
B3	8*f*	0.5185(13)	0.0750(12)	0.4788(6)	1	0.0244(33)

B4	8*f*	0.5820(13)	0.2307(13)	0.5214(6)	1	0.0240(33)
B5	8*f*	0.6149(13)	0.3736(13)	0.4800(7)	1	0.0256(34)
B6	4*e*	0.5	-0.1216(14)	0.25	1	0.0072(34)
B7	8*f*	0.6537(11)	0.2268(11)	0.2513(6)	1	0.0122(26)
B8	4*e*	0.5	-0.4290(16)	0.25	1	0.0170(41)

all 2601 independent reflections. The final atomic coordinates, occupancy factors and temperature factors are listed in table 2. All sites, except for the B6 and B8 sites that have the 4*e* site symmetry, are general positions (8*f*); thus, the number of atoms that accommodate at the B6 and B8 sites is half of others and unit chemical formula can be described as $Y_{10+x}B_7C_{10-x}$ (x≈0.1). There are eight formula units per unit cell. The occupancy ratio of C:Y at the C/Y mixed occupancy site is approximately 0.91:0.09.

The interatomic distances are summarized in table 3. The Y-C bond distances range from 0.224 to 0.274 nm. As will be discussed later, $Y_{10+x}B_7C_{10-x}$ has a layer structure in which layers are composed of Y and C, span parallel to the *a-b* plane and stack along the *c*-axis. The Y-C bond distances within the layers are approximately 0.25 nm, which is comparable to that (0.248 nm) in the binary Y_2C compound [13]. On the other hand, the Y-C distances perpendicular to the Y-C layers vary largely

Table 3. Interatomic distances (described in $\times 10^{-1}$ nm=Å unit)

Bonds between Y and C					
Y1-C7	2.491(11)	Y2-C5	2.265(11)	Y3-C1	2.512(10)
-C8	2.497(11)	-C/Y	2.447(7)	-C2	2.522(10)
-C5	2.510(10)	-C6	2.507(11)	-C5	2.527(10)
-C2	2.547(10)	-C4	2.524(10)	-C7	2.547(11)
-C6	2.736(12)	-C3	2.568(10)	-C/Y	2.628(7)
Y4-C1	2.255(11)	Y5-C1	2.498(10)	Y6-C2	2.244(11)
-C/Y	2.439(8)	-C8	2.514(11)	-C/Y	2.437(8)
-C3	2.532(10)	-C7	2.518(10)	-C9	2.532(11)
-C4	2.540(10)	-C2	2.525(10)	-C6	2.535(11)
-C9	2.571(10)	-C9	2.737(11)	-C4	2.563(10)
Y7-C5	2.515(10)	Y8-C2	2.495(10)	Y9-C7	2.258(11)
-C8	2.520(11)	-C5	2.526(10)	-C/Y	2.466(7)
-C7	2.522(11)	-C8	2.534(11)	-C9	2.511(10)
-C1	2.527(10)	-C1	2.545(10)	-C3	2.513(10)
-C3	2.708(11)	-C4	2.743(11)	-C6	2.571(11)
Y10-C8	2.357(10)				
-C3	2.518(10)				
-C6	2.527(11)				
-C4	2.532(11)				
-C9	2.536(11)				
Direct bond between Y-C double layers					
Y10-C/Y	2.926(8)				
Bonds in C-B-B-C chain					
C1-B4	1.461(17)	B1-B2	2.127(21)	B4-B5	1.933(20)
C2-B2	1.467(18)	B1-B5	2.095(20)		
C5-B3	1.454(18)	B2-B2	1.925(29)		

C7-B5	1.482(18)	B3-B3	2.022(28)		
C8-B1	1.474(16)	B3-B4	2.083(20)		
Bonds in C-B-C chain					
C3-B7	1.451(16)	C4-B6	1.444(11)	C6-B7	1.468(16)
C9-B8	1.450(11)				

from 0.224 to 0.274 nm due to the irregular thickness of the Y-C layers. $Y_{10+x}B_7C_{10-x}$ has two chain structures of C-B-C and C-B-B-C. The C-B bond distances within both the C-B-B-C and C-B-C chains range from 0.1444 to 0.1482 nm and are normal to a C-B bond distance. The B-B distances range from 0.1925 to 0.2127 nm, which is rather large compared with typical B-B bond distances of 0.17-0.19 nm.

3.2. Structure description

The obtained crystal structure viewed close to the [110] direction is shown in figure 1(a). As expected, $Y_{10+x}B_7C_{10-x}$ has a layered structure. Y-C double layers, in which Y and C alternatively bond like the NaCl structure, are dominant. The Y-C double layers spread parallel to the (001) plane and stack along the *c*-axis. A Y-C double layer is shown in figure 1(b).; it is connected to neighbor Y-C double layers via C-B-C chains or C-B-B-C chains.

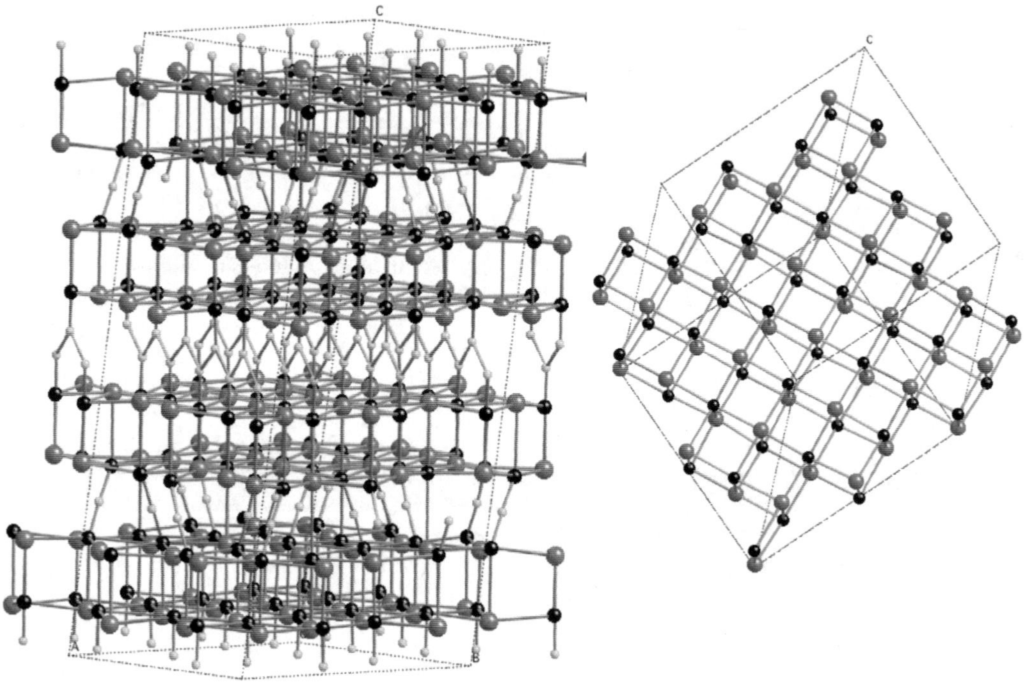

(a) (b)

Figure 1. (a) Crystal structure of $Y_{10+x}B_7C_{10-x}$ viewed close to the [110] direction. Red, green and black spheres correspond to Y, B and C atoms, respectively. Blue arrow indicates one of the C/Y mixed occupancy sites. (b) Y-C double layer viewed close to the [111] direction.

A Y-C double layer connected by the C-B-C chains to another Y-C double layer mutually rotates 45° along the *c*-axis. Distance between the Y-C double layers is approximately 0.293 nm. C atoms in the Y-C double layer face to C atoms in the counter Y-C layer; thus, they can form a C-B-C chain. The C/Y mixed occupancy site should be originally a C site to maintain the alternative bonding scheme between Y and C atoms, as shown in figure 1(a) by a blue arrow. However, the counter part of the C/Y mixed occupancy site is not a C site but a Y site, which does not allow forming a C-B-C chain and changes it from the C site to the C/Y mixed occupancy site. On the contrary, the Y-C double layers connected by the C-B-B-C chains are parallel each other along C and Y atoms alignment, thus, the C-B-B-C chains regularly line up forming an infinite zigzag B chain, as can be seen in figure 1(a). Distance between the double layers connected by the C-B-B-C chains is much longer than that connected by the C-B-C chains and is approximately 0.36 nm.

C/Y mixed occupancy sites or vacancy sites in (rare earth)-boron-carbon compounds often exhibit ordering. We examined the possibility of the ordering of the present C/Y mixed occupancy site, which may appear by solving the crystal structure using a lower symmetry space group than *C*2c (No.15). Extinction rule allowed us to use an alternative space group of *C*c (No. 9). The structure refinement did not show any evidences of the ordering, thus we concluded that C and Y atoms are randomly distributed over the C/Y mixed occupancy sites.

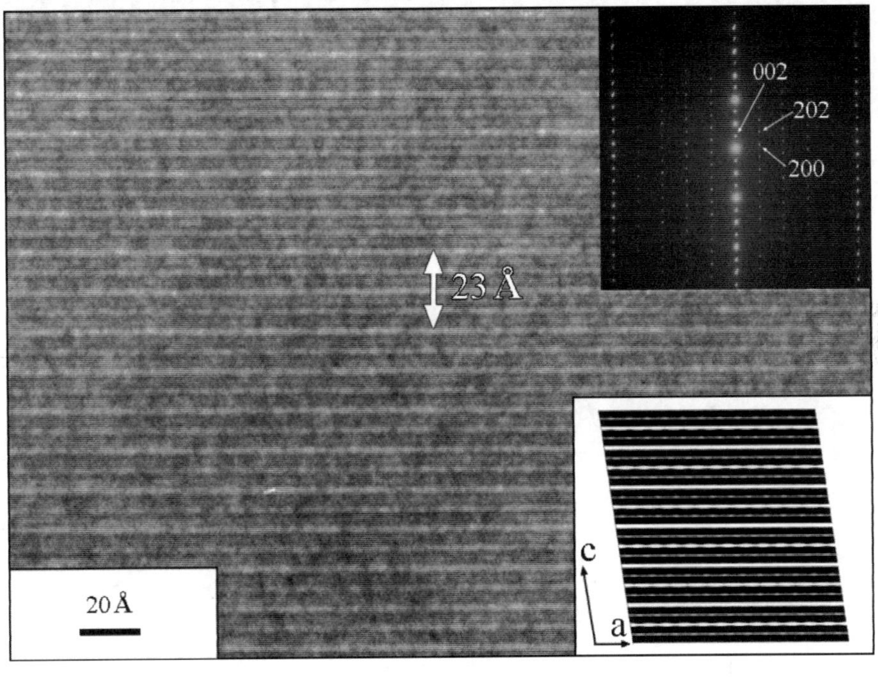

Figure 2. HRTEM image of $Y_{10+x}B_7C_{10-x}$ taken for the incident beam parallel to the [010] direction. The right-top inset indicates an ED pattern taken along this direction, and a computer simulation image of the proposed structure is shown in the bottom-right inset.

Structural similarity can be seen in $Ho_5B_2C_{6+x}$ [14] and $Th_3B_2C_3$ [15]. $Ho_5B_2C_{6+x}$ has a tetragonal structure (*P*4), where Ho and C form a double layer. In that layer, Ho and C bond alternatively forming a tetragonal network and incorporating C_2 dumbbells that can be seen in CaC_2. The Ho-C double layer stacks along the *c*-axis and connects each other by C-B-C chains. $Th_3B_2C_3$ has a monoclinic structure (*P*2/m), where Th and C form a NaCl-like triple layer that stacks along the *c*-

axis. The Th-C triple layers are connected by the C-B-B-C chains. Thus, $Y_{10+x}B_7C_{10-x}$ structure can be understood as an admixture of $Ho_5B_2C_{6+x}$ and $Th_3B_2C_3$ structures or, in a broad sense, to be homologous to them. This makes the crystal structure of $Y_{10+x}B_7C_{10-x}$ more complicated, with a much larger unit cell size than those other two structures.

Very recently, Babizhetsky [16] has refined crystal structure of $RE_{10}B_7C_{10}$ (RE = Gd - Er) that appears to be isotypic to $Y_{10+x}B_7C_{10-x}$ because they all have a monoclinic crystal structure with same space group and approximately same lattice constants. This similarity is plausible because Y and Gd - Er have similar chemical nature and atomic radii.

3.3. HRTEM and Ed observations

Figure 2 presents an HRTEM lattice image for the incident electron beam close to the [010] direction. The image clearly reveals the layered structure of the compound.

In the computer-simulated image, the black double layer separated by a thin white line corresponds to the Y-C double layer. The bright, continuous white line corresponds to the C-B-B-C chain layer and the weaker white line with periodical intensity variation corresponds to the C-B-C chain layer. The observed lattice image is reproduced well by the simulation. Bundles with four black double layers are separated by 2.3 nm, which is consistent with the lattice constant c = 2.36 nm obtained by XRD. All these facts confirm the $Y_{10+x}B_7C_{10-x}$ crystal structure proposed by the XRD structure analysis. Note that ED analysis along the [010], [100], [110] and [001] axes indicated no evidence of the ordering of the C/Y mixed occupancy site.

3.4. Magnetization measurement

The measured magnetic susceptibility is shown in figure 3. Superconducting transition can be seen at 7.5 K; however, the observed magnetic susceptibility is of the 10^{-6} emu/g order, which is too small for bulk superconductivity. We attributed this superconductivity to an unidentified impurity phase and concluded that $Y_{10+x}B_7C_{10-x}$ does not exhibit superconductivity above 2 K.

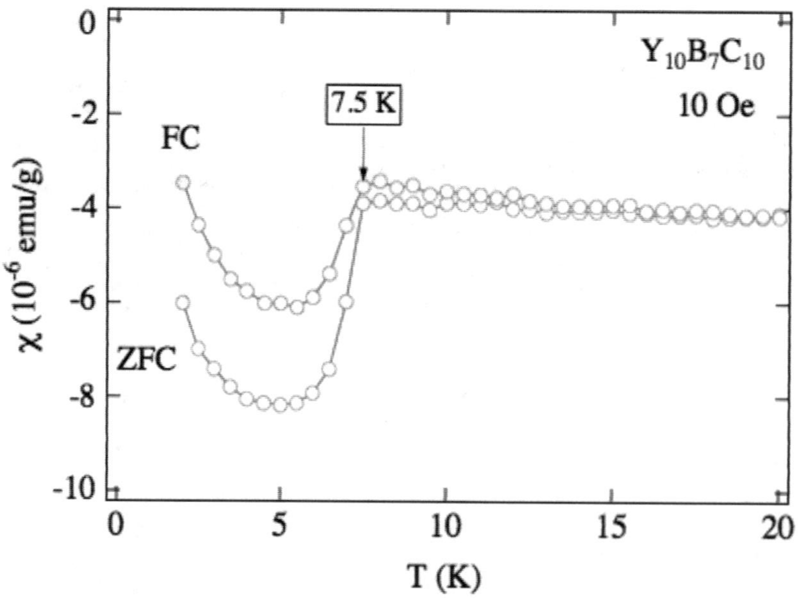

Figure 3. Temperature dependence of magnetic susceptibility of $Y_{10+x}B_7C_{10-x}$.

4. Conclusions

We established a new yttrium borocarbide $Y_{10+x}B_7C_{10-x}$ ($x \approx 0.1$) that has previously [8] been reported as Y_2BC_2. Crystal structure of $Y_{10+x}B_7C_{10-x}$ is monoclinic (space group *C*2/c (No.15) and lattice constants a = 1.1273(14) nm, b = 1.1159(14) nm, c = 2.3566(30) nm and β = 98.15(2)°), and it is "homologous" to $Ho_5B_2C_{6+x}$ and $Th_3B_2C_3$. $Y_{10+x}B_7C_{10-x}$ does not exhibit superconductivity above 2 K.

This result does not completely deny our expectation that a ternary Y-B-C compound isotypic to $Sc_2B_{1.1}C_{3.2}$ may exist. During the course of this investigation we noticed an unidentified compound whose composition is close to $Y_{10+x}B_7C_{10-x}$. Electron diffraction analysis revealed that the compound has a hexagonal structure with a = 0.94 nm. The compound may be responsible for the superconductivity shown in figure 3.

Acknowledgement

T. T. thanks Mr. E. Bannai, NIMS for technical support on the arc-melting and floating zone crystal growth processes.

References

[1] Rogl P, 1998 *Phase Diagrams of the Ternary Metal-Boron-Carbon Systems* (ed. Effenberg G) ASM International, Materials Park, OH
[2] Amano G, Akutugawa S, Muranaka T, Zenitani Y and Akimitsu J 2004 *J. Phys. Soc. Jpn.* **73** 530
[3] Akutagawa S and Akimitsu J 2006 *Sci. Technol Adv. Mater.* **7** 2
[4] Ozawa C T and Kauzlarich S M 2008 *Sci. Technol. Adv. Mater.* 9 0300xx
[5] Shi Y, Leithe-Jasper A, Bourgeois L, Bando Y and Tanaka T, 1999 *J. Solid State Chem.* **148** 442
[6] Onoda M, Shi Y, Leithe-Jasper A and Tanaka T, 2001 *Acta Crystallogr., Sect. B: Struct. Sci.* **57** 449
[7] Mori T, Shi Y and Tanaka T, 2000 *Mol. Cryst. Liq. Cryst.* **340** 83
[8] Bauer J and Nowotony H, 1971 *Monatshefte Chem.* **102** 1129
[9] Altomare A, Cascarano G, Giacovazzo C, Guagliardi A, Burla M, Polidori G and Camalli M, 1994 *J. Appl. Crystallogr.* **27** 435
[10] Sheldrick G M, 1997 *SHELX97: a program for the solution and refinement of crystal structure*, Universität Göttingen, Göttingen, Germany
[11] Palmer D, *CrystalMaker Version 5.0.7, CrystalMaker software*, Bicester, Oxfordshire, OX6 7BS, UK
[12] Kimoto K, Asaka T, Nagai T, Saito M, Matsui Y and Ishizuka K, 2007 *Nature* **450** 702
[13] Atoji M and Kikuchi M 1969 *J. Chem. Phys.* **51** 3863
[14] Bauer J, Vennegués P and Vergneau J L 1985 *J. Less-Commn Met.* **110** 295
[15] Rogl P 1979 *J. Nucl. Mater.* **79** 154
[16] Babizhetsky V 2008 private communication

Crystal structure of $R_3Pd_{25-x}B_{8-y}$, R = La, Ce

Oksana Sologub[1], Peter Rogl[1], Andriy Grytsiv[1] and Gerald Giester[2]
[1]Institut für Physikalische Chemie, Universität Wien, A-1090 Wien, Währingerstrasse 42, Austria
[2]Institut für Mineralogie und Kristallographie, Universität Wien, A-1090 Wien, Althanstrasse 14, Austria
peter.franz.rogl@univie.ac.at

Abstract. New ternary compounds $La_3Pd_{25-x}B_{8-y}$ (x=1.75, y=0.07) and $Ce_3Pd_{25-x}B_{8-y}$ (x=1.25, y=2.00) have been synthesized by arc melting. Their crystal structures were determined from X-ray single crystal diffraction data. Both compounds crystallize with a new monoclinic structure type, space group $P2_1/c$ with a=1.17567(2) nm, b=1.05757(2) nm, c=1.60664(3) nm, β =102.1788(10) deg, R_F^2=0.030 for 7427 structure factors and 289 variable parameters for $La_3Pd_{25-x}B_{8-y}$ and a=1.17453(3) nm, b=1.04571(3) nm, c=1.59759(5) nm, β=102.132(2) deg, R_F^2=0.035 for 2805 structure factors and 283 variable parameters for $Ce_3Pd_{25-x}B_{8-y}$. Rietveld refinement for the X-ray powder diffraction patterns for both compounds confirmed the structure models determined from X-ray single crystal diffraction data.

1. Introduction

Exceptional physical properties discovered for $CePt_3Si$ [1, 2, 3] prompted us to expand our investigations towards related systems with boron. In continuation of these studies, a series of novel ternary compounds $M_2Pd_{14+x}B_{5-x}$ (M= Th, Ce, Pr, Nd, Sm, Eu, Gd) was reported recently [4] of which the crystal structures of $Nd_2Pd_{14+x}B_{5-x}$ and $Th_2Pd_{14+x}B_{5-x}$ were determined from X-ray single crystal data [4] and found to be closely related to the structure type of $Sc_4Ni_{29}Si_{10}$ [5]. The evaluation of X-ray powder diffraction data of the light rare earths containing members (from Ce to Gd) of this series revealed isotypism with the $Nd_2Pd_{14+x}B_{5-x}$ and $Th_2Pd_{14+x}B_{5-x}$ structure. The $La_2Pd_{14}B_5$ sample, initially prepared as non-magnetic counterpart to the magnetic rare earth compounds, however, showed a hitherto unknown X-ray powder diffraction pattern. In the course of our studies of the Ce-Pd-B system at 850 °C, similar X-ray diffraction patterns were observed for alloys within the concentration range 70-75 at.% Pd. Detailed structural studies of these novel phases became the subject of the present paper.

2. Experimental

$La_2Pd_{14}B_5$ and $Ce_3Pd_{25}B_8$ samples with a total amount of about 1g, were prepared by argon arc-melting elemental pieces of lanthanum and cerium (purity >99.9 mass%, Auer Remy, D), palladium foil (99.9%, Ögussa, A) and crystalline boron (98%, Alfa Aesar, D). For homogeneity, the samples were re-melted several times without significant weight losses (< 1 mass%). The alloy buttons were wrapped in Mo-foil, sealed in an evacuated quartz tube and annealed at 850 °C for 240 h before quenching in cold water. X-ray powder diffraction data were collected employing a Guinier-Huber image plate system with monochromatic Cu $K\alpha_1$ radiation (8°<2θ<100°). Precise lattice parameters were calculated by least squares fits to the indexed 2θ values calibrated with Ge as internal standard (a_{Ge} =0.565791 nm). Single crystals were mechanically isolated from the crushed as cast alloys. Inspections on an AXS-GADDS texture goniometer assured high crystal quality, unit cell dimensions and monoclinic Laue symmetry of the specimens prior to the X-ray intensity data collections on a four-circle Nonius Kappa diffractometer equipped with a CCD area detector employing graphite monochromated Mo K_α radiation (λ=0.071069 nm). Orientation matrices and unit cell parameters werederived using the program DENZO [6] (Table 1). No absorption corrections were performed because of the rather regular crystal shapes and small dimensions of the investigated specimens. The structures were solved by direct methods and were refined with the SHELXL-97 program [7, 8]. Quantitative Rietveld refinements of the X-ray powder diffraction data were performed with the FULLPROF program [9].

Table 1. Structural data for $R_3Pd_{25-x}B_{8-y}$; R=La, Ce (space group $P2_1/c$; No. 14).

Alloy composition	$La_{9.5}Pd_{67}B_{23.5}$	$Ce_9Pd_{73}B_{18}$
Formula from refinement	$La_{12}Pd_{93.09}B_{31.75}$	$Ce_{12}Pd_{95.09}B_{24}$
Data collection	Nonius KappaCCD	Nonius KappaCCD
Radiation	MoK_α	MoK_α
2θ range	$2.32 < 2\theta < 72.60$	$2.34 < 2\theta < 55.00$
Crystal size	$18x18x35\mu m^3$	$15x15x28\mu m^3$
a [nm]	1.17567(2)	1.17453(3)
b [nm]	1.05757(2)	1.04571(3)
c [nm]	1.60664(3)	1.59759(5)
β [deg]	102.1788(10)	102.132(2)
Reflections in refinement	7427 $F_o > 4\sigma(F_o)$ of 9344	2805 $F_o > 4\sigma(F_o)$ of 8593
Mosaicity	<0.35	<0.45
Number of variables	289	283
$R_F^2 = \Sigma\|F_0^2-F_c^2\|/\Sigma F_0^2$	0.030	0.035
R_{Int}		
GOF	1.030	1.036
Extinction (Zachariasen)	0.00015(1)	0.00007(2)
Residual electron density; max; min in [electrons/Å^3]	3.31; -3.49	4.13; -3.95

3. Results

3.1. X-ray single crystal diffraction

Single crystal X-ray intensity patterns indicated for both crystals Laue class *2/m* and systematic extinctions consistent with space group $P2_1/c$ (no. 14). Intensity statistics unambiguously confirmed the probability of a centrosymmetric space group ($|E^2-1|=0.976$) [10].

3.1.1. $La_3Pd_{25-x}B_{8-y}$ structure. Initial atomic positions for heavy atoms La and Pd were found by direct methods whereas eight boron atoms were located from difference Fourier synthesis. The Pd2 and Pd25 sites exhibited an increased ADP (atom displacement parameter) indicating partial occupancy. This structure model was refined to a residual value $R_F^2 = 0.035$ with anisotropic thermal displacement parameters for the metal atoms but isotropic temperature factors for the boron atoms, resulting in the chemical formula $La_6Pd_{47-x}B_{16}$. An additional peak (13.7 e/Å^3) in the difference electron density map at the distance of 0.065 nm from Pd23 and at 0.145 nm from B8 was assigned to a partially occupied Pd24 site. When allowed to refine freely, the occupancy of Pd23/Pd24 and B8 sites converged to 0.923(2)/0.077(2) and 0.923(2), respectively. Replacement of Pd24 by a boron atom resulted in 100(1)% occupancy of this atomic position whereas the occupancy of Pd23 and B8 sites remained unchanged. In case the occupancy parameters of positions filled by mixture of Pd23 and B atoms and only boron were constrained to 1.0, the refinement was unsatisfactory revealing unacceptably small ADP's of boron atom and two additional peaks of electron density 17 e/Å^3 and 13 e/Å^3 at the distance 0.012 nm and 0.074 nm respectively from Pd23. This procedure did not improve the distances as well:

still short Pd23-boron atom and boron - boron contacts (0.065 nm and 0.145 nm) prevented the coexistence of all three atoms in the unit cell. The analysis of the coordination figure for this atomic site in case of its occupancy by boron atom showed that due to the smaller atomic radii, the B atom could contact only 4 neighbouring atoms arranged in form of an unusual almost flat and distorted tetrahedron with the central B atom itself staying aside the polyhedron. All these reasons prompted us to remain with the solution of the split sites Pd23/Pd24 with a filling level of 92.3/7.7%. Attempts to attain an ordered structure model in space groups of lower symmetry were unsuccessful. Thus we kept the structure model in space group $P2_1/c$ finally arriving at the formula $La_6Pd_{49-x}B_{16-y}$ with a reliability factor R_F^2=3.14%. To verify our results and to check on the atom order, X-ray intensity data were collected for a second crystal specimen: the structure solution arrived at an identical arrangement of atomic sites and split sites but refinement converged to a slightly higher value of R_F^2 = 3.54%.

3.1.2. $Ce_3Pd_{25-x}B_{8-y}$ structure. The structure solution in $P2_1/c$ space group by direct methods revealed 3 atomic positions of cerium and 24 atomic positions of palladium. Although isotypism with the structure model for the La-compound was obvious, the structure refinement (residual value R_F^2=12.58% adopting isotropic thermal parameters for all atoms) yielded a significant residual electron density of 69 e/Å^3 in (1/2,0,0) at a distance of 0.236 nm from Pd26 and large displacement parameters for Pd10 and Pd26 (the atomic numbers correspond to the ones given in table 2). A corresponding atomic site was attributed to Pd28 and occupancy parameters were refined for all three atoms Pd10, Pd26 and Pd28. This procedure improved the R factor to 10.69% for a partial occupancy of Pd10, Pd26 and Pd28 sites. Six boron atoms were located from difference Fourier syntheses and the structure was refined to R_F^2=7.21% adopting anisotropic thermal parameters for heavy atoms and isotropic displacement parameters for borons. Due to a rather significant residual electron density of 22.93 e/Å^3 observed at (0.2700, 0.4187, 0.4868) in the vicinity of Pd10, this atomic site had also to be refined as split position ($d_{Pd10\text{-}Pd27}$=0.089 nm). The following refinement of ADP's resulted in a filling level of 80(1)% Pd10 + 20(1)% Pd27 (R_F^2=5.28%). Because of unacceptably large displacement parameters for Pd26, i.e. 0.246(9), the Pd26 atomic position was split into (0.6543, 0.1594, 0.0283) (Pd26) and (0.6055, 0.0746, 0.0116) (Pd23). This procedure reduced the residual electron density to acceptable 4.13 e/Å^3 and a final structure refinement yielded R_F^2=3.51% suggesting occupancies as 50(1)% Pd26/Pd28, 30(1)% Pd23 + 20(1)% Pd26.

Comparing the atomic positions of the two structures, we observe that the structure models of $Ce_3Pd_{25-x}B_{8-y}$ and $La_6Pd_{49-x}B_{16-y}$ differ only by the occupancy of Pd positions within a domain with partially filled atomic sites in the Ce structure, and by the presence of two additional boron positions in the La structure (Table 2). After structural refinement for $Ce_3Pd_{25-x}B_{8-y}$ was accomplished, we reconsidered the La structure model assuming that the partially occupied atomic sites observed in the Ce structure are not filled in the La structure. Due to a slightly increased ADP of Pd10 a refinement of the occupancy parameter for this atomic site showed only 93.1%. As a result, the final difference Fourier map revealed residual peaks as low as 3.31 e/Å^3 at the distance 0.048 nm from Pd23 and the R value was reduced to R_F^2=3.00%.

Journal of Physics: Conference Series **176** (2009) 012007 doi:10.1088/1742-6596/176/1/012007

Table 2. Atomic coordinates and equivalent isotropic displacement parameters in [10^2 nm^2] for the atoms in $R_3Pd_{25-x}B_{8-y}$ [a]

Atom	Wyckoff notation	x		y		z		Occupancy		U_{iso}	
		$La_3Pd_{25-x}B_{8-y}$	$Ce_3Pd_{25-x}B_{8-y}$	$La_3Pd_{25-x}B_{8-y}$	$Ce_3Pd_{25-x}B_{8-y}$	$La_3Pd_{25-x}B_{8-y}$	$Ce_3Pd_{25-x}B_{8-y}$	$La_3Pd_{25-x}B_{8-y}$	$Ce_3Pd_{25-x}B_{8-y}$	$La_3Pd_{25-x}B_{8-y}$	$Ce_3Pd_{25-x}B_{8-y}$
R1	4(e)	0.13374(3)	0.13011(6)	0.56675(3)	0.56995(8)	0.34969(2)	0.34940(5)	1	1	0.01016(6)	0.0134(2)
R2	4(e)	0.24480(3)	0.24889(7)	0.37282(3)	0.37295(7)	0.00077(2)	0.00027(5)	1	1	0.01251(6)	0.0130(2)
R3	4(e)	0.63178(3)	0.62792(7)	0.41148(3)	0.41846(8)	0.33042(2)	0.33996(6)	1	1	0.00967(6)	0.0177(2)
Pd1	4(e)	0.00199(3)	0.00085(8)	0.33761(4)	0.33575(10)	0.23095(3)	0.22983(7)	1	1	0.00917(7)	0.0111(2)
Pd2	4(e)	0.03801(4)	0.04532(9)	0.31167(5)	0.30660(10)	0.41038(3)	0.40981(7)	0.916(2)	1	0.0137(1)	0.0127(2)
Pd3	4(e)	0.05183(3)	0.05574(8)	0.23589(4)	0.23852(10)	0.08490(3)	0.08312(7)	1	1	0.00966(7)	0.0115(2)
Pd4	4(e)	0.05714(4)	0.06089(8)	0.08773(4)	0.08460(10)	0.32382(3)	0.32240(7)	1	1	0.01097(8)	0.0125(2)
Pd5	4(e)	0.06264(3)	0.06414(8)	0.59388(4)	0.59214(10)	0.04148(3)	0.04177(6)	1	1	0.00883(7)	0.0112(2)
Pd6	4(e)	0.12355(3)	0.12618(9)	0.50450(4)	0.50487(10)	0.67239(3)	0.67016(7)	1	1	0.00951(7)	0.0113(2)
Pd7	4(e)	0.16254(3)	0.15969(9)	0.72379(4)	0.72332(11)	0.19057(3)	0.18952(7)	1	1	0.00851(7)	0.0116(2)
Pd8	4(e)	0.17168(4)	0.16304(10)	0.82557(4)	0.82687(11)	0.02820(3)	0.03174(7)	1	1	0.01067(8)	0.0171(2)
Pd9	4(e)	0.19143(4)	0.19173(9)	0.45483(4)	0.45523(10)	0.17853(3)	0.17910(7)	1	1	0.01002(7)	0.0120(2)
Pd10	4(e)	0.21592(4)	0.2255(1)	0.08348(4)	0.0826(1)	0.02277(3)	0.0166(1)	0.931(2)	0.80(1)	0.01528(9)	0.0177(3)
Pd11	4(e)	0.22250(3)	0.22249(9)	0.28869(4)	0.29117(11)	0.33131(3)	0.32778(7)	1	1	0.01158(8)	0.0126(2)
Pd12	4(e)	0.27557(3)	0.27867(9)	0.20246(4)	0.20492(10)	0.18001(3)	0.17735(7)	1	1	0.00980(7)	0.0120(2)
Pd13	4(e)	0.30879(4)	0.30976(9)	0.03434(4)	0.03844(10)	0.32564(3)	0.32335(7)	1	1	0.00963(7)	0.0122(2)
Pd14	4(e)	0.32107(3)	0.31968(9)	0.62460(4)	0.63043(11)	0.09332(3)	0.08927(7)	1	1	0.00955(7)	0.0119(2)
Pd15	4(e)	0.38022(3)	0.37758(9)	0.48256(4)	0.49367(12)	0.32652(3)	0.33002(7)	1	1	0.00958(7)	0.0157(2)
Pd16	4(e)	0.43730(4)	0.44466(9)	0.17115(4)	0.18241(12)	0.07630(3)	0.08311(8)	1	1	0.01006(8)	0.0190(2)
Pd17	4(e)	0.4394284)	0.43687(9)	0.39686(4)	0.40812(11)	0.16881(3)	0.17425(7)	1	1	0.01048(8)	0.0145(2)
Pd18	4(e)	0.45221(3)	0.44672(9)	0.25328(4)	0.25726(11)	0.40847(3)	0.41328(7)	1	1	0.00937(7)	0.0137(2)
Pd19	4(e)	0.50137(3)	0.50008(9)	0.14704(4)	0.15603(11)	0.26224(3)	0.26656(7)	1	1	0.00898(7)	0.0125(2)
Pd20	4(e)	0.56116(3)	0.56123(9)	0.40029(4)	0.40224(11)	0.03436(3)	0.03899(7)	1	1	0.00892(7)	0.0122(2)
Pd21	4(e)	0.66042(3)	0.65775(9)	0.25038(4)	0.26336(12)	0.16713(3)	0.18023(8)	1	1	0.00976(8)	0.0210(2)
Pd22	4(e)	0.82049(3)	0.81828(9)	0.37011(4)	0.36598(11)	0.08916(3)	0.08715(7)	1	1	0.00941(7)	0.0118(2)
Pd23	4(e)	0.60608(5)	0.6055(4)	0.07331(7)	0.0746(4)	0.01000(3)	0.0116(2)	0.923(2)	0.30(1)	0.0112(2)	0.0355(6)
Pd24	4(e)	0.6381(7)		0.1306(9)		0.0192(4)		0.077(2)		0.0115(1)	
Pd25	2(a)	0	0	0	0	0	0	0.712(1)	0.50(1)	0.0117(19)	0.0132(5)
Pd26	4(e)		0.6543(1)		0.1594(2)		0.0283(1)		0.70(1)		0.0355(6)
Pd27	4(e)		0.2700(7)		0.4187(6)		0.4868(5)		0.20(1)		0.0177(3)
Pd28	2(b)		½		0		0		0.50(1)		0.0369(14)
B1	4(e)	0.0088(5)	0.0024(12)	0.4315(6)	0.4296(15)	0.0987(4)	0.1002(9)	1	1	0.012(1)	0.013(3)
B2	4(e)	0.1172(5)	0.1249(13)	0.1833(6)	0.1858(16)	0.2207(4)	0.2222(9)	1	1	0.011(1)	0.017(3)
B3	4(e)	0.3122(5)	0.3130(12)	0.6134(5)	0.6147(1)	0.2276(4)	0.2208(9)	1	1	0.010(1)	0.011(3)
B4	4(e)	0.3395(6)		0.3877(7)		0.4522(5)		1		0.018(1)	
B5	4(e)	0.3820(6)	0.3761(20)	0.3001(6)	0.3059(25)	0.2750(4)	0.2824(16)	1	1	0.015(1)	0.057(6)
B6	4(e)	0.4938(5)	0.4936(12)	0.0525(6)	0.0598(16)	0.3947(4)	0.3993(10)	1	1	0.012(1)	0.015(3)
B7	4(e)	0.8117(5)	0.8125(14)	0.3737(6)	0.3740(17)	0.2184(4)	0.2183(11)	1	1	0.011(1)	0.023(4)
B8	4(e)	0.6855(5)		0.2518(7)		0.0204(5)		0.923(2)		0.011(1)	

[a]crystal structure data are standardized using the program Structure Tidy [11].

Analysis of interatomic distances in $Ce_3Pd_{25-x}B_{8-y}$ showed that the proximity of the positions Pd23 and Pd27 to those of Pd26, Pd28 and Pd10, respectively, precludes all these sites from being simultaneously occupied. For the La structure, short contacts of 0.648(7) Å and 1.449(1) Å were observed for Pd23/Pd24 (92.3%/7.7%) and B8/Pd24 (92.3%/7.7%), respectively.

3.2. X-ray powder diffraction

Full profile Rietveld refinements, performed on powder X-ray diffraction data of alloys $La_2Pd_{14}B_5$ (Figure 1) and $Ce_3Pd_{25}B_8$, converged with residual values as low as R_F=2.56 and R_F=2.91, respectively thus confirming the structure models obtained from single crystals. A small amount of a secondary phase in the La sample was consistent with a new compound $La_2Pd_{14+x}B_{5-x}$ ($I4_1/amd$ space group, a=0.8612(1) nm, c=1.6643(3) nm, R_F=3.65) close to the $Sc_4Ni_{29}B_{10}$ derivative structure [5] and most probably isotypic with the recently determined structure of $Nd_2Pd_{14+x}B_{5-x}$ [4].

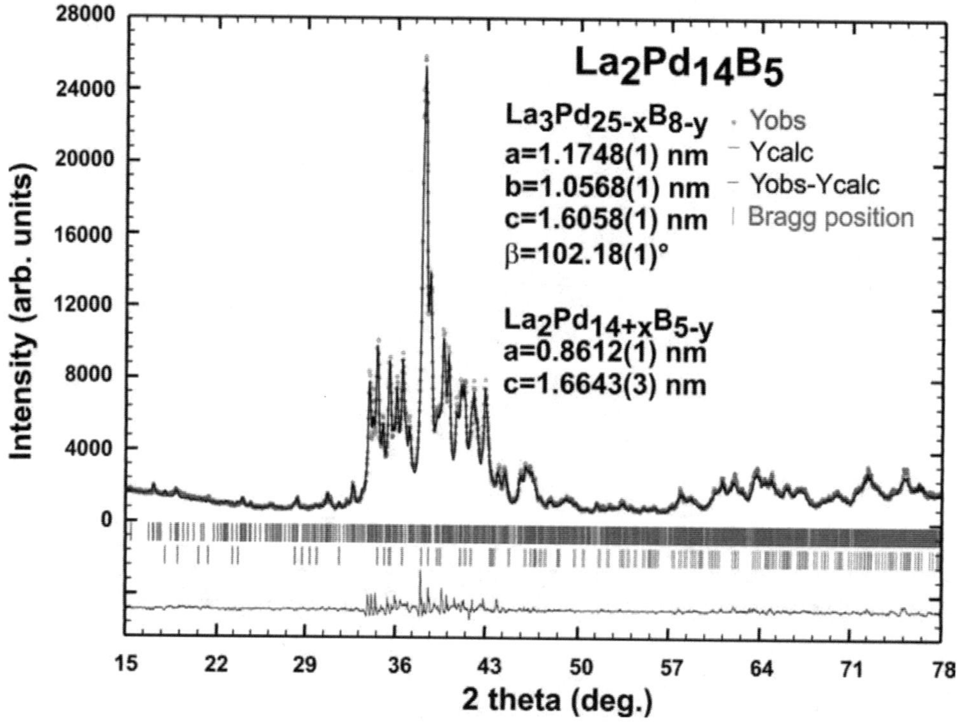

Figure 1. X-ray diffraction pattern of $La_2Pd_{14}B_5$. The solid line derives from the Rietveld refinement. $Y_{obs.}$-$Y_{calc.}$ is the intensity difference between exprimental data and Rietveld calculations. Upper pattern of Bragg positions corresponds to $La_3Pd_{25-x}B_{8-y}$ structure and lower pattern shows Bragg positions of the $La_2Pd_{14}B_5$ structure.

3.3. Structural analysis

Both La and Ce structures exhibit the same structural arrangement of fully occupied RE, palladium and boron atomic sites with slight shifts in atomic coordinates. Three defect positions of palladium atoms in $Ce_3Pd_{25-x}B_{8-y}$ do not appear to be filled in $La_3Pd_{25-x}B_{8-y}$. And vice versa, two atomic sites of boron in the La structure are vacant in the structure with Ce. Several empty polyhedra exist in both structures (see figures 2a and 2b), which show sufficiently large distances from center to the surrounding atoms to eventually host additional boron and/or palladium atoms. From the common architecture both structures can be attributed to the same $R_3Pd_{25-x}B_{8-y}$ structural type with the La compound as the Pd-deficient and the Ce compound as the B-deficient members.

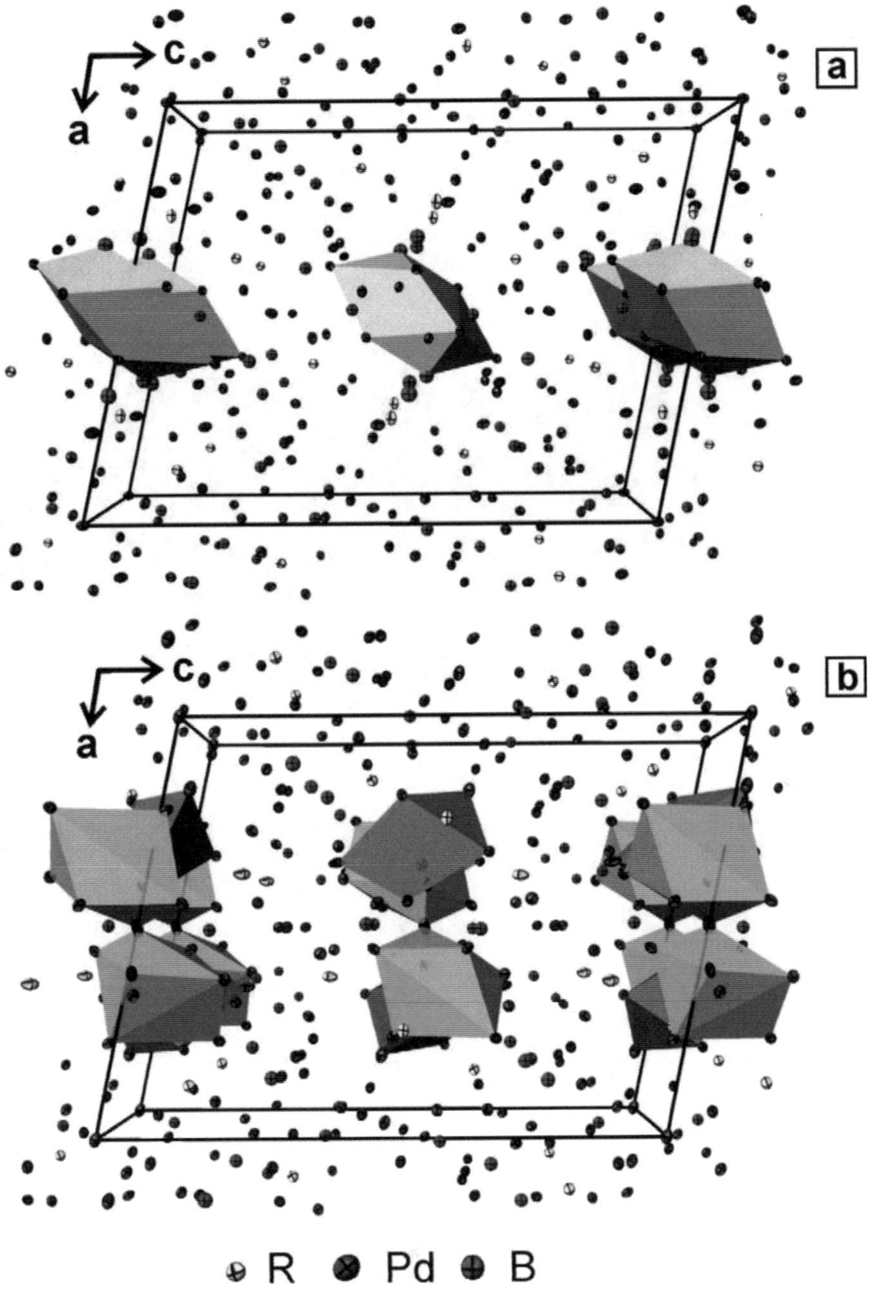

Figure 2. Perspective views of $La_3Pd_{25-x}B_{8-y}$ (a) and $Ce_3Pd_{25-x}B_{8-y}$ (b) along [101] axis with outlined vacant polyhedra.

Perspective views of both structures $La_3Pd_{25-x}B_{8-y}$ and $Ce_3Pd_{25-x}B_{8-y}$ along the b direction are given in figures 2a-2b; the coordination polyhedra for atoms are shown in detail in figure 3 to 6 for only $La_3Pd_{25-x}B_{8-y}$ as a representative member of the structure. Each La atom possesses a large coordination sphere formed by palladium and boron atoms. Five boron atoms are found in slightly distorted monocapped and tricapped trigonal prisms whereas B3 and B7 are coordinated by tetragonal antiprisms. The environment of B8 located in the site with partial occupancy has the shape of a distorted heptagonal bi-pyramid with one additional atom (Figure 6).

As the consequence of symmetry, the arrangement of polyhedra in the structure shows a large variety: polyhedra isolated and coupled by common face, edges or vertices; complexes of certain Pd atoms polyhedra arranged along x direction (Figure 7); corrugated (due to the mutual shift for a half of translation unit) columns of connected by common edges coordination units of palladium atoms stretched along z direction, etc. All palladium atoms except Pd2 and Pd25 participate in the formation of coordination spheres for the boron atoms. Boron trigonal prisms axes are oriented in various ways with respect to x, y and z directions (Figure 8). In agreement with the classification of boride structure types with respect to the type of boron-boron aggregation as a function of metal-boron ratio [12], no boron-boron contact was observed in the structure (Table 3).

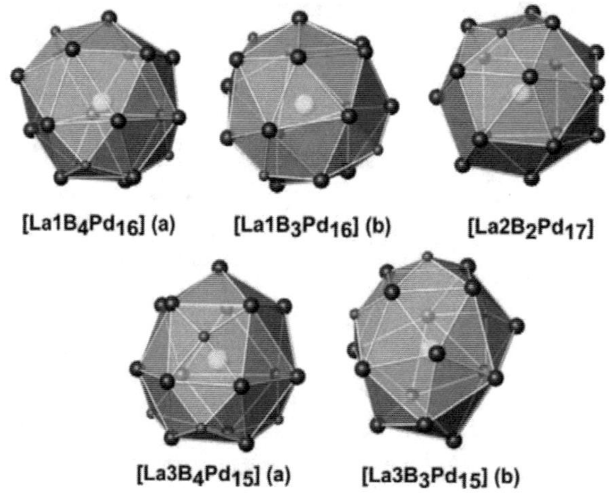

Figure 3. Coordination polyhedra of lanthanum atoms in $La_3Pd_{25-x}B_{8-y}$. La1 and La3 have two different coordinations depending on which one of the neighbouring sites is occupied: La1(La3) for Pd23/B8 contact (a) and La1(La3) for Pd24 contact (b).

Table 3. Interatomic distances (nm) for boron atoms in $La_3Pd_{25-x}B_{8-y}$

Atoms	Distance	Atoms	Distance	Atoms	Distance	Atoms	Distance
B1 – Pd5	0.2108(6)	B3 – Pd7	0.2092(7)	B5 - Pd12	0.2034(7)	B7 - Pd22	0.2101(6)
B1 – Pd3	0.2153(6)	B3 – Pd15	0.2131(7)	B5 - Pd15	0.2106(7)	B7 - Pd7	0.2135(6)
B1 - Pd5	0.2249(6)	B3 – Pd14	0.2187(6)	B5 - Pd19	0.2178(7)	B7 - Pd6	0.2177(6)
B1 – Pd9	0.2270(6)	B3 - Pd19	0.2189(6)	B5 - Pd18	0.2184(6)	B7 - Pd21	0.2224(6)
B1 - Pd22	0.2281(6)	B3 - Pd21	0.2194(7)	B5 - Pd17	0.2219(7)	B7 – Pd1	0.2231(6)
B1 - Pd4	0.2299(6)	B3 – Pd9	0.2236(6)	B5 - Pd11	0.2247(7)	B7 - Pd13	0.2233(6)
B1 – Pd1	0.2361(6)	B3 – Pd17	0.2997(7)	B5 – Pd9	0.2940(7)	B7 – Pd4	0.2895(6)
		B3 – La1	0.3199(7)	B5 – Pd13	0.3090(7)	B7 – La3	0.3082(6)
B2 – Pd12	0.2108(6)			B5 – La3	0.3120(7)		
B2 - Pd6	0.2145(6)	B4 - Pd10	0.2054(7)			B8 - Pd23	0.2096(7)
B2 – Pd1	0.2146(6)	B4 - Pd23	0.2116(8)	B6 - Pd20	0.2145(6)	B8 - Pd22	0.2140(7)
B2 – Pd4	0.2182(6)	(or Pd24)	0.262(1)	B6 - Pd17	0.2176(6)	B8 - Pd8	0.2156(7)
B2 – Pd3	0.2228(6)	B4 – Pd18	0.2149(8)	B6 - Pd18	0.2197(6)	B8 - Pd20	0.2189(7)
B2 – Pd11	0.2232(6)	B4 - Pd16	0.2167(8)	B6 - Pd13	0.2231(6)	B8 - Pd14	0.2230(8)
B2 – Pd13	0.2965(6)	B4 – Pd11	0.2375(8)	B6 – Pd20	0.2266(6)	B8 - Pd21	0.2443(8)
B2 – Pd9	0.3112(6)	B4 – Pd15	0.2390(8)	B6 – Pd14	0.2279(6)	B8 – Pd18	0.2941(8)
B2 - La1	0.3182(6)	B4 – Pd23	0.3094(8)	B6 - Pd19	0.2376(6)	B8 – La1	0.3296(7)
		B4 – La2	0.3131(9)				
		B4 – La1	0.3246(9)				

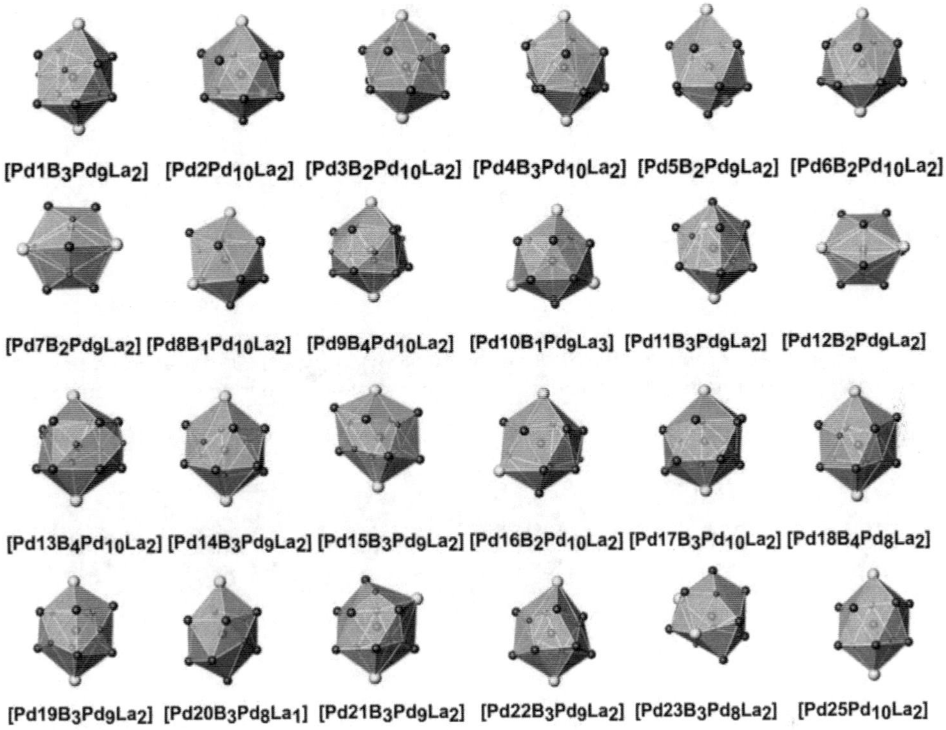

Figure 4. Coordination polyhedra of palladium atoms in $La_3Pd_{25-x}B_{8-y}$: Pd23/B8 sites are occupied.

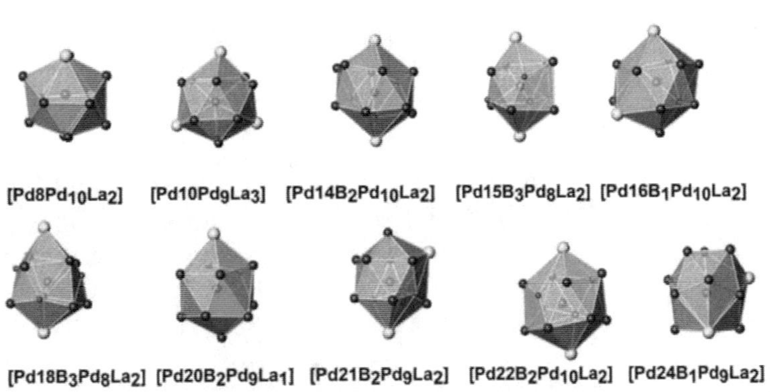

Figure 5. Coordination polyhedra of palladium atoms in $La_3Pd_{25-x}B_{8-y}$: Pd24 site is occupied.

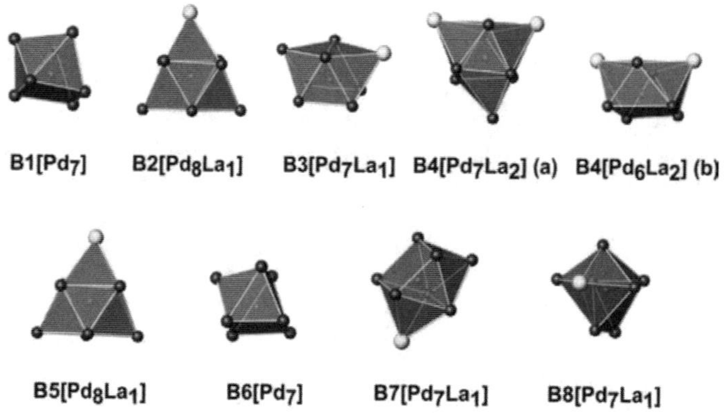

Figure 6. Coordination polyhedra of boron atoms in $La_3Pd_{25-x}B_{8-y}$ structure. B4 atoms have two different polyhedra depending on which one of neighbouring sites is occupied, Pd23 (a) or Pd24 (b).

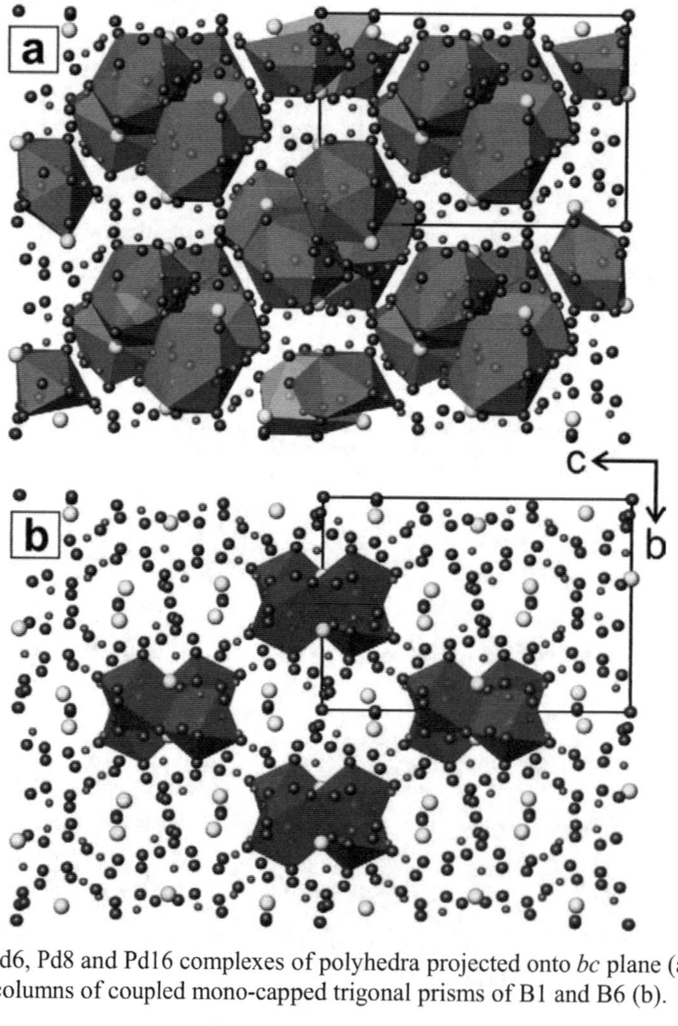

Figure 7. Pd2, Pd6, Pd8 and Pd16 complexes of polyhedra projected onto *bc* plane (a) separated by columns of coupled mono-capped trigonal prisms of B1 and B6 (b).

Journal of Physics: Conference Series **176** (2009) 012007 doi:10.1088/1742-6596/176/1/012007

3.4. Structural relationships

A comparative analysis of Pd- and Ni-rich ternary rare earth systems is given in figure 9; the concentration range for the formation of new compounds with palladium is shown in the inset. Several structures are found to form within a rather narrow concentration field of R 8-10 at.% - T 60-80 at.% - B 33-10 at.%.

Based on the $Sc_4Ni_{29}B_{10}$ atomic model, first determined by Kuzma *et al* (1988) for $R_4Ni_{29}B_{10}$ R=Sc, Tm, Yb, Lu [5], three new $Sc_4Ni_{29}B_{10}$ derivative structural arrangements have been reported more recently: $Y_2Pd_{14}B_5$ [13], $ErNi_7B_3$ [14] and $Nd_2Pd_{14+x}B_{5-x}$ [4]; all of them differ from the parent structure by the mode of occupancy of coupled tetragonal antiprisms with Pd and B atoms. All the coordination spheres for boron atoms are tetragonal antiprisms and mono-capped trigonal prisms and the structures show no boron-boron contacts.

Two more structures were found to form with the heavy rare earth ternary borides: $Ho_2Ni_{15}B_6$ [15] and $Ho_2Ni_{10}B_5$ [16]. The coordination polyhedra for boron atoms in $Ho_2Ni_{15}B_6$ ($P2_1/c$, a=1.4223 nm, b=1.0672 nm, c=0.9582 nm, β=94.23°) all are tetragonal antiprisms; one B-B pair was observed besides seven isolated boron atoms. For the $Ho_2Ni_{10}B_5$ – type (*Pbca*, a=0.9001 nm, b=0.9477 nm, c=1.7588 nm), 2 boron-boron pairs form besides three isolated borons. The atomic environments for boron atoms are mono-capped tetragonal antiprisms and bi-capped trigonal prisms.

The most common structural arrangements for the metal rich ternary rare earth borides were observed to be an ordered version of $CaCu_5$ type, $CeCo_3B_2$ type and its derivatives with the trigonal prisms as the coordination polyhedra for boron atoms [17].

With respect to the type of boron atom coordination (mono-capped and tri-capped trigonal prisms and tetragonal antiprisms), the $La_3Pd_{25-x}B_{8-y}$ and $Ce_3Pd_{25-x}B_{8-y}$ structures are related to $Sc_4Ni_{29}B_{10}$ – and $CeCo_3B_2$ - derivative structures.

Because of the similarities in composition and unit cell dimensions of $R_3Pd_{25-x}B_{8-y}$ and $ErNi_{7.9}B_2$ ($C2/c$ space group, a=1.662 nm, b=0.952 nm, c=1.068 nm, β=117.7 deg) [18], a test for isotypism among both structures was performed. The primitive unit cell of $Ce_3Pd_{25-x}B_{8-y}$ was transformed to a C-centered cell with lattice parameters a=1.77288 nm, b=1.04571 nm, c=1.17453 nm, β=118.23 deg applying the (-a-c, b, a) transformation matrix, however the structure solution in space group $C2/c$ revealed significant violations on systematic absences *hkl* ($h+k$=2n), *0kl* (k=2n) and *h0l* (h=2n, l=2n). These facts clearly show that $Ce_3Pd_{25-x}B_{8-y}$ and $ErNi_{7.9}B_2$ belong to different structure types.

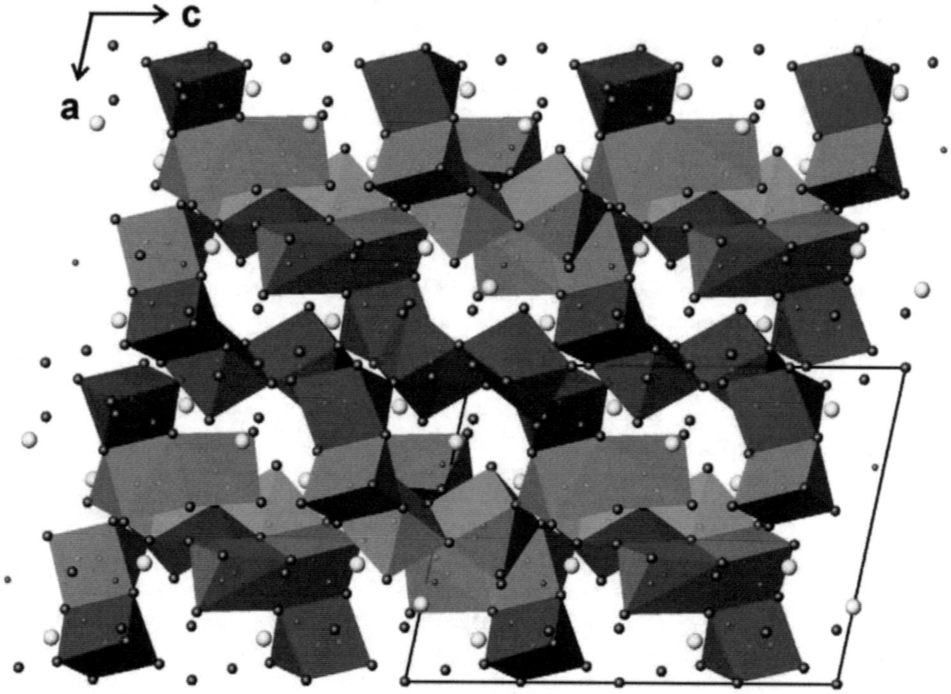

Figure 8. B trigonal prism aggregation in the $La_3Pd_{25-x}B_{8-y}$ structure.

4. Summary

The crystal structures of new ternary compounds $La_3Pd_{25-x}B_{8-y}$ (x=1.75, y=0.07) and $Ce_3Pd_{25-x}B_{8-y}$ (x=1.25, y=2.00) have been determined from X-ray single crystal data. Both compounds were found to form new monoclinic structural arrangements in space group $P2_1/c$ with a=1.17567(2) nm, b=1.05757(2) nm, c=1.60664(3) nm, β=102.1788(10) deg, R_F^2=0.0300 for 7427 structure factors and 289 variable parameters for $La_3Pd_{25-x}B_{8-y}$ and a=1.17453(3) nm, b=1.04571(3) nm, c=1.59759(5) nm, β =102.132(2) deg, R_F^2=0.0495 for 2805 structure factors and 283 variable parameters for $Ce_3Pd_{25-x}B_{8-y}$. Rietveld refinement for the X-ray powder diffraction patterns for both compounds confirmed the structure models determined from X-ray single crystal data.

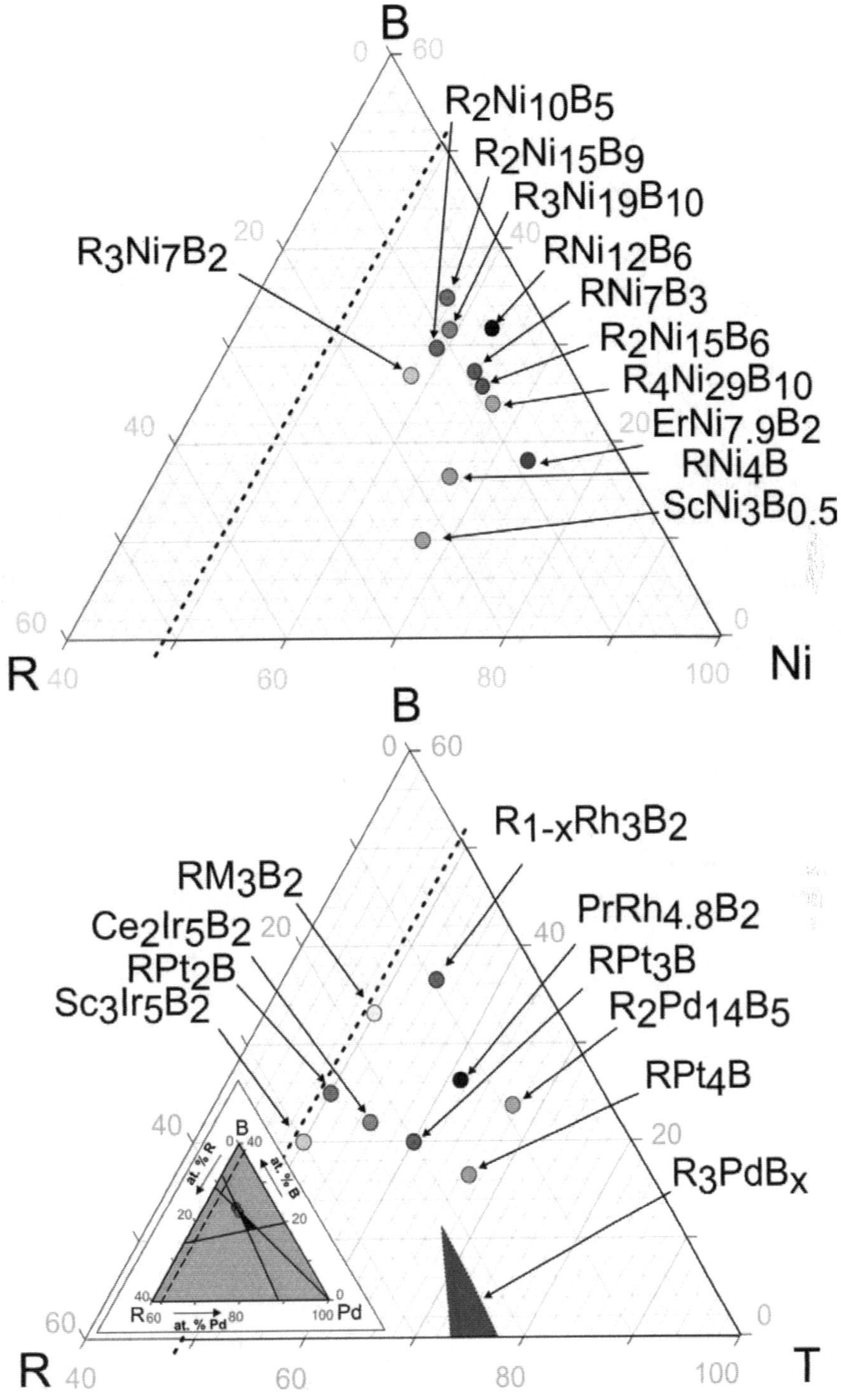

Figure 9. Formation of compounds in platinum metal subgroup and Ni ternary rare earth boride systems; inset figure shows the range of existence for $R_3Pd_{25-x}B_{8-y}$ and $Nd_2Pd_{14+x}B_{5-x}$ compounds.

Acknowledgements. Research supported by Austrian National Science Foundation FWF project 18054 and M1067-N20.

References

[1] Bauer E, Hilscher G, Michor H, Paul C, Scheidt E W, Gribanov A, Seropegin Y, Noel H, Sigrist M and Rogl P 2004 *Phys. Rev. Lett.* **92** 027003 1

[2] Yogi M, Kitaoka K, Hashimoto S, Yasuda T, Settai R, Matsuda T D, Haga Y, Onuki Y, Rogl P and Bauer E 2004 *Phys. Rev. Lett.* **93** 027003 1

[3] Bauer E, Lackner R, Hilscher G, Michor H, Sieberer M, Eichler A, Gribanov A, Seropegin Y and Rogl P 2005 *J. Phys.: Cond. Mat.* **17** 1

[4] Royanian E, Bauer E, Kaldarar H, Galatanu A, Hilscher G, Michor H, Reissner M, Rogl P, Sologub O, Giester G and Pereira-Goncalves A, submitted

[5] Kuz'ma Y B, Dub O M, Bruskov V A, Chaban N and Zavalii L V 1988 *Kristallografiya* **33** 841

[6] Nonius Kappa CCD, Program Package COLLECT, DENZO, SCALEPACK, SORTAV, Nonius Delft, The Netherlands.

[7] Sheldrick G M 2008 *Acta Crystallogr.* **A64** 112

[8] Sheldrick G M 1997 *SHELXS-97, Program for Crystal Structure Refinement* University of Göttingen, Germany Windows version by McArdle, Natl. Univ. Ireland, Galway.

[9] Roisnel T and Rodriguez-Carvalaj J 2001 *Materials-Science-Forum* **378-381(1)** 118

[10] Marsh R E 1995 *Acta Crystallogr.* **B51** 897

[11] Parthé E, Gelato L, Chabot B, Penzo M, Censual K and Gladyshevskii R 1994 *TYPIX – Standardized Data and Crystal Chemical Characterization of Inorganic Structure Types* Berlin, Heidelberg, Springer-Verlag

[12] Rogl P 1991 "Existence and Crystal Chemistry of Borides" in: *Inorganic Reactions and Methods* Ed. J. J. Zuckerman VCH-Publications Inc. **13(6)** 85-167

[13] Salamakha P S, Goncalves A P, Sologub O L and Almeida M 2003 *J. Alloys Comp.* **360** 61

[14] Kuz'ma Yu B, Babizhetskyj V, Veremchuk I and Chaban N 2004 *J. Solid State Chem.* **177** 425

[15] Glovyak T, Oryshchin S V, Gubich I B, Chaban N F and Kuzma Yu B 1989 *Kristallografiya* **34** 602

[16] Kuzma Yu B, Zavalii P Yu, Gubich I B and Bruskov V A 1991 *Kristallografiya* **36** 269

[17] Veremchuk I V, Babizhetskyy V S, Chaban N F and Kuzma Yu B 2004 *Neorg. Mater.* **40** 697

Growth and physical properties of Sc_2AlB_6 crystals

Shigeru Okada[1], Takao Mori[2], Kunio Kudou[3], Toetsu Shishido[4] and Takaho Tanaka[2]

[1] Department of Science and Engineering, Kokushikan University, 4-28-1 Setagaya, Setagaya-ku, Tokyo 154-8515, Japan

[2] National Institute for Materials Science, 1-1 Namiki, Tsukuba 305-0044, Japan

[3] Department of Mechanical Engineering, Kanagawa University, 3-27-1 Rokkakubashi, Kanagawa-ku, Yokohama 221-8686, Japan

[4] Institute for Materials Research, Tohoku University, 2-1-1 Katahira, Aoba-ku, Sendai 980-0812, Japan

sokada@kokushikan.ac.jp

Abstract. Single crystals of Sc_2AlB_6 were produced in Al self-flux, under Ar atmosphere, at soaking temperature of 1550 °C for 5 h, from Sc compounds Sc_2O_3, $ScCl_3$, ScF_3, $Sc(CH_3COO)_3$ or $Sc_2(CO_3)_3$ and crystalline boron. Sc_2AlB_6 crystals grown from Sc_2O_3 or ScF_3 had respective maximum sizes of ~0.4×0.4×4.2 and 0.2×0.2×2.1 mm^3. Sc_2AlB_6 samples prepared from $ScCl_3$, $Sc(CH_3COO)_3$ or $Sc_2(CO_3)_3$ were not large enough for most measurements or were not single crystals. Physical properties of the isoelectronic compounds Sc_2AlB_6 and Lu_2AlB_6 were investigated. Superconductivity was not observed down to 1.8 K. The magnetic susceptibilities of both samples were temperature independent and paramagnetic. The Pauli paramagnetism of Sc_2AlB_6 is sizably larger than that for Lu_2AlB_6. The electronic specific heat coefficient γ = 3.3 mJ mol^{-1} K^{-2} and Debye temperature θ_D = 757 K were deduced from the measured specific heat of Lu_2AlB_6.

1. Introduction

RE_2AlB_6 (RE=Yb and Lu), which are expected to have the Y_2ReB_6-type structure (orthorhombic *Pbam*) [1], have been synthesized in the past [2, 3]. Recently, we prepared single crystals of isostructural Sc_2AlB_6 from the high-temperature Al solution, using amorphous boron and Sc or Sc_2O_3 powders as starting materials [4]. As shown in figure 1, the Sc_2AlB_6 structure is built up by a two-dimensional boron network sandwiched between metal layers. These boron atoms exist in the interstitial sites of trigonal prisms formed by the Sc and Al atoms. However, there is little information on the physical and chemical properties of Sc_2AlB_6. In our previous work, we presented results on micro-hardness, electrical resistivity, and oxidation resistance at high temperatures in air [5].

This paper reports optimization of the growth parameters of Sc_2AlB_6 aiming at synthesis of large single crystals. The growth occurs in molten aluminum metal, under argon atmosphere, from crystalline boron and a Sc compound, such as Sc_2O_3, $ScCl_3$, ScF_3, $Sc(CH_3COO)_3$ or $Sc_2(CO_3)_3$. Those compounds are a more suitable than Sc metal because of their high chemical stability in air and low reactivity with alumina crucibles at high temperatures. The morphology and crystallographic parameters of the crystals were evaluated, and the magnetic susceptibility and specific heat were measured at low temperatures.

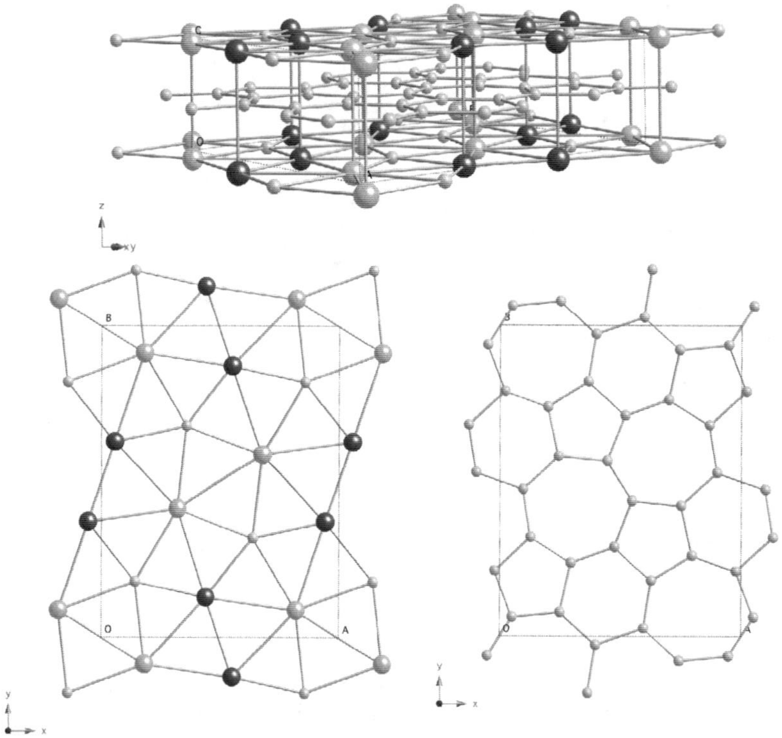

Figure 1. The crystal structure of Sc_2AlB_6 [4] viewed from the [1-10] direction (above), the metal (bottom, left) and boron (bottom right) layers viewed from the [001] direction, where the largest spheres (violet) are the Sc (Sc2 is darker) atoms, the middle-size spheres are the Al (light blue) and the smallest spheres (light green) are the B stoms.

2. Experimental details

We used the following reagents: Sc_2O_3, $ScCl_3$, ScF_3, $Sc(CH_3COO)_3$ and $Sc_2(CO_3)_3$ (purity 99.9%; Wako Pure Chemical Industries Ltd.)), crystalline boron powders (99%; Mitsuwa's Pure Chemicals) and aluminum metal chips (99.99%; Nippon Light Metal Co. Ltd). The B/Sc ratio ranged between 1.0 and 6.0, and Al metal was added to each mixture at mass ratio of 1:15-1:20. The powders were placed in a dense alumina crucible (99%; Nikkato Corporation), heated up at a rate of 300 K h^{-1}, held at soaking temperature of 1550 °C for 5 h, and cooled down at a rate of 5 K h^{-1} to 1200 °C. Then, the furnace was rapidly cooled down to room temperature. The crystals were removed from the solidified melt by dissolving it in 6-mol hydrochloric acid; they were then rinsed in water and dried. Figure 2 shows the flow chart of the synthesis. Sc_2AlB_6 crystals were selected under a stereomicroscope for the X-ray diffraction analysis. Magnetic susceptibilities of Sc_2AlB_6 and Lu_2AlB_6 were measured with a superconducting quantum interference device (SQUID) magnetometer in the temperature range 300 - 1.8 K. We used Lu_2AlB_6 as a comparison for Sc_2AlB_6 and investigated its physical properties as well. Specific heat was measured using a transient heat pulse method with a small temperature increase of 2 % relative to the system temperature. Samples were attached to an alumina plate holder using Apiezon N grease. The temperature was varied from 60 to 1.8 K. The specific heat of Lu_2AlB_6 was obtained by subtracting the blank run data from the data with the sample.

3. Results and discussion

3.1. Synthesis of Sc_2AlB_6 crystals

Crystals of Sc_2AlB_6, ScB_2, α-AlB_{12} and ß-AlB_{12}-type compound simultaneously formed from the aluminum flux, Sc reagents and crystalline boron. The maximum dimensions of Sc_2AlB_6 crystals grown from Sc_2O_3 and ScF_3 were ~0.4×0.4×4.2 mm^3 and 0.2×0.2×2.1 mm^3, respectively. However the Sc_2AlB_6 samples prepared from $ScCl_3$, $Sc(CH_3COO)_3$ or $Sc_2(CO_3)_3$ were not large enough for most measurements or were not single crystals.

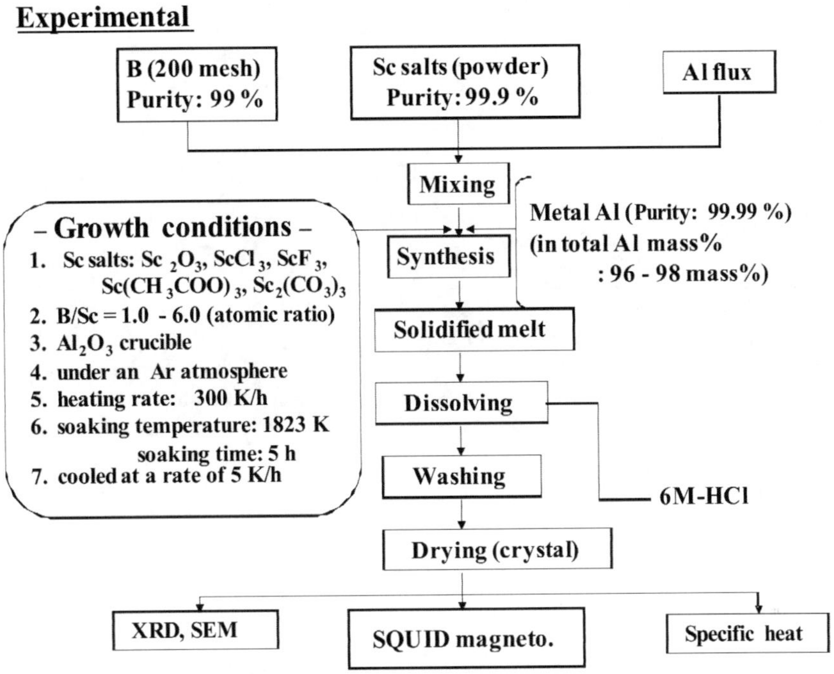

Figure 2. Flow chart for the growth of Sc_2AlB_6 crystals.

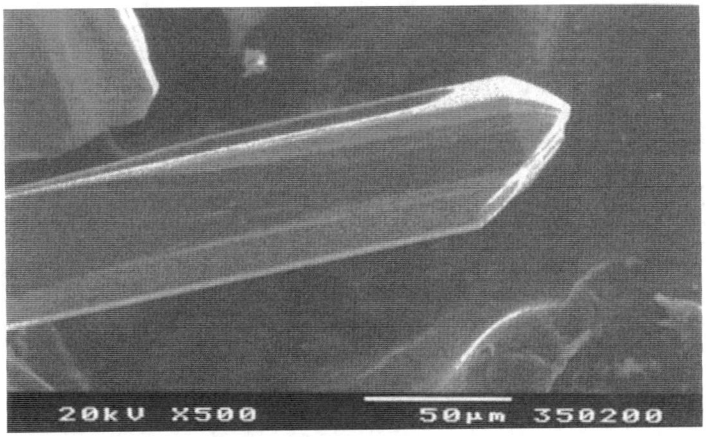

Figure 3. Sc_2AlB_6 crystal grown from ScF_3.

Low solubility of $ScCl_3$, $Sc(CH_3COO)_3$ and $Sc_2(CO_3)_3$ in the Al flux at high temperature could be blamed for poor quality of the corresponding samples. Sc_2AlB_6 crystals were shaped as prisms extending along the *b*-axis. A Sc_2AlB_6 crystal grown from ScF_3 is shown in figure 3. Unit cell parameters for Sc_2AlB_6 are summarized in table 1; they are rather close to the previously published data for Yb_2AlB_6, Lu_2AlB_6 and Y_2ReB_6 [1-3]. The unit cell parameters correlate to ionic radii (Sc (+3), 0.075 nm; Yb (+3), 0.099 nm; Lu (+3), 0.098 nm) of coordination number 6 or 8 [6].

Table 1. Crystallographic parameters of Sc_2AlB_6 and Y_2ReB_6-type crystals.

Formula unit	Sc_2AlB_6	Yb_2AlB_6	Lu_2AlB_6	Y_2ReB_6
Crystal system	ortho.	ortho.	ortho.	ortho.
a (nm)	0.8936(2)	0.9127(5)	0.8987(1)	0.9175(5)
b (nm)	1.1228(3)	1.146(1)	1.1334(1)	1.155(1)
c (nm)	0.3432(1)	0.3584(4)	0.3633(1)	0.3673(4)
V (nm^3)	0.3443(1)	0.3749(1)	0.3701(1)	0.3892(1)
Space group	*Pbam*	*Pbam*	*Pbam*	*Pbam*
Z	4	4	4	4
Ref.	This work	[2]	[3]	[1]

3.2. Magnetic susceptibilities of Sc_2AlB_6 and Lu_2AlB_6

The attention to rare earth aluminum boride system RE-Al-B is increasing due to the recent discoveries of its interesting physical properties. For example, multiple magnetic transitions occur in $TmAlB_4$ at low temperatures below the antiferromagnetic transition temperature T_N [7]. Interestingly, the low-temperature anomalies originate from intrinsic lattice defects, which are generated by the presence of two closely related structures, $YCrB_4$-type and $ThMoB_4$-type [8]. The Y_2ReB_6-type structure has similarities to the $TmAlB_4$ structure – both are layered compounds composed of two-dimensional boron polygon layers and metal atom layers. Although Sc_2AlB_6 does not have any elements of large magnetic moment, it is worthwhile to measure its magnetic susceptibility, expecting anomalies caused by lattice defects and occurrence of superconductivity.

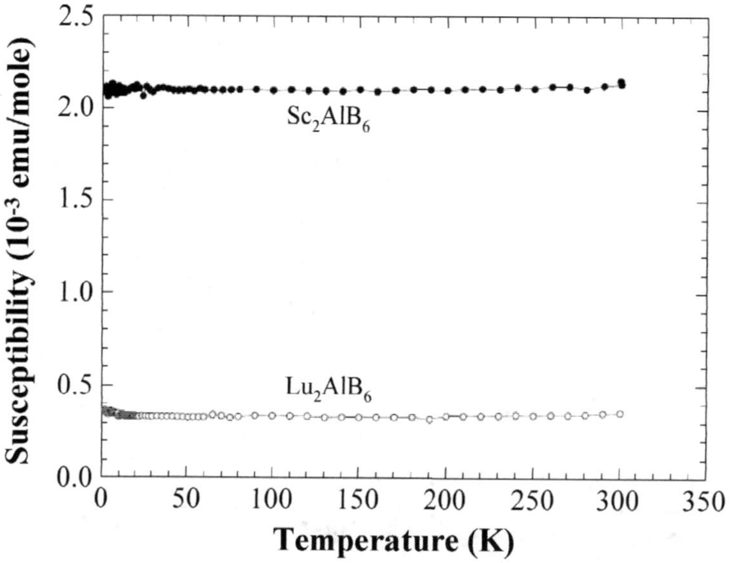

Figure 4. Magnetic susceptibilities of Sc_2AlB_6 and Lu_2AlB_6.

The magnetic susceptibilities of Sc_2AlB_6 and Lu_2AlB_6 are shown in figure 4. Measurements in low magnetic field were repeated in 70 kOe. Superconductivity was not observed in either compound down to 1.8 K. The susceptibilities of both samples are temperature independent and paramagnetic. The absolute values are 2.1×10^{-3} emu/mole and 3.3×10^{-4} emu/mole for Sc_2AlB_6 and Lu_2AlB_6, respectively. The difference between Sc_2AlB_6 and Lu_2AlB_6 is significantly larger than expected from the differences in the core diamagnetism. Thus, we can conclude that the Pauli paramagnetism of Sc_2AlB_6 is significantly larger than that of Lu_2AlB_6. This in turn indicates that the density of states at the Fermi level is sizably different between these isoelectronic compounds. Estimates of orbital susceptibility are ambiguous and thus we did not attempt to evaluate the Pauli paramagnetism from our data.

3.3. Specific heat of Lu_2AlB_6

The Lu_2AlB_6, but not Sc_2AlB_6, crystals were large enough for specific heat measurement. The low-temperature specific heat C of Lu_2AlB_6 is presented in figure 5 in a conventional plot of C/T versus T^2. A linear relation is obtained over the whole temperature range, indicating that the specific heat can be well described as the sum of a linear electronic specific heat term and a phonon term.

$$C = \gamma T + \beta T^3 \qquad (1)$$

An electronic specific heat coefficient was deduced as $\gamma = 3.3$ mJ mol^{-1} K^{-2}. The finite value of γ indicates a finite density of states at the Fermi level $D(E_F)$, which agrees with our expectation of Lu_2AlB_6 to be a metal. This γ value corresponds to $D(E_F) = 8.3\times10^{23}$ states eV^{-1}mole^{-1}.

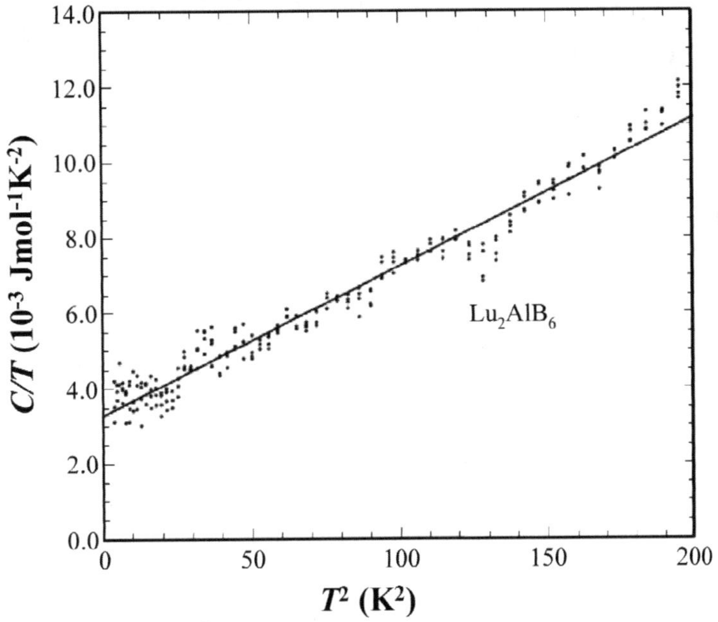

Figure 5. Low-temperature specific heat of Lu_2AlB_6. The solid line is a fit by equation 1.

From the phonon term of the specific heat, the Debye temperature θ_D of Lu_2AlB_6 was estimated as 757 K. Interestingly, Lu_2AlB_6 has the much higher Debye temperature than $LuAlB_4$ (θ_D = 415 K [8]) although both compounds have a similar structure.

4. Conclusions

Single crystals of Sc_2AlB_6 were grown in Al self-flux, under Ar atmosphere, from a Sc compound (Sc_2O_3, $ScCl_3$, ScF_3, $Sc(CH_3COO)_3$ or $Sc_2(CO_3)_3$) and crystalline boron. The maximum crystal size was ~0.4×0.4×4.2 and 0.2×0.2×2.1 mm^3 for Sc_2O_3 and ScF_3 starting materials, respectively.

Sc_2AlB_6 crystals were prisms extending in the *b*-axis direction. The Sc_2AlB_6 samples prepared from $ScCl_3$, $Sc(CH_3COO)_3$ or $Sc_2(CO_3)_3$ were too small for measurements or were not single crystals. The magnetic susceptibilities of Sc_2AlB_6 and Lu_2AlB_6 were temperature independent and paramagnetic. The absolute values are 2.1×10^{-3} emu/mole and 3.3×10^{-4} emu/mole for Sc_2AlB_6 and Lu_2AlB_6, respectively. An electronic specific heat coefficient $\gamma = 3.3$ mJ/mol/K^2 and Debye temperature of 757 K were deduced for Lu_2AlB_6. This γ value indicates that Lu_2AlB_6 is a metal with the density of states at Fermi level $D(E_F) = 8.3\times10^{23}$ states eV^{-1} mole^{-1}.

Acknowledgments
The authors would like to thank Drs. Iizumi K and Sawada Y of Tokyo Polytechnic University for help with the experiments.

References

[1] Kuz'ma Yu B and Svaricheskaya S I 1972 *Kristallografiya* **17** 658
[2] Mikhalenko S I, Kuz'ma Yu B, Korsukova M M and Gurin V N 1980 *Izv. Akad. Nauk SSSR Neorg Mater.* **16** 1941
[3] Okada S, Yu Y, Lundström T, Kudou K and Tanaka T 1996 *Jpn. J. Appl. Phys.* **35** 4718
[4] Okada S, Tanaka T, Leithe-Jasper A, Michiue Y and Gurin V N 2000 *J. Solid State Chem.* **154** 49
[5] Okada S, Kudou K, Tanaka T, Shishido T, Gurin V N and Lundström T 2004 *J. Solid State Chem.* **177** 547
[6] Lide D R 1996 *CRC Handbook of Chemistry and Physics – Ionic Radii in Crystals* ed H T Evans (New York: CRC Press) p12
[7] Mori T, Okada S and Kudou K 2005 *J. Appl. Phys.* **97** 10A910
[8] Mori T, Borrmann H, Okada S, Kudou K, Leithe-Jasper A, Burkhardt U and Grin Yu 2007 *Phys. Rev.* B **76** 064404 1

Rim region growth and its composition in reaction bonded boron carbide composites with core-rim structure.

S. Hayun, A.Weizmann, H. Dilman, M. P. Dariel , and N. Frage

Department of Materials Engineering, Ben-Gurion University of the Negev,

P. 0. Box 653, Beer-Sheva 84105, Israel.

hayuns@bgu.ac.il

Abstract. Aluminum was detected in reaction-bonded boron carbide that had been prepared by pressureless infiltration of boron carbide preforms with molten silicon in a graphite furnace under vacuum. The presence of Al_2O_3 in the heated zone, even though not in contact with the boron carbide preform, stands behind the presence of aluminium in the rim region that interconnects the initial boron carbide particles. The composition of the rim corresponds to the $B_x(C,Si,Al)_y$ quaternary carbide phase. The reaction of alumina with graphite and the formation of a gaseous aluminum suboxide (Al_2O) accounts for the transfer of aluminum in the melt and, subsequently in the rim regions. The presence of Al increases the solubility of boron in liquid silicon, but with increasing aluminum content the activity of boron decreases. These features dominate the structural evolution of the rim-core in the presence of aluminum in the melt.

1. Introduction

Light ceramics are particularly attractive for personal, land and airborne vehicle armor. High values of hardness are by common consensus of crucial importance for good ballistic resistance. Thus, boron carbide, which has very high hardness and low mass per volume ratio [1], has long been considered as a choice armor material candidate. However, difficulties and costs of fabrication, mainly by hot pressing, are factors of paramount importance in determining whether a particular material is considered for armor applications and limit the extended use of boron carbide based armor.

An alternative way for the fabrication of a fully dense boron carbide composite is the so-called "reaction bonding" process [2,3]. According to this method, a green body of boron carbide with or without free carbon is infiltrated with molten silicon. The reaction of molten silicon with the boron carbide particles or with free carbon leads to the formation of silicon carbide. The final composite consists mostly of the initial boron carbide grains, a newly formed $B_{12}(B,C,Si)_3$ phase, which surrounds the initial boron carbide particles, the SiC phase and some unreacted residual silicon. A typical microstructure of the infiltrated boron carbide composites is shown in Fig 1. Electrochemical etching allowed to reveal the presence of the initial boron carbide particles (core regions) surrounded by the newly formed carbide phase (rim regions) (Fig. 1 insert). The formation of this morphology was attributed to the dissolution-precipitation reactions in the boron carbide - silicon system and has been discussed previously [4].

Fig. 1. SEM micrograph of a typical microstructure of the infiltrated boron carbide composites. SEM micrograph of a electrochemically etched composite (insert) reveals the presence of the initial boron carbide particles (core regions) surrounded by the newly formed carbide phase (rim regions).

The composite material was prepared by pressureless infiltration of boron carbide preforms with molten silicon in a graphite furnace under vacuum. Some alumina parts were present in the heated zone, but none in any contact with the boron carbide preform. Neither the starting boron carbide powder nor the high purity silicon used for infiltration contained any aluminum. Nevertheless, substantial amounts of aluminum were detected by WDS analysis in the rim regions of the composite (Fig.2).

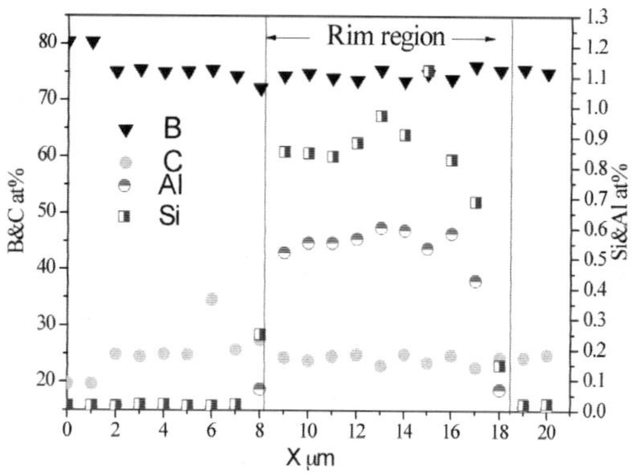

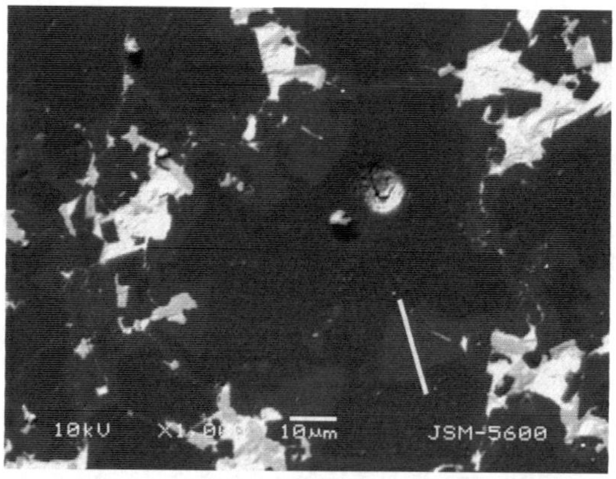

Fig. 2. Line-scan along the straight white line in the backscatter electron image of boron carbide composite.

The presence of aluminum affects to a considerable extent the morphology and the content of the rim region in the composite. The present work was undertaken in order to gain insight in the mechanism whereby aluminum is transported and makes its appearance in the rim region and in order to study its effect on the rim growth kinetics.

2. Experimental procedure

Two sets of ceramic preforms (20 mm diameter and 3 mm height) were made (compacting under uniaxial pressure of 25MPa.) from coarse boron carbide powder (average size of about 106μm, "Mudan Jiang" Chinese company) and infiltrated with molten silicon (Alfa Aesar 99.9999%).

The infiltration was conducted in a vacuum furnace (10^{-5} torr) at 1753K and the holding at the infiltrating temperatures varied from 10 to 240 min. The infiltration was carried out by placing an appropriate silicon lump on the top of the porous preforms. In the first set of the experiments, the preforms were placed on a SiC substrate (*type-A*), in the second set (*type-B*) Al_2O_3 granules had been placed beneath the SiC substrate. No direct contact whatsoever took place between the alumina granules and the boron carbide preform (Fig. 3).

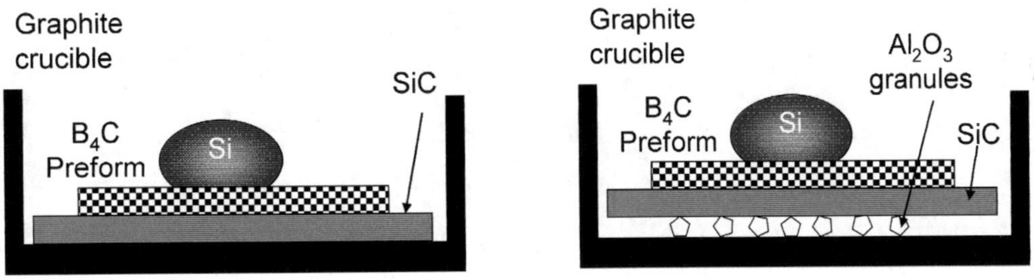

Fig. 3. Schematic illustration of the infiltration setup a) in the case of free Al_2O_3 and b) in the case of Al_2O_3 present.

The microstructure of the samples was studied by optical microscopy (OM, Zeiss Axiovert 25), scanning electron microscopy (SEM, JEOL-35) in conjunction with an energy-dispersive spectrometer (EDS) and a wavelength-dispersive spectrometer (WDS). The phase composition and the structure of the samples was analyzed by X-Ray diffraction (XRD), using a Rigaku RINT 2100 diffractometer with Cu Kα radiation. The scanning angle 2θ was increased in a step of 0.02°.

3. Result and discussion

3.1. Phase composition

The XRD patterns of the *type-A* and *type-B* composites, as a function of holding time at the infiltrating temperature, are presented in Fig. 4. According to the XRD results, the *type-A* specimens that have been infiltrated in Al_2O_3-free conditions, consist of four phases namely, boron carbide (B_xC), a ternary B-C-Si carbide phase, β–SiC and residual silicon. It is noteworthy that the position and the shape of the peaks corresponding to the ternary carbide do not change with the holding time. These peaks correspond to the $B_{12}(B,C,Si)_3$ phase [5,6,7].

For *type-B* specimens that had been infiltrated in the presences of Al_2O_3, the evolution of the phase composition is different. First, an additional relatively broad peak (2θ=38.5) appeared after a 30 min heat treatment. This peak may be attributed to an aluminum based solid solution. The intensity of the peaks corresponding to the boron carbide phase decreases, while the peaks attributed to the new formed carbide phase $B_x(C,Si,Al)_y$ increase with time. These changes suggest that substantial transformation of the initial boron carbide particles to the newly formed carbide phase took place. After a four hour-long heat treatment, the boron carbide peaks disappear, indicating an almost complete transformation into a new structure. In contrast to *type-A* composites, in which the position and the shape of the peaks corresponding to the ternary carbide do not change, in *type-B* samples, the peaks corresponding to the new formed carbide phase shift towards lower angles. Furthermore, the shape of these peaks indicates that the rim region consists of several layers with varying lattice parameters, layers that were formed during the holding time (Fig. 4.).

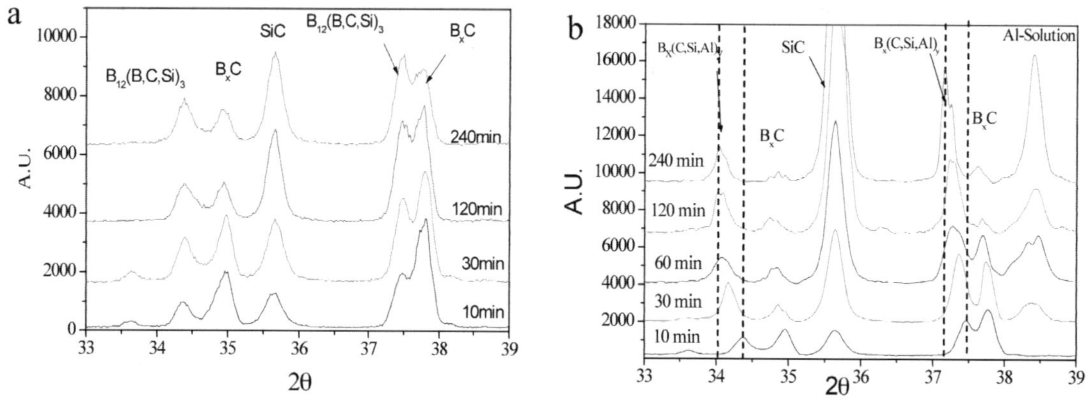

Fig. 4. XRD patterns as function of holding time at the infiltration temperature of a) *type-A* and b) *type-B* composites.

3.2. Origin of aluminium in the melt.

The presence of aluminum in the composite having been unequivocally demonstrated, it was necessary to account for its transport from the alumina particles into the composite. The starting point of the model is based on the chemical interaction between alumina and graphite that is expressed by the following equations:

$$Al_2O_{3(s)} + C_{(s)} = Al_2O_{(g)} + CO_{2(g)} \tag{1}$$

$$Al_2O_{(g)} = 2[Al] + 0.5O_{2(g)} \tag{2}$$

$$C_{(s)} + CO_{2(g)} = 2CO_{(g)} \tag{3}$$

$$CO_{(g)} + 0.5O_{2(g)} = CO_{2(g)} \tag{4}$$

where *s* and *g* denote solid and gaseous phases, respectively, and the square brackets denote that the element is in a liquid solution.

In the quaternary Si-Al-O-C system with four phases, namely the liquid solution, alumina, graphite and the gaseous phase are in equilibrium, the degree of freedom at constant temperature (F= 4+1-4) is equal to 1. The partial pressure of Al_2O is therefore a function of the aluminum content in the liquid solution. This function may be derived from the equilibrium constants of the reactions (1-4)

$$K_1 = P_{CO_2} P_{Al_2O},\ K_2 = \frac{P_{O_2}^{\ 0.5} a_{AL}^{\ 2}}{P_{Al_2O}},\ K_3 = \frac{P_{CO}^{\ 2}}{P_{CO_2}},\ K_4 = \frac{P_{CO_2}}{P_{CO} P_{O_2}^{\ 0.5}} \tag{5}$$

$$\ln P_{Al_2O} = 4\ln a_{Al} + \ln\left(\frac{K_3}{K_1}\left(\frac{K_4}{K_2}\right)^2\right) \tag{6}$$

where a_{Al} stands for the activity of aluminum in the melt. The equilibrium constants were calculated according to thermodynamic data, presented in Table 1.

The calculated values of P_{Al_2O} as a function of aluminum activity in the melt at 1753K are presented in Fig. 5. The horizontal line corresponds to the vacuum level in the furnace, which is about $1.3 \cdot 10^{-8}\,atm$.

Table 1. Standard Gibbs free energy of reaction 1-4 [8]

Reaction	$\Delta G^0(1)$	$\Delta G^0(2)$	$\Delta G^0(3)$	$\Delta G^0(4)$
ΔG^0, kJ/mole	$1066.03-377.72T$	$226.73+0.049T$	$169-0.172T$	$-282.69+0.086T$

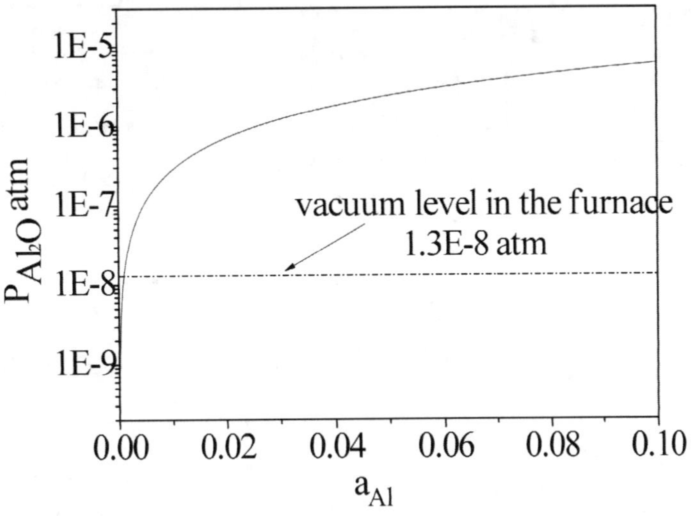

Fig. 5. Al_2O partial pressure (atm) as a function of aluminium activity. The total pressure in the chamber is $1.3 \cdot 10^{-8}\,atm$.

Since the equilibrium partial pressure of P_{Al_2O} is higher than the total pressure in the furnace, reaction 1 takes place, alumina decomposes and aluminum is being transferred via the gaseous phase into the melt. It has to be noted that this process may occur without direct contact between graphite and alumina. Thus, according to the results of the thermodynamic analysis, the interaction between graphite and alumina stands at the origin of the presence of aluminum in the melt.

3.3. Rim growth kinetics.

The microstructural evolution of the *type-A* and *type-B* composites in the course of the heat treatment for various durations may be followed using the SEM micrographs shown in Fig 6. All the samples were electro-chemically etched in KOH solution in order to emphasize the difference between rim and core regions.

 doi:10.1088/1742-6596/176/1/012009

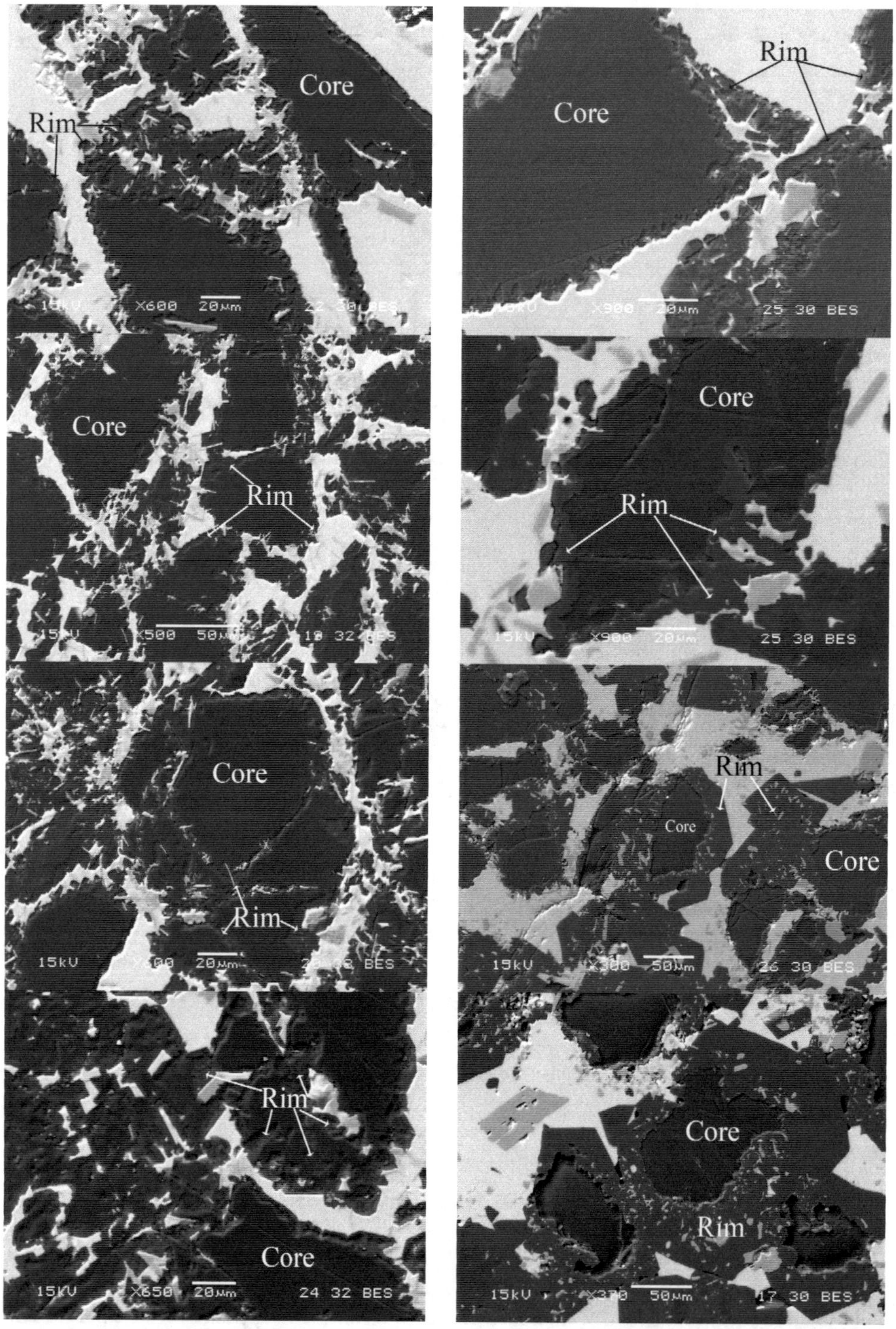

Fig. 6. SEM micrographs of *type-A* and *type-B* composites that electro-chemically etched in KOH solution in order emphasize the differences between rim and core regions.

For *type-A* composites, the thickness of the rim layer shows a very weak dependence on the holding time, while for *type-B* composites substantial widening of this region takes place. In *type-B* specimens, after 60 min most initial boron carbide particles are transformed into the quaternary carbide phase $B_x(C,Si,Al)_y$ [9]. These microstructural observations are in a good agreement with the results of the XRD analysis. According to the microstructural observations, it appears that in *type-A* composites the dissolution precipitation process operates only during the initial stages of the heat treatment. The rim region has a coherent boundary with the initial boron carbide particles [4] and prevents the continuous dissolution of boron carbide. After about 10 min the transformation of boron carbide to $B_{12}(B,C,Si)_3$ and the rim regions growth are, probably, controlled by a diffusion process through the newly formed layer.

The presence of Al significantly changes the rate of growth of the rim region. We suggest that a reason of this feature is the effect of Al on the boron solubility in liquid Si. According to thermodynamic data [8] Al increases the solubility of boron in liquid Si and decreases its activity. Thus, when Al is absorbed in the melt, enhanced dissolution of boron carbide and $B_{12}(B,C,Si)_3$, which is formed during the initial stage of infiltration (where aluminum is absent in the melt) takes place. With increasing Al content, the thermodynamic equilibrium requirements dictate additional boron carbide phase dissolution in parallel with the precipitation of the Al and Si containing carbide phase. As a result of this process the rim region consists of a sequence of quaternary carbide layers with various Al contents, as was detected by XRD analysis.

An interesting additional feature was revealed during the electro-chemical etching procedure. In *type-A* samples, the rim regions are always attacked preferably, while for *type-B* samples, after 30 min of holding, the core regions were more likely than the rim regions to be attacked. This feature is attributed to the increased aluminum content within the rim regions that promotes the stability of the quaternary carbide phase in the KOH solution.

4. Conclusion

The presence of alumina parts within a furnace with graphite heating elements leads to the presence of aluminum in the composites fabricated by infiltration of porous boron carbide preforms with liquid Si at 1753K under vacuum. The presence of aluminum in the boron carbide–silicon system significantly enhances the transformation of the initial boron carbide particles into a newly formed quaternary carbide phase and the kinetics of the rim-region growth.

5. References

[1] F. Thévenot, "Boron carbide - a comprehensive review" J. Eur. Cer. Soc. , 6 205-225 (1990).

[2] M. K. Aghajanian, B. N. Morgan, J. R. Singh, J. Mears and R. A. Wolffe, "A New Family of Reaction Bonded Ceramics for Armor Applications"; pp. 527–539 in Ceramic Transactions, vol 134,Ceramic Armor Material by Design Edited by J. W. McCauley, A. Crowson, American Ceramic Society 2001

[3] K.M. Taylor, R.J. Palicke , "Dense Carbide Composite for Armor and Abrasives". U.S. Pat. No. 3 765 300, Oct.16 1973

[4] S. Hayun, N. Frage and M. P. Dariel,, J. Sol. St. Chem., 179(9) 2875-2879 (2006).

[5] A. Lipp, M. Röder "Verbindungen im System B-C-Si bzw. B-Si Z. Anorg. Allg. Chem. 344 (1966) 225.

[6] B, Morosin, T. L. Aselage, R. S. Feigelson,"Crystal structure refinements of rhombohedral symmetry materials containing boron-rich icosahedra." Mater. Res. Soc. Symp. Proc. 97 145-149 (1987).

[7] B. Kasper, Ph. D. Thesis, University of Stuttgart, Stuttgart, 1996

[8] SGTE Substances Database , Ver. 4, Scientific Groupe Thermodata Europe, Grenoble Campus, 1001 Avenue centrale BP66, F-38402 Saint Martine D'Heres, France.

[9] P. Dorner, " Ph. D. Thesis, University of Stuttgart, Stuttgart, 1982

High–temperature crystallization of novel rare–earth borate materials: single crystals and thin films

Nikolay Leonyuk[1], Victor Maltsev[1], Elena Volkova[1], Elizaveta Koporulina[1], LarisaNekrasova[1],
NikolayTolstik[2] and Nikolay Kuleshov[2]
[1]Department of Crystallography and Crystallochemistry, Geological Faculty, Moscow State University, 119991/GSP1 Moscow, Russian Federation
[2]Institute for Optical Materials and Technologies, Belarus Technical University, Minsk, Belarus

E-mail: leon@geol.msu.ru

Abstract. Phase formation has been studied in the complex system $YbAl_3(BO_3)_4$ - $K_2Mo_3O_{10}$ - B_2O_3 - Yb_2O_3 within the stability region of $YbAl_3(BO_3)_4$ (YbAB). The obtained data were compared with those for $YAl_3(BO_3)_4$ as another end member of the YbAB-YAB solid solutions. Visually transparent (Er,Yb):$YAl_3(BO_3)_4$ single crystals with a typical size up to $10\times10\times15$ mm^3 and crystalline layers were obtained from $K_2Mo_3O_{10}$ based fluxed melts. The 1.5 μm emission spectra of (Er,Yb):$YAl_3(BO_3)_4$ thin films as well as bulk crystals were measured using CW Ti:sapphire laser tuned to 976 nm as a pump source.

1. Introduction

Development of small optical components requires highly efficient single crystals and crystalline layers. They can be used, for example, as compact disk lasers or as active waveguides with different pumping configurations in laser and amplifier applications. In this regard, borate crystals with a general formula $R Al_3(BO_3)_4$ (R = Y; Pr–Lu) are potentially interesting because of their wide isomorphous substitutions [1]. In this family, $YAl_3(BO_3)_4$ (YAB) crystal is a well-known laser host for Yb and Nd ions [2-3]. YAB bulk crystals and thin layers, co-doped with Yb and Er, also demonstrate promising properties for telecommunications, medical and range-finding applications [4-6]. Recently, (Er,Yb):YAB was proposed as a laser material, which can potentially generate multi-watt powers under diode-laser pumping at ~980 nm. CW output power of 1 W with a slope efficiency of 35% was obtained, but the output power was limited by the available pump source [4,5]. Mode-locked regime has been realized with pulse duration of about 4 ps and average output power of 270 mW [6].

Flux growth technology of YAB crystals is well known and is based on the pseudo-quaternary YAB - $K_2Mo_3O_{10}$ - B_2O_3 - Y_2O_3 system [7]. As for other end member of (Yb,Y)$Al_3(BO_3)_4$ solid solutions, i.e., $YbAl_3(BO_3)_4$ (YbAB), its crystallization conditions from similar complex high-temperature solutions needs to be studied.

Here, we discuss recent results on phase formations in the complex system $YbAl_3(BO_3)_4$ - $K_2Mo_3O_{10}$ - B_2O_3 - Yb_2O_3, as well as on the growth of (Er,Yb)–codoped YAB bulk crystals and (Er,Yb):YAB crystalline layers. We also present the 1.5 μm emission spectra of thin films and bulk crystals.

2. Experimental details

Phase relationships in the pseudo-quaternary YbAB - $K_2Mo_3O_{10}$ - B_2O_3 - Yb_2O_3 system were studied in the temperature range 1000-1150 °C by a spontaneous nucleation approach for two cross sections corresponding to 15 and 20 wt% of YbAB. Formation of solid phases was revealed for regions of YbAB stability and existence of the melt in the studied system. The data are presented as projections on the temperature level at 1000 °C.

The complex solvent, containing potassium trimolybdate $K_2Mo_3O_{10}$ with an excess of boron and rare earth oxides, was used in all runs. Starting chemicals (at least 99.99% and 99.999% purity for rare earths and other materials, respectively) were Y_2O_3, Yb_2O_3, Er_2O_3, Yb_2O_3, Al_2O_3 and B_2O_3, but $K_2Mo_3O_{10}$ was previously sintered from K_2MoO_4 and H_2MoO_4 at 650 °C according to the reaction:

$$K_2MoO_4 + 2H_2MoO_3 = K_2Mo_3O_{10} + 2H_2O\uparrow.$$

The starting charge was mixed and placed into a platinum crucible and homogenized at 1150 °C in a high-temperature furnace for 24 – 48 h. The temperature at the bottom of the crucible was kept 2–3 °C higher than at the melt surface. The flux components were varied from 30 to 85 mol% ($K_2Mo_3O_{10}$), 10 to 70 mol% (B_2O_3) and 0 to 20 mol% (Yb_2O_3). Then, the temperature was lowered at a rate of 0.5-1 °C/h to 1000 °C (15 wt% YbAB) or to 1050 °C (20 wt% YbAB). After this procedure, the crucibles were quickly cooled down to 300 °C and removed from the furnace. The solid phases were released from the solidified melt by dissolution in hot diluted hydrochloric acid, in order to identify them by X–ray powder diffraction.

Temperature dependence of YbAB solubility was studied in the range of 8 - 14 mol% for the fluxes (I) 55 mol% $K_2Mo_3O_{10}$, 45 mol% B_2O_3 and (II) 55 mol% $K_2Mo_3O_{10}$, 40 mol% B_2O_3, 5 mol% Yb_2O_3 by a probe technique. The saturation temperatures were found from the experimental data on changes in weight and micro-relief of the probe seeds after soaking for a period from 10 min to several hours in fluxed melts, depending on the deviation from their equilibrium points.

Single crystals of $Er_xYb_yY_{1-x-y}Al_3(BO_3)_4$ with $x = 0.005$ to 0.015 and $y = 0.07$ to 0.15 in the starting crystalline substances, as well as pure $YAl_3(BO_3)_4$ single crystals, were obtained by dipping seeded solution growth (DSSG). The fraction of YAB and (Er,Yb):YAB crystalline substances in the starting solutions was 17 wt%. Before DSSG, the saturation temperatures of fluxed melts were determined by a probe technique as 1060–1080 °C depending on the dopant concentration. A "point" YAB seed of $0.3\times0.3\times1.0$ mm^3 size was dipped into fluxed melt. During the crystal growth, supersaturation was kept within the temperature range 1080–1000 °C by the cooling of fluxed melts at a rate 0.2–5 °C/day following the experiments on the solubility and crystallization kinetics. At the end of the growth process, the crystal was pulled out and cooled to the room temperature within several days.

Thin films were grown under similar conditions, using liquid phase epitaxy (LPE). At the beginning of epitaxial growth, the YAB substrate was slowly dipped into the solution at the temperature 1 °C higher than saturation point, T_{sat}. After 10–20 min, the temperature was decreased to 3–12 °C below T_{sat} and kept at this level for 9–36 h in order to grow the (Er,Yb):YAB layers depending on supercooling of the fluxed melt and on expected thickness of the layer, as it follows from their growth kinetics. Then, the substrate with deposited film were slowly pulled out, cooled to the room temperature and released from the crystallized melt by dissolution in hydrochloric acid, distilled water and ethanol.

Substrate choice is crucial for LPE growth. It should match the material to be deposited both in the lattice parameters and in the thermal expansion coefficient. Otherwise strains are introduced into the film and cracks occur if the mismatch is large. Furthermore, differences in refractive indices between the substrate and the active layer are important because YAB crystalline thin layers doped with Er^{3+} and Yb^{3+} can be applied in active waveguide devices, such as planar and channel waveguide lasers. In homoepitaxy, the lattice parameters, melting temperatures, and thermal expansion coefficients are matched between the substrate and film. However, it is not suitable for making waveguides as the film should have higher refractive index than the substrate. According to ref. [8], substitution of Y^{3+} by Yb^{3+} up to 7 at% increases the ordinary and extraordinary refractive indices by ~0.002 and 0.001, respectively. In this respect, single crystals of YAB are an optimal substrate for the LPE of (Er, Yb):YAB layers.

The substrate wafers with sizes up to 0.7–10 mm^2 and thickness of 0.5–1.5 mm were cut from undoped DSSG–grown YAB single crystals parallel to rhombohedron face and mechanically polished. The X-ray powder diffraction patterns were obtained using DRON-1UM and STOE STADIMP diffractometers. The composition, homogeneity and external morphology of single-crystal layers were studied with an analytical scanning electron microscope (ASEM) JSM-5300 + Link ISIS. Microprobe analysis of unpolished samples was performed with an accuracy of 0.2-0.3 wt%. A Cameca analyzer was used for study layers with a minor dopant concentration. The distribution coefficient was calculated as $K_d = C_{cryst}/C_{dissRAYB}$, where C_{cryst} is the content of element R (R = Y; Pr–Lu) in RYAB grown epilayers and $C_{dissRAYB}$ is the R concentration in the borate crystalline of the fluxed melt.

Room-temperature polarized absorption spectra of LPE-grown (Er,Yb):YAB thin films in the 1 μm spectral region were measured using a Cary-5000 spectrophotometer. The 1.5 μm emission spectra of Er,Yb:YAB thin film as well as of bulk crystal were obtained using CW Ti:sapphire laser tuned to 976 nm as a pump source. The laser beam was focused using wide-aperture 10× microscope objective. The fluorescence was collected by the same objective, reflected by the selective mirror and focused on the slits of a 1-m monochromator. It was detected by a Si photodiode and processed by a lock-in amplifier. The spectral response of the system was calibrated using a stabilized tungsten lamp. The surface and morphology of the films were examined with Leica DM 4000 M optical microscope equipped with a Leica DFC-290 camera. The optical magnification was varied from 40× to 500×. Digital interference contrast (DIC) technique was used to improve the contrast of the small surface irregularities.

3. Results and discussion

No YbAB single-phase crystallization was found in the pseudo-quaternary 20 wt% $YbAl_3(BO_3)_4$ - 80 wt% ($K_2Mo_3O_{10}$-B_2O_3-Yb_2O_3) system in the temperature range 1150-1050 °C. YbAB and $YbBO_3$ orthoborates co-crystallize in region II (figure 1*a*). Only small hexagonal $YbBO_3$ plates and needle-shaped $Al_5(BO_3)O_6$ orthotriborate crystals were observed in the regions III and IV, respectively.

As for the 15 wt% YbAB cross-section of the above complex system, single-phase YbAB crystallization was revealed upon lowering the temperature from 1100 to 1000 °C (figure 1*b*). Its field is limited by variations of the Yb_2O_3 and B_2O_3 contents in the flux from 5 to 10 mol% and 20 to 40 mol%, respectively. In the region I, isometric YbAB crystals exhibit well-developed $\{10\bar{1}1\}$, $\{11\bar{2}0\}$ and $\{2\bar{1}\bar{1}0\}$ faces. In this case, the saturation points of YbAB insignificantly increase with an excess of ytterbium oxide in the complex $K_2Mo_3O_{10}$-B_2O_3-Yb_2O_3 solvent (figure 2*a*). On the other hand, these temperatures decrease substantially with increasing B_2O_3 concentration in the $K_2Mo_3O_{10}$-B_2O_3 flux (figure 2*b*). Only $YbBO_3$ orthoborate crystallizes in the regions III (figure 1b), but increasing Yb_2O_3 content up to 20 mol% in the solvent leads to a considerable increase in the melting temperature of starting load (figure 2b, region V).

Comparative analysis of phase relationships in the YbAB (15 wt%) and YAB (17 wt%) based systems indicates that R_2O_3 concentration in the $K_2Mo_3O_{10}$-B_2O_3-R_2O_3 solvent seems to be not higher than 10 mol%. The regions of single phase YbAB and YAB crystallization coincide in a narrow range (overlapping field in figure 3).

The visually transparent YAB and (Er,Yb):YAB crystals were obtained with a typical size up to 10×10×15 mm^3 and 8×8×12 mm^3, respectively. YAB crystals are colorless, but the (Er, Yb):YAB samples have the characteristic for erbium light pink color. The average Er and Yb effective distribution coefficient is 0.84 as a consequence of minor differences in the sizes of Y^{3+}, Er^{3+} and Yb^{3+} cations.

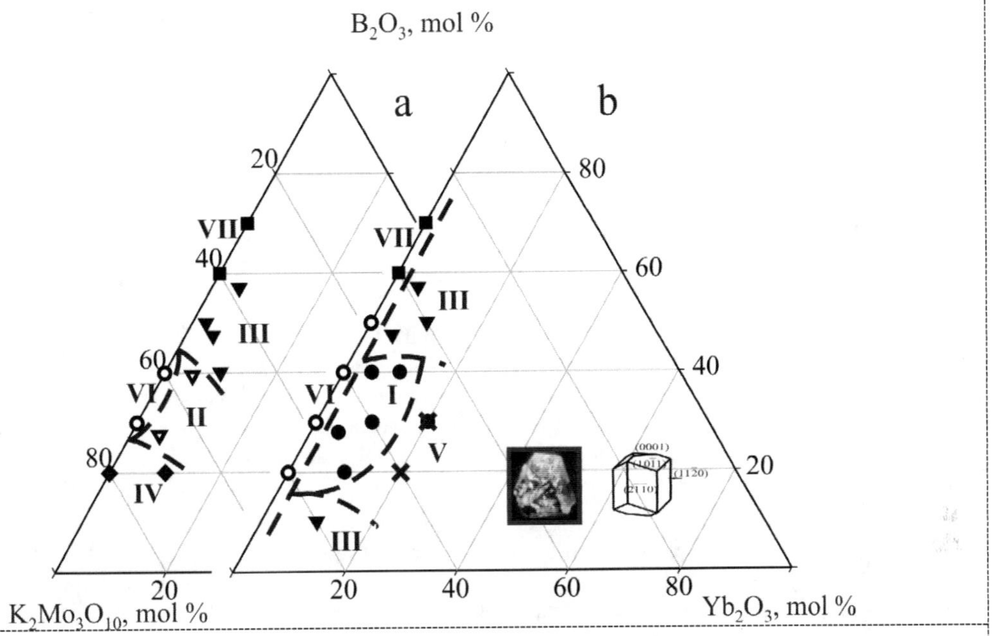

Figure 1. Phase relationships in the systems 20 wt% YbAB–80 wt% ($K_2Mo_3O_{10}$-B_2O_3-Yb_2O_3) (*a*) and 15 wt% YbAB – 85 wt% ($K_2Mo_3O_{10}$-B_2O_3-Yb_2O_3) (*b*) in the temperature range 1000-1150 °C: (I) ● - YbAB; (II) ▽– YbAB+$YbBO_3$; (III) ▼ – $YbBO_3$; (IV)◆ – $Al_5(BO_3)O_6$; (V)✖ - YbAB, as a result of solid state reactions (✖ and **x** loads do not melt at 1150 °C); (VI)○ – glass forming melt; (VII)■ – separating melt.

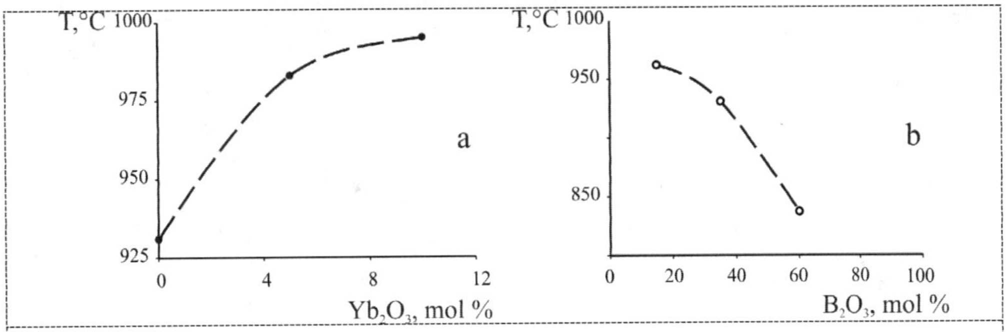

Figure 2. YbAB saturation points in 15 wt% fluxed melt depending on the Yb_2O_3 (*a*) and B_2O_3 (*b*) concentrations in the complex solvents.

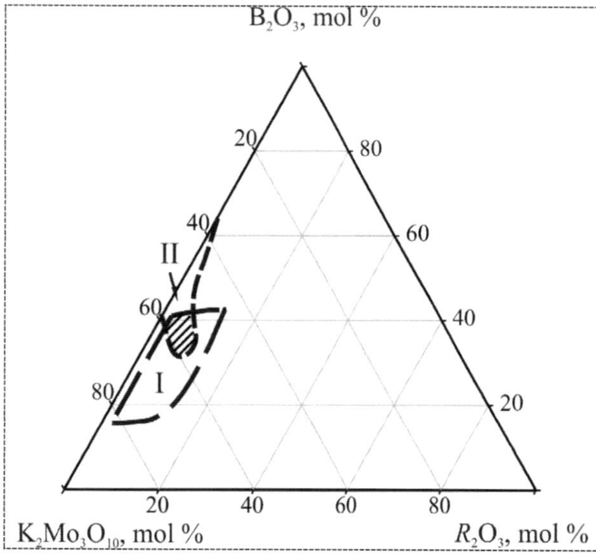

Figure 3. Schematic comparison of single-phase YbAB (I) and YAB (II) crystallization in $K_2Mo_3O_{10}$ based fluxed melts, as projections to their 1000 °C (I) and 900 °C (II) cross sections.

The observed growth rates were similar for all the studied $Er_{0.015}Yb_{0.11}Y_{0.875}Al_3(BO_3)_4$ and $Yb_{01}Y_{0.9}Al_3(BO_3)_4$ single-crystal layers, because the presence of a small amount of dopant does not change the development of the layers. Thus the LPE growth can be considered as almost homoepitaxial. The surface morphology was quite good and flat over large areas. There were no visible inclusions in the layers. Though the polished edge looks optically good, the as-grown surface has some imperfections and cracks. All the surface irregularities are smaller than 50 µm (figure 4). The border between the doped film and the undoped substrate is clearly visible in polarized light (figure 5).

Figure 4. 50-µm thick as-grown (Er,Yb):YAB layer on the $(10\bar{1}1)$ oriented YAB substrate

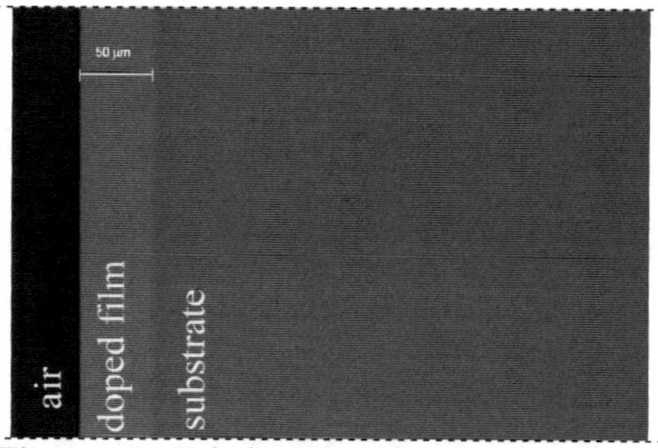

Figure 5. Polarized optical image of the polished cross-section of the (Er,Yb):YAB layer on the YAB substrate

Room-temperature polarized absorption and emission spectra of (Er,Yb):YAB bulk crystal and thin film are shown in figures 6 and 7, respectively. The spectra look very similar; neither absorption nor emission peaks of the LPE-grown layers are broadened in comparison with the bulk crystal. The similarity of the absorption and fluorescence spectra of (Er,Yb):YAB bulk crystals and LPE-grown thin films indicates good crystallinity of the latter.

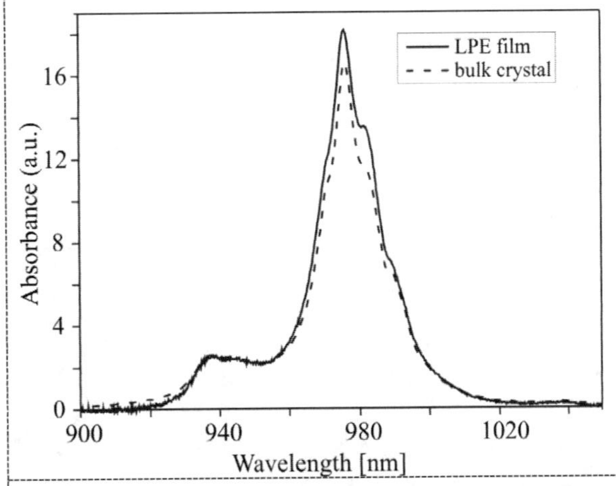

Figure 6. Room-temperature absorption spectra of (Er,Yb):YAB bulk crystal and LPE-grown film in σ-polarization.

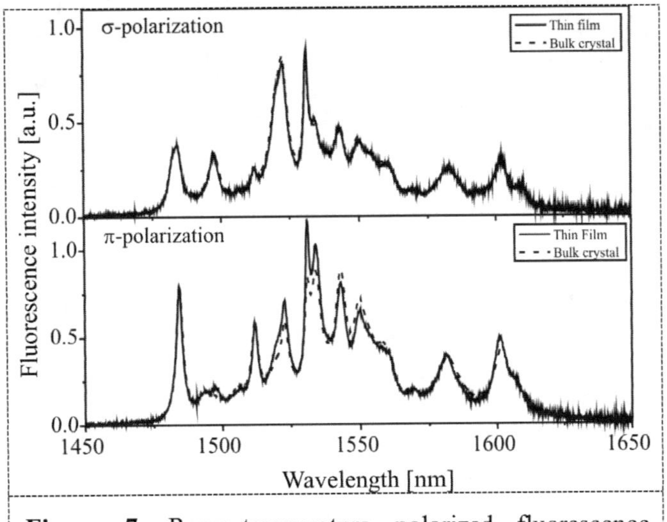

Figure 7. Room-temperature polarized fluorescence spectra of (Er,Yb):YAB bulk crystal and thin film.

5. Conclusions

The region of primary YbAB crystallization was revealed for 15 wt% of YbAB cross section in YbAB-$K_2Mo_3O_{10}$-B_2O_3-Yb_2O_3 pseudo-quaternary system. The saturation points of YbAB containing fluxed insignificantly increase with increasing ytterbium oxide and decreasing B_2O_3 concentration in the flux. Phase relationships are similar in both ytterbium and yttrium systems. The regions of single-phase YbAB and YAB crystallization somewhat overlap each other in the temperature range 1000 – 1150 °C. Most preferable fluxes for crystal growth of YbAB - YAB solid solutions appear as: (I) 55 mol% $K_2Mo_3O_{10}$, 45 mol% B_2O_3 and (II) 55 mol% $K_2Mo_3O_{10}$, 40 mol% B_2O_3, 5 mol% R_2O_3. Optically transparent YAB and (Er,Yb):YAB crystals with sizes up to 10×10×15 mm^3 were obtained by the DSSG technique. (Er,Yb):YAB thin film growth was performed under similar conditions, using liquid phase epitaxy. The 1.5 μm emission spectra of thin films and bulk crystals were measured using CW Ti:sapphire laser tuned to 976 nm.

Acknowledgments
This research was supported by the RFBR grants ## 07-05-00680-a, 08-05-12038_ofi and 08-05-90010_bel.

References

[1] Hong H Y-P and Dwight K 1974 *Mater. Res. Bull.* **9** 1661
[2] Dekker P, Dawes J M, Piper J A, Liu Y and Wang J 2001 *Opt. Commun.* **195** 431
[3] Jaque D, Capmany J and Sole J G 1999 *Appl. Phys. Lett.* **75** 325
[4] You W, Lin Y, Chen Y, Luo Z and Huang Y 2004 *J. Cryst. Growth* **270** 481
[5] Tolstik N A, Kurilchik S V, Kisel V E, Kuleshov N V, Maltsev V V, Pilipenko O V, Koporulina E V and Leonyuk N I 2007 *Opt. Lett.* **32** 3222
[6] Lagatsky A A, Kisel V E, Troshin A E, Tolstik N A, Kuleshov N V, Leonyuk N I, Zhukov A E, Rafailov E U and Sibbett W 2008 *Opt. Lett.* **33** 83
[7] Azizov A V, Leonyuk N I, Timchenko N I and Belov N V 1979 *Sov. Phys. Dokl.* **24** 313
[8] Filippov V V, Bondar' I T, Kuleshov N V, Leonyuk N I, Mal'tsev V V and Pilipenko O V 2007 *J. Opt. Technol.* **74** 717

Synthesis of crystalline boron nanoribbons and calcium hexaboride nanowires by low pressure chemical vapor deposition

Panchatapa Jash and Michael Trenary
Department of Chemistry, University of Illinois at Chicago, IL 60607, USA
Email: mtrenary@uic.edu

Abstract
Low pressure chemical vapor deposition was used to synthesize crystalline boron nanoribbons of 16 nm thickness by both uncatalyzed and nickel-catalyzed pyrolysis of diborane at relatively low temperatures. The nanoribbons were characterized with scanning electron microscopy, transmission electron microscopy, X-ray photoelectron spectroscopy, Auger electron spectroscopy, and Raman spectroscopy. Similar procedures were used to synthesize and characterize calcium hexaboride nanowires. The nanostructures reported here are compared with recent literature results for boron nanoribbons and CaB_6 nanowires grown by the same method.

1. Introduction

Boron and boron-rich solids possess unusual geometric and electronic structures that make them interesting from a fundamental perspective as well as uniquely suited to certain applications [1-3]. For this reason and the fact that boron and carbon share many properties, there has been much interest in recent years in exploring the properties of the boron analogs of fullerenes and carbon nanotubes. Based on an Aufbau principle, Boustani has predicted that novel layered, tubular, and quasi-crystalline boron nanostructures built from elemental subunits should be relatively stable [4-10]. Moreover, theoretical predictions indicate that boron-related nanostructures might be better candidates for applications as nanoscale interconnects than are carbon nanotubes. Unlike carbon nanotubes, boron nanoribbons do not roll-up to multiwall tubular forms and therefore the dangling bonds at their surfaces may be available for bonding to other elements, including hydrogen. For this reason, boron and boride nanostructures are of interest for hydrogen storage applications.

Calcium hexaboride (CaB_6) has a density of 2.45 g/cm^3, a melting point of 2235 °C, a resistivity of 222 μΩ-cm, and a Young's modulus of 451 GPa [11]. Its electronic properties have been the focus of many recent studies, which have addressed whether CaB_6 is a high temperature ferromagnet [12-15], a semiconductor [16,17], or a semimetal [18-20]. Applications of CaB_6 as a high temperature material, a surface protector, and as a wear resistant material have all been considered [2,11]. Various methods of synthesising CaB_6 have been reported [21-24], but because high temperatures where calcium has a high vapor pressure are used, it is challenging to prepare the pure compound with a precise stoichiometery [25,26]. Recently, it was shown that a mixture of CaB_6 and CaH_2 will react with H_2 to produce $Ca(BH_4)_2$ [27]. Since the reverse reaction, the dehydriding of $Ca(BH_4)_2$ to produce CaB_6 and CaH_2, releases 9.6 wt% H_2, $Ca(BH_4)_2$ is of interest as a material that can reversibly store hydrogen at high gravimetric and volumetric densities. It is likely that the high surface-to-volume ratio of CaB_6 nanowires will greatly facilitate its reaction with H_2 to form $Ca(BH_4)_2$, thus making studies of the reaction of hydrogen with CaB_6 nanowires of great interest to the hydrogen storage problem. As a

Journal of Physics: Conference Series **176** (2009) 012011 doi:10.1088/1742-6596/176/1/012011

prelude to such studies, methods to synthesize the appropriate boron and boride nanostructures must be established and the synthesized products well characterized.

2. Experimental set-up

A low-pressure chemical vapor deposition (LPCVD) system was designed and assembled for synthesis of boron-related nanostructures. It follows the design of the apparatus described by Xu *et al* for the synthesis of boron nanoribbons [28] and CaB_6 nanowires [29]. In this work, the nanostructures were deposited onto silicon (Si) substrates with a one-micron thick thermally grown SiO_2 layer (University Wafer). The Si substrates were cut from the wafers and ultrasonically cleaned first using H_2O_2 and concentrated H_2SO_4 and then with acetone and ethanol (time for each cleaning: 15 min). For the uncatalyzed growth of the boron nanoribbons, the cleaned substrates were simply loaded into a quartz boat and placed in the 1-inch diameter quartz tube, which in turn was placed into a tube furnace. As the ends of the quartz tube extend beyond the tube furnace, there is a temperature gradient from the middle of the furnace, where the temperature is highest, to the ends of the tube, which are near room temperature. Due to this temperature gradient, it was found that different results were obtained depending on the exact placement of the samples along the length of the quartz tube. For the CaB_6 synthesis, a thermal evaporator was used to deposit a Ni layer (6 nm) onto a randomly distributed powder of CaO (Alfa Aesar; 99.998% purity) placed on a Si substrate. For the Ni-catalyzed growth of the boron nanoribbons, the Ni layer was evaporated onto the bare Si substrates. The chamber was ramped up to 925 °C (center position temperature) in 45 min under a continuous flow of argon (BOC gases). A gas mixture of 1% diborane in argon (Matheson Trigas) at a flow rate of 20 sccm (standard cubic centimeter per minute) was then introduced into the chamber for 120 min. The total pressure for each run was ~ 400 mtorr. After reaction, several dark gray "puffy balls" (some visible to the eye) were found randomly deposited on the substrates, which were positioned in the 630-750 °C temperature zone region of the furnace. The substrates were removed, and the deposits were characterized by various techniques, including scanning electron microscopy (SEM), transmission electron microscopy (TEM) including energy dispersive X-ray spectroscopy (EDX) and selected area electron diffraction (SAED), X-ray photoelectron spectroscopy (XPS), and Raman spectroscopy.

3. Results and discussion

3.1. Uncatalyzed growth of boron nanoribbons

The SEM micrographs in figure 1 were recorded using a Nova NanoSEM 600 and JEOL JSM 6320 FE-SEM and show the different types of morphologies observed for nanostructures grown in the different temperature regions of the furnace. These morphologies can be designated as nanoribbons or nanoscrolls, as suggested by Xu *et al* [28]. The widths of these structures, including of the nanoscrolls calculated as if they could be unrolled, are more than a hundred nm with a thickness range of 15-20 nm. The inset in the right-hand panel of figure 1 shows the cross-section of one ~16 nm thick scroll. The "grass-like" nanoribbons, which are most common, display various features. For example, some are twisted or have zigzag edges and the tips (or ends) of some have slots in the middle, forming a "fork-like" structure. At the relatively higher temperature of formation of the wider nanoribbons, some break into narrower structures. Also, as seen in the inset of the upper left-hand panel, new ribbons seem to grow from the edge of a wide ribbon. This morphology was not described by Xu *et al* [28].

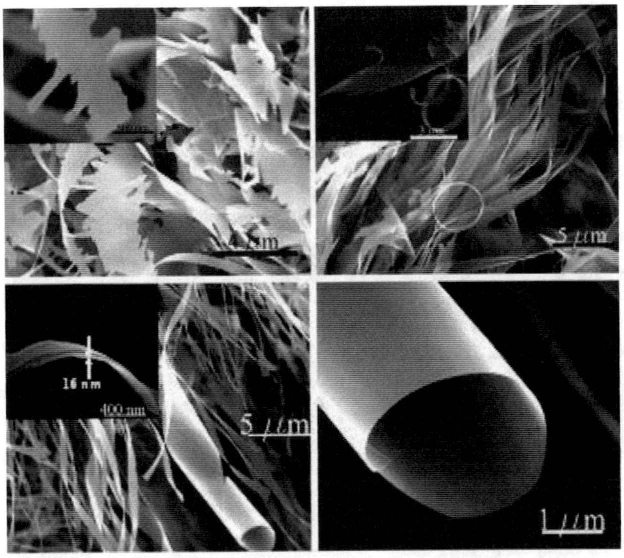

Figure 1. SEM images of pure boron nanoribbons and nanoscrolls with thicknesses of about 16 nm.

To characterize the boron nanoribbons with transmission electron microscopy (JEOL JEM-3010 TEM), some puffy balls were scraped off the substrates and ultrasonically dispersed in ethanol for 20 min. A drop of suspension was placed on a copper TEM grid covered with a holey carbon film (SPI Supplies). Figure 2 shows high resolution TEM micrographs of twisted nanoribbons. The spacing between atomic layers is about 0.26 nm (left) and 0.39 nm (right). The 0.39 nm value is a good match (within 10%) for the spacing between the (202) or (113) lattice planes of α-tetragonal boron, while the 0.26 nm spacing likely corresponds to a higher index set of planes, such as (215). Selected area electron diffraction shows that the nanoribbons are crystalline. Measured d-spacings from the SAED pattern (figure 3) are well matched by those in JDPDS file # 77-1275, which confirms that our product is α-tetragonal boron, a conclusion also reached by Xu *et al* [28].

Figure 2. HRTEM images showing 0.26 nm (left) and 0.39 nm (right) atomic layer spacings.

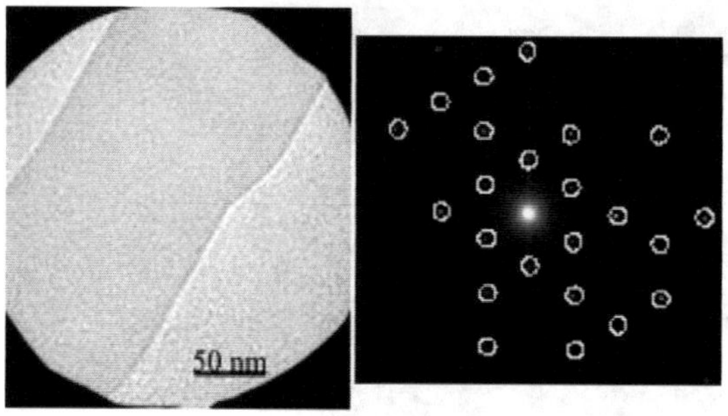

Figure 3. Low resolution TEM image of a nanoribbon (left) and selected area electron diffraction pattern (right) indicating that the nanoribbons are crystalline with the α-tetragonal boron structure (JDPDS file # 77-1275). The circles indicate the position of all observed diffraction spots, some of which are quite faint.

Further information on the nature of the boron nanoribbons is provided by the Raman spectra in figure 4. Table 1 compares the peaks reported by Xu *et al* [28] with those observed here. The peak at 520 cm^{-1} is due to the silicon substrate. All of the peaks reported by Xu *et al* [28] are within a few cm^{-1} of the peaks that we observe, with the exception of their peaks at 705 and 746 cm^{-1}, which are not seen in our spectra. Also, we observe additional peaks that are not apparent in their spectra, indicating that additional boron phases are present in our samples. However, the peaks we observe do not coincide exactly with those in the Raman spectra of α- or β-rhombohedral boron [30]. The preparation of pure elemental boron is difficult, because boron is very reactive toward other elements at the elevated temperatures required for reduction of its compounds. Moreover, the crystallinity of the final boron product is highly temperature dependent [2,3]. For example, amorphous boron powders are usually made at 600-800 °C; vitreous boron pieces are formed at 800-1100 °C; and various crystalline boron allotropes are commonly obtained above 1000 °C. Therefore it is not surprising to have peaks from different forms of elemental boron, some of which may be stabilized by impurities.

Table 1. Summary of Raman peak positions of boron nanoribbons

Materials	Raman Shift (cm^{-1})										
Xu *et al*	366	490	642		705		746	795	910	1100	
This work	--	493	646	683	--	761	--	789	911	1117	1170

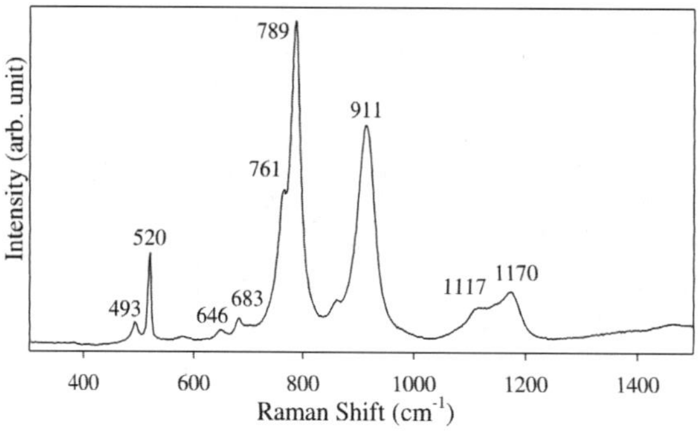

Figure 4. Raman spectrum of the boron nanoribbons.

The chemical composition of the nanoribbons was analyzed with EDX (figure 5) using a JEOL JEM 3010 TEM, which shows that they consist mainly of boron with only a small amount of oxygen detected. The oxygen peak is likely due to an amorphous oxide surface layer as boron is superficially oxidized in air at room temperature [2,3,31].

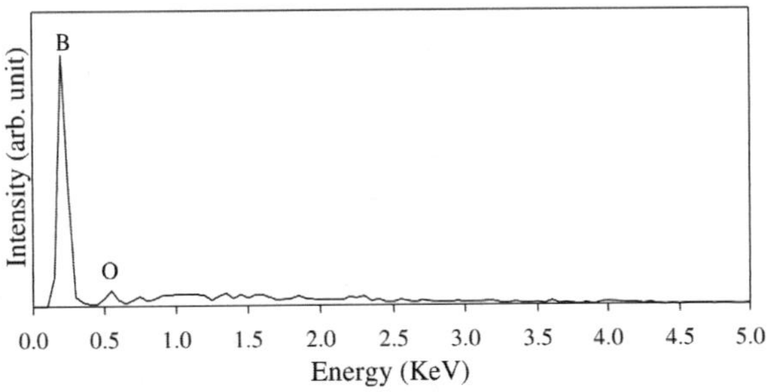

Figure 5. EDX spectrum shows that the boron nanoribbons contain oxygen.

The X-ray photoelectron spectra in figure 6 show that in addition to the boron and oxygen peaks, a peak due to carbon and a very weak nitrogen peak are also present. The possible source of the carbon and nitrogen impurities might be the gas precursor, which contains a small amount of CH_4, O_2, N_2, and CO_2 (all less than 10 ppm) in the ultra high purity diborane and argon used. Quantitative analysis revealed that within the XPS interrogation volume, the nanoribbons consist of ~70 at.% boron, ~14 at.% oxygen, ~14 at.% carbon, and trace amounts of other impurities including nitrogen. Like the oxygen, the carbon is likely present mainly at the surface. Since XPS is a surface analysis tool, it is not surprising that it yields higher apparent oxygen content than implied by EDX, if the oxygen is present only in a surface oxide layer. This observation can be confirmed from the Auger electron spectra of figure 7. Quantitative analysis of the AES data using the appropriate atomic sensitivity factors shows that the percentages of B, O and C in the nanoribbon sample are ~43%, ~13%, ~44%, respectively. As AES is slightly more surface sensitive than XPS, the higher carbon content from AES suggests that the carbon is present mainly in the topmost atomic layers.

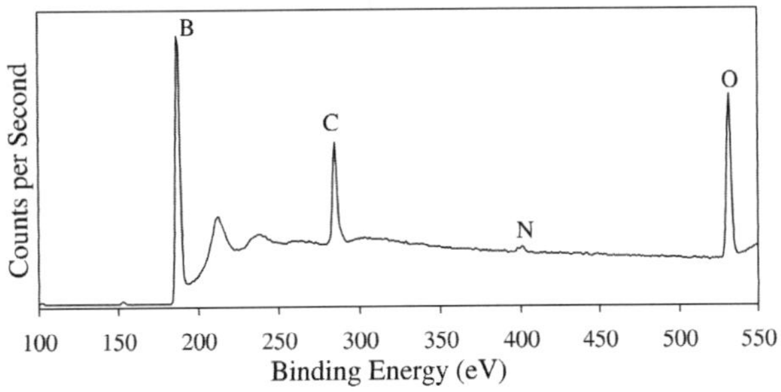

Figure 6. XPS spectrum of a "puffy ball" showing carbon and nitrogen contamination in addition to a surface oxide.

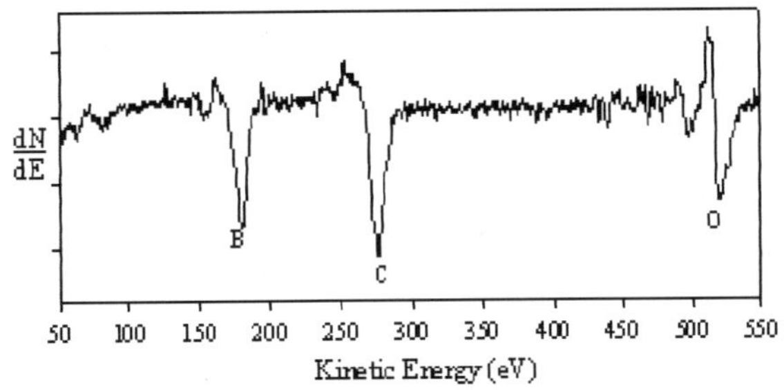

Figure 7. Auger spectrum showing a higher carbon

concentration than detected with XPS.

The B 1s, C 1s, and O 1s regions of the XPS spectrum of the boron nanoribbons are displayed in figure 8. The B 1s peak shows two components at 187.0 and 188.2 eV. The larger and lower binding energy component is assigned to boron atoms bonded only to other boron atoms [31], whereas the less intense component at 188.2 eV is assigned to boron bonded to carbon [32,33]. The C 1s core-level spectrum is broad and asymmetric. It is composed of a peak at 285 eV due to the C-C bond from hydrocarbon contamination of the surface. The C 1s component at 286.6 eV was assigned as due to either a C-O bond or C-OH bond [34]. A corresponding O 1s component associated with the C-O (or C-OH) bond appeared at 531.7 eV. The main O 1s component is assigned to boron oxide. These assignments are generally in agreement with those of Xu *et al* [28].

There are several polymorphs of crystalline boron, including α-rhombohedral, β-rhombohedral, α-tetragonal, β-tetragonal, and others [2,3]. Among them, the first form to be discovered was α-tetragonal boron [35]. This form has a unit cell with a large number of holes, which in principle can accommodate both additional boron and impurity atoms [3]. A quantum mechanical calculation by Longuet-Higgins and Roberts [36] using the concept of a closed shell of boron icosahedra showed that the proposed B_{50} structure is unstable. The tetragonal framework is stabilized through introduction of electron donor atoms, such as carbon, nitrogen, beryllium or aluminium [25,2,37]. A concentration of electron donor atoms of only 1 at.% can stabilize the structure [18]. Even though most of the carbon is concentrated on the surface, the XPS results in the B 1s region shows some evidence of B-C bonding. This carbon may serve to stabilize the α-tetragonal form of the nanoribbons, which is the crystalline form implied by the SAED results.

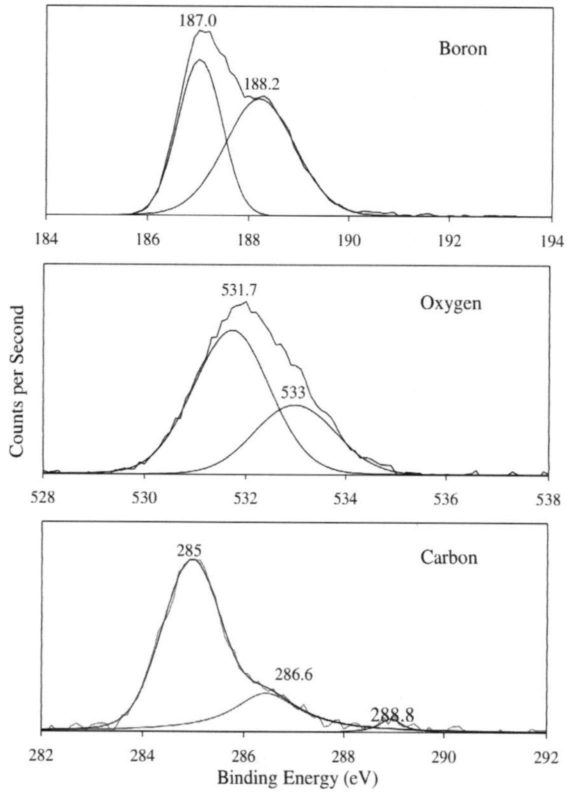

Figure 8. XPS of the boron nanoribbons shows that different components of the B, O and C 1s peaks can be resolved and assigned to B-O, B-C, C-H and C-O bonds.

3.2. Catalyzed growth of boron nanoribbons

By using nickel as a catalyst, boron crystalline nanoribbons and boron nanowire-catalyst-nanotube hybrid structures were synthesized. Although the growth mechanism of these nanostructures has not been fully explored, it is believed that the vapor-liquid-solid (VLS) growth mechanism plays an important role as catalyst particles can always be observed by SEM (figure 9) at the end of the nanowire. Xu *et al* have worked with Au and Pt/Pd alloy catalysts and by reducing the thickness of the catalytic metal film they have synthesized nanostructures with diameters of 8-10 nm [38]. Boron nanowires formed by the nickel catalyzed reaction are about 100 nm in diameter with bright metal particles at the tips. Some nanoribbons are also formed (figure 10), which have similar thicknesses of those formed by the non-catalyzed reaction. A Raman spectrum corresponding to the nanowires of figure 9, is shown in figure 11 and with only two wide peaks at 901 and 1106 cm^{-1} it is quite different from the Raman spectrum shown in figure 5 of the crystalline boron nanoribbons. This suggests that these nanostructures are composed of amorphous boron.

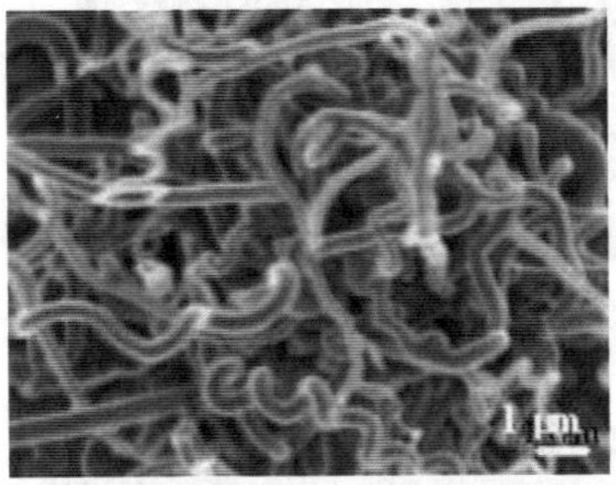

Figure 9. SEM images of the Ni-catalyzed boron nanotubes shows bright metallic particles on the tips. Diameters of the tubes are about 200 nm, which is controlled by the thickness of the deposited catalyst layer.

Figure 10. SEM images show that the Ni-catalyzed growth also leads to the formation of boron nanoribbons.

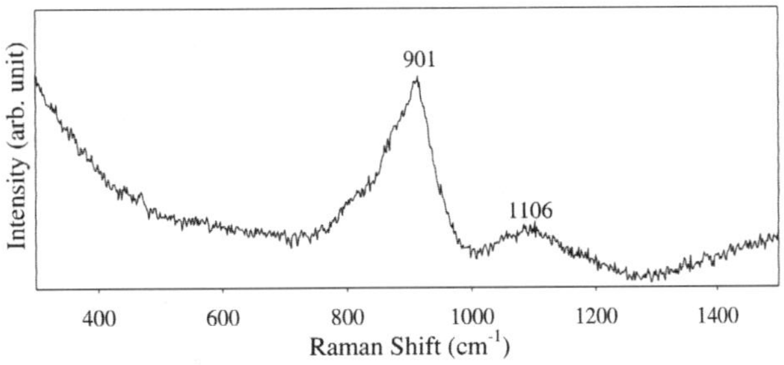

Figure 11. Raman spectrum encompassing all of the structures shown in figures 9 and 10.

3.3. Synthesis and characterization of CaB_6 nanowires

As noted in the experimental section, with CaO placed on the substrate, wire-like CaB_6 nanostructures of various morphologies, depending on temperature region of the furnace, were observed to form. The nanowires are ~15-50 nm in diameter, and ~1-10 µm in length. Catalyst particles were observed on the tips of most of the nanowires. Figure 12 is a low-magnification SEM image of the as-synthesized nanowires which shows that similar to pure boron nanoribbons, CaB_6 also appears in the form of "puffy balls". But unlike for pure boron, these balls are not visible to the naked eye. The higher magnification image in figure 13 shows the presence of catalyst particles on the tips.

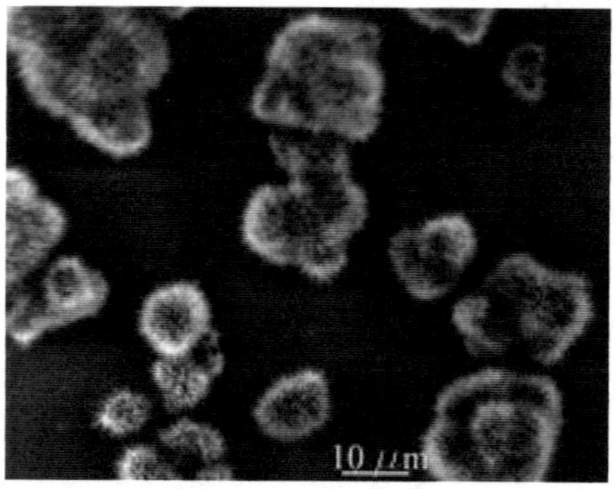

Figure 12. SEM image of "puffy balls" of CaB_6 nanowires.

Figure 13. High magnification SEM image of the CaB_6 deposits shows straight stick-like wires with catalyst particles at the tips.

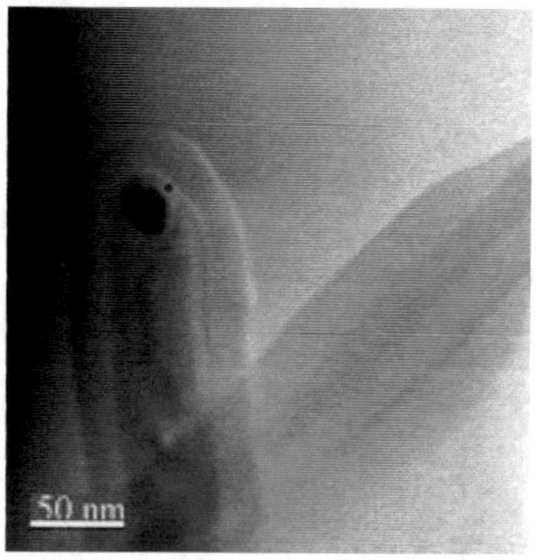

Figure 14. Low resolution TEM micrograph shows the growth of an amorphous layer around the crystalline CaB_6 nanowire (diameter: 45 nm). As CaB_6 is easily oxidized, the amorphous layer is likely an oxide.

A low resolution TEM micrograph, figure 14, shows a portion of a nanowire with a catalyst particle on the tip. Figure 15 shows an EDX spectrum obtained in STEM mode with a 1 nm probe size, along with a TEM image showing a catalyst particle at the tip of the nanowires. The EDX spectrum shows the presence of only B, Ca, and Ni. The HRTEM image in figure 16 shows that the spacing between atomic layers is 0.27 nm. According to JDPDS file # 31-0254, the corresponding plane is (110), while growth is preferentially along the CaB_6 [001] direction.

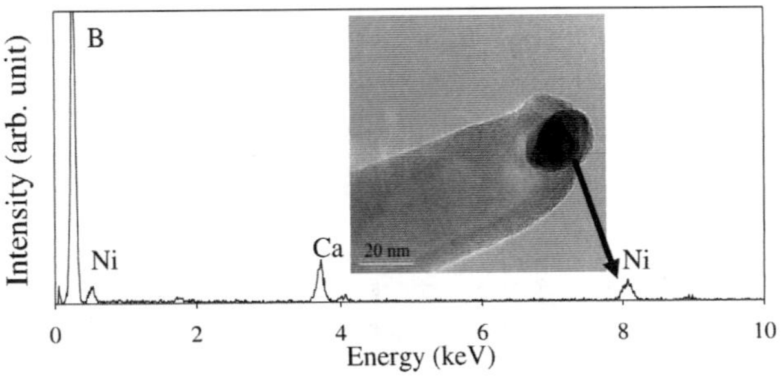

Figure 15. EDX spectrum of CaB_6 with a low resolution TEM image in the inset.

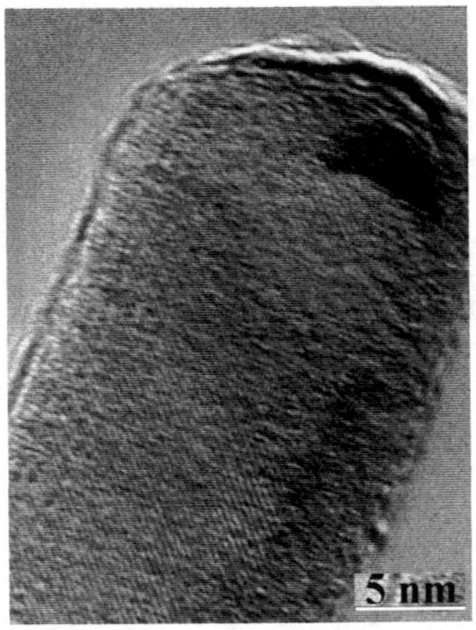

Figure 16. HRTEM image of crystalline nanowires with atomic layer spacing of 0.27 nm and nickel particle at the tip.

Further information on the elemental composition was obtained with EELS line scans (STEM mode; probe size 0.2 nm). Figure 17 is a typical EELS spectrum, which shows the characteristic B and Ca ionization edges at ~188 and ~346 eV, respectively. Quantitative analysis of these results confirms the Ca:B ratio of 1:6.

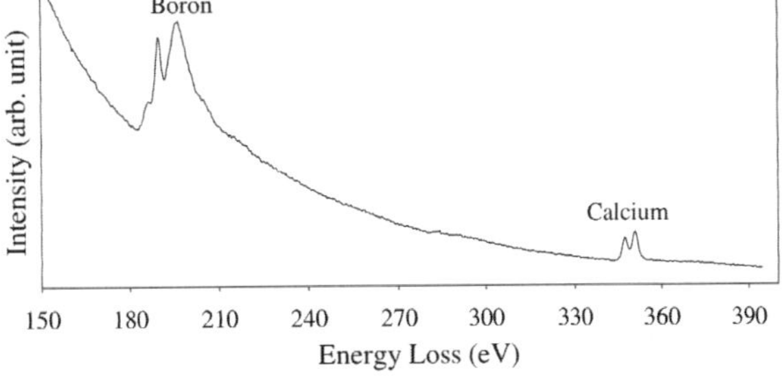

Figure 17. EEL spectrum shows that the ratio of calcium to boron is 1:6.

Finally, figure 18 shows the Raman spectrum of the as-synthesized nanowires acquired at room temperature in ambient. As indicated by table 2, the peaks at 778, 1135, and 1280 cm^{-1} match the Raman active modes A_{1g}, E_g and T_{2g}, respectively, of CaB_6 as reported by Xu *et al* [29]

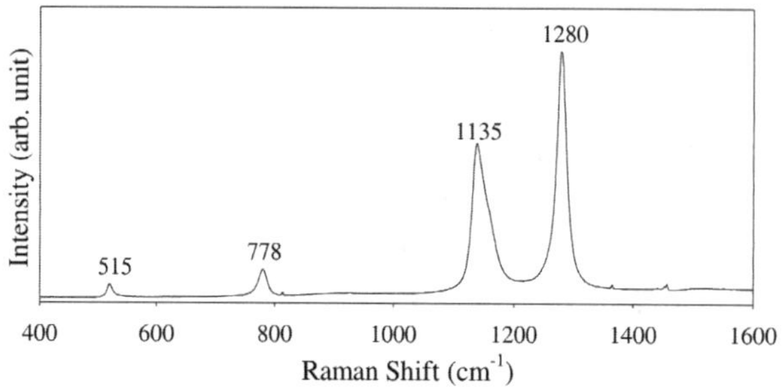

Figure 18. Raman spectrum of crystalline CaB_6 nanowires.

Table 2. Summary of Raman Peak Positions of CaB_6

Materials	Peak Position (cm^{-1})		
	A_{1g}	E_g	T_{2g}
Single crystal CaB_6 nanowires (this work)	1280	1135	778
Single crystal CaB_6 nanowires [29]	1260	1121	767

As mentioned by Xu, *et al* different morphologies were observed as a function of reaction chamber temperature.[29] As the CaB_6 nanowires grew in the presence of certain metal catalysts and catalytic particles were observed on the tip of most nanowires at a relatively low reaction temperature (860-900 °C), the growth probably occurs by the VLS mechanism [39]. Xu *et al* [29] speculated on the details of CaB_6 growth by this mechanism. The presence of a bulbous tip in some nanowires and bullet-like tips in other cases tells us that possibly both root-growth and tip-growth VLS took place. In addition to the VLS mechanism, it is also possible that growth occurred by the vapor-solid (VS) growth mechanism. The combination of different mechanisms may be responsible for the various morphologies observed.

4. Summary

The same LPCVD apparatus and general synthesis method involving the pyrolysis of diborane was used for the growth of both boron nanoribbons and CaB_6 nanowires. The uncatalyzed growth of boron nanoribbons led to a variety of structures with the appearance of blades of grass, ribbons, and rolled up scrolls. These different morphologies all feature locally flat boron sheets only 15-20 nm thick. The use of a Ni catalyst in the growth process gave similar structures but in addition yielded boron nanowire-catalyst-nanotube hybrid structures. The addition of CaO to the system resulted in the growth of CaB_6 nanowires with diameters of ~15-50 nm, and lengths of ~1-10 μm. Recently, we have extended the method to the growth of SrB_6 nanowires [40], with similar characteristics to those of the CaB_6 nanowires described here.

Acknowledgements

We thank R. S. Ruoff and T. Xu for advice and assistance in setting up the LPCVD apparatus and for help in interpreting our results. This work was supported by a grant from the US Department of Energy, grant no. DE-FG02-05ER15726.

References

[1] Greenwood NN and Earnshaw A. 1997. *Chemistry of the Elements*. UK: Reed Educational and Professional Publishing Ltd.
[2] Matkovich VI, editor. 1977. *Boron and Refractory Borides*. Berlin: Springer-Berlag.
[3] Newkirk AE. 1964. *Boron Metallo-Boron Compounds and Boranes*. Adams RM, editor. New York: Interscience Publishers.

[4] Boustani I 1997 *Phys. Rev. B* **55** 16426
[5] Boustani I 1997 *Surf. Sci.* **370** 355
[6] Boustani I 1997 *J. Solid State Chem.* **133** 182
[7] Boustani I and Quandt A 1997 *Europhys. Lett.* **39** 527
[8] Boustani I, Quandt A, Hernandez E and Rubio A 1999 *J. Chem. Phys.* **110** 3176
[9] Boustani I, Quandt A and Rubio A 2000 *J. Solid State Chem.* **154** 269
[10] Boustani I, Rubio A and Alonso JA 1999 *Chem. Phys. Lett.* **311** 21
[11] Post B. 1964. Refractory Binary Borides. In: Adams RM, editor. *Boron, Metallo-Boron Compounds and Boranes*. New York: Interscience Publishers.
[12] Murakami S, Shindou R, Nagaosa N and Mishchenko AS 2002 *J. Phys. Chem. Solids* **63** 1285
[13] Otani S and Mori T 2002 *J. Phys. Soc. Jpn.* **71** 1791
[14] Young DP, Hall D, Torelli ME, Fisk Z, Sarrao JL, Thompson JD, Ott HR, Oseroff SB, Goodrich RG and Zysler R 1999 *Nature* **397** 412
[15] Zhitomirsky ME, Rice TM and Anisimov VI 1999 *Nature* **402** 251
[16] Souma S, Komatsu H, Takahashi T, Kaji R, Sasaki T, Yokoo Y and Akimitsu J 2003 *Phys. Rev. Lett.* **90** 027202
[17] Tromp HJ, van Gelderen P, Kelly PJ, Brocks G and Bobbert PA 2001 *Phys. Rev. Lett.* **87** 016401
[18] Kino H, Aryasetiawan F, van Schilfgaarde M, Kotani T, Miyake T and Terakura K 2002 *J. Phys. Chem. Solids* **63** 1595
[19] Hasegawa A and Yanase A 1979 *J. Phys. C* **12** 5431
[20] Massidda S, Continenza A, dePascale TM and Monnier R 1997 *Z. Phys. B: Condens. Matter* **102** 83
[21] Dutta SK 1975 *Am. Ceram. Soc. Bull.* **54** 727
[22] Johnson RW and Daane AH 1963 *J. Chem. Phys.* **38** 425
[23] Zheng S, Min G, Zou Z, Yu H and Han J 2001 *J. Am. Ceram. Soc.* **84** 2725
[24] Otani S and Mori T 2003 *J. Mater. Sci. Lett.* **22** 1065
[25] Greenwood NN. 1997. *Chemistry of the elements* / N.N. Greenwood and A. Earnshaw. Oxford ; Boston Butterworth-Heinemann
[26] Post B, editor. 1964. *Boron, Metallo-Boron, Compounds and Boranes*. New York: Interscience Publishers.
[27] Ronnebro E and Majzoub EH 2007 *J. Phys. Chem. B* **111** 12045
[28] Xu TT, Zheng JG, Wu NQ, Nicholls AW, Roth JR, Dikin DA and Ruoff RS 2004 *Nano Letters* **4** 963
[29] Xu TT, Zheng JG, Nicholls AW, Stankovich S, Piner RD and Ruoff RS 2004 *Nano Letters* **4** 2051
[30] Werheit H 1988 *Prog. Cryst. Growth Char.* **16** 179
[31] Foo WC, Ozcomert JS and Trenary M 1991 *Surf. Sci.* **255** 245
[32] Moulder J, Stickle W, Sobel P and Bomben E, editors. 1995. *Handbook of X-ray Photoelectron Spectroscopy*. Eden Prairie: Physical Electronics.
[33] Ronning C, Schwen D, Eyhusen S, Vetter U and Hofsäss H 2002 *Surf. Coat. Technol.* **158-159** 382
[34] Vickerman JC, editor. 1997. *Surface Analysis - The Principle Techniques*. Chichester, England: John Wiley & Sons.
[35] Laubengayer AW, Hurd DT, Newkirk AE and Hoard JL 1943 *J. Am. Chem. Soc.* **65** 1924
[36] Longuet-Higgins HC and Roberts MdV 1955 *Proc. Roy. Soc. London, A* **230** 110
[37] Will G and Ploog K 1974 *Nature* **251** 406
[38] Xu TT, Nicholls AW and Ruoff RS 2006 *NANO: Brief Reports and Reviews* **1** 1
[39] Wagner RS and Ellis WC 1964 *Appl. Phys. Lett.* **4** 89
[40] Jash P, Nicholls AW, Ruoff RS and Trenary M 2008 *Nano Lett.* **8** 3794

Binary immiscible metal systems for preparation of borides

Vladimir N Gurin[1,2], Ulrich Burkhardt[2] and Yuri Grin[2]

[1]Ioffe Physical-Technical Institute, Russian Academy of Sciences, 194021 St Petersburg, Russia
[2]Max-Planck-Institut für Chemische Physik fester Stoffe, 01187 Dresden, Germany

E-mail: vladimir.gurin@mail.ioffe.ru

Abstract. The binary systems with the immiscibility in the liquid state are considered as a tool for synthesis and crystallization of borides. New preparation method using two solvents (immiscibility gap method) is compared with the conventional solution-melt method using one solvent (flux method). The appropriate systems are analyzed with respect to the temperature and concentration limits of the immiscibility gap, density ratio of solvents and reagents, interaction of reagents, formation and crystallization of compounds, and allocation of products in the metallic matrix.

1. Introduction

Recently, we reported first results on the new route of processing and investigation of interaction of elements in binary systems with immiscibility in the liquid state (immiscibility gap method) with respect to the synthesis and crystallization from two immiscible solvents L1 and L2 [1, 2]. In the present work, the immiscibility method was applied as the tool for synthesis of metal borides. Differences are discussed between the new and the conventional solution-melt method using one solvent L (flux method [3]). Experimental part has been already described in [2].

Analysis of the known phase diagrams [4] reveals five binary systems most suitable for experiments under ambient conditions: Ga/Pb, Cd/Ga, Al/Pb, Al/Cd and Pb/Zn. The system Ga/Pb has an immiscibility gap, but the upper immiscibility temperature (610 °C) is too low for synthesis of refractory compounds. The system Cd/Ga is unsuitable for immiscibility gap method because of very low temperatures (9 °C) and narrow concentration range of immiscibility (52.25 at% Ga). Remaining three - Al/Pb, Al/Cd and Pb/Zn - possess wide temperature and concentration ranges of immiscibility (table 1) and are the most suitable for synthesis and crystallization of refractory compounds. The basic differences between the flux and immiscibility gap methods are related to:

(i) temperature-concentration boundary conditions for the liquid phase;
(ii) presence of the immiscibility range bound to a binodal;
(iii) density ratios of the solvents and the reagents;
(iv) conditions for the chemical reaction of the reagents and formation of the compounds;
(v) conditions of crystallization and allocation of the products in the metallic matrix.

Journal of Physics: Conference Series **176** (2009) 012012 doi:10.1088/1742-6596/176/1/012012

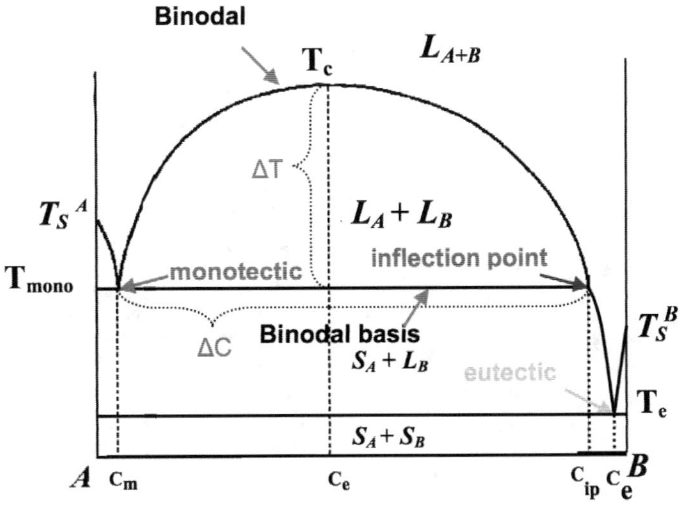

Figure 1. Generalized phase diagram with an immiscibility gap.

Table 1. Temperature and concentration limits of the immiscibility gap in selected systems

Binary system	ΔT^{L1+L2}=T_c-T_{mono} (°C)	ΔC^{L1+L2}=$C_{mono,ip}$-$C_{ip,mono}$ (at% of one component)
Al/Cd	370 = 1020 - 650	93.14 = 94.9 - 1.76 (Cd)
Al/Pb	907 = 1566 - 659	97.71 = 97.91 - 0.19 (Pb)
Cd/Ga	9 = 291 - 282	52.25 = 74.95 - 22.7 (Ga)
Ga/Pb	297.4 = 610 - 312.6	91.9 = 93.7 - 1.8 (Pb)
Zn/Pb	380.2 = 798 - 417.8	93.7 = 94 - 0.3 (Pb)

2. Temperature-concentration limits of the immiscibility gap

This condition is crucial taking into account the fact, that in such systems the reagents are dissolved in different solvents. They react at the immiscibility border of the solvents. The interaction is supported by the counter diffusion and convection of solvated particles of reagents within immiscible layers.

In a generalized phase diagram (figure 1) the immiscibility area is bound to a dome-shaped curve (a binodal) and its basis (monotectic horizontal). The temperature interval ΔT is defined by the monotectic temperature (T_{mono}) and the critical point of immiscibility (T_c). There are two types of the phase diagrams with an immiscibility range: systems with monotectic and inflection points and systems with two inflection points [5]. Thus, the concentration range ΔC is defined either by the monotectic point C_{mono} and an inflection point C_{ip}, or by two points of an inflection [5]. For the elements with the melting temperature lower than T_{mono}, an eutectic (T_e, C_e) should be located below the monotectic reaction. In such systems preparation of compounds and their crystallization is possible in the temperature-concentration areas of $\Delta T = T_c - T_{mono}$ and $\Delta C = C_{mono,ip} - C_{ip,mono}$. The larger are ΔT and ΔC, the more variable is the given system for synthesis in respect to the temperatures, solvent ratios L1 and L2, amounts of the dissolved reagents, and conditions for crystallization of products (e.g., more time for growth of the larger crystals with less defects).

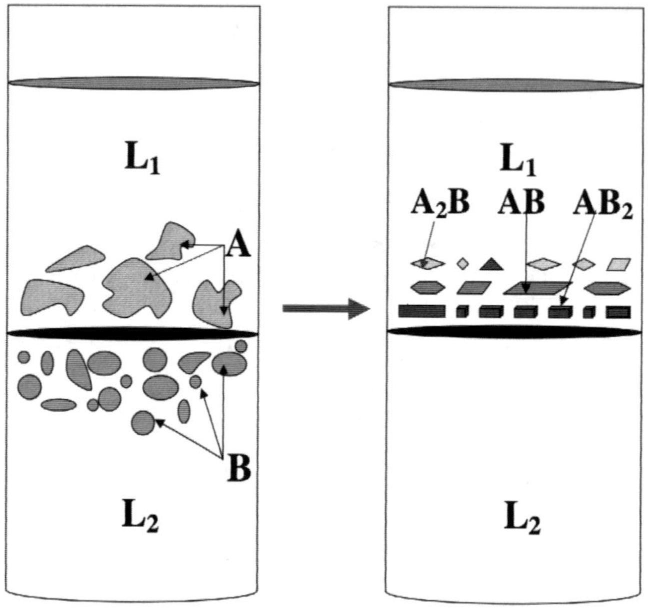

Figure 2. Arrangement of the reagents (a) and reaction products (b) in the system with an immiscibility gap.

3. Density ratio of the dissolved reagents and solvents

In addition to the immiscibility gap, selected solvents and reagents have to fulfill the following condition. The reagent dissolved in the upper layer should sink in the top liquid solvent, but not in the bottom liquid solvent. The bottom reagent should emerge in the bottom layer, but not in the top layer (figure 2a). Such density ratio allows the transport of the reagents and supports chemical reaction at the liquid phase boundary by counter diffusion and convection.

For the immiscibility gap preparation of the borides the key point is the selection of an appropriate solvent for boron. E.g., metals with low melting temperatures, such as Al (density 2.70 g cm^{-3}), Zn (7.14) and Cd (8.64), are heavier than B (2.34). Boron dissolves only partially in Al, and is almost insoluble in Zn or Cd. Therefore, the system Ca (1.55)/Ce(6.78) was chosen for synthesis of TiB_2 (figure 3a). Here Ca is lighter than B and dissolves boron. Titanium (4.50) is soluble and emerges in liquid cerium, reacting with boron at phase boundary and forming refractory boride TiB_2.

Syntheses in the system Ca(+B)/Ce(+Ti) were carried out in a sealed tantalum crucible. The obtained ingots were cut in two parts: one was used for metallographic characterization, another was dissolved in strongly diluted acids to isolate the crystalline products. Different shapes of crystals were observed: needles (figure 4a), isometric formations (figure 4b) or staples of plates (figure 4c). Please note that TiB_2 prepared by usual flux method is usually obtained in form of thin hexagonal plates [3]. More detailed investigations on the origin of the different crystal shapes are in progress.

Another solvent for boron can be obtained by reducing of the density of the aluminium melt by adding a lighter metal, e.g., Mg (1.74). The Mg/Al ratio should be selected so that the density of the mixture is lower than the density of boron. So, for the target density of 2.2 g cm^{-3} the mass

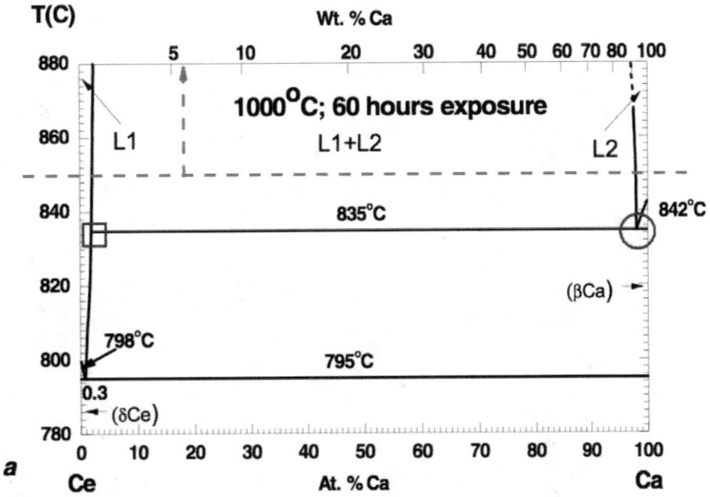

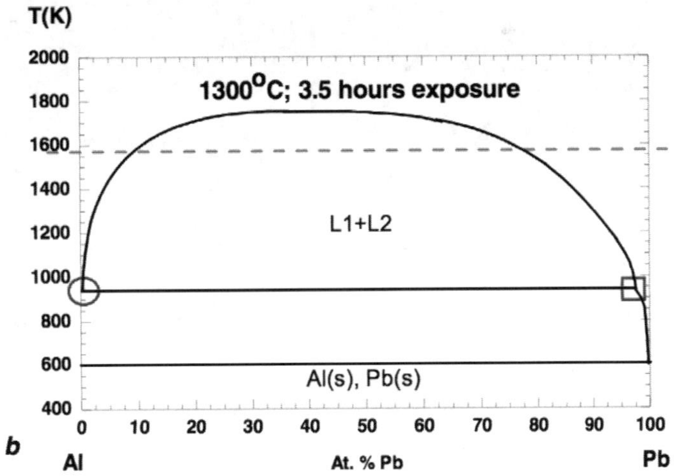

Figure 3. Phase diagrams Ca/Ce and Al/Pb with immiscibility gaps, used for synthesis and crystallization of borides.

ratio for the mixture is Al : Mg = 10 : 7. The so-obtained Al/Mg solvent forms an immiscibility gap system with Pb. It is possible to perform synthesis and crystallization of hexaborides of the rare-earth metals in this system despite formation of intermetallic compounds between Al and Mg and between Mg and Pb. These compounds decompose at the temperatures between 450 and 550 °C and do not hinder the formation of refractory rare-earth hexaborides at elevated temperatures. Based on these considerations, single crystals of solid solutions $La_{1-x}Nd_xB_6$ were obtained from the immiscibility system (Al,Mg)(+B)/Pb (+La,Nd) in form of isometric crystals, needles and plates (figure 5).

Use of the mixed solvents in immiscibility systems has special aspects. Ideally, the component added to one solvent should form also a phase diagram with an immiscibility gap with the solvent

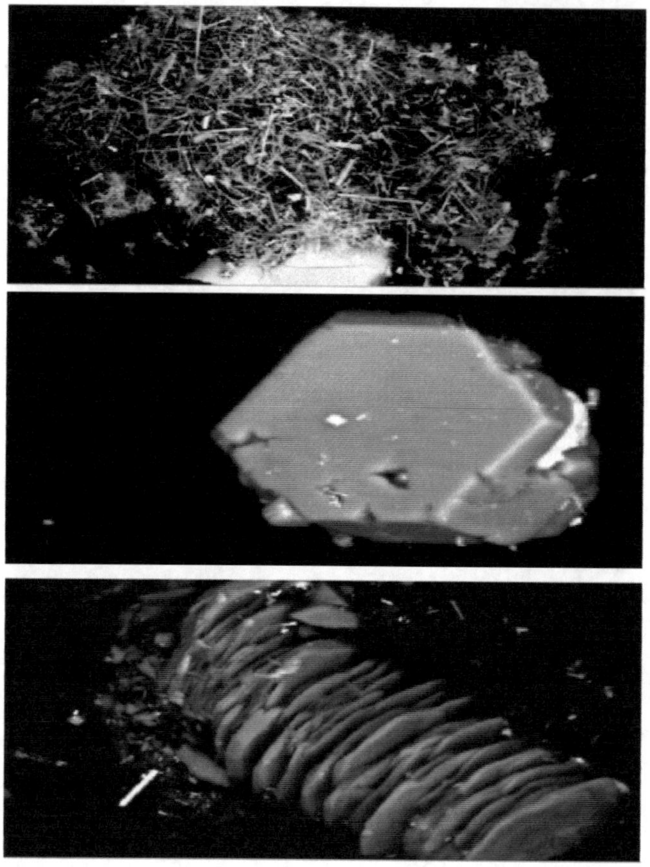

Figure 4. Crystals of TiB_2 prepared in the system Ca(+B)/Ce(+Ti) with different shape: needles (a), isometric formations (b) and staples of plates (c).

of the second layer. So, addition of well-soluble Ga to Al in the system Al/Pb will be in favor, because both Ga and Al form with Pb immiscibility systems having large ΔT and ΔC. For the same reasons, an addition of Al to Ga in the immiscibility system Ga/Pb will be also in favor. To the best of our knowledge, this paper is the first report on the use of the mixed solvents for preparation and crystallization with the immiscibility gap method.

4. Allocation of the products in immiscibility systems

In general, several variants of synthesis and crystallization of compounds in immiscibility systems may be considered. In the first scenario, the reagents are soluble in different layers and the product is soluble in the top layer. The dissolved reagents meet at the immiscibility border and react, forming a compound with a density lower than the density of the bottom solvent and higher than the density of the top solvent. The product moves to the top layer. Under further cooling, the crystallization occurs close to the immiscibility border.

In the second scenario, one of the reagents is almost insoluble in the bottom solvent, but emerges in it (e.g., Co in Pb). In this case, the insoluble particles move to the immiscibility border by convection and react with the dissolved top reagent. The possibility of the crystal growth depends here on the solubility of the products in one of the layers. If they are insoluble

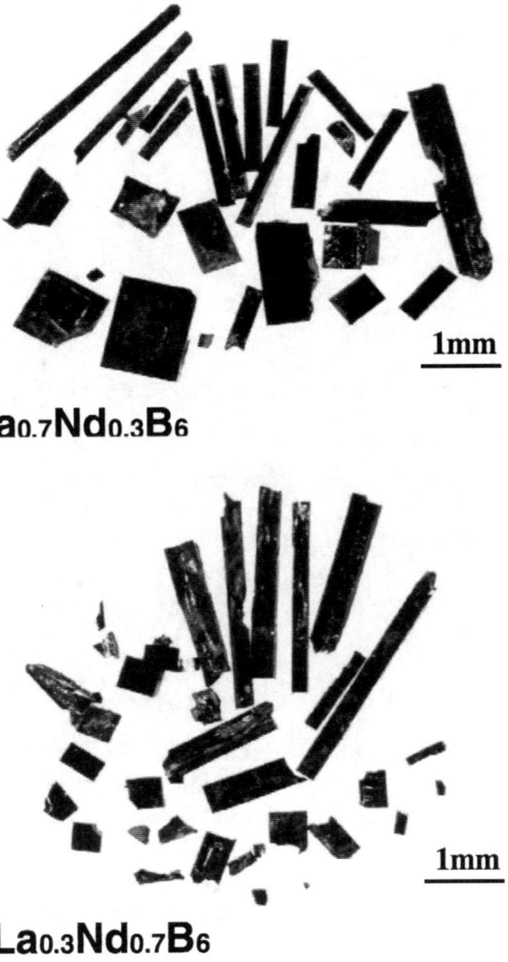

Figure 5. Single crystals of $La_{1-x}Nd_xB_6$ in different shapes, obtained from the system (Al,Mg)(+B)/Pb(+La,Nd).

in both solvents, a polycrystalline product is formed immediately at the immiscibility border. If the product is soluble in one of the layers, it crystallizes in the vicinity of the immiscibility border in the according layer. This was observed in the system Pb(+Co)/Zn(+Ge): insoluble in Pb cobalt reacts with Ge from the top layer, forming binary compound $Co_{61}Ge_{39} \approx Co_3Ge_2$ (figure 6). In this case, the reaction is complicated by the formation of the ternary byproduct $Co_{10}Zn_{87+x}Ge_x$ with the solvent [2].

5. Separation of products of synthesis and crystallization

In accordance with the first scenario described above, the products in the form of crystals with different shapes concentrate in the bottom part of the top layer (figure 2). Their separation in many cases is easier and faster than in case of the usual flux method. The top layer of the solidified ingot can be cut off and dissolved in an appropriate solvent which removes the matrix and does not attack the target crystals. This reduces processing time and costs.

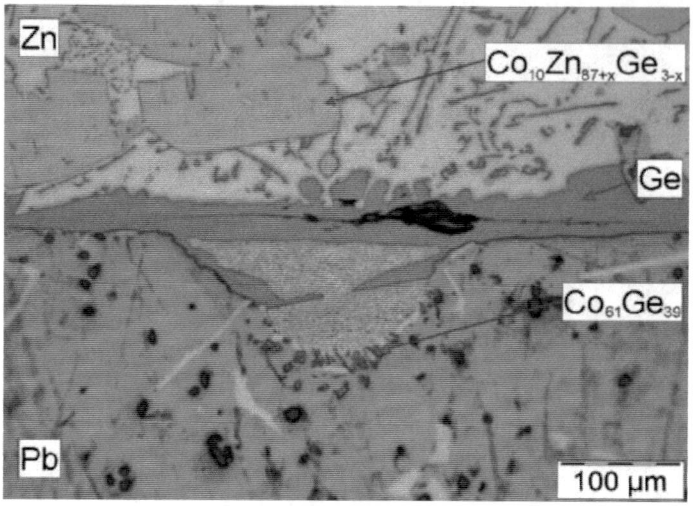

Figure 6. The microstructure of the ingot after immiscibility gap preparation in the system Pb(+Co)/Zn(+Ge). Formation of intermetallic compounds is observed in the vicinity of the immiscibility border in both layers together with the excess of elemental Ge crystallized by cooling from the Zn melt.

6. Conclusions

An appropriate choice of the binary systems with the immiscibility gap in the liquid state allows performing chemical reactions and crystallization of the products at the immiscibility border. Key requirements for this immiscibility gap method are solubility of the reagents in different layers and reasonable density ratio between the reagents and the solvents. The feasibility of the new preparation route is shown on example of synthesis and crystallization of refractory borides TiB_2 and $La_{1-x}Nd_xB_6$.

Acknowledgments

The authors would like to express their gratitude to Drs. P. Hohn, M. Schmidt and A. Leithe-Lasper, as well as to M. Eckert and P. Scheppan for valuable discussions and support in performing experiments.

References

[1] Burkhardt U, Boström M, Schnelle W, Hui Z, Grin Yu and Gurin V 2003 *Proc. 9th Euro. Conf. Solid State Chem.* (Stuttgart, Germany) p 204
[2] Burkhardt U, Gurin V N and Grin Yu 2006 *Develop. Instit. Sci. Rep.* (MPI CPfS, Dresden, Germany) p 246
[3] Gurin V N and Korsukova M M 1983 *Prog. Cryst. Growth Charact.* **16** 59
[4] Massalski T B (ed) 1996 *Binary Alloy Phase Diagrams* 2nd ed (ASM International, The Materials Information Society) Vol. 1-3
[5] Anosov V Ya, Ozerova M I and Fialkov Yu Ya 1976 *Principles of physical chemical analysis* (Moscow, Nauka) p 140 (in Russian)

Journal of Physics: Conference Series **176** (2009) 012013 doi:10.1088/1742-6596/176/1/012013

On quasi-classical estimations of boron nanotubes ground-state parameters

Levan Chkhartishvili

Department of Physics, Georgian Technical University, 77 M Kostava Avenue, Tbilisi, 0175, Georgia

E-mail: chkharti2003@yahoo.com

Abstract. The explicit expressions in term of B–B bond length are obtained for atomic sites coordinates and intersite distances in boron nanotubes of regular geometry. Using the charge and potential distributions in boron atom parameterized within initial quasi-classical approximation, the boron–boron pair potential curve is constructed. It leads to the boron–boron interatomic distance, binding energy, and relative vibration frequency equilibrium values which only slightly are deviated from averaged bond length in structural units of boron and boron-rich solids, dissociation energy, and oscillatory quantum of the diboron molecule, respectively. This curve together with obtained formulas should be useful for calculating of boron nanotubes geometrical and other ground-state parameters.

1. Introduction

As it is well known, variety of boron crystalline modifications and boron-rich solids mainly are constructed from the interconnected boron icosahedra [1]. The same short-range structure is characteristic for amorphous boron [2]. Then, most of B atoms usually are members of regular atomic triangles. This circumstance leads to possibility of the new kind on boron-based materials in form of planar or rolled boron surfaces with triangular two-dimensional lattices.

Such geometries for finite B_n clusters experimentally can be estimated based on measurements of apparent ionization potentials and fragmentation patterns for collision-induced dissociation [3]. Theoretically nanostructural formation of boron clusters was proposed for the first time by Boustani [4]. According to local-spin-density calculations (including nonlocal corrections), most of structures were found to prefer planar or quasi-planar arrangements (the latter is generated from planar one by forcing the boron atoms out of plane, up and down, and in periodic fashion) and can be considered as fragments of a surface or as segments of a sphere.

Experimentally the atomic and electronic structures and chemical bonding of small boron clusters up to 20 atoms were probed by the photoelectron spectroscopy combined with appropriate quantum calculations. In particular, clusters B_5^- and B_5 were investigated [5] using anion photoelectron spectroscopy. The global minimum of B_5^- was found to have a planar structure with a closed-shell ground-state. Excellent agreement was observed between ab initio detachment energies and experimental spectra, firmly establishing the ground-state-structures for both B_5^- and B_5. Hepta- and octa-coordinated boron atoms in molecular wheels B_8 and B_9 were observed and confirmed in [6]. It was also reported [7] about experimental and theoretical evidences that small boron clusters prefer planar structures and exhibit aromaticity and antiaromaticity according the Hückel rules. Aromatic boron clusters posses more circular shapes, whereas antiaromatic ones are elongated. Thus, planar boron clusters can be considered as only series of molecules representing a new dimension of boron chemistry. Most of the small boron clusters B_n^+, B_n, and B_n^- with $n = 2-15$ as individual species in the gas phase have been reviewed in [8]. It was emphasized that free small boron clusters were characterized using photoelectron spectroscopy and ab initio calculations, which established their planar or quasi-planar shapes. Experimentally quasi-planarity of the small boron clusters also were confirmed in [9,10].

Experimental investigations and computational simulations revealed [11] that boron clusters, which favour planar structures up to 18 atoms, prefer three-dimensional structures beginning at 20 atoms. Neutral cluster B_{20} has found to have a double-ring tubular structure, even though the B_{20}^- remains planar. As for the anion B_{20}^- itself, the tubular structure was shown to be isoenergetic to two-dimensional structures, which were observed and confirmed by photoelectron spectroscopy. The two-dimensional – three-dimensional transition observed at B_{20} suggests that it may be considered as the embryo of the thinnest single-walled all-boron nanotubes. In [12], it was reported first synthesis of pure boron single-walled nanotubes by a chemical reaction over a catalyst with parallel, uniform-radius cylindrical pores. According to recent communication, the multi-walled boron nanotubes also have been synthesized (see reference in [13]).

A systematic theoretical study on B_n clusters carried out applying first-principles quantum-chemical methods to determine their geometric structures show that they are composed of two fundamental units: either of hexagonal or pentagonal pyramids [14]. The resulting quasi-planar and convex structures can be considered as fragments of planar surfaces and segments of nanotubes or hollow spheres, respectively [15,16]. According to the Hartree–Fock self-consistent-field geometry optimization performed in [17] using Slater-type orbitals basis set, fragment of the quasi-planar boron structure is more stable than its isomers in form of isolated unit cell of the real α-rhombohedral boron crystal or predicted boron quasi-crystals. The nanotubular isomers composed of parallel-lying (staggered) atomic rings in turn are shown to be even more stable. As it was mentioned, both predicted quasi-planarity of boron clusters [18] and boron tubular forms [19] were confirmed experimentally.

Recently, density-functional-theory has been employed [20] to calculate the electronic and geometric structures, total and binding energies, harmonic frequencies, point symmetries, and highest-occupied-molecular-orbital – lowest-unoccupied-molecular-orbital gaps of small boron clusters B_n ($n = 2-12$). The linear, planar, convex, quasi-planar, cage, and open-cage structures have been found. The first and second energy differences have been used to obtain most stable sizes of clusters. None of the lowest-energy-structures and their isomers is shown to have an inner atom. Later, the studies have been extended on B_n clusters with $n = 13-20$ (see [21]). They are found to form additional diverse structural motifs, ultimately can yield distinct nanostructures. Within whole range of $n = 2-20$, almost all physical properties of B_n clusters are size-dependent. In particular, vibration frequencies are increasing with size. The double-ring tubular form of B_{20} cluster has found to have the highest binding energy per atom.

First-principles calculations reported in [22] also reveal that boron can form wide variety of matastable planar and tubular forms. The preferred planar structure is a buckled triangular lattice that breaks the threefold ground-state degeneracy of the flat triangular plane. When the plane is rolled into a tube, this degeneracy leads to a strong chirality dependence of the binding energy. The achiral $(n,0)$-tubes derive their structure from the flat triangular plane, while the achiral (n,n)-tubes arise from the buckling plane, and have large cohesion energies and different structure as a result. Density functional theory calculations carried out [23] to obtain geometric structure of boron sheets and nanotubes show that buckled boron sheet with certain buckling height is more stable than the flat sheet. But, boron nanotubes formed by rolling up of a boron sheet exhibit buckling surfaces if helicity allows for the formation of alternating up and down B-atoms rows in the surface. In all other cases, boron nanotubes exhibit only flat surfaces. Hence, all the $(n,0)$-tubes have to prefer buckled geometry, while not all (n,n)-tubes do same.

2. Methods

For analysis of the boron nanotubular structures and future purposeful designing devices based on nanotubular boron it is important to be able to predict reliably the sizes of nanotubes with given indexes. Here this task is solved for the mostly regular forms with equal B–B lengths, which exhibit flat surfaces. Namely, the quasi-classical B–B pair interatomic binding energy curve is constructed, and then the expressions of radii of the zigzag and armchair boron nanotubes in terms of B–B bond length are obtained considering geometries of such structures.

Previously, it had been demonstrated [24,25] that an atom can be considered as a quasi-classical electron system in sense of Maslov criterion and, on this basis, the quasi-classical approach to the construction of interatomic pair potentials had been elaborated. Within the initial quasi-classical approximation, using quadratic polynomial approximation for the electron screening factor standing in the effective atomic potential the binding energy $E = E(d)$ between pair of the identical (e.g. boron) atoms was expressed as

$$-4E(d) = \sum_{i,j=0}^{i,j=s} (n_i U_j + n_j U_i)(W(R_i, R_j, d) + W(R_{i+1}, R_{j+1}, d) - W(R_i, R_{j+1}, d) - W(R_{i+1}, R_j, d)) + \\ + 2A_0 \sum_{i=0}^{i=s} (A_i / d + B_i + C_i d)\theta(d - R_i)\theta(R_{i+1} - d). \quad (1)$$

Here d is the interatomic distance; n_i, U_i, A_i, B_i, and C_i are the known quasi-classical parameters of the electron-density and potential distributions in the interacting atoms; R_i is the radius of effective electron cloud affecting an electron of the ith subshell, while s is the total number of subshells in an atom. $W(R_i, R_j, d)$ is a known algebraic function expressing dependence of two spheres' intersection region volume upon their radii R_i and R_j, and centre-to-centre distance d; and $\theta(R_i - d)$ is a Heaviside step-function. These parameters and electron cloud radii can be obtained by fitting quasi-classical electron energy levels and mean orbital radii of the electrons to their first-principles, e.g. Hartree–Fock, values.

As for the procedure of the determination of boron nanotubes radii, it is following.

If $d_{(n,0)}$ is the B–B bonds lengths in a zigzag (*n*,0) nanotube (figure 1), the height of the equilateral triangle built up from the boron atoms in vertexes equals to $\delta_{(n,0)} = \sqrt{3}\, d_{(n,0)} / 2$. On the other hand, the regular $2n$-gon with sides of $\delta_{(n,0)}$ is inscribed into the circle with radius coincident the tube radius $r_{(n,0)}$. Consequently, $r_{(n,0)} = \delta_{(n,0)} / 2 \sin \pi / 2n$ and

$$r_{(n,0)} = \frac{\sqrt{3}\, d_{(n,0)}}{4 \sin \dfrac{\pi}{2n}}. \quad (2)$$

Let $d_{(n,n)}$ and $r_{(n,n)}$, respectively, denote the B–B bond length and radius in an armchair (*n*,*n*) nanotube (figure 2). It is clear that the regular n-gon with sides of $d_{(n,n)}$ is inscribed into the circle with radius of $r_{(n,n)}$. Consequently,

$$r_{(n,n)} = \frac{d_{(n,n)}}{2 \sin \dfrac{\pi}{n}}. \quad (3)$$

Nanotube index $n = 1,2,3,...$ determines the number of atoms because the nanotube unit cell consists of $2n$ boron atoms. At $n = 1$ zigzag nanotube degenerate in zigzag atomic chain, while armchair nanotube degenerate in straight atomic chain. Correspondingly, formula obtained for $r_{(n,n)}$ in general case, does not works; instead it should be assumed $r_{(1,1)} = 0$.

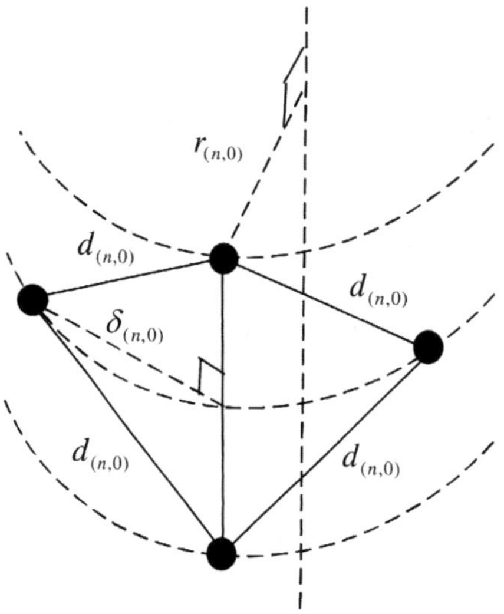

Figure 1. For calculation of boron zigzag nanotube radius (black circles – B atoms, solid legs – B–B bonds).

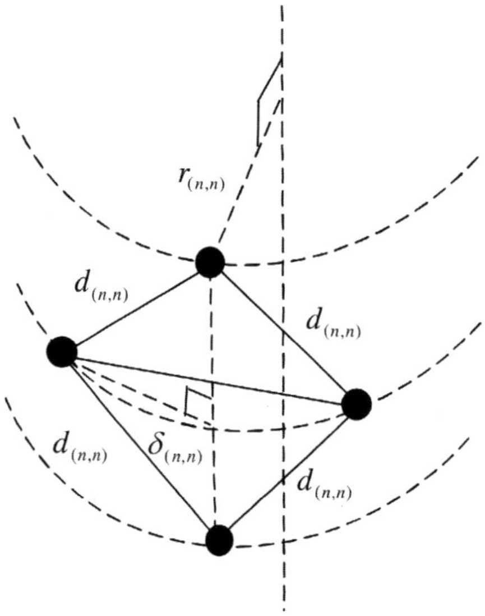

Figure 2. For calculation of boron armchair nanotube radius (black circles – B atoms, solid legs – B–B bonds).

Now about the detailed regular geometries of the zigzag and armchair boron nanotubes, which we will describe using cylindrical coordinates $\vec{r}\,(\rho,\varphi,z)$ useful in tubular binding energy and electronic structure calculations.

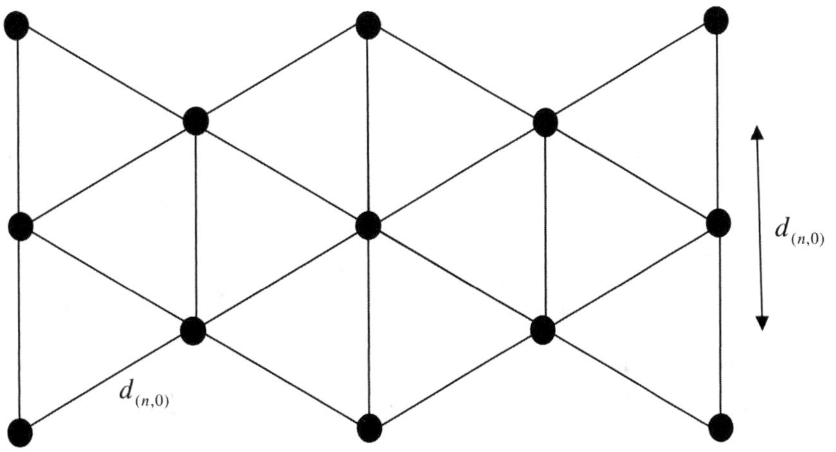

Figure 3. Lattice constant of boron zigzag nanotube (black circles – B atoms, solid legs – B–B bonds).

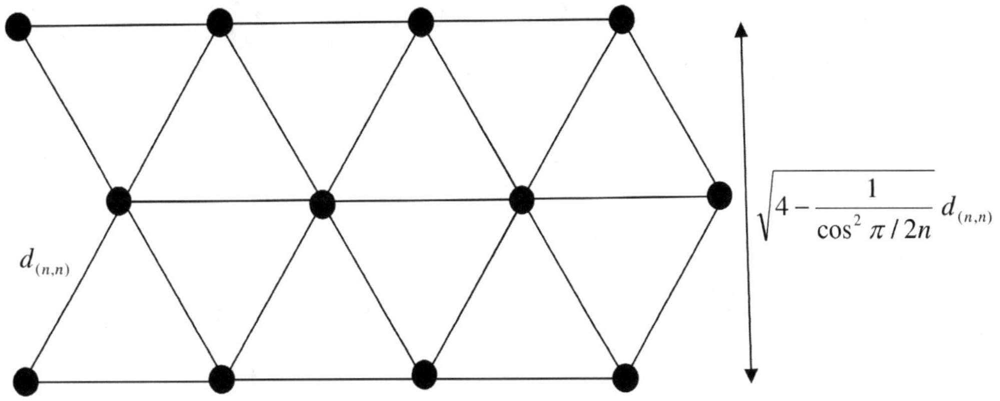

Figure 4. Lattice constant of boron armchair nanotube (black circles – B atoms, solid legs – B–B bonds).

Constant of the zigzag $(n,0)$ nanotube one-dimensional lattice equals to $d_{(n,0)}$ (figure 3). Its unit cell consists of two equidistant atomic rings in parallel planes perpendicular to the axis, each with n boron atoms. Evidently, cylindrical coordinate ρ for all atomic sites equals to tube radius:

$$\rho = r_{(n,0)} \,. \tag{4}$$

As for the coordinates φ and z, they equal to

$$\varphi = \frac{(4k-3-(-1)^l)\pi}{2n}, \tag{5}$$

$$z = \frac{l\, d_{(n,0)}}{2}. \tag{6}$$

Here $k = 1,2,3,\ldots,n$ and $l = 0,\pm1,\pm2,\ldots$ number atomic sites in given plane and atomic planes, respectively.

If $d_{(n,n)}$ is the B–B bond length in an armchair (n,n) nanotube, the side $\delta_{(n,n)}$ of the regular $2n$-gon inscribed into the circle with tube radius of $r_{(n,n)}$ is determined as $\delta_{(n,n)} = 2r_{(n,n)} \sin \pi / 2n$. One-dimensional lattice constant of the armchair nanotube (figure 4) equals to the doubled leg of the right triangle with another leg of $\delta_{(n,n)}$ and hypotenuse of $d_{(n,n)}$, i.e., according to the Pythagorean theorem, $2\sqrt{d_{(n,n)}^2 - \delta_{(n,n)}^2} = \sqrt{4 - 1/\cos^2 \pi / 2n}\, d_{(n,n)}$. Its unit cell also consists of two atomic rings in parallel planes perpendicular to the tube axis and, from its part, each ring consists of n boron atoms. Coordinate ρ for all atomic sites again equals to tube radius:

$$\rho_{\mathrm{B}} = r_{(n,n)}, \tag{7}$$

while the rest cylindrical coordinates equal to

$$\varphi = \frac{(4k - 3 - (-1)^l)\pi}{2n}, \tag{8}$$

$$z = \sqrt{1 - \frac{1}{4\cos^2 \frac{\pi}{2n}}}\; l\, d_{(n,n)}, \tag{9}$$

where again $k = 1,2,3,\ldots,n$ and $l = 0,\pm1,\pm2,\ldots$.

On the basis of above obtained relations, it was found the squared distances between arbitrary atomic site $\vec{r}\,(\rho,\varphi,z)$ and site $\vec{r}_0\,(\rho,0,0)$ of the so-called central atom with $l = 0$ and $k = 1$, i.e., with $\varphi = 0$ and $z = 0$, $(\vec{r} - \vec{r}_0)^2 = 4\rho^2 \sin^2 \varphi/2 + z^2$, in zigzag

$$\frac{({}^{kl}_{(n,0)}\mathrm{B} - {}^{00}_{(n,0)}\mathrm{B})^2}{d_{(n,0)}^2} = \frac{3\sin^2 \dfrac{(4k - 3 - (-1)^l)\pi}{4n}}{4\sin^2 \dfrac{\pi}{2n}} + \frac{l^2}{4}, \tag{10}$$

and armchair tubes

$$\frac{({}^{kl}_{(n,n)}\mathrm{B} - {}^{00}_{(n,n)}\mathrm{B})^2}{d_{(n,n)}^2} = \frac{\sin^2 \dfrac{(4k - 3 - (-1)^l)\pi}{4n}}{\sin^2 \dfrac{\pi}{n}} + \left(1 - \frac{1}{4\cos^2 \dfrac{\pi}{2n}}\right) l^2. \tag{11}$$

3. Results and discussion

Using quasi-classical B–B interatomic pair potential and based on above given description of the boron nanotubes' geometry it is possible to calculate easily their ground state parameters, like the equilibrium bond length, equilibrium binding energy per atom, zero-point vibration frequencies etc. By them the absolute and relative stabilities of regular boron nanotubes can be estimated.

Thereto the quasi-classical B–B binding energy $E = E(d)$ in dependence on interatomic distance d seems to be useful. Part of its parameters, like the coefficients of harmonic and anharmonic terms $c \approx 0.12$ au and $g \approx 4.0$ au in the corresponding interatomic potential $U(d)$, which in the vicinity of equilibrium distance $d = d_0$ is approximated by the function $U(d_0) + c(d - d_0)^2 - g(d - d_0)^3$, and atomic core charge of $A_0 \approx 3.151$ (which is near the boron atom valence of 3) were successfully applied to interpret isotopic composition effect on boron structural [26] and melting parameters [27], respectively. Analysis of the corresponding binding energy curve (1) (converted from potential energy one using virial theorem for system total energy) based on the parameters obtained equating quasi-classical electron energy levels and mean orbital radii of the electrons with their Hartree–Fock values (table 1) leads to the equilibrium B–B distance of 3.37 au (1.78 Å), binding energy at that distance of

0.103 au (2.80 eV), and relative vibration frequency of 0.00480 au (0.131 eV). These values slightly, only by few percents are deviated from the averaged experimental B–B bond length in main structural units of boron and boron-rich compounds – icosahedra of 3.40 au (1.80 Å), dissociation energy of 0.099 au (2.69 eV), and oscillatory quantum of 0.00479 au (0.130 eV) of diboron molecule B_2, respectively.

Table 1. Quasi-classical parameters of the electron-density and potential distributions in B atom.

i	R_i, au	A_i, au	$-B_i$, au	C_i, au	$-n_i$, au	U_i, au
0	0	3.151	2.078	0.3864	0.03982	0.6342
1	2.317	1.351	0.523	0.0509	0.00213	0.0332
2	4.964	0.333	0.113	0.0096	0.00028	0.0005
3	5.890					

Table 2. Tabulated boron–boron binding energy function.

d, au	$E(d)$, au
3.20	0.101408
3.21	0.101656
3.22	0.101887
3.23	0.102100
3.24	0.102297
3.25	0.102478
3.26	0.102642
3.27	0.102791
3.28	0.102925
3.29	0.103043
3.30	0.103147
3.31	0.103236
3.32	0.103311
3.33	0.103372
3.34	0.103420
3.35	0.103454
3.36	0.103475
3.37	0.103483
3.38	0.103479
3.39	0.103462
3.40	0.103433
3.41	0.103393
3.42	0.103341
3.43	0.103278
3.44	0.103203
3.45	0.103118
3.46	0.103023
3.47	0.102917
3.48	0.102802
3.49	0.102677
3.50	0.102542

Journal of Physics: Conference Series **176** (2009) 012013 doi:10.1088/1742-6596/176/1/012013

The binding energy curve obtained in initial quasi-classical approximation for interacting boron atoms are expressed by a piece-wise analytical function. Evidently, for numerical calculations the tabulated form would be more appropriate. Here such form (in atomic units) is given (table 2) over the range from $a = 3.20$ up to 3.50 au (i.e. in the vicinity of equilibrium interatomic distance of 3.37 au) with step of argument of 0.01 au, what seems sufficient to provide accurate energy computations.

On basis of above obtained formulas, sizes of the small single-walled B-nanotubes were estimated (table 3) assuming bond lengths as $d_0 \approx 1.78$ Å, what is the equilibrium interatomic distance value according to the quasi-classical B–B pair potential. Actually, $d_{(n,0)}$ and $d_{(n,n)}$ depend on n; and their exact values can not be determined geometrically. In general, this task requires solving of the physical problem of the binding energy maximization for given nanotubular structure with respect B–B bond length. But, these dependences evidently have to be weak and in all cases in satisfactiry approximation $d_{(n,0)}$ and $d_{(n,n)}$ can be substituted by the mentioned value of 1.78 Å.

Table 3. Estimated radii of small-sized boron zigzag and armchair nanotubes.

(n,0) or (n,n)	$r_{(n,0)}$ or $r_{(n,n)}$, Å
(1,1)	0.00
(1,0)	0.77
(2,2)	0.89
(3,3)	1.03
(2,0)	1.09
(4,4)	1.26
(5,5)	1.51
(3,0)	1.54
(6,6)	1.78
(4,0)	2.01
(7,7)	2.05
(8,8)	2.33
(5,0)	2.49
(9,9)	2.60
(10,10)	2.88
(6,0)	2.98
(11,11)	3.16
(12,12)	3.44
(7,0)	3.46

Analogously, one can estimate molar binding energy of a boron nanotube. On the one hand, any constituent boron atom has 6 nearest neighbours. On the other hand, binding energy between pair of boron atoms is about 2.80 eV. So, taking into account interaction only between nearest neighbours the tube binding energy per atom may be calculated as $2.80 \text{ eV} \times 6 / 2 = 8.40$ eV. Of course, it is the estimation: actually binding energy should depend on the tube index.

4. Summary

Thus, using the electron-density and potential distributions in a boron atom parameterized within the initial quasi-classical approximation, it is possible to tabulate boron–boron pair binding energy curve, which leads to the boron–boron equilibrium interatomic distance, equilibrium binding energy, and relative vibration frequency values only slightly deviated from averaged bond length in structural units of boron and boron-rich solids, dissociation energy, and oscillatory quantum of the diboron molecule, respectively. On this basis, the B–B bond length and molar binding energy in a boron nanotube are predicted to be about 1.78 Å and 8.40 eV, respectively. Besides, the explicit expressions in term of B–B bond length are found for the flat zigzag and armchair boron nanotubes radii, atomic sites coordinates and intersite distances. There are made numerical estimations for the mostly small-sized boron nanotubes.

The proposed B–B binding energy curve together with obtained explicit formulas describing tubular geometries should be useful for calculating not only geometrical, but also other ground-state parameters of the boron nanotubes. From their part, these results can serve as a relevant basis for further investigations of the boron nanotubes electron structure.

References

[1] Tsagareishvili G V, Antadze M E and Tavadze F N 1991 *Production and Structure of Boron* Tbilisi (Metsniereba)
[2] Volkov V V, Yur'ev G S, Myakishev K G and Il'inchik E A 2006 *Vacuum Nanotechnology and Equipment* Neklyudov I M and Shulayev V M (Eds) Kharkiv (KhPTI) **1** 336
[3] Hanley L, Whitten J L and Anderson S L 1988 *J. Phys Chem.* **92** 5803
[4] Boustani I 1994 *J. Quant. Chem.* **52** 1081
[5] Zhai H-J, Wang L-Sh, Alexandrova A N and Boldyrev A I 2002 *J. Chem. Phys.* **117** 7917
[6] Zhai H-J, Alexandrova A N, Birch K A, Boldyrev A I and Wang L-Sh 2003 *Angew. Chem. Int. Ed.* **42** 6004
[7] Zhai H J, Kiran B, Li J and Wang L-Sh 2003 *Nat. Mat.* **2** 827
[8] Alexandrova A N, Boldyrev A I, Zhai H-J and Wang L-Sh 2006 *Coord. Chem.* **250** 2811
[9] Perkins C L, Trenary M and Tanaka T 1998 *Phys. Rev.* B **58** 9980
[10] Cao L M, Sun L L, Gao C X, He M, Wang Y Q, Li Y C, Zhang X Y, Li G, Zhang J and Wang W K 2001 *Adv. Mater.* **13** 1701
[11] Kiran B, Bulusu S, Zhai H-J, Yoo S, Zeng X Ch and Wang L-Sh 2005 *Proc. Natl. Acad. Sci. (USA)* **102** 962
[12] Ciuparu D, Klie R F, Zhu Y and Pfefferle L 2004 *J. Phys. Chem.* B **108** 3967
[13] Mukhopadhyay S, Pandey R, Yap Y K and Boustani I 2008 *16th Int. Symp. Boron, Borides & Rel. Mat.* Matsue 62
[14] Boustani I 1997 *Phys. Rev.* B-II **55** 16426
[15] Boustani I 1997 *J. Solid State Chem.* **133** 182
[16] Boustani I and Quandt A 1997 *Euro Phys. Lett.* **39** 527
[17] Boustani I, Quandt A and Rubio A 2000 *J. Solid State Chem.* **154** 269
[18] Boustani I 1997 *Surf. Sci.* **370** 355
[19] Boustani I, Quandt A, Hernández E and Rubio A 1999 *J. Chem. Phys.* **110** 3176
[20] Atiş M, Özdoğan C and Güvenç Z B 2007 *Int. J. Quan. Chem.* **107** 729
[21] Atiş M, Boyukata M, Özdoğan C, Yildirim E and Güvenç Z B 2008 *16th Int. Symp. Boron, Borides & Rel. Mat.* Matsue 33
[22] Evans M H, Joannopoulos J D and Pantelides S T 2005 *Phys. Rev.* B **72** 045434
[23] Cabria I, Alonso J A and López M J 2006 *Phys. Status Solidi* A **203** 1105
[24] Chkhartishvili L, Lezhava D, Tsagareishvili O and Gulua D 1999 *Trans. AMIAG*, **1** 295
[25] Chkhartishvili L, Lezhava D and Tsagareishvili O 2000 *J. Solid State Chem.* **154** 148
[26] Chkhartishvili L S, Gabunia D L and Tsagareishvili O A 2008 *Powd. Metal. & Met. Cer.* **47** 616
[27] Chkhartishvili L S, Gabunia D L and Tsagareishvili O A 2007 *Inorg. Mater.* **43** 594

Journal of Physics: Conference Series **176** (2009) 012014 doi:10.1088/1742-6596/176/1/012014

Boron nitride nanosystems of regular geometry

Levan Chkhartishvili

Department of Physics, Georgian Technical University, 77 M Kostava Avenue, Tbilisi, 0175, Georgia

E-mail: chkharti2003@yahoo.com

Abstract. The explicit expressions in term of B–N bond length are obtained for atomic site coordinates and intersite distances in regular boron nitride nanotubes and fullerenes. The radii of single-walled BN nanotubes and single-shelled BN fullerenes are estimated, and their most stable associations in form of double-walled nanotubes and double-shelled fullerenes are predicted. The differences between radii of regular boron nitride fullerenes with indexes of $(n+3)$ and (n) are almost equal to the interlayer distance in layered boron nitride structures. Description made for the boron nitride nanosystems of regular geometries may serve as basis for further ground state and electron structure calculations.

1. Introduction

Boron nitride (BN) can be found in forms of one-dimensional diatomic molecule, two-dimensional nanotubular and fullerene surfaces and three-dimensional crystals, like the layered hexagonal h-BN and rhombohedral r-BN, cubic zinc-blende c-BN and wurtzite w-BN, etc. Boron and nitrogen atoms are tetrahedrally coordinated in both c-BN and w-BN crystals. However, unlike the zinc-blende case, in the wurtzite structure there are two different types of nearest-neighbor bonds (both bond lengths are almost equal, whereas the tetrahedral angles are distorted).

Any constituent atom of the h-BN crystal, which corresponds to the boron nitride ground state, may be considered as 3-coordinated because the strong covalent bonding occurs only within the layers, whereas weak van der Waals forces are mostly responsible for interlayer interactions. The h-BN crystal posses a "graphitic" structure with two-layer stacking sequence, while r-BN is characterized by a three-layer stacking. The layers in these layered structures consist of regular hexagons, i.e., 6-membered atomic rings, with vertexes alternatively occupied by B and N atoms. In h-BN crystal, B atoms are placed directly above N atoms and vice versa. Hence, it would be possible to think that the interlayer bonding has some ionicity. However, the electrostatic component is insignificant due to the large interlayer distance. Same is argued by the existence of layered r-BN crystal, in which each subsequent layer is turned, and also by the isolated plane defects and their bundles presented in real layered BN crystals, in which any given atom can be placed above the like atom. Besides, it is possible to obtain turbostratic t-BN and amorphous structures in form of mixtures of the various boron nitride crystalline phases, and multi-walled nanotubular and multi-shelled fullerene-like BN structures. Strong chemical bonding between atoms in given layer and weak interlayer interaction in layered boron nitrides specify an opportunity of the physical and chemical intercalations by various atoms and molecules.

Based upon the similarity of structures of the specified boron nitride phases with those of carbon, it was assumed [1] that alongside with C nanotubes and fullerenes, BN nanotubes and fullerenes should be stable. The latter are hexagonal or mixed BN layers wrapped into cylinders or spheres, respectively. A modified-neglecting-of-diatomic-overlap (MNDO) study of BN analogues of C fullerenes was presented in [2] and $B_{30}N_{30}$ cage-molecule was predicted to be stable and, consequently, relatively simple to synthesize. In fact, by means of arc-discharge synthesis, BN nanosystems had been obtained both from carbon nanotubes [3] and in carbon-free plasma [4]. Like the bulk boron nitrides, the nanoscale ones are the materials with a diversity of technical and industrial applications. There are various synthesis methods and applications for the BN nanotubes and fullerenes. For those and also for electronic structure and physical properties of BN nanosystems see, for example, reviews [5-7] and references therein.

Design of devices based on nanostructured BN requires knowledge of the sizes of the nanotubes and fullerenes with given indexes. Here this task is solved for the most stable forms. Namely, in next sections, the geometries of regular BN nanotubes and fullerenes are presented. Note that in the figures below, B and N atoms are shown by the black and white circles, respectively; B–N bonds are shown by the solid lines, while auxiliary lines and arcs are dashed.

2. Geometry of regular boron nitride nanotubes

Primarily, the expressions of radii of the zigzag and armchair BN nanotubes in terms of B–N bond length have been obtained considering geometries of such structures (figures 1 and 2).

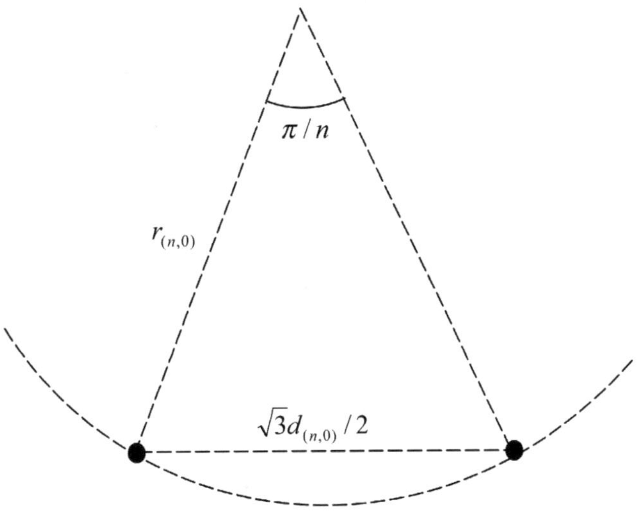

Figure 1. Calculation of the zigzag nanotube radius.

The formulas for the radii of $(n,0)$ and (n,n) BN nanotubes $r_{(n,0)}$ and $r_{(n,n)}$ are:

$$\frac{r_{(n,0)}}{d_{(n,0)}} = \frac{\sqrt{3}}{4\sin \pi / 2n}, \tag{2.1}$$

$$\frac{r_{(n,n)}}{d_{(n,n)}} = \frac{\sqrt{5+4\cos\pi/2n}}{4\sin\pi/2n}. \tag{2.2}$$

Here $d_{(n,0)}$ and $d_{(n,n)}$ are the B–N bond lengths in zigzag and armchair nanotubes, respectively. Nanotube index $n = 1,2,3,...$ determines the number of atoms because its unit cell consists of $2n$ BN formula units.

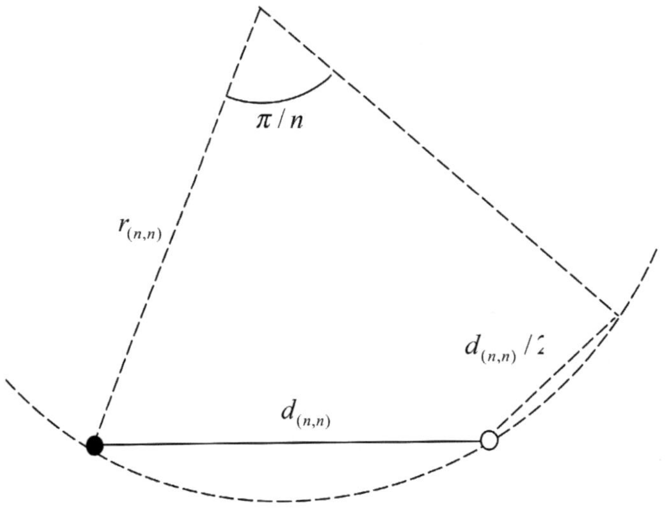

Figure 2. Calculation of the armchair nanotube radius.

On this basis, radii of the small single-walled BN nanotubes from (1,0) to (10,10) were estimated (table 1). Also, the most stable double-walled nanotubes from (1,0)@(9,0) to (5,5)@(10,10) were predicted (table 2) assuming bond lengths and inter-wall distances as $d \approx 1.4457$ Å and $c/2 \approx 3.3306$ Å, which are the experimental values of the B–N bond length in layers and interlayer distance in h-BN layered crystals, respectively. Actually, $d_{(n,0)}$ and $d_{(n,n)}$ depend on n. Therefore, their equilibrium values can not be determined geometrically and can be found by maximizing binding energy of a given nanotube with respect to the B–N bond length. Evidently these dependences have to be weak, and in all cases, $d_{(n,0)}$ and $d_{(n,n)}$ can be substituted by the value $d \approx 1.4457$ Å. Analogously, inter-wall distances in all double-walled nanotubes can by approximated by $c/2 \approx 3.3306$ Å.

Table 1. Estimated radii $r_{(n,0)}$ and $r_{(n,n)}$ of small-sized boron nitride zigzag and armchair nanotubes.

Nanotubes	Radii, Å
(1,0)	0.63

(1,1)	0.81
(2,0)	0.89
(3,0)	1.35
(2,2)	1.43
(4,0)	1.64
(5,0)	2.03
(3,3)	2.11
(6,0)	2.42
(4,4)	2.79
(7,0)	2.82
(8,0)	3.21
(5,5)	3.47
(9,0)	3.61
(10,0)	4.00
(6,6)	4.16
(11,0)	4.40
(12,0)	4.80
(7,7)	4.85
(13,0)	5.20
(8,8)	5.54
(14,0)	5.59
(15,0)	5.99
(9,9)	6.23
(16,0)	6.39
(17,0)	6.79
(10,10)	6.91

Table 2. Predicted stable double-walled boron nitride nanotubes built up from regular single-walled ones.

Double-walled nanotubes	Deviations of difference between radii
(1,0) @ (9,0)	– 10.5 %
(1,0) @ (10,0)	+ 1.4 %
(1,1) @ (10,0)	– 4.3 %
(1,1) @ (6,6)	+ 0.5 %
(2,0) @ (6,6)	– 1.8 %
(2,0) @ (11,0)	+ 5.6 %
(3,0) @ (11,0)	– 5.4 %
(3,0) @ (12,0)	+ 6.5 %
(2,2) @ (11,0)	– 10.8 %
(2,2) @ (12,0)	+ 1.1 %
(4,0) @ (7,7)	– 3.7 %
(4,0) @ (13,0)	+ 6.9 %
(5,0) @ (13,0)	– 4.8 %
(5,0) @ (8,8)	+ 5.4 %

(3,3) @ (13,0)	− 7.2 %
(3,3) @ (8,8)	+ 3.0 %
(6,0) @ (14,0)	− 4.8 %
(6,0) @ (15,0)	+ 7.2 %
(4,4) @ (15,0)	− 3.8 %
(4,4) @ (9,9)	+ 3.3 %
(7,0) @ (15,0)	− 4.7 %
(7,0) @ (9,9)	+ 2.4 %
(8,0) @ (16,0)	− 4.7 %
(8,0) @ (17,0)	+ 7.4 %
(5,5) @ (17,0)	− 0.5 %
(5,5) @ (10,10)	+ 3.3 %

Estimated radii of the single-walled BN nanotubes can be used for a prediction of their most probable combinations in multi-walled structures. Listed pairs as most suitable for formation of the stable double-walled BN nanotubes have been selected from the above calculated single-walled nanotubes. Remaining small divergences in sizes of the neighboring regular nanotubes can be compensated by defects and small chiral distortions. Such transformations of the zigzag or armchair nanotubes into chiral one will be accompanied, respectively, by increase or decrease in their radii. Hence, if the difference in radii between regular nanotubes are more (less) than $c/2$, the nanotube is favored in which the internal wall is zigzag (armchair) and the external wall has armchair (zigzag) structure.

Let us describe the geometries of the zigzag and armchair BN nanotubes using cylindrical coordinates (ρ, φ, z).

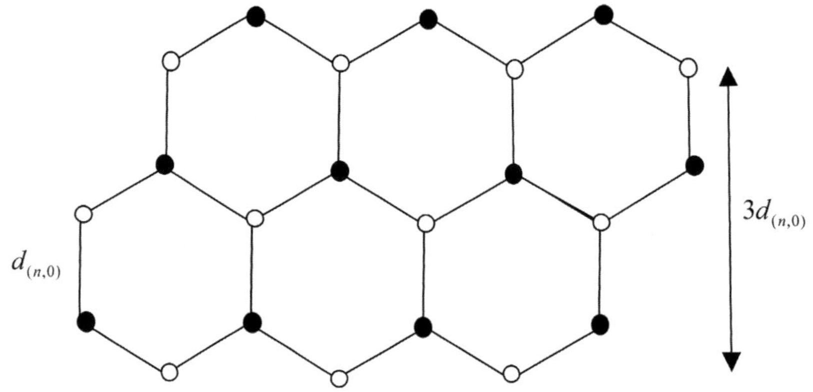

Figure 3. Lattice constant of the zigzag nanotube.

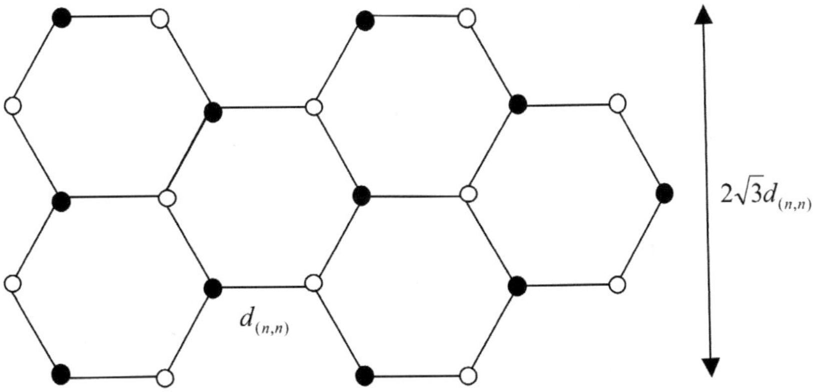

Figure 4. Lattice constant of the armchair nanotube.

The constant of zigzag (n,0) nanotube one-dimensional lattice equals to $3d_{(n,0)}$ (figure 3). Its unit cell consists of 4 atomic rings in parallel planes perpendicular to the axis. There are 2 pairs of rings, each consisting of 2 planes with n boron or n nitrogen atoms. Evidently, cylindrical coordinate ρ for all atomic sites equals to the tube radius:

$$\rho_B = \rho_N = r_{(n,0)}. \quad (2.3)$$

As for the coordinates φ and z in the first and second pairs of atomic rings, they equal to

$$\varphi_B = \varphi_N = 2l\pi / n, \quad (2.4)$$

$$z_B = (3m + 1/2)d_{(n,0)}, \quad (2.5)$$

$$z_N = (3m - 1/2)d_{(n,0)}, \quad (2.6)$$

and

$$\varphi_B = \varphi_N = (2l + 1)\pi / n, \quad (2.7)$$

$$z_B = (3m - 1)d_{(n,0)}, \quad (2.8)$$

$$z_N = (3m + 1)d_{(n,0)}, \quad (2.9)$$

respectively. Here $l = 0,1,2,...,n-1$ and $m = 0,\pm1,\pm2,...$ number atomic pairs in given pair of the atomic rings and these rings themselves, respectively.

The constant of the armchair (n,n) nanotube one-dimensional lattice equals to $2\sqrt{3}d_{(n,n)}$ (figure 4). Its unit cell consists of 2 atomic rings in parallel planes perpendicular to the tube axis. Each ring consists of n boron and n nitrogen atoms. Coordinate ρ for all atomic sites again equals the tube radius:

$$\rho_B = \rho_N = r_{(n,n)}, \quad (2.10)$$

The other cylindrical coordinates in the first and second atomic rings, respectively, equal to

$$\varphi_B = \varphi_1 + 2l\pi / n, \quad (2.11)$$

$$\varphi_N = -\varphi_1 + 2l\pi / n, \quad (2.12)$$

$$z_B = z_N = \sqrt{3}md_{(n,n)}, \quad (2.13)$$

and

$$\varphi_B = -\varphi_1 - 2\varphi_2 + 2l\pi / n, \quad (2.14)$$

$$\varphi_N = \varphi_1 + 2\varphi_2 + 2l\pi / n, \quad (2.15)$$

$$z_{\mathrm{B}} = z_{\mathrm{N}} = \sqrt{3}(m + 1/2)d_{(n,n)}. \tag{2.16}$$

Here

$$\sin\varphi_1 = \frac{2\sin\pi/2n}{\sqrt{5+4\cos\pi/2n}}, \tag{2.17}$$

$$\sin\varphi_2 = \frac{\sin\pi/2n}{\sqrt{5+4\cos\pi/2n}}, \tag{2.18}$$

$l = 0,1,2,\ldots,n-1$ and $m = 0,\pm1,\pm2,\ldots$ number B or N atoms in atomic rings and these rings themselves, respectively. The geometry of angles φ_1 and φ_2 is shown in figures 5 and 6.

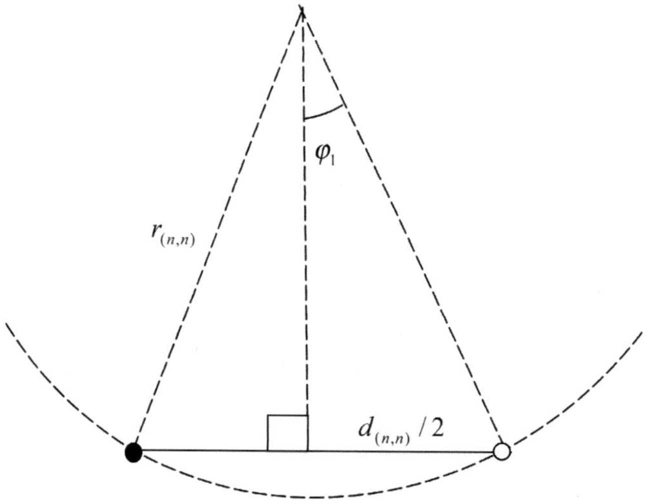

Figure 5. Geometry of angle φ_1.

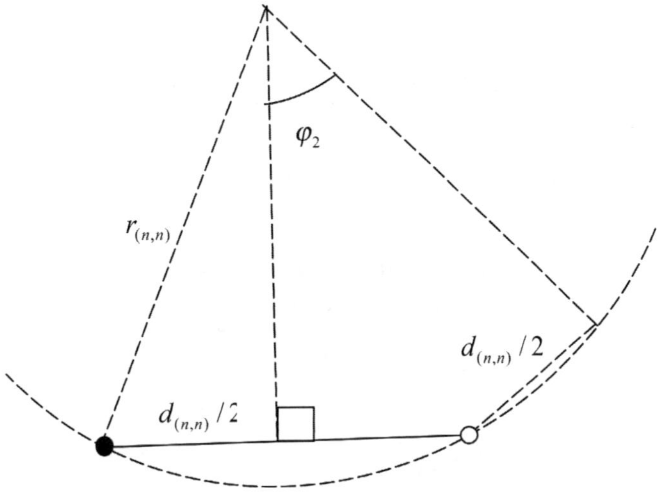

Figure 6. Geometry of angle φ_2.

On this basis, the distances between given atomic site and sites in so-called central atomic pairs are expressed for zigzag tubes ($l = m = 0$: $\varphi_B = \varphi_N = 0$, $z = d_{(n,0)}/2$, and $z = -d_{(n,0)}/2$):

$$\frac{({}^{lm}_{(n,0)}\mathrm{B1} - {}^{00}_{(n,0)}\mathrm{B1})^2}{d^2_{(n,0)}} = \frac{3\sin^2 l\pi/n}{4\sin^2 \pi/2n} + 9m^2, \tag{2.19}$$

$$\frac{({}^{lm}_{(n,0)}\mathrm{B2} - {}^{00}_{(n,0)}\mathrm{B1})^2}{d^2_{(n,0)}} = \frac{3\sin^2 (2l+1)\pi/2n}{4\sin^2 \pi/2n} + \frac{9(2m-1)^2}{4}, \tag{2.20}$$

$$\frac{({}^{lm}_{(n,0)}\mathrm{N1} - {}^{00}_{(n,0)}\mathrm{B1})^2}{d^2_{(n,0)}} = \frac{3\sin^2 l\pi/n}{4\sin^2 \pi/2n} + (3m-1)^2, \tag{2.21}$$

$$\frac{({}^{lm}_{(n,0)}\mathrm{N2} - {}^{00}_{(n,0)}\mathrm{B1})^2}{d^2_{(n,0)}} = \frac{3\sin^2 (2l+1)\pi/2n}{4\sin^2 \pi/2n} + \frac{(6m+1)^2}{4}, \tag{2.22}$$

$$\frac{({}^{lm}_{(n,0)}\mathrm{B1} - {}^{00}_{(n,0)}\mathrm{N1})^2}{d^2_{(n,0)}} = \frac{3\sin^2 l\pi/n}{4\sin^2 \pi/2n} + (3m+1)^2, \tag{2.23}$$

$$\frac{({}^{lm}_{(n,0)}\mathrm{B2} - {}^{00}_{(n,0)}\mathrm{N1})^2}{d^2_{(n,0)}} = \frac{3\sin^2 (2l+1)\pi/2n}{4\sin^2 \pi/2n} + \frac{(6m-1)^2}{4}, \tag{2.24}$$

$$\frac{({}^{lm}_{(n,0)}\mathrm{N1} - {}^{00}_{(n,0)}\mathrm{N1})^2}{d^2_{(n,0)}} = \frac{3\sin^2 l\pi/n}{4\sin^2 \pi/2n} + 9m^2, \tag{2.25}$$

$$\frac{({}^{lm}_{(n,0)}\mathrm{N2} - {}^{00}_{(n,0)}\mathrm{N1})^2}{d^2_{(n,0)}} = \frac{2\sin^2 (2l+1)\pi/2n}{4\sin^2 \pi/2n} + \frac{9(2m+1)^2}{4}; \tag{2.26}$$

and for armchair ($l = m = 0: \varphi_B = \varphi_1$, $\varphi_N = -\varphi_1$, and $z_B = z_N = 0$) tubes:

$$\frac{({}^{lm}_{(n,n)}\mathrm{B1}-{}^{00}_{(n,n)}\mathrm{B1})^2}{d^2_{(n,n)}}=\frac{(5+4\cos\pi/2n)\sin^2 l\pi/n}{4\sin^2\pi/2n}+3m^2, \tag{2.27}$$

$$\frac{({}^{lm}_{(n,n)}\mathrm{B2}-{}^{00}_{(n,n)}\mathrm{B1})^2}{d^2_{(n,n)}}=\frac{(5+4\cos\pi/2n)\sin^2(2l-1)\pi/2n}{4\sin^2\pi/2n}+\frac{3(2m+1)^2}{4}, \tag{2.28}$$

$$\frac{({}^{lm}_{(n,n)}\mathrm{N1}-{}^{00}_{(n,n)}\mathrm{B1})^2}{d^2_{(n,n)}}=\frac{(2\sin(2l-1)\pi/2n+\sin l\pi/n)^2}{4\sin^2\pi/2n}+3m^2, \tag{2.29}$$

$$\frac{({}^{lm}_{(n,n)}\mathrm{N2}-{}^{00}_{(n,n)}\mathrm{B1})^2}{d^2_{(n,n)}}=\frac{(\sin(2l+1)\pi/2n+2\sin l\pi/n)^2}{4\sin^2\pi/2n}+\frac{3(2m+1)^2}{4}, \tag{2.30}$$

$$\frac{({}^{lm}_{(n,n)}\mathrm{B1}-{}^{00}_{(n,n)}\mathrm{N1})^2}{d^2_{(n,n)}}=\frac{(2\sin(2l+1)\pi/2n+\sin l\pi/n)^2}{4\sin^2\pi/2n}+3m^2, \tag{2.31}$$

$$\frac{({}^{lm}_{(n,n)}\mathrm{B2}-{}^{00}_{(n,n)}\mathrm{N1})^2}{d^2_{(n,n)}}=\frac{(\sin(2l-1)\pi/2n+2\sin l\pi/n)^2}{4\sin^2\pi/2n}+\frac{3(2m+1)^2}{4}, \tag{2.32}$$

$$\frac{({}^{lm}_{(n,n)}\mathrm{N1}-{}^{00}_{(n,n)}\mathrm{N1})^2}{d^2_{(n,n)}}=\frac{(5+4\cos\pi/2n)\sin^2 l\pi/n}{4\sin^2\pi/2n}+3m^2, \tag{2.33}$$

$$\frac{({}^{lm}_{(n,n)}\mathrm{N2}-{}^{00}_{(n,n)}\mathrm{N1})^2}{d^2_{(n,n)}}=\frac{(5+4\cos\pi/2n)\sin^2(2l+1)\pi/2n}{4\sin^2\pi/2n}+\frac{3(2m+1)^2}{4}. \tag{2.34}$$

3. Geometry of regular boron nitride fullerenes

Let us introduce the concept of a regular boron nitride fullerene. The structure of the BN fullerene may be considered regular if it meets certain conditions: (1) structure consists of equal numbers of B and N atoms; (2) only 3-coordinated chemical bonds are realized (i.e. as in layers of h-BN, r-BN, and t-BN crystals, as well as in regular BN nanotubes, but with distinguished valence angles); (3) atomic rings represent plane or broken (along diagonals between B and N sites) regular polygons with even number of vertexes alternatively occupied by B and N atoms. Given definition yields equal B–N bond lengths. Let us denote this only structural parameter for (n)-fullerene ($n=1,2,3,\ldots$) by $d_{(n)}$.

In general, regular boron nitride (n) fullerene consists of $n+1$ parallel atomic rings in northern hemisphere and $n+1$ parallel atomic rings in southern hemisphere. If the fullerene index n equals an integer power of 2 (i.e., $2^0=1$, $2^1=2$, $2^2=4,\ldots$) then a pair of parallel atomic planes adjacent to the equatorial plane coincides in that plane. There are $2n$ atoms, n B atoms and n N atoms, in each plane. Then, the chemical formula of a regular BN fullerene should be $B_{2n(n+1)}N_{2n(n+1)}$. Let us assume that index $m=\pm1,\ldots,\pm(n+1)$ numbers atomic rings in parallel planes. Atomic sites in m-plane with $m\neq\pm1$ serve as vertexes of the regular $2n$-gon inscribed in circle with sides alternatively equal to certain parameters $d_{\mathrm{B}\,n,m}$ or $d_{\mathrm{N}\,n,m}$, which can be calculated as diagonals between certain B and N vertexes in regular $2k$-gon ($k=1,2,3,\ldots$) with vertexes alternatively occupied by B and N atoms and side of $d_{(n)}$. In particular, when $m=\pm(n+1)$, i.e., for polar circles, $d_{\mathrm{B}\,n,m}=d_{\mathrm{N}\,n,m}\equiv d_{(n)}$. In other words, polar circles present atomic rings in form of regular $2n$-gons with side of $d_{(n)}$. For planes with $m=\pm1$, which are adjacent or coincide with equatorial plane, only one from these two parameters equals to $d_{(n)}$, while the remaining parameter designated as $D_{(n)}$ should be determined separately. Constructions carried out for small (n) allows us to suggest that there is always a unique assembly of planes or broken (along diagonals between B and N sites) regular polygons with even number of vertexes, alternatively occupied by B and N atoms covering regular BN-fullerene surface without overlaps or gaps. As the regular fullerene is inscribed in sphere, the spherical or, more precisely, geographical (r, φ, λ) coordinates are useful to describe atomic site positions. Evidently, the geographical coordinate r for all atomic sites equals to fullerene radius $r_{(n)}$,

$$r_{\mathrm{B}} = r_{\mathrm{N}} = r_{(n)} . \tag{3.1}$$

The latitudes φ are determined as

$$\varphi_{\mathrm{B}} = \varphi_{\mathrm{N}} = \operatorname{sgn} m \arccos(r_{n,m} / r_{(n)}) , \tag{3.2}$$

while longitudes λ are

$$\lambda_{\mathrm{B}} = \arcsin(d_{\mathrm{B}\,n,m} / r_{n,m}) + \pi((4l - 3 - (-1)^m)/2n - 1) , \tag{3.3}$$

$$\lambda_{\mathrm{N}} = \arcsin(d_{\mathrm{N}\,n,m} / r_{n,m}) + \pi((4l - 3 - (-1)^m)/2n - 1) . \tag{3.4}$$

Here index $l = 1, \ldots, n$ numbers B and N atoms in planes.

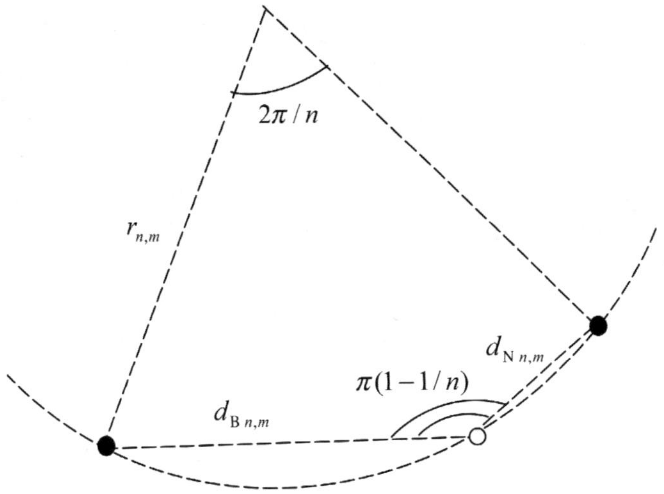

Figure 7. Calculation of the radius of atomic ring in a parallel plane.

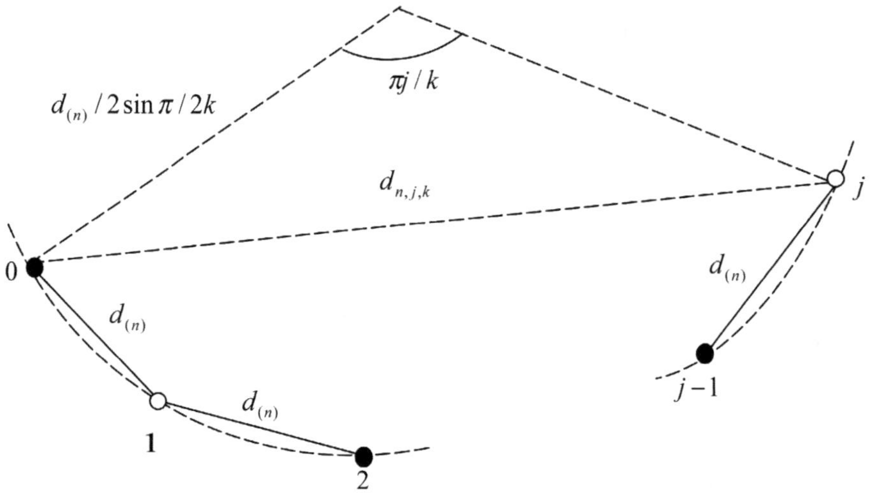

Figure 8. Calculation of the diagonal $d_{n,k,j}$ between certain B and N vertexes in regular $2k$-gon with vertexes alternatively occupied by B and N atoms and side of $d_{(n)}$.

Radius $r_{n,m}$ of the atomic ring in m-plane can be calculated from the following relation (figure 7),

$$r_{n,m}^2 = \frac{d_{\mathrm{B}\,n,m}^2 + 2d_{\mathrm{B}\,n,m}d_{\mathrm{N}\,n,m}\cos\pi/n + d_{\mathrm{N}\,n,m}^2}{4\sin^2\pi/n}. \qquad (3.5)$$

For planes with $m \neq \pm 1$, $d_{\mathrm{B}\,n,m}$ and $d_{\mathrm{N}\,n,m}$ are certain interatomic distances, which equal to a diagonals $d_{n,k,j}$ (figure 8) between certain B and N vertexes in regular $2k$-gon ($k = 1,2,3,\ldots$ and $j = 1,\ldots,k$) with vertexes alternatively occupied by B and N atoms and side $d_{(n)}$:

$$\frac{d_{n,k,j}}{d_{(n)}} = \frac{\sin\pi j/2k}{\sin\pi/2k}. \qquad (3.6)$$

If $m = \pm 1$, i.e., in planes adjacent or coincident with equatorial plane, one from the parameters $d_{\mathrm{B}\,n,m}$ and $d_{\mathrm{N}\,n,m}$ equals to $d_{(n)}$, while another equals to parameter $D_{(n)}$ expressed as

$$\frac{D_{(n)}}{d_{(n)}} = \frac{\sin\pi j_{\mathrm{B}}/2k_{\mathrm{B}}}{\sin\pi/2k_{\mathrm{B}}} + \frac{\sin\pi j_{\mathrm{N}}/2k_{\mathrm{N}}}{\sin\pi/2k_{\mathrm{N}}} - 1, \qquad (3.7)$$

where j_{B} and j_{N} are the j-values in planes with $m = \pm 2$. Stated relation is obtained taking into account that distance between closest atoms in planes with $m = \pm 2$ equals to square diagonal $\sqrt{2}d_{(n)}$ (corresponding drawing is analogous to that for fullerene radius case given below).

And finally, radius $r_{(n)}$ of regular (n) fullerene is determined as (figure 9)

$$\frac{2r_{(n)}^2}{d_{(n)}^2} = 1 + \frac{1}{8\sin^2\pi/n}\left(\frac{\sin\pi j_{\mathrm{B}}/2k_{\mathrm{B}}}{\sin\pi/2k_{\mathrm{B}}} + \frac{\sin\pi j_{\mathrm{N}}/2k_{\mathrm{N}}}{\sin\pi/2k_{\mathrm{N}}}\right)^2. \qquad (3.8)$$

In particular, in case of $n = 2^0, 2^1, 2^2, \ldots$, i.e., for fullerenes B_4N_4, $B_{12}N_{12}$, $B_{24}N_{24}$ etc, planes with $m = \pm 1$ coincide with the equatorial one, and one can obtain the simplified formula (figure 10):

$$\frac{r_{(n)}^2}{d_{(n)}^2} = \frac{1}{\sin^2 \pi / 2n}\left(\frac{3}{4} + \frac{\cos \pi / 2n}{\sqrt{2}}\right). \tag{3.9}$$

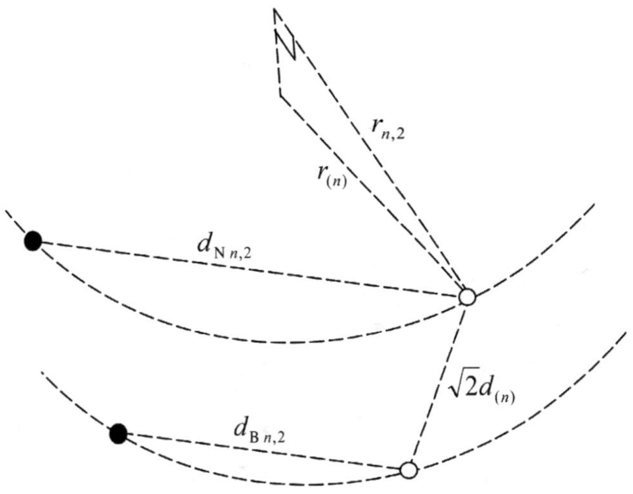

Figure 9. Calculation of the regular fullerene radius (general case).

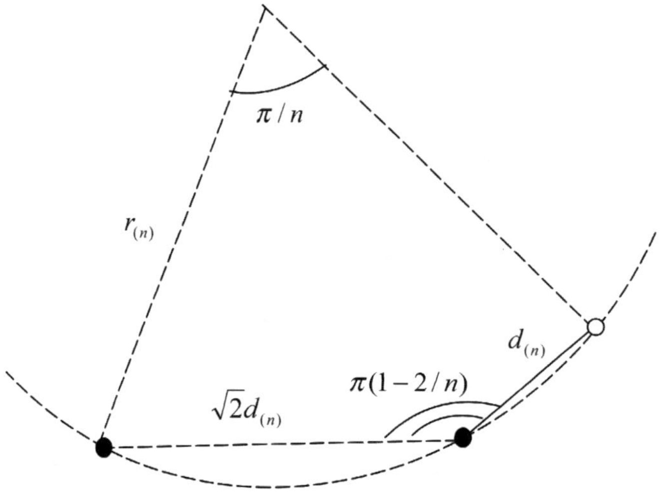

Figure 10. Calculation of the regular fullerene radius (if $n = 2^0, 2^1, 2^2, \ldots$).

Extrapolating this simplified formula over all regular BN fullerenes and again assuming $d_{(n)} \approx 1.4457$ Å, their radii have been calculated (table 3). Their most stable aggregates in form of double-shelled fullerenes predicted for inter-shell distances $c/2 \approx 3.3306$ Å are shown in table 4.

Table 3. Estimated radii $r_{(n)}$ of regular boron nitride fullerenes.

Regular fullerenes	Radii, Å
(1)	1.25
(2)	2.29
(3)	3.37
(4)	4.48
(5)	5.58
(6)	6.69
(7)	7.79
(8)	8.90
(9)	10.01
(10)	11.12
(11)	12.23
(12)	13.34
(13)	14.45
(14)	15.56
(15)	16.67

(16)	17.78
(17)	18.90
(18)	20.00
(19)	21.12
(20)	22.23

Table 4. Predicted stable double-shelled boron nitride fullerenes.

Double-walled fullerenes	Deviations of difference between radii
(1) @ (4)	– 3.0 %
(2) @ (5)	– 1.2 %
(3) @ (6)	– 0.3 %
(4) @ (7)	– 0.6 %
(5) @ (8)	– 0.3 %
(6) @ (9)	– 0.3 %
(7) @ (10)	– 0.0 %
(8) @ (11)	± 0.0 %
(9) @ (12)	± 0.0 %
(10) @ (13)	± 0.0 %
(11) @ (14)	± 0.0 %
(12) @ (15)	+ 0.0 %
(13) @ (16)	+ 0.0 %
(14) @ (17)	± 0.0 %
(15) @ (18)	– 0.0 %
(16) @ 19)	+ 0.3 %
(17) @ (20)	+ 0.3 %

From table 4 one can find that for regular fullerenes with indexes of (n) and $(n+3)$, deviations of the difference between radii from the expected inter-shell distance value are almost zero. Explanation for that is at $n \to \infty$ the radius of regular BN fullerene $r_{(n)} \to \sqrt{3+2\sqrt{2}}nd_{(n)}/\pi$ and then $r_{(n+3)} - r_{(n)} \to 3\sqrt{3+2\sqrt{2}}d_{(n)}/\pi$. Substituting $d_{(n)} \to 1.4457$ Å we get $r_{(n+3)} - r_{(n)} \to 3.3329$ Å which is the almost interlayer distance $c/2 \approx 3.3306$ Å in layered BN crystals. This result does not seem accidental. Maybe the obtained formula for fullerene radius reveals general relation between intralayer and interlayer structural parameters a and c in all layered boron nitride structures, such as layered crystals, multi-walled nanotubes, and multi-shelled fullerenes, $c/a \approx 2\sqrt{3(3+2\sqrt{2})}/\pi \approx 2.6621$ (in real h-BN layered crystals $c/a \approx 2.6602$).

On the basis of the stated relations, the explicit expressions (in term of B–N bond length $d_{(n)}$) of the interatomic distances in regular boron nitride fullerene have been obtained:

$$(\mathrm{B}\ n,m',l' - \mathrm{N}\ n,m'',l'')^2 =$$
$$= 2r_n^2 - 2\,\mathrm{sgn}\,m'm''\sqrt{(r_{(n)}^2 - r_{n,m'}^2)(r_{(n)}^2 - r_{n,m''}^2)} -$$

$$
\begin{aligned}
&-\left(d_{\mathrm{B}\,n,m'}d_{\mathrm{N}\,n,m''}+d_{\mathrm{N}\,n,m'}d_{\mathrm{B}\,n,m''}\right)\times\\
&\times\frac{\cos\left(4(l'-l'')+(-1)^{m'}+(-1)^{m''}\right)\pi/2n}{4\sin^2\pi/2n}+\\
&+\left(d_{\mathrm{B}\,n,m'}d_{\mathrm{B}\,n,m''}-d_{\mathrm{N}\,n,m'}d_{\mathrm{N}\,n,m''}\right)\times\\
&\times\frac{\sin\left(4(l'-l'')+(-1)^{m'}+(-1)^{m''}\right)\pi/2n}{2\sin\pi/n},
\end{aligned}
\tag{3.10}
$$

$$
\begin{aligned}
&(\mathrm{N}\,n,m',l'-\mathrm{B}\,n,m'',l'')^2=\\
&=2r_n^2-2\,\mathrm{sgn}\,m'm''\sqrt{(r_{(n)}^2-r_{n,m'}^2)(r_{(n)}^2-r_{n,m''}^2)}-\\
&-\left(d_{\mathrm{N}\,n,m'}d_{\mathrm{B}\,n,m''}+d_{\mathrm{B}\,n,m'}d_{\mathrm{N}\,n,m''}\right)\times\\
&\times\frac{\cos\left(4(l'-l'')-(-1)^{m'}-(-1)^{m''}\right)\pi/2n}{4\sin^2\pi/2n}+\\
&+\left(d_{\mathrm{N}\,n,m'}d_{\mathrm{N}\,n,m''}-d_{\mathrm{B}\,n,m'}d_{\mathrm{B}\,n,m''}\right)\times\\
&\times\frac{\sin\left(4(l'-l'')-(-1)^{m'}-(-1)^{m''}\right)\pi/2n}{2\sin\pi/n},
\end{aligned}
\tag{3.11}
$$

$$
\begin{aligned}
&(\mathrm{B}\,n,m',l'-\mathrm{B}\,n,m'',l'')^2=\\
&=2r_n^2-2\,\mathrm{sgn}\,m'm''\sqrt{(r_{(n)}^2-r_{n,m'}^2)(r_{(n)}^2-r_{n,m''}^2)}-\\
&-\left(d_{\mathrm{B}\,n,m'}d_{\mathrm{B}\,n,m''}+d_{\mathrm{N}\,n,m'}d_{\mathrm{N}\,n,m''}\right)\times\\
&\times\frac{\cos\left(4(l'-l'')+(-1)^{m'}-(-1)^{m''}\right)\pi/2n}{4\sin^2\pi/2n}+\\
&+\left(d_{\mathrm{B}\,n,m'}d_{\mathrm{N}\,n,m''}-d_{\mathrm{N}\,n,m'}d_{\mathrm{B}\,n,m''}\right)\times\\
&\times\frac{\sin\left(4(l'-l'')+(-1)^{m'}+(-1)^{m''}\right)\pi/2n}{2\sin\pi/n},
\end{aligned}
\tag{3.12}
$$

$$
\begin{aligned}
&(\mathrm{N}\,n,m',l'-\mathrm{N}\,n,m'',l'')^2=\\
&=2r_n^2-2\,\mathrm{sgn}\,m'm''\sqrt{(r_{(n)}^2-r_{n,m'}^2)(r_{(n)}^2-r_{n,m''}^2)}-\\
&-\left(d_{\mathrm{N}\,n,m'}d_{\mathrm{N}\,n,m''}+d_{\mathrm{B}\,n,m'}d_{\mathrm{B}\,n,m''}\right)\times\\
&\times\frac{\cos\left(4(l'-l'')-(-1)^{m'}+(-1)^{m''}\right)\pi/2n}{4\sin^2\pi/2n}+\\
&+\left(d_{\mathrm{N}\,n,m'}d_{\mathrm{B}\,n,m''}-d_{\mathrm{B}\,n,m'}d_{\mathrm{N}\,n,m''}\right)\times\\
&\times\frac{\sin\left(4(l'-l'')-(-1)^{m'}+(-1)^{m''}\right)\pi/2n}{2\sin\pi/n}.
\end{aligned}
\tag{3.13}
$$

4. Summary

From geometrical considerations, we have deduced the explicit expressions for the following parameters of zigzag and armchair BN nanotubes: radii, atomic sites coordinates and intersite distances in terms of B–N bond length and index. We have also introduced concept of the regular BN fullerene and derived analogous expressions for them. In particular, the differences between radii of regular fullerenes with indexes of $(n+3)$ and (n) are close to the interlayer distance in layered BN crystals. It seems that formula of a regular fullerene radius reveals general relation between intralayer and interlayer structural parameters in all layered boron nitride structures. Using obtained expressions, we made numerical estimations for the small-sized boron nitride single-walled nanotubes and single-shelled fullerenes and then, based on these estimations, predicted double-walled nanotubes and double-shelled fullerenes.

How can we solve the same task for multi-walled nanotubes or multi-shelled fullerenes? In such cases, it will be necessary to calculate radii of nanotubes or fullerenes with high indexes aiming at selection of the sequences of nanotubes or fullerenes, whose radii are close to an arithmetical progression with arithmetic ratio close to the interlayer distance in layered boron nitride structures. However, only geometrical consideration will be insufficient because in multi-walled or multi-shelled nanosystems there are intermediate layers as well. For this reason, the choice of the most stable structure should be made by comparing change in the binding energy caused by deviation from the equilibrium interlayer distance with that caused by the distortions of the regular structures.

Description made for the regular geometries of the boron nitride nanosystems may serve as basis for further ground state and electron structure calculations. In particular, quasi-classical approach may be used for this purpose (physical theory see in [8,9], and for related mathematical problems like the computing of matrix elements and solving of secular equation see [10-13]), which have already been applied to other structural modifications of boron nitride – diatomic molecule, isolated plane sheet, hexagonal h-BN, cubic c-BN, and wurtzite-like w-BN crystals [14-22]. Recently the relative stability of the ultra-small BN nanotubes has been studied by this approach, and a complex dependence of the molar energy on the radius was obtained [23]. Similar calculations for BN fullerenes should give answers to such questions as: whether the spheroid form or the flattened fullerenes containing defects are energetically more preferable; what are the maximal sizes of regular BN fullerenes, etc. In addition to the geometrical factors, such consideration needs to take into account features characterizing BN nanosystems in view of general equations derived [24] for fluctuations in energy of the small completely open (incompressible) systems. Those equations show that the fluctuations may be large, because there are no restraints on the size of the system; and, besides, fluctuations in total number or partial numbers of atoms in binary systems indirectly contribute to the fluctuations in their energy.

References

[1] Rubio A, Corkill J L and Cohen M L 1994 *Phys. Rev.* B **49** 5081
[2] Xia X, Jelski D A, Bowsner J R and George Th F 1992 *J. Am. Chem. Soc.* **114** 6493
[3] Weng-Seih Z, Cherrey K, Chopra N G, Blasé X, Miyamoto Y, Rubio A, Cohen M L, Louie S G, Zettl A and Gronsky R 1995 *Phys. Rev.* B **51** 11229
[4] Chopra N G, Luyken R J, Cherrey K, Crespi V H, Cohen M L, Louie S G and Zettl A 1995 *Science* **269** 966
[5] Chkhartishvili L S 2006 *Thin Films in Optics and Nanoelectronics* Neklyudov I M and Shulayev V M (Eds) Kharkiv (KhPTI) **2** 367
[6] Chkhartishvili L S 2008 *Nanomaterials* Neklyudov I M and Shulayev V M (Eds) Kharkiv (KhPTI) **2** 23
[7] Chkhartishvili L 2009 *Boron Based Solids* Chkhartishvili L (Ed) Trivandrum (Research Signpost) (in press)
[8] Chkhartishvili L 2004 *Quasi-Classical Theory of Substance Ground State* Tbilisi (GTU)
[9] Chkhartishvili L 2006 *Quasi-Classical Method of Calculation of Substance Structural and Electronic Energy Spectrum Parameters* Tbilisi (TSU)
[10] Chkhartishvili L S 2001 *Math. Notes* **69** 421
[11] Chkhartishvili L S 2004 *Trans. GTU* **452** 15
[12] Chkhartishvili L S 2005 *Math. Notes* **77** 273
[13] Chkhartishvili L S 2009 *Math. Notes* (in press)
[14] Chkhartishvili L, Lezhava D, Tsagareishvili O and Gulua D 1999 *Trans. AMIAG* **1** 295
[15] Chkhartishvili L, Lezhava D and Tsagareishvili O 2000 *J. Solid State Chem.* **154** 148
[16] Chkhartishvili L and Lezhava D 2001 *Trans. GTU* **439** 87
[17] Chkhartishvili L 2002 *Proc. 1st Internat. Boron Symp.* Erarslan K (Ed) Kütahya (DU) 139
[18] Chkhartishvili L S 2004 *Phys. Solid State* **46** 2056

[19] Chkhartishvili L 2004 *J. Solid State Chem.* **177** 395
[20] Chkhartishvili L 2006 *Proc. TSU (Phys.)* **40** 130
[21] Chkhartishvili L 2006 *Mater. Sci.: An Ind. J.* **2** 18
[22] Chkhartishvili L S 2006 *Phys. Solid State* **48** 846
[23] Chkhartishvili L S 2009 *Nanostruct. Mater. Sci.* **1** 33
[24] Hill T L and Chamberlin R V 2002 *Nano Lett.* **2** 609

Journal of Physics: Conference Series **176** (2009) 012015 doi:10.1088/1742-6596/176/1/012015

Structure refinement of quaternary RE-B-C-Si compounds: $Y_{3-x}(B_{12})_3(CSi)Si_8$ (x ≈ 0.96) and $Dy_{3-x}(B_{12})_3(CSi)Si_8$ (x ≈ 0.90)

Takaho Tanaka[1], **Akira Sato**[1] **and FuXiang Zhang**[2]

[1] National Institute for Materials Science, 1-2-1 Sengen, Tsukuba, Ibaraki 305-0047, Japan

[2] Department of Georogical Science, University of Michigan, Ann Arbor, MI 48109, USA

E-mail: TANAKA.Takaho@nims.go.jp

Abstract. We refined the crystal structure of quaternary RE-B-C-Si (RE = Y or Dy) compounds $Y_{3-x}(B_{12})_3(CSi)Si_8$ (x ≈ 0.96) and $Dy_{3-x}(B_{12})_3(CSi)Si_8$ (x ≈ 0.90). Both compounds have a trigonal crystal structure with space group $R\overline{3}m$ (No. 166). They have previously been assigned as ternary RE-B-Si compounds, but we incorporated carbon to the new structure refinements. The refinement for $Y_{3-x}(B_{12})_3(CSi)Si_8$ achieved a final $R1 = 0.0288$ for 1178 independent reflections with $F_O > 4\sigma(F_O)$ and 0.0431 for all 1484 reflections, and the refinement for $Dy_{3-x}(B_{12})_3(CSi)Si_8$ achieved a final $R1 = 0.0382$ for 2009 independent reflections with $F_O > 4\sigma(F_O)$ and 0.0385 for all 2022 reflections. The new structure refinements suggest the presence of $(B_{12})_3$-Si-C-$(B_{12})_3$ units, which, together with two other units, B_{12} icosahedron cluster and Si_8 ethane-like cluster, stabilize the $RE_{3-x}(B_{12})_3(CSi)Si_8$ crystal structure.

1. Introduction

Rare-earth boron-rich borides have unique crystal structures in which boron framework structure consists mainly of boron icosahedra (B_{12}). Because the B_{12} cluster has 5-fold symmetry, simple arrangement of the B_{12} clusters cannot achieve periodical long-range order. Thus, the boron framework is rather complicated and the unit cell is large. This framework contains many voids that accommodate the rare earth atoms. Recently cluster-compounds have attracted much attention from both scientific and practical points of view because of their unique crystal structures, chemical and physical properties and potential thermoelectric and superconducting applications [1, 2, 3, 4]. Knowing their crystal structure is crucial for understanding their properties.

We have reported the synthesis and structure analysis of a ternary Y-B-Si compound $YB_{17.6}Si_{4.6}$ [5], which has a trigonal structure with space group $R\overline{3}m$ (No.166) and lattice constants $a = 1.00841(4)$ and $c = 1.64714(5)$ nm. We described the crystal structure of $YB_{17.6}Si_{4.6}$ based on three units of a B_{12} icosahedron, a Si_8 ethane-like cluster and a bonding configuration $(B_{12})_3$-Si-Si-$(B_{12})_3$. In the latter, we have first assigned the Si site as a split position, i.e., Si3a and Si3b sites, but then re-assigned the Si3b site to the B site in the structure refinement of $Dy_{0.7}B_{12.33}Si_3$ [6] that is isotypic to $YB_{17.6}Si_{4.6}$. Lattice constants of $Dy_{0.7}B_{12.33}Si_3$ are $a = b = 1.00782(3)$ and $c = 1.64651(4)$ nm.

Meanwhile Salvador *et al* [7] reported the crystal structure of $Tb_{3-x}C_2Si_8(B_{12})_3$ as a quaternary boron-rich phase containing B_{12} icosahedra. Both structures appear isotypic.

Journal of Physics: Conference Series **176** (2009) 012015 doi:10.1088/1742-6596/176/1/012015

Salvador *et al* reported that $Tb_{3-x}C_2Si_8(B_{12})_3$ could only be synthesized if carbon is introduced into the growth. In the crystal structure of $Tb_{3-x}C_2Si_8(B_{12})_3$ they found a bonding scheme of $(B_{12})_3$-C-C-$(B_{12})_3$ in addition to B_{12} icosahedra and Si_8 ethane-like clusters. On the contrary, we did not incorporate carbon in the crystal structure because we have used carbon-free raw materials and crucibles in our growth of single crystals by high temperature solution method using Si flux. Further, we have synthesized powder samples of $REB_{17.6}Si_{4.6}$ (RE= Y, Gd-Lu) by solid-state reaction [8]. We undertook Rietveld structure analysis of Tb system compound using powder X-ray diffraction (XRD) data and have assigned only Si to the Si3 site with a partial occupancy of 71%, because the electron density of the site was obviously too high for a C site and we did not detect carbon with either chemical analysis or electron probe microanalysis. However, we have used graphite for a heating element and heat insulation in both synthesis experiments, which might contaminate the samples. Thus, we conducted synthesis experiments of $YB_{17.6}Si_{4.6}$ using a carbon-free furnace to confirm whether or not $YB_{17.6}Si_{4.6}$ can be synthesized without carbon. The result suggested that in absence of carbon, formation of $YB_{17.6}Si_{4.6}$ is difficult, as reported by Salvador *et al.* This finding motivated us to reinvestigate the crystal structure of both Y and Dy compounds.

In this study, we report the structure refinement of both compounds incorporating carbon, which resulted in a new structure description of $RE_{3-x}(B_{12})_3(CSi)Si_8$ (RE=Y and Dy, $x \approx 0.96$ for Y and 0.90 for Dy) and a new bonding scheme of $(B_{12})_3$-Si-C-$(B_{12})_3$.

2. Phase formation

The powder synthesis of ternary Y-B-Si compounds was carried out using an RF-induction-heating furnace. To exclude carbon, a composite consisting of BN and TiB_2 (BN-composite-EC, DENKA, Japan) was used as a heating element. The heating element was centered in a BN container, which was filled with BN granules for heat-insulation. Raw materials were YB_4 powder (3N, carbon content 0.1 wt%, Shin-Nihon Kinzoku, Japan), amorphous B powder (3N, carbon content 0.1 wt%, SB-Boron, USA) and Si powder (5N, Koujundokagaku, Japan). The synthesis was carried out at approximately 1750 °C under vacuum, for 10 h. The expected reaction equation was $YB_4 + 14B + 5Si = YB_{18}Si_5$ with minor losses of B and Si. The obtained sample was analyzed by powder XRD. The XRD pattern showed contributions of yttrium higher boride (YB_{12}) and boron silicide, which indicated that the phase formation of $YB_{17.6}Si_{4.6}$ without carbon is difficult. Whereas if small amount of carbon was added into the system, the XRD pattern of the phase previously assigned as $YB_{17.6}Si_{4.6}$ appeared. Thus, we concluded that carbon is important for the phase formation, which stimulated us to refine the structure incorporating carbon into it.

3. Structure refinement

3.1. Structure refinement of $Y_{3-x}(B_{12})_3(CSi)Si_8$

The structure refinement of $Y_{3-x}(B_{12})_3(CSi)Si_8$ was carried out using previously collected single-crystal XRD data and SHELXL97 program [9]. The result was visualized using the crystallographic program CrystalMaker [10]. The previous atomic coordinates were used as an initial solution [5]. Eight atomic sites have previously been assigned to one yttrium atom Y, four boron atoms B1-B4 and three silicon atoms Si1-Si3, in which Si3 site was assigned to the split sites Si3a and Si3b with partial occupancies of 0.50(7) and 0.16(8), respectively, as mentioned above. All remaining sites except yttrium site are fully occupied. Considering the electron density of the sites, carbon incorporation is only possible to the Si3b site; thus, in turn, the Si3b site was re-assigned to the C3 site. The final refinement of 38 parameters for 1178 independent reflections with $F_0 > 4\sigma(F_0)$ resulted in an *R1* value of 2.88% and 4.31% for all 1484 independent reflections. The *R1* values are slightly better than those 2.96% and 4.41% obtained previously

[5]. The final atomic coordinates, occupancy factors and temperature factors are listed in table 1.

Table 1. Atomic coordinates, site occupancy and isotropic displacement parameters of $Y_{3-x}(B_{12})_3(CSi)Si_8$

Atom	Site	x/a	y/b	z/c	Occ.	$U_{eq}(\text{Å}^2\times10^3)$
Y	9*e*	1/6	1/3	1/3	0.68(1	6.1(1)
B1	36*i*	0.4916(1)	0.1556(1)	0.1353(1)	1.0	4.5(1)
B2	36*i*	0.3671(1)	0.0400(1)	0.2181(1)	1.0	4.6(1)
B3	18*h*	0.4838(2)	0.2419(1)	0.2307(1)	1.0	7.4(2)
B4	18*h*	0.2900(2)	0.1450(1)	0.2697(1)	1.0	5.1(2)
C3	6*c*	2/3	1/3	0.2666(12)	0.58(5)*	2.9(5)
Si1	6*c*	1/3	2/3	0.2379(1)	1.0	2.7(1)
Si2	18*h*	0.4648(1)	0.5352(1)	0.2730(1)	1.0	4.1(1)
Si3	6*c*	2/3	1/3	0.2917(3)	0.42(2)*	1.1(2)

*The sum of those values was fixed at 1.0.

3.2. Structure refinement of $Dy_{3-x}(B_{12})_3(CSi)Si_8$

The structure refinement of $Dy_{3-x}(B_{12})_3(CSi)Si_8$ was carried out in a similar fashion, using previously collected single-crystal XRD data and SHELXL97 program [9], and the previous atomic coordinates were used as an initial solution [6]. The B(5) site was re-assigned to the C3 site. The final refinement of 34 parameters for 2009 independent reflections with $F_0 > 4\sigma(F_0)$ resulted in an *R1* value of 3.82% and 3.85% for all 2022 independent reflections. The final atomic coordinates, occupancy factors and temperature factors are listed in table 2.

Table 2. Atomic coordinates, site occupancy and isotropic displacement parameters of $Dy_{3-x}(B_{12})_3(CSi)Si_8$

Atom	Site	x/a	y/b	z/c	Occ.	$U_{eq}(\text{Å}^2\times10^3)$
Dy	9*e*	1/6	1/3	1/3	0.70(0)	10.4(1)
B1	36*i*	0.4916(1)	0.1556(1)	0.1354(1)	1.0	4.9(2)
B2	36*i*	0.3669(1)	0.0397(2)	0.2180(1)	1.0	5.1(2)
B3	18*h*	0.4840(1)	0.2420(1)	0.2306(1)	1.0	8.1(2)
B4	18*h*	0.2900(1)	0.1450(1)	0.0636(1)	1.0	5.6(3)
C3	6*c*	2/3	1/3	0.2642(3)	0.50(1)*	2.9(5)
Si1	6*c*	1/3	2/3	0.2378(1)	1.0	6.2(1)
Si2	18*h*	0.4647(1)	0.5353(1)	0.2731(1)	1.0	6.0(1)
Si3	6*c*	2/3	1/3	0.2909(3)	0.50(1)*	3.2(3)

*The sum of those values was fixed at 1.0.

4. Discussion

The distances between the C3 and Si3 sites and the surrounding sites for $Y_{3-x}(B_{12})_3(CSi)Si_8$ and $D_{3-x}(B_{12})_3(CSi)Si_8$ are summarized in table 3. The shortest distances between C3 and Si3 split positions are only 0.413 and 0.438 Å for $Y_{3-x}(B_{12})_3(CSi)Si_8$ and $Dy_{3-x}(B_{12})_3(CSi)Si_8$, respectively, as shown in figure 1(a). Therefore, locally in such a pair positions, either the C3 or the Si3 position, but not both, is occupied.

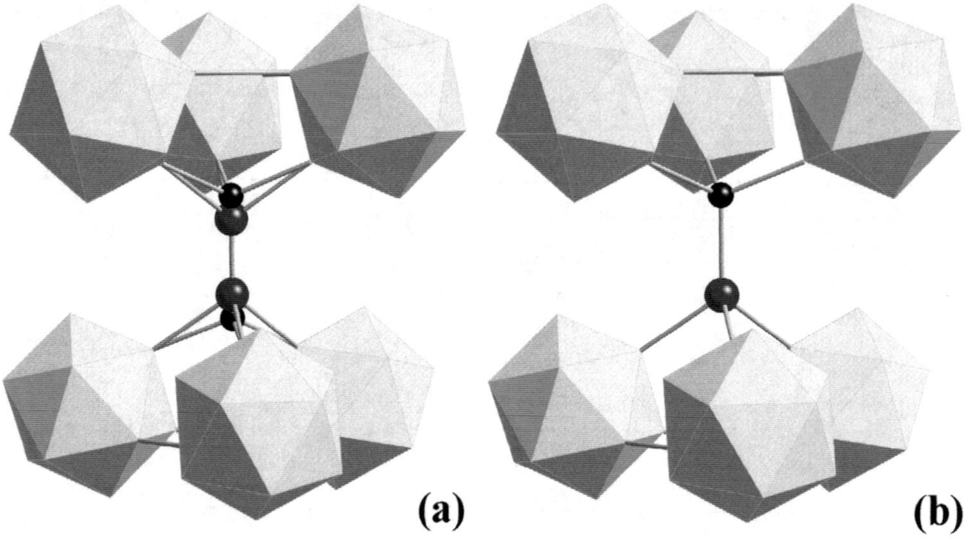

Figure 1. (a) C3 and Si3 split positions are simultaneously shown, and (b) $(B_{12})_3$-Si-C-$(B_{12})_3$ bonding scheme (green: B_{12} icosahedron, black: carbon, blue: silicon).

Table 3. Interatomic distances between the listed sites. Left and right values correspond to $Y_{3-x}(B_{12})_3(CSi)Si_8$ and $Dy_{3-x}(B_{12})_3(CSi)Si_8$, respectively

Atoms	Distance (Å)	Atoms	Distance (Å)
C3-B3	1.703(7)/1.688(3)	C3-C3	2.198(6)/2.268(4)
Si3-B3	1.887(3)/1.878(2)	Si3-C3	0.413(3)/0.438(2)
C3-Si3	1.786(24)/1.838(6)	Si3-Si3	1.373(10)/1.398(4)

Three combinations of C3 and Si3 sites are possible: Si-C, C-C and Si-Si. The latter two are unlikely because of too large C-C atom separation of 2.198/2.268 and too short Si-Si separation of 1.373/1.398 Å. Because the Si-C bond length in SiC is 1.89 Å, the separation of 1.786/1.838 Å more likely originates from Si-C than C-C or Si-Si pairs. Formation of Si-C pair at the C3 and Si3 sites agrees well with the fact that the occupancies of both the C3 and Si3 sites are close to 50% for $Y_{3-x}(B_{12})_3(CSi)Si_8$ and equal to 50% for $Dy_{3-x}(B_{12})_3(CSi)Si_8$. Then, we can expect that a new bonding scheme of $(B_{12})_3$-Si-C-$(B_{12})_3$ is available in the $RE_{3-x}(B_{12})_3(CSi)Si_8$ crystal structure and it can effectively stabilize the crystal structure. Thus the carbon incorporation must be important for the phase formation of $RE_{3-x}(B_{12})_3(CSi)Si_8$. The $(B_{12})_3$-Si-C-$(B_{12})_3$ unit and overall crystal structure are shown in figures 1(b) and 2, respectively.

In their analysis of $Tb_{3-x}C_2Si_8(B_{12})_3$, Salvador *et al* [7] found only carbon atoms at the Si3/C3 site, and they did not detect split positions. Thus, their bonding scheme is $(B_{12})_3$-C-C-$(B_{12})_3$, where the C-C distance is 1.695 Å. This value is much shorter than the C3-Si3 distances of $Y_{3-x}(B_{12})_3(CSi)Si_8$ and $Dy_{3-x}(B_{12})_3(CSi)Si_8$. Meanwhile, in our Rietveld structure analysis using powder XRD data of the Tb system compound we assigned only Si atoms to the Si3 site with a partial occupancy of 71% [8]. The electron density at this site is obviously too high for carbon. Considering the importance of the Si3 site in bridging the B_{12} icosahedron framework, it is unlikely that vacancies are available on 29% of the Si3 site. Instead, we think that the introduction of Si/C mixed occupation with 100% occupancy to the site is

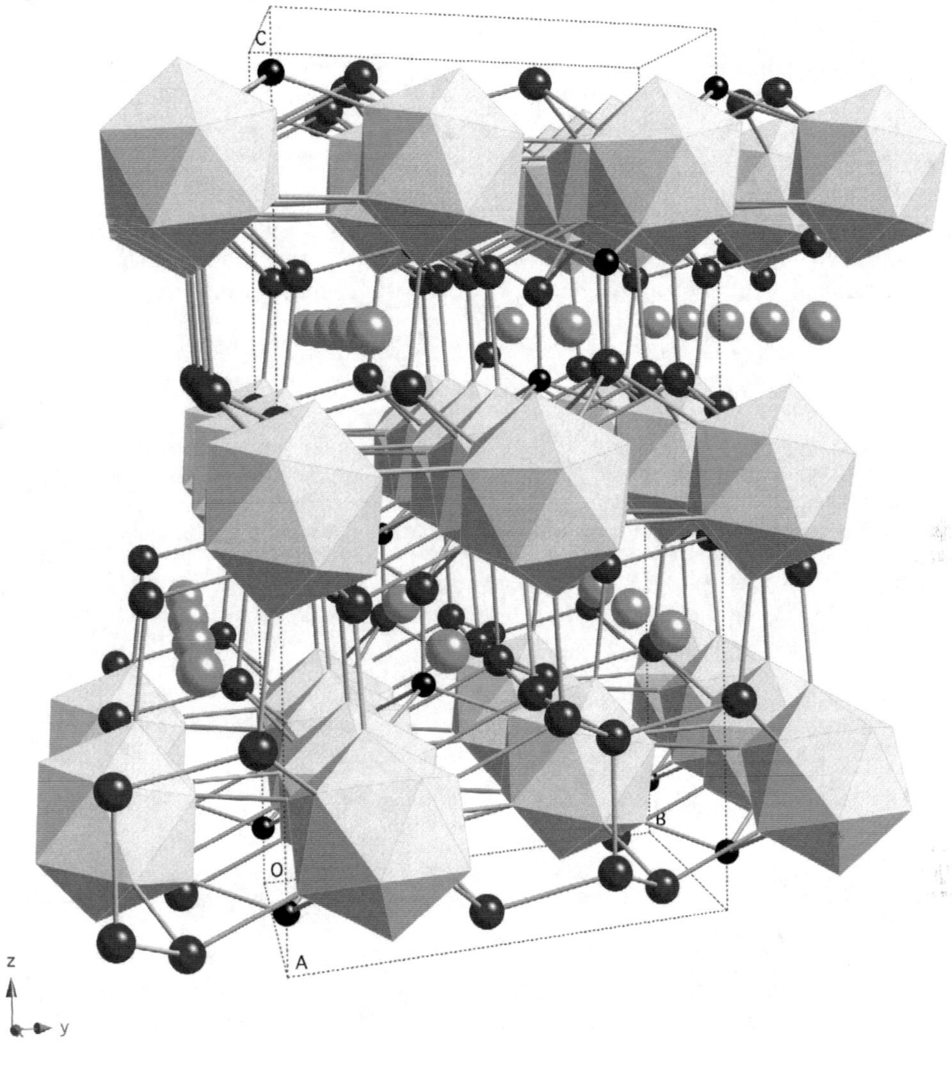

Figure 2. Crystal structure of $RE_{3-x}(B_{12})_3(CSi)Si_8$ (RE=Y or Dy) viewed along the direction close to [100]. Red, black and blue spheres indicate the Y or Dy, C and Si atoms, respectively. Vacancies at the Y or Dy site are ignored. Green polyhedra indicate the B_{12} icosahedra.

more reasonable, which immediately results in equal occupancy of Si and C to the site. This result is consistent with that obtained for $Y_{3-x}(B_{12})_3(CSi)Si_8$ and $Dy_{3-x}(B_{12})_3(CSi)Si_8$. Thus the composition should be $Tb_{3-x}(B_{12})_3(CSi)Si_8$ ($x \approx 0.96$) and it should include $(B_{12})_3$-Si-C-$(B_{12})_3$ units. However, we did not detect a splitting for the site. This is probably due to a lower resolution of the Rietveld method compared to that of single-crystal XRD method. The structure refinement by Salvador *et al* indicated the existence of $(B_{12})_3$-Si-C-$(B_{12})_3$ units, probably because they intentionally added carbon during the growth, and thus their crystals had higher carbon content than ours; carbon incorporation to our sample was accidental.

Existence of both bonding schemes of $(B_{12})_3$-Si-C-$(B_{12})_3$ and $(B_{12})_3$-C-C-$(B_{12})_3$ suggests a solid-solubility region of carbon on the site from 50 to 100%. On the other hand, $(B_{12})_3$-Si-Si-

$(B_{12})_3$ bonding scheme is unlikely because of too short Si-Si distance, which suggests that the minimum carbon occupancy at the site is 50%. Some B atoms may replace C atoms at the C3 site, as previously assigned to the B site [6]. However, the C occupation is more likely because the site is tetrahedrally coordinated, and the B occupation at the site needs extra electron transfer to complete tetrahedral bonding. Thus, carbon is indispensable for the phase formation of this group of compounds.

5. Conclusion

Crystal structures of $Y_{3-x}(B_{12})_3(CSi)Si_8$ (x ≈ 0.96) and $D_{3-x}(B_{12})_3(CSi)Si_8$ (x ≈ 0.90) were refined. The newly found bonding scheme $(B_{12})_3$-Si-C-$(B_{12})_3$, together with the B_{12} icosahedron and the Si_8 ethane-like cluster, plays an important role in phase formation of this type of rare-earth borocarbosilicide.

References

[1] Zhang F X, Xu F F, Leithe-Jasper A,Mori T, Tanaka T, Xu J, Sato A, Bando Y and Matsui Y 2001 *Inorg. Chem.* **40** 6948
[2] Mori T, Nishimura T, Yamaura K and Takayama-Muromachi E 2007 *J. Appl. Phys.* **101** 093714
[3] Yamanaka S, Horie H -O, Nakano H and Ishikawa M 1995 *Fullerene Sci. Technol.* **3** 21
[4] Kawaji H, Horie H -O, Yamanaka S and Ishikawa M 1995 *Phys. Rev. Lett.* **74** 1427
[5] Zhang F X, Sato A and Tanaka T 2002 *J. Solid State Chem.* **164** 361
[6] Zhang F X and Tanaka T 2003 *Z. Kristallogr. NCS* **218** 26
[7] Salvador J R, Bilc D, Mahanti S D and Kanatzidis M G 2002 *Angew. Chem. Int. Ed.* **41** 844
[8] Zhang F X, Xu F F, Mori T, Liu Q L and Tanaka T 2003 *J. Solid State Chem.* **170** 75
[9] Sheldrick G M, 1997 *SHELX97: a program for the solution and refinement of crystal structure* Universitt Gttingen, Gttingen, Germany
[10] Palmer D *CrystalMaker Version 5.0.7, CrystalMaker software* Bicester, Oxfordshire, OX6 7BS, UK

Journal of Physics: Conference Series **176** (2009) 012016 doi:10.1088/1742-6596/176/1/012016

Crystal structure and physical properties of new boride $Ca_{1+\varepsilon}Co_4B_4$

Yukari Katsura, Hiraku Ogino, Yutaka Matsumura, Shigeru Horii, Jun-ichi Shimoyama and Kohji Kishio

Department of Applied Chemistry, School of Engineering, the University of Tokyo, 7-3-1 Hongo, Bunkyo-ku, Tokyo, 113-8656, Japan

tt67114@mail.ecc.u-tokyo.ac.jp

Abstract. We report the crystal structures and physical properties of a novel boride $Ca_{1+\varepsilon}Co_4B_4$. This is a $Gd_{1+\varepsilon}Fe_4B_4$-type compound, composed of tetrahedral chains of CoB and infinite atomic chains of Ca, with independent translation periods c_{Ca} and c_{Co}. This is the first boride of this structural type that has a divalent metal at the rare earth site. The very low electrical resistivity showing linear decrease with decreasing temperature indicated a highly metallic nature of $Ca_{1+\varepsilon}Co_4B_4$. The very small magnetization suggested that $Ca_{1+\varepsilon}Co_4B_4$ is paramagnetic. First-principles calculations supported this paramagnetism, and revealed the nature of chemical bonding and electric conduction in this compound.

1. Introduction

Ternary metal borides RM_xB_y where R is an electropositive metal (usually rare earth metals) and M is a transition metal, have a wide variety of compositions and crystal structures. Alkaline earth elements are known to have similar electronegativity and ionic radii with rare earth elements. However, ternary borides containing alkaline earth elements have much extent for exploration, compared to the well-studied rare earth elements.

There exist three structural prototypes for ternary borides that has the composition RM_4B_4: $CeCo_4B_4$ [1], $LuRh_4B_4$ [2] and $NdCo_4B_4$ [3] structures. These structures are all composed of M_4 tetrahedra and B_2 dimers in different configurations. The structure of the newly synthesized $Ca_{1+\varepsilon}Co_4B_4$ is presented in figure 1. This structure belongs to $Gd_{1+\varepsilon}Fe_4B_4$ structure[4-9], which is an incommensurate variation of $NdCo_4B_4$ structure.

$NdCo_4B_4$ structure belongs to a tetragonal space group $P4_2/n$. Among above three structures, $NdCo_4B_4$ is the only structure with edge-sharing Co_4 tetrahedra forming one-dimensional tetrahedral chains of Co. These chains are bonded to adjacent chains by B_2 dimers, forming channels of CoB. A single atomic chain of R atoms is formed inside each channel. R atoms form a body-centered tetragonal lattice with neighboring R chains. The c-axis length of the R lattice (c_R) is equal to the c-axis length of the MB lattice (c_M).

In $Gd_{1+\varepsilon}Fe_4B_4$ structure, the spacing of R atomic chain is 5-15% shorter than the period of M_4 tetrahedral chains, so that c_R and c_M are independent. Therefore stoichiometry of these compounds can not be described with simple integers, and the value of ε varies from 0.06 to 0.18. These structures can be expressed as pseudo-commensurate supercells $R_p(M_4B_4)_q$ where p and q are integers.

It is difficult to clarify why some of the borides prefer commensurate structures and why others prefer incommensurate ones. The commensurate $NdCo_4B_4$ structure are observed in RCo_4B_4 (R = La, Pr, Nd) [3], $LaRu_4B_4$ [11], ROs_4B_4 (R = La-Nd, Sm, Eu, Th) [12-14] and RIr_4B_4 (R = Y, La-Nd, Sm-Tb, Th) [12-14]. On the other hand, $Pr_{1+\varepsilon}Mn_4B_4$ [15], $R_{1+\varepsilon}Fe_4B_4$ (R = Sc, Y, La-Nd, Sm, Gd-Tm, Lu) [6,8,10] and $Sm_{1+\varepsilon}Co_4B_4$ [16] are known to have the incommensurate $Gd_{1+\varepsilon}Fe_4B_4$ structure. No RM_4B_4 or $R_{1+\varepsilon}M_4B_4$ phases with alkaline earths for R have been reported so far.

The c-axis lengths of R lattices (c_R) and a-axis lengths show strong dependence on the ionic radii of R^{3+}. The c-axis lengths of MB lattices (c_M) remain almost fixed, regardless of R. The relatively short a-axis length and c_R in $Ce_{1+\varepsilon}M_4B_4$ (M = Fe [8], Os [12]) suggest that the high valence state above +3 is favored for R when M has 8 valence electrons.

Physical properties of $NdCo_4B_4$-type and $Gd_{1+\varepsilon}Fe_4B_4$-type borides have not been studied except for magnetic properties. Lack of magnetism in Fe atoms in $R_{1+\varepsilon}Fe_4B_4$ is observed from ^{57}Fe-Mössbauer studies of $R_{1+\varepsilon}Fe_4B_4$ (R = Nd, Gd). Ferromagnetism in $R_{1+\varepsilon}Fe_4B_4$ is observed only when R ions have magnetic moments due to f-electrons (R = Pr, Nd, Gd-Ho) [8,17]. Few studies were carried out on the electrical resistivity of $R_{1+\varepsilon}M_4B_4$. Electronic structure of these compounds has been completely unknown.

In this study, we discovered $Ca_{1+\varepsilon}Co_4B_4$ as the first $R_{1+\varepsilon}M_4B_4$ phase with alkaline earth element at the R site. We investigated its crystal structure, electrical and magnetic properties, and its electronic structure thus far.

2. Experimental

A mixture of Co and B powder with a molar ratio of 1:1.1 was pressed into pellet and heated in a quartz ampoule at 900-1100°C for 48-72 h to obtain CoB precursor. Boron was placed in excess to prevent formation of ferromagnetic impurities. CoB was crushed into powder, and filled in a closed Nb tube with Ca granules. The tube was pressed and then heated in a quartz ampoule at 1100°C for 48-120 h, so that liquid Ca could penetrate and diffuse into CoB.

Distribution of impurities and composition were analyzed by EDX spectra of polished surface through SEM (*KEYENCE VE7800*) equipped with EDX spectrometer (*EDAX PV7750/75ME*). Powder XRD patterns of crushed sample washed by water were obtained by using *Rigaku Ultima-IV* with Cu-K_α radiation. *RIETAN-2000* code [19] was used for Rietveld simulation.

Bright-field TEM images and electron diffraction patterns were obtained using *JEOL JEM-2010F* (200 kV) from a thin specimen prepared by *JEOL Ion Slicer*. The script on *EMS on Line* [20] was used for simulation of TEM images. Electrical resistivity of polycrystalline bulk was measured by the four-point-probe method under magnetic field of 0-5 T. Magnetization of the bulk was measured by using SQUID magnetometer (*Quantum design MPMS-XL5s*).

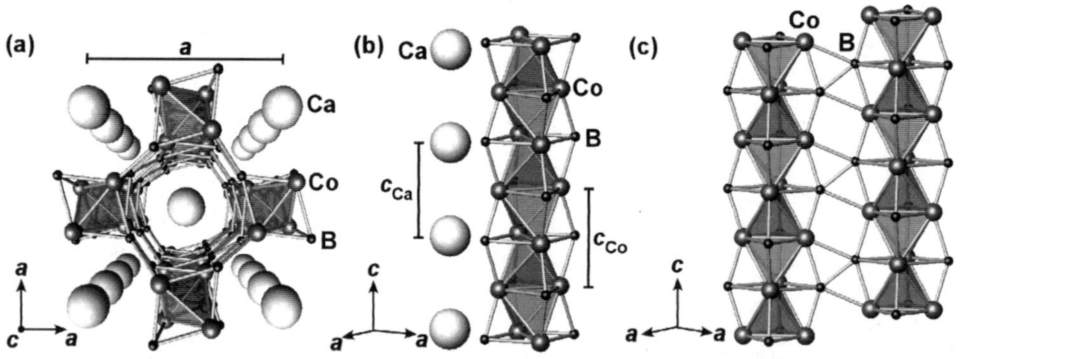

Figure 1. Crystal structure of $Ca_{1+\varepsilon}Co_4B_4$. (a) View from [001] direction. (b) Illustration of two c- axis parameters c_{Ca} and c_{Co}. (c) Connection between two tetrahedral chains. The bonds between atoms are presented for visual aid, and they do not necessarily mean presence of chemical bonds.

Journal of Physics: Conference Series **176** (2009) 012016 doi:10.1088/1742-6596/176/1/012016

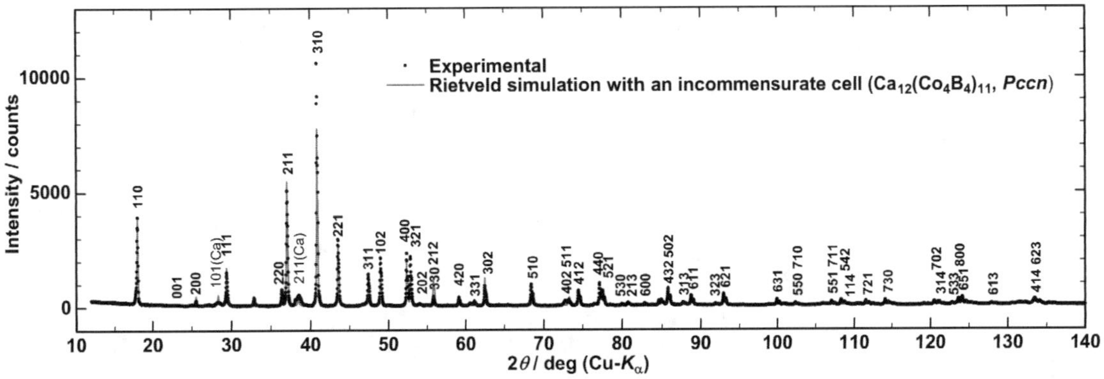

Figure 2. (colour online) Experimental and simulated powder XRD patterns of $Ca_{1+\varepsilon}Co_4B_4$.

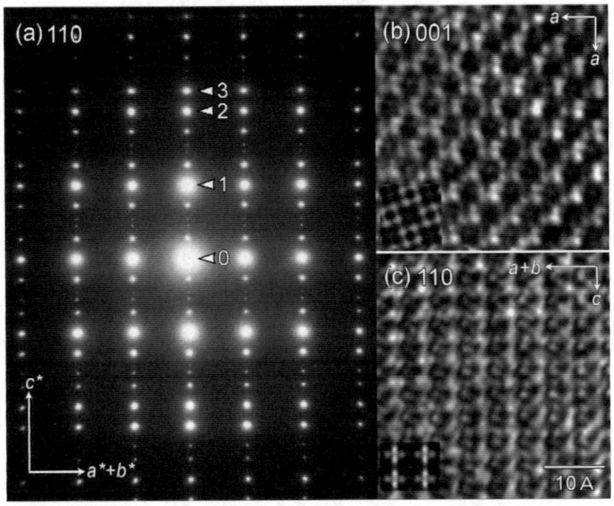

Figure 3. (a) Electron diffraction pattern from [110] direction. The diffraction spots marked with 0, 1, 2 and 3 correspond to 000, 001 and 002 reflections from CoB sublattice, and 002 reflection from Ca sublattice, respectively. Other satellite spots are produced by multiple scattering over the two sublattices.
(b)(c) Bright field TEM images from [001] and [110] directions with simulated images for commensurate $CaCo_4B_4$.

Spin-polarized first-principles calculation was carried out by FLAPW method using *WIEN2k* code [21] with GGA (PBE96) electron correlation potential. 8×8×14 *k*-point mesh was employed. Energy and charge were converged within 0.0001 Ry and 0.002*e*. Muffin-tin radii for Ca, Co and B were 2.50, 2.00 and 1.63 Bohr, respectively. Electron isosurface and Fermi surface were visualized by XCrySDen [22].

3. Results and discussion

Dark gray polycrystalline bulk was obtained by removal of Nb tube. From powder XRD pattern shown in figure 2, the phase was characterized by a tetragonal lattice with a = 6.992 Å and c = 3.861 Å. These values were similar to the lattice parameters for $NdCo_4B_4$ (a = 7.070 Å, c = 3.822 Å) [3] and the CoB sublattice in $Sm_{1+\varepsilon}Co_4B_4$ (a = 7.051 Å, c_{Co} ~ 3.871 Å) [16], implying that these peaks originate from CoB lattice in $CaCo_4B_4$ or $Ca_{1+\varepsilon}Co_4B_4$. The strong 310 reflection is a typical feature for powder diffraction patterns of $NdCo_4B_4$-type and $Gd_{1+\varepsilon}Fe_4B_4$-type compounds. Most of the peaks were successfully reproduced by Rietveld simulation of both commensurate cell $CaCo_4B_4$ and incommensurate cell $Ca_{12}(Co_4B_4)_{11}$.The presence of Ca sublattice (body-centered tetragonal, c_{Ca} ~ 3.50-3.53 Å) was implied from the diffraction peaks such as 2θ = 28.3°(101) and 38.5°(211). Although the peaks from Ca sublattice were much broader than the ones from CoB sublattice, this could be explained by low periodicity of Ca atoms in the chain.

A more convincing evidence for the presence of incommensurate sublattice was obtained from selected area electron diffraction pattern from $Ca_{1+\varepsilon}Co_4B_4$. Diffraction spots from two sublattices with different c-axis parameters c_{Co} and c_{Ca} are clearly observed in figure 3(a). Satellite spots due to multiple scattering by the two sublattices were observed along c^*-axis. Such pattern was consistent with TEM studies of $R_{1+\varepsilon}Fe_4B_4$ [7,10]. Bright-field TEM images in figures 3(b) and 3(c) from [001] and [110] directions of $Ca_{1+\varepsilon}Co_4B_4$, show a good agreement with simulated patterns.

From SEM observations of fractured surface in figure 4, grain sizes were 5-10 μm. Standardless EDX analysis on the grains showed the composition of Ca to Co ~ 1.0-1.1: 4.0, supporting that the resultant phase is $CaCo_4B_4$ or $Ca_{1+\varepsilon}Co_4B_4$. The grains are well connected, and the volume fractions of unreacted Ca metal and impurities were small enough for resistivity measurements. No contamination due to Nb tube was observed in the samples. The density of the bulk was 4.36 gcm^{-3}, which was 77% of the calculated density for $Ca_{1+\varepsilon}Co_4B_4$ (5.68 gcm^{-3}).

As shown in figure 5, the $Ca_{1+\varepsilon}Co_4B_4$ bulk exhibited highly metallic resistivity, with a slight increase in resistivity under magnetic field. Magnetic susceptibility of the best $Ca_{1+\varepsilon}Co_4B_4$ bulk was 3.0×10^{-4} (2 K, 1 T) and 1.7×10^{-4} (340 K, 1 T), suggesting that $Ca_{1+\varepsilon}Co_4B_4$ is paramagnetic. However, the gradual decrease in susceptibility at higher temperatures implied the presence of small amount of ferromagnetic impurities with Curie temperatures higher than 340 K.

First-principles calculations were carried out on a commensurate cell $CaCo_4B_4$, with addition of 0.2 electrons per formula unit to reproduce the Fermi level in the incommensurate cell. Electron isosurface of $CaCo_4B_4$ shown in figure 6(a) reveals the presence of rod-like Co-B-B-Co covalent bonds, which connect the tetrahedral chains of Co. CoB channel and Ca appear to be ionically bonded. Density of states (DOS) in figure 6(b) shows that $CaCo_4B_4$ is almost paramagnetic. Pauli paramagnetic susceptibility is estimated to be 2.0×10^{-4} at 0 K, however, this value should be lower in the sample because the DOS in the actual cell would be decreased due to lowered symmetry. The Fermi level lying in Co d-band indicates that the metallic bonds in tetrahedral chains of Co are responsible for electric conduction. Figure 6(c) shows four sets of Fermi sheets stacked along c^*-axis, suggesting anisotropy in conductivity with high conductivity along c-axis.

4. Conclusion

We investigated the crystal structure and physical properties of a novel ternary boride $Ca_{1+\varepsilon}Co_4B_4$, characterized by an incommensurate $Gd_{1+\varepsilon}Fe_4B_4$-type structure. The presence of incommensurate substructures along c-axis was confirmed by the powder XRD pattern and electron diffraction patterns of $Ca_{1+\varepsilon}Co_4B_4$. Among all $NdCo_4B_4$-based borides, this is the first compound that has a divalent metal at the rare earth site. Such displacement of the rare earth sites in ternary borides will be one of the powerful methods for finding novel ternary borides. From resistivity and magnetization measurements, $Ca_{1+\varepsilon}Co_4B_4$ was implied to be a paramagnetic metal. A preliminary first-principles calculation on the commensurate cell showed that $CaCo_4B_4$ is paramagnetic. It was also found that the tetrahedral chains of Co atoms are responsible for electric conduction, and Ca atoms are ionically bonded to the CoB channel structure. Single crystal growth of $Ca_{1+\varepsilon}Co_4B_4$ is needed for refinement of atomic positions and further studies on physical properties, especially the magnetization and anisotropic physical properties.

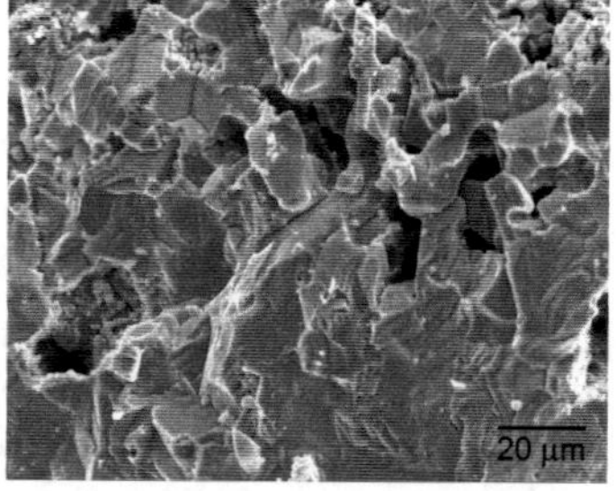

Figure 4. Secondary electron image of a fractured surface of $Ca_{1+\varepsilon}Co_4B_4$ bulk.

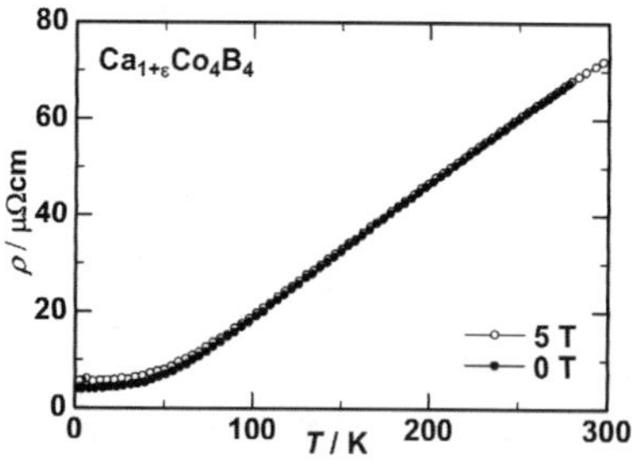

Figure 5. Temperature dependence of resistivity of $Ca_{1+\varepsilon}Co_4B_4$ bulk.

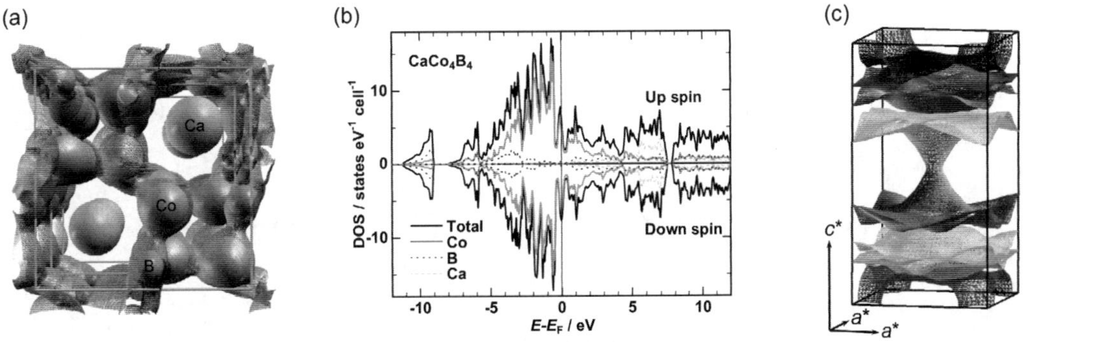

Figure 6. (colour online) (a) Electron isosurface of $CaCo_4B_4$ (b) Spin-polarized density of states of $CaCo_4B_4$ (c) Fermi surface of $CaCo_4B_4$, calculated by FLAPW method.

Acknowledgments
The authors thank T. Itoh, Y. Kakegawa, K. Ibe and H. Tsunakawa (the University of Tokyo) for their assistance on TEM observations. This work was performed under the inter-university cooperation research program of the Institute for Material Research, Tohoku University. This study was partially supported by National Institute for Materials Science (NIMS) and by the Japan Society for the Promotion of Science under the Grant-in-Aid for JSPS Fellows, and the Grant-in-Aid for Scientific Research of Priority Areas (MEXT) No.16080103.

References
[1] Kuz'ma Yu B and Bilonizhko N S 1971 *Kristallografiya* **16** 1030
[2] Yvon K and Johnston D C 1982 *Acta Crystallogr. B* **38** 247
[3] Kuz'ma Yu B and Bilonizhko N S 1978 *Dopov. Akad. Nauk Ukr. RSR, Ser. A* 275
[4] Chaban N F, Kuz'ma Yu B, Bilonizhko N S, Kachmar O O, Petriv N V 1979 *Dopov. Akad. Nauk Ukr. RSR Ser. A* **10** 873
[5] M. Sagawa M, Fujimura S, Yamamoto H, Matsuura Y and Hiraga K 1984 *IEEE Trans. Magn.* **20** 1584
[6] Givord D, Moreau J M and Tenaud P 1985 *Solid State Commun.*, **55** 303
[7] Bezinge A, Braun H F and J. Muller J 1987 *Phys. Rev. B* **36** 1406
[8] Niarchos D, Zouganelis G, Kostikas A and Simopoulos A 1986 *Solid State Commun.* **59** 389
[9] Zhao Z B and Hong S Y 1992 *J. Phys.: Condens. Matter* **4** 9221
[10] Österreicher K and Österreicher H 1987 *J. Solid State Chem.* **70** 313
[11] Grüttner A and Yvon K 1979 *Acta Cryst. B* **35** 451
[12] Rogl P 1979 *Monatsh. Chem.* **110** 235
[13] Hiebl K, Rogl P and Sienko M J 1982 *Inorg. Chem.* **21** 1128
[14] Rogl P 1980 *Monatsh. Chem.* **111** 517
[15] Zavalij P Yu, Mykhalenko S I and. Kuz'ma Yu B 1994 *J. Alloys Compd.* **203** 55
[16] Zavalij P Yu,. Kuz'ma Yu B and Zavalii L V 1990 *Kristallografiya* **35** 638
[17] Givord D, Li H S, Moreau J M and Tenaud P, *J. Magn. Magn. Mat.* 1986 **54-57** 445
[18] Buschow K H J, de Mooij D B, Daams J L C and van Noort H M 1986 *J. Less-Common Metals* **115** 357
[19] Izumi F and Ikeda T 2000 *Mater. Sci. Forum* **321-324** 198
[20] Jouneau P -H and Stadelmann P 1985 Centre Interdépartemental de Microscopie Electronique, EPFL, Lausanne
[21] Blaha P, Schwarz K, Madsen G, Kvasnicka D and Luitz J 2001 *WIEN2k, An Augmented Plane Wave+ Local Orbitals Program for Calculating Crystal Properties* (Vienna University of Technology)
[22] Kokalj A 2003 *Comp. Mater. Sci.* **28** 155

Effect of surface energy on the growth of boron nanocrystals

Wataru Hayami and Shigeki Otani

Advanced Nanomaterials Laboratory, National Institute for Materials Science, 1-1 Namiki, Tsukuba, Ibaraki 305-0044, Japan

e-mail : hayami.wataru@nims.go.jp

Abstract. The surface energies of α-rhombohedral (α-rh), β-rhombohedral (β-rh), α-tetragonal (α-t), and β-tetragonal (β-t) boron were calculated from first principles to investigate their role in nano-scale crystal growth. Equilibrium shapes of boron crystals were obtained using Wulff's theorem. Our results shows that α-t boron, despite its low cohesive energy, is more stable than the other structures as a result of its low surface energy when the number of atoms is less than about 216. Since the nanowire of α-t boron that was obtained experimentally was larger than this, it was probably in a metastable state. The difference between the surface energy of the *ab*-plane and that of the *ac*-plane explains why the α-t nanowires grow in the *c* direction.

1. Introduction

Boron is an element in Group IIIB, and is generally a semiconductive material with covalent bonds. Whereas boron forms compounds that have a variety of structures, elemental boron itself has mainly four polymorphs: α-rhombohedral (α-rh), β-rhombohedral (β-rh), α-tetragonal (α-t), and β-tetragonal (β-t). Of these, α-rh and β-rh boron have been well studied because they are fairly stable and easy to synthesize. α-rh boron is believed to be the most stable at low temperatures. It transforms into β-rh boron at about 1400 K [1]. α-t and β-t boron are less studied than α-rh and β-rh boron.

The synthesis of α-t boron was first reported in 1943 by Laubengayer *et al* [2] and later its structure was precisely determined as B_{50} by Hoard *et al* [3]. However, in 1971, studies by Amberger *et al* concluded that pure α-t boron could not be synthesized in the absence of carbon or nitrogen as an impurity [4-6]. These studies asserted that what was assumed to be α-t boron was actually $B_{50}C_2$ or $B_{50}N_2$ and the structures of these compounds were proposed. In 1992, Lee *et al* [7] showed by first-principles calculations that α-t boron could be made more stable by the inclusion of carbon, and along with their calculations of lattice parameters, they concluded that the α-t boron obtained by Hoard *et al* [3] was probably $B_{50}C_2$. It is now thought that α-t boron does not exist in pure form.

Nevertheless, we believe that this assertion is not fully confirmed. First, the original synthesis experiments [2-6] were conducted using chemical vapor deposition (CVD), and it is not known what will happen if other methods are applied. Second, the work by Lee *et al* [7] indicates only that $B_{50}C_2$ is more stable than pure B_{50}, and does not directly disprove the existence of B_{50}.

Since 2001, the synthesis of wire-like structures from boron has been reported [8-15], with some structures adopting α-t structures [11, 12, 14, 15] (table 1). These are called nanowires [11, 14], nanobelts [12], or nanoribbons [15]. For the sake of simplicity, the term "nanowire" will be used throughout this paper to refer to these structures. Yang *et al* [14] and Xu *et al* [15] reported that α-t nanowires contain impurities, a view that is consistent with the earlier results. Zhang *et al* [11], on the other hand, suggested that the α-t nanowire that they produced consisted solely of boron atoms, and

Wang *et al* [12] insisted that they did not observe any impurities in their nanowire. These two groups used the laser ablation (LA) method to create the nanowires, whereas CVD was used in the other preparations. If pure α-t boron actually exists, this is a significant result, since it has long been regarded as an impossibility.

Table 1. Experimental and calculated lattice constants of α-tetragonal boron. LA: laser ablation, CVD: chemical vapor deposition.

a (Å)	*c* (Å)	synthesis method	impurities	reference
8.84	5.00	LA without catalyst	undetected	Wang *et al* [12]
8.75	5.06	LA with catalyst	undetected	Zhang *et al* [11]
8.73	5.03	CVD	C, Cu	Yang *et al* [14]
8.79	5.09	CVD	C, O	Xu *et al* [15]
8.75	5.06	CVD	not available	Hoard *et al* [3]
8.753	5.093	CVD	C ($B_{50}C_2$)	Will *et al* [6]
8.634	5.128	CVD	N ($B_{50}N_2$)	Will *et al* [6]
8.838	4.908	calculation	none	Lee *et al* [7]
8.81	5.01	calculation	none	this work
8.76	5.03	calculation	C ($B_{50}C_2$)	this work

For the growth of nano-sized crystals, it is more likely that the surface energy makes as large a contribution as the bulk energy. When we looked at the crystal structure of α-t boron, we noticed that its (100) surface has fewer bonds than the other surfaces. In fact, on the α-t nanowire obtained by Wang *et al*, the (100) surface appears very large, suggesting that it has much lower energy than the other surfaces. This prompted us to wonder whether the α-t crystal could be stabilized by its low surface energy, even though its cohesive energy is small, suggesting instability.

There have been few studies on the surface energy of boron. This is partly because the unit cells of boron crystals are large and the optimization of the geometry of the bulk and surfaces requires prolonged computation. We performed first-principle scalculations of the surface energies of four polymorphs of boron and investigated whether a nano-sized α-t crystal could be stable [16, 17].

2. Computational Details

The calculation of the total energies and geometry optimization were performed using the CPMD code, version 3.9.1 [18, 19]. This code is based on the density functional theory with plane waves and pseudopotentials [20-22]. The norm-conserving Troullier-Martins-type pseudopotentials [23] were used. The generalized gradient approximation was included by means of the functional derived by Becke [24] and by Lee, Yang, and Parr [25]. An energy cutoff of 50 Ry for the plane-wave expansion was sufficient to provide a convergence for total energies and geometries. We have already confirmed in a previous study [26] that the total energies converged at a smaller cutoff of 40 Ry for this pseudopotential. Because the unit cell is large, the *k*-point sampling in the total energy calculation was performed using Monkhorst-Pack sampling [27] of a $(2 \times 2 \times 2)$ mesh. The results were compared with those of a finer mesh, and it was found that the difference in the total energy per atom was less than 6×10^{-4} eV. In some cases, a larger unit cell was required to construct a model of a particular surface. For example, the (111) surface of a rhombohedral structure needs a hexagonal lattice that contains three times as many atoms as the primitive unit cell. For these larger cells, the calculations were performed only at the Γ point, and a good convergence was confirmed by comparison with a $(2 \times 2 \times 2)$ mesh. Calculations were performed on a parallel computer (Hitachi SR11000) using a message-passing interface. Typically, 16 CPUs were used and the computational times for a geometry optimization were a few days to a few weeks.

To evaluate the surface energy, the total energy of the bulk was first calculated. A space of about 7 Å was then inserted at a certain cross section to generate a surface and the total energy was recalculated. The geometries were optimized in both calculations. The difference between the two total energies divided by the surface area was defined as the surface energy. The space of 7 Å was

Journal of Physics: Conference Series **176** (2009) 012017 doi:10.1088/1742-6596/176/1/012017

sufficiently wide to render the interaction between surfaces negligible. We assumed that no structural defects like twining occur.

There are an infinite number of ways of inserting surfaces, and no strict proofs exist that a given set of surfaces is sufficient to reach the minimum surface energy. However, the experience of crystallography tells us that surfaces with high Miller indices rarely appear. In addition, surfaces that cut fewer bonds have inherently less surface energy, and those surfaces that make the crystal shape round also reduce the surface area. From these considerations, we lay down the following criteria for choosing surfaces: (i) surfaces with Miller indices of up to two; (ii) surfaces cutting fewer bonds; and (iii) surfaces that make the crystal shape round. The selected surfaces (cross sections) are indicated by red lines in figure 1.

Aside from the α-rh structure, fractional occupation sites exist for boron atoms. Because the unit cells for such structures are large, the fractional occupation does not have much effect on the surface energy. We assumed the number of atoms in a unit cell to be 105 for β-rh, 50 for α-t, and 196 for β-t.

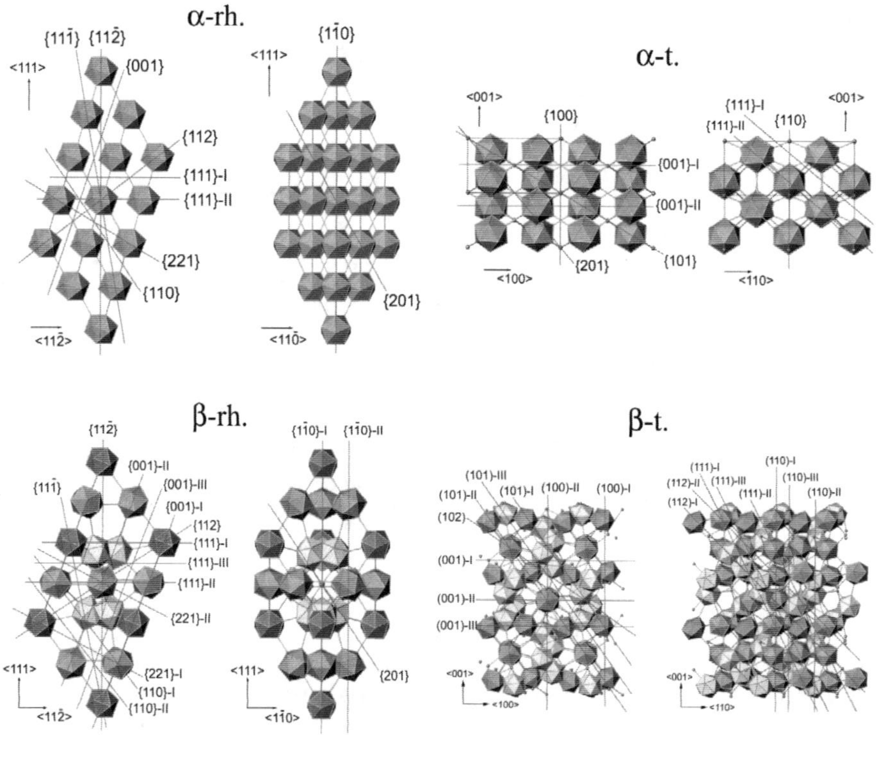

Figure 1. Atomic structures and surfaces of α-rh, β-rh, α-t and β-t boron.

3. Results and Discussion

First, we calculated the lattice constants of α-t boron and compared them with the experimental values listed in table 1. Our calculation indicates that pure α-t boron B_{50} has a longer *a* and a shorter *c* value than *a* and *c* in $B_{50}C_2$. The calculation by Lee *et al* [7] also showed a similar trend. Among the experimental lattice constants listed in table 1, the results of Wang *et al* [12], who insist that their nanowire included no impurities, show a longer *a* and a shorter *c* than the other results. This means that their α-t nanowire is almost pure, or at least much purer than the others. Their experiment adopted LA without using any catalyst, giving little probability that impurities had contaminated their sample.

The calculated surface energies are exhibited in figure 2. The minimum-energy surfaces in the same Miller index group are excerpted from references [16, 17]. It is seen that α-t boron has the lowest average surface energy, as we had expected. Notably, the four main surfaces of (100), (001), (101), and (110) are the lowest of all the surfaces. β-t boron is the second lowest on average, followed by β-rh and α-rh boron. The surface energies of β-rh and α-rh have certain similarities, probably due to structural parallels. Both have a rhombohedral symmetry, and the orientations of the B_{12} icosahedra are the same, except that β-rh boron has additional B_{28} subunits at the center.

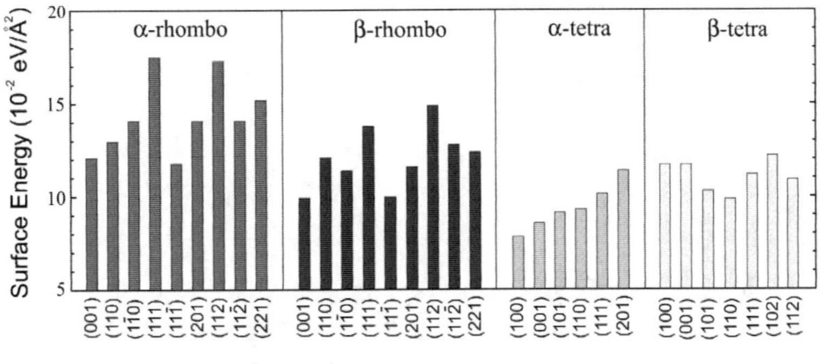

Figure 2. Calculated surface energies of boron.

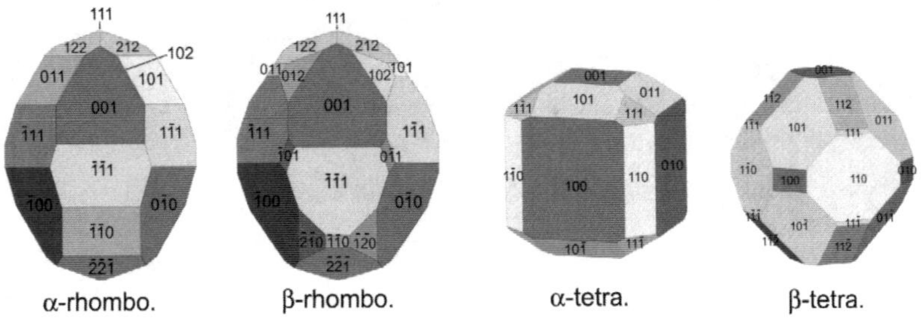

Figure 3. Equilibrium shapes of boron crystals calculated from Wulff's theorem. The surfaces are labeled with Miller indices and equivalent surfaces in a crystal are the same color.

When all the surface energies of a crystal are known, the crystal shape that minimizes the total surface energy at a constant volume is given by Wulff's theorem [28, 29]. The theorem states that the distance from the crystal center to a surface is proportional to its surface energy per unit area. Low-energy surfaces are closer to the center and, as a result, exhibit a larger surface area. The crystal shapes of the four boron polymorphs obtained by Wulff's theorem are shown in figure 3. The surfaces are labeled by their Miller indices. Equivalent surfaces are given the same color. It is interesting that the (100) surface appears very large in α-t boron, as was observed in the experiment by Wang *et al* [12]. Its rectangular shape would appear to make it suitable for nanowire growth. β-t boron has a round shape which results from its isotropic surface energies. The structure of β-t boron is rather complicated, since it includes 196 atoms per unit cell, which means that number of cut bonds does not differ very much between each surface. The shapes of β-rh and α-rh boron are similar to each other as would be expected from their surface energies. β-rh boron has additional small faces because its B_{28} subunits make the surface energy more isotropic than that of α-rh boron.

Once the crystal shape has been determined, it is easy to calculate the total surface energy. When the number of atoms is N, the surface area and energy are proportional to $N^{2/3}$, since the surface area is proportional to $v^{2/3}$, where v is the crystal volume, and v is proportional to N. The total energy of the crystal E_{tot} can be expressed as

$$E_{tot} = -C_v N + C_s N^{2/3} . \quad (1)$$

The first term is the bulk part and C_v is equal to the cohesive energy per atom. The second term is the surface term. C_s can be calculated from the surface energy, crystal volume, and atom density. The coefficients C_v and C_s are listed in table 2. A comparison of C_v values shows α-rh boron to be the most stable and α-t boron the most unstable. On the other hand, a comparison of C_s values shows α-t boron to have an extremely low (stable) value, with α-rh being the most unstable.

Table 2. Calculated coefficients C_v, C_s and atom densities used for the calculations.

	C_v (eV/atom)	C_s (eV/atom$^{2/3}$)	atom density (atoms/Å^3)
α-rh	5.98	2.48	0.137
β-rh	5.97	2.16	0.128
α-t	5.91	1.78	0.129
β-t	5.96	2.08	0.130

Table 3. Critical sizes of boron crystals calculated with equation (2).

	α-t	β-t	β-rh	α-rh
α-t	---	216	254	1000
β-t	0	---	512	8000
β-rh	0	0	---	32800
α-rh	0	0	0	---

The total energy of the crystal is dominated by C_v when N is large, and by C_s when N is small. When there are two phases of α and β satisfying $C_v^\alpha > C_v^\beta$, $C_s^\alpha > C_s^\beta$, where the superscripts α and β correspond to each phase, phase α is more stable for large N and phase β for small N. The critical size, N_c, at which the total energies of the two phases become equal is calculated from equation (1) as

$$N_c = \{ (C_s^\alpha - C_s^\beta) / (C_v^\alpha - C_v^\beta) \}^3. \quad (2)$$

The critical sizes for each boron polymorph are listed in table 3. For α-t boron, the N_c against β-t (216) is lower than that against β-rh (254) and α-rh (1000). As a result, α-t boron is the most stable when N is between 0 ~ 216. In the same way, β-t, β-rh and α-rh boron are the most stable between 216 ~ 512, 512 ~ 32800 and 32800 ~, respectively. β-t boron has a rather narrow region, which would make it hard to synthesize. β-rh and α-rh boron have a wide stable region, suggesting that they are easy to synthesize.

The thickness of nanowires obtained in the experiment by Wang *et al* [12] was several tens of nanometers. This is much greater than the critical size of the α-t crystal, even if the estimated value includes some degree of error. Therefore, if the nanowire is pure α-t boron, it must be in a metastable state. In an equilibrium state, according to Wulff's theorem, the aspect ratio *c/a* of an α-t crystal shape is equal to the ratio between the surface energies of the (001) and the (100) surfaces, and is estimated from the values in figure 2 to be 1.09. In real nanowires, the aspect ratio is much larger than this estimated value, which makes it likely that α-t nanowires grow via a non-equilibrium process. Below we propose some possible growth mechanisms for α-t nanowires.

The first possibility is that α-t nanowires grow via a vapor-solid (VS) process. In this case, the growth rate is dominated by the sticking probability of vapor atoms to the surface rather than by the surface energy itself. The difference between the sticking probabilities of the (001) and the (100) surfaces could be greater than the difference between their surface energies, with the result that nanowires should grow more rapidly in the <001> direction than in the <100> direction. The second possibility is that α-t nanowires grow via a vapor-liquid-solid (VLS) process. In this case, the nanowires would grow in the direction normal to the surface on which a liquid droplet preferably remains. The binding energy between the droplet and the surface determines which surface is preferred. It is difficult to calculate the binding energy because the droplet is in a liquid state. It is, however, logical to expect binding energy to be correlated to the surface energy, since the surface structures are identical. A surface with a higher surface energy would have a higher binding energy to the droplet, so that the droplet would prefer the (001) surface to the (100) surface.

In summary, the following processes are suggested as the growth mechanism of pure α-t nanowires.

(i) First, an equilibrium nucleation takes place of an α-t boron crystal that is smaller than $N_c = 216$.
(ii) Second, based on this nano-sized nucleus, the crystal, in a metastable state, grows more rapidly in the <001> direction than in the <100> direction. This may be due to the difference between sticking probabilities (VS process) or between the binding energies of the droplet on the surface (VLS process).

In the experiment by Wang *et al* [12], the dimensions of the nanowire were not actually tetragonal, i.e., *a* *b*. The reason for this cannot be explained by our calculations, since the surface energies of the *a* and *b* surfaces in pure α-t boron are identical. We speculate that the lack of tetragonal dimensions may be due to the presence of traces of impurities below the detectable level or to the experimental conditions, such as the angle of the laser beam.

The reason that pure α-t is formed only in gas-phase syntheses, such as LA, is that in the gas phase, the fluctuations in the atom density are large, allowing small ($N < N_c$) clusters to exist as isolated entities. We speculate that LA, where a high temperature and a low density are achieved, is superior to CVD for the synthesis of α-t boron crystals.

4. Conclusions

It has been proven that pure α-t boron, despite its low cohesive energy, can exist as a nano-sized crystal because of its low surface energy. The calculated critical (maximum) number of atoms in the α-t crystal is 216. Since the α-t nanowire obtained in the experiment is much larger than this, it is probably in a metastable state. The difference in the surface energies explains why the nanowire grows in the *c* direction. The calculated lattice constants of pure α-t boron and α-t $B_{50}C_2$ also agree with those determined experimentally by Wang *et al* [12], in which pure α-t has a longer *a* and a shorter *c* than α-t $B_{50}C_2$. We propose a growth mechanism for α-t boron from the viewpoint of surface energies. Because boron clusters need to be isolated to form an equilibrium α-t nucleus, laser ablation, in which the fluctuation of density is large, would favor the synthesis of α-t boron.

References

[1] Runow P 1972 *J. Mater. Sci.* **7** 499
[2] Laubengayer A W, Hurd D T, Newkirk A E and Hoard J L 1943 *J. Am. Chem. Soc.* **65** 1924
[3] Hoard J L, Hughes R E and Sands D E 1958 *J. Am. Chem. Soc.* **80** 4507
[4] Amberger E and Ploog K 1971 *J Less. Common. Metals* **23** 21
[5] Ploog K, Schmidt H, Amberger E, Will G and Kossobutzki K H 1972 *J. Less. Common. Metals* **29** 161
[6] Will G and Kossobutzki K H 1976 *J. Less. Common. Metals* **47** 33.
[7] Lee S, Bylander D M, Kim S W and Kleinman L 1992 *Phys. Rev. B* **45** 3248
[8] Wu Y, Messer B and Yang P 2001 *Adv. Mater.* **13** 1487
[9] Cao L, Zhang Z, Sun L, Gao C, He M, Wang Y, Li Y, Zhang X, Li G, Zhang J and Wang W 2001 *Adv. Mater.* **13** 1701
[10] Otten C J, Lourie O R, Yu M, Cowley J M, Dyer M J, Ruoff R S and Buhro W E 2002 *J. Am. Chem. Soc.* **124** 4564.
[11] Zhang Y, Ago H, Yumura M, Komatsu T, Ohshima S, Uchida K and Iijima S 2002 *Chem. Comm.* **2002** 2806
[12] Wang Z, Shimizu Y, Sasaki T, Kawaguchi K, Kimura K and Koshizaki N 2003 *Chem. Phys. Lett.* **368** 663
[13] Wang Y Q and Duan X F 2003 *Appl. Phys. Lett.* **82** 272
[14] Yang Q, Sha J, Xu J, Ji Y J, Ma X Y, Niu J J, Hua H Q and Yang D R 2003 *Chem. Phys. Lett.* **379** 87
[15] Xu T T, Zheng J G, Wu N, Nicholls A W, Roth J R, Dikin D A, Ruoff R S 2004 *Nano Lett.* **4** 963
[16] Hayami W and Otani S 2007 *J. Phys. Chem. C* **111** 688
[17] Hayami W and Otani S 2007 *J. Phys. Chem. C* **111** 10394
[18] CPMD, Copyright IBM Corp. 1990-2004, Copyright MPI fur Festkorperforschung Stuttgart 1997-2001, http://www.cpmd.org/
[19] Marx D and Hutter J 2000 *NIC Series* Vol.1 John von Neumann Institute for Computing: Jülich

pp301-449
[20] Hohenberg P and Kohn W 1964 *Phys. Rev.* **136** B864
[21] Kohn W and Sham L J 1965 *Phys. Rev.* **140** A1133
[22] Car R and Parrinello M 1985 *Phys. Rev. Lett.* **55** 2471
[23] Troullier N and Martins J L 1991 *Phys. Rev. B* **43** 1993
[24] Becke A D 1988 *Phys. Rev. A* **38** 3098
[25] Lee C, Yang W and Parr R G 1988 *Phys. Rev. B* **37** 785
[26] Hayami W, Tanaka T and Otani S 2005 *J. Phys. Chem. A* **109** 11975
[27] Monkhorst H J and Pack J D 1976 *Phys. Rev. B.* **13** 5188
[28] Wulff G Z. 1901 *Kristallogr.* **34** 449
[29] Benson G C and Patterson D 1955 *J. Chem. Phys.* **23** 670

Journal of Physics: Conference Series **176** (2009) 012018 doi:10.1088/1742-6596/176/1/012018

Si-doping effect on bonding nature and elasticity of $AlMB_{14}$ with M=Li, Mg, and Na

Ryoji Sahara[1], Toetsu Shishido[1], Akiko Nomura[1], Kunio Kudo[2], Shigeru Okada[3], Vijay Kumar[4], Kazuo Nakajima[1] and Yoshiyuki Kawazoe[1]

[1]Institute for Materials Research (IMR), Tohoku University, Aoba-ku, Sendai 980-8577, Japan
[2]Faculty of Engineering, Kanagawa University, Kanagawa-ku, Yokohama 221-8686, Japan
[3]Faculty of Engineering, Kokushikan University, Tokyo 154- 8515, Japan
[4]Dr. Vijay Kumar Foundation, 1969 Sector 4, Gurgaon 122001, Haryana, India, Institute for Materials Research (IMR), Tohoku University, Aoba-ku, Sendai 980-8577, Japan

E-mail: `sahara@imr.edu`

Abstract. We study the effects of Si doping on the bonding nature and elastic properties of $AlMB_{14}$ compounds having B_{12} icosahedral clusters with M = Li, Mg, and Na, by means of *ab initio* calculations. We find that Si substitution leads to a decrease in the cohesive energy as well as in the values of the bulk, shear, and Young's moduli.

1. Introduction

Boron-rich compounds containing B_{12} icosahedral clusters are of great technological interest because of their high hardness. A variety of compounds have been made by doping metal/carbon species in the void space between the boron icosahedra. Among these, the $AlMB_{14}$-type compounds are simple, and many experimental studies have been carried out in recent years [1, 2, 3] after the discovery [4] of $AlMgB_{14}$. However, theoretical studies of these compounds are few[5, 6]. One reason for the slow progress has been the large number of atoms (64) in the unit cell of this compound as shown in figure 1. This makes *ab initio* calculations of such materials computationally very demanding. Also there are no experimental reports of elastic properties because it is difficult to make single crystals of these compounds. With the development of faster computers, calculations of such large pure and doped systems can be done now routinely. Of particular interest is the doping of Si that has been reported to enhance hardness of $AlMgB_{14}$ compound [7]. However, it is not yet understood what leads to an increase in the hardness. Here we report calculations on the effects of Si doping on the structural, electronic, and the elastic properties of such boron rich compounds.

The unit cell of $AlMB_{14}$ compound (figure 1) has four formula units (64 atoms) in an orthorhombic structure with space group *Imma*. Each B_{12} icosahedron is constructed of four crystallographically independent boron atom sites, namely two B1, four B3, four B4, and two B5 sites. There are eight boron sites in the unit cell denoted as B2 type[3] that lie in between the B_{12} icosahedra. Each boron atom on an icosahedron is connected to another boron atom either on a neighboring icosahedron or with a bridging B2 atom. Changing the M atoms such as Li or Mg, as well as the doping of Si, affects the number of valence electrons in the unit cell and possibly the volume. It could also affect the local bonding nature particularly around the Si sites.

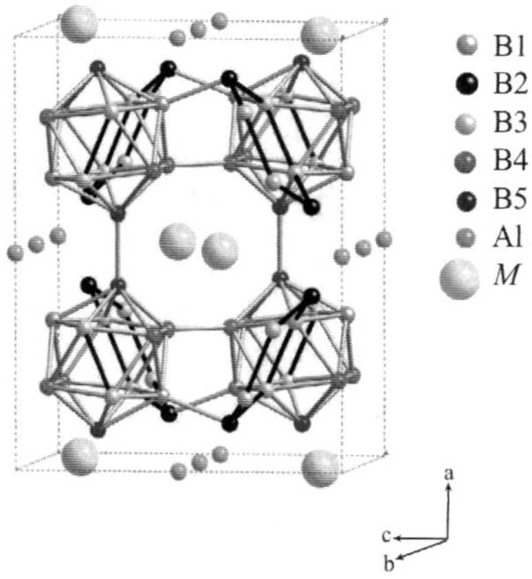

Figure 1. Unit cell of $AlMB_{14}$.

In an earlier paper, we have reported the results of *ab initio* calculations on the atomic and electronic structures as well as the elastic properties of $AlMB_{14}$, M = Li, Na, and Mg compounds. We showed that Mg compounds are metallic while Li and Na compounds are semiconducting with quite different band gaps. It was also shown that there exists an asymmetric covalent bonding between B_{12} icosahedra that, together with a void along the b axis, leads to an asymmetric dependence of the elastic properties when M is varied. We also explored the changes in the properties when Al or M atoms were replaced by Si [8]. *Ab initio* studies of Si doping in $AlMgB_{14}$ by other researchers [7] have been carried out using the local density approximation, and it has been shown that addition of Si decreases the elastic properties. This contradicts to the experimental results that have shown an increase in hardness by addition of Si. In these studies by Cook *et al.* Si substitution was done on Al sites[7] and the possibility of substitution on other sites was not explored.

The aim of this paper is to understand the effects of Si doping on the bonding character in boron-rich $AlMB_{14}$ compounds with M = Li, Mg, and Na and to explore substitution of Si atoms on Al or M or B1, B2, or B3 sites. Furthermore we also varied the number of Si atoms in the unit cell up to a maximum of sixteen atoms and analyzed the cohesive energy, the density of states, and the charge density to understand the changes due to Si substitution. We also explored Si atom in an interstitial site as there is a void in the unit cell. In Sec. 2, we discuss our calculation procedure. The results are given in Sec. 3 and a summary in Sec. 4.

2. Calculation

We follow the procedure of ref. [8] and use the projector augmented wave method [9, 10] as implemented in the Vienna *ab initio* simulation package [12, 11]. The exchange-correlation energy is calculated within the generalized gradient approximation [13]. The free energy is minimized over the degrees of freedom of both the electron density and the ionic positions using conjugate-gradient iterative minimization technique. The cut-off energy for the plane wave expansion is taken as 477.97 eV. The Brillouin zone integrations are performed using a set of $8 \times 8 \times 8$ **k**-points in all cases. Direct optimization of the cell gave the lattice constants, a, b, and c. The Murnaghan equation of state [14] is used to determine the bulk modulus, B_M, for a single crystal by fitting the total energy at various volumes of the cell. For orthorhombic crystals, we calculated nine elastic constants, C_{11}, C_{22}, C_{33}, C_{44}, C_{55}, C_{66}, C_{12}, C_{13}, and C_{23}, and the derivative of the energy as a function of the lattice strain [15] δ which is taken as 0.005 and 0.01 in all cases.

For a statistically isotropic polycrystalline single phase material, the bulk modulus B and the shear modulus, G are related to the elastic constants following the Voigt approximation [16] or the Reuss approximation [17]. The Voigt approximation assumes a uniform strain and gives the bulk modulus, B_v, and the shear modulus, G_v, as $B_v = 1/9(C_{11}+C_{22}+C_{33})+2/9(C_{12}+C_{23}+C_{13})$ and $G_v = 1/15(C_{11}+C_{22}+C_{33}) - 1/15(C_{12}+C_{13}+C_{23}) + 1/5(C_{44}+C_{55}+C_{66})$. The Reuss approximation assumes a uniform stress and gives the bulk modulus, B_R, and the shear modulus, G_R, as $B_R = 1/((s_{11}+s_{22}+s_{33}) + 2(s_{12}+s_{13}+s_{23}))$ and $G_R = 15/(4(s_{11}+s_{22}+s_{33}) - 4(s_{12}+s_{13}+s_{23}) + 3(s_{44}+s_{55}+s_{66}))$. Here s_{ij} are the compliance constants. These equations give the theoretical maximum (Voigt approximation) and minimum (Reuss approximation) values of the average isotropic elastic moduli of a polycrystalline sample using the anisotropic single crystal elastic constants. Finally we arithmetically averaged the values obtained from Voigt approximation and Reuss approximation following Hill *et al.* [18] to estimate the elastic properties of a polycrystalline sample. Accordingly the bulk modulus and shear modulus of a polycrystalline sample are obtained from $B = (B_v + B_R)/2$ and $G = (G_v + G_R)/2$, respectively. The Young's modulus, Y, and the Poisson ratio, ν, are estimated from $Y = 9BG/(3B+G)$ and $\nu = (3B-2G)/2(3B+G)$.

In order to understand the effects of Si doping, we considered model systems in which Si is substituted on Al, M, B_1, B_2, or B_3 sites. We explored substitution on Al sites in detail taking $Al_{4-x}Si_xM_4B_{56}$ compounds with x = 1-4. For x = 1 and 2 we further studied different possible Al sites in order to find the minimum energy configuration. For Al and Li sites, the maximum number of Si atoms considered is four, whereas for Si on boron sites (B_1, B_2, or B_3), the maximum number of Si atoms is sixteen. Furthermore, Si doping at interstitial site (0.5 0.5 0.5) is also treated in the case M = Mg. In each case, the structure is reoptimized. For comparison, the cases without Si doping are also shown by taking results from ref. [8].

3. Results and discussion

Tables 1 - 3 show the calculated cohesive energy, optimized lattice constants, a, b, and c, and the bulk modulus, B_M for M = Li, Mg, and Na, respectively, obtained using Murnaghan equation of state for some configurations. Note that there is slight lattice distortion when adding one Si atom.

From table 1, it is clear that in all cases, Si substitution lowers the cohesive energy and with increasing x, it decreases more significantly. By comparing the cases of Si substitution on either bridging B2 or B1/B3 sites that construct boron icosahedral cluster and connect with bridging B2 sites, we find that among these boron sites, the B3 site is energetically most favorable. However, these are not the best sites for Si doping. Substitution on M = Li or Mg site costs the least energy, while for M = Na, substitution on Al site is energetically the best. Furthermore, by comparing the energies in the cases of one Si substitution on Al sites, we find that the cohesive energy is nearly the same for all Al sites. However, when two Si atoms replace two Al atoms, substitution on two neighboring Al sites in the unit cell ($Al_2Li_4Si_2B_{56}$-2) is most favorable. In the case of substitution on Li sites, we considered the compound $Al_4Li_{4-x}Si_xB_{56}$ with x = 1. The cohesive energy decreases compared with the undoped case. The substitution of a Si atom on a Li site is favorable by 0.003 eV/atom as compared to substitution on an Al site. The lattice constants a, b, and c for substitution on an Li site are 10.37, 5.88, and 8.14 Å, respectively as compared to the values 10.35, 5.85, and 8.14 Å for the undoped case. Therefore, there is a slight expansion of the lattice by Si substitution on Li sites, while Si substitution on Al sites decreases the lattice volume compared with the undoped case. However, the volume change is quite small for substitution on Al or M sites, and therefore such sites are the likely candidates for

Table 1. Cohesive energy, lattice constants, and the bulk modulus, in the case of M=Li

number and site of Si substitution	configuration	cohesive energy (eV/atom)	a (Å)	b (Å)	c (Å)	B_M (GPa)
0	$Al_4Li_4B_{56}$ [8]	6.100	10.350	5.850	8.140	211
one Al-site	$Al_3Li_4Si_1B_{56}$-1	6.079	10.336	5.856	8.138	-
one Al-site	$Al_3Li_4Si_1B_{56}$-2	6.079	10.336	5.856	8.138	-
one Li-site	$Al_4Li_3Si_1B_{56}$	6.082	10.373	5.882	8.136	191
one B1-site	$Al_4Li_4Si_1B_{55}$	6.021	10.410	5.898	8.173	-
one B2-site	$Al_4Li_4Si_1B_{55}$	6.030	10.427	5.939	8.161	-
one B3-site	$Al_4Li_4Si_1B_{55}$	6.032	10.411	5.914	8.153	-
two Al-sites	$Al_2Li_4Si_2B_{56}$-1	6.060	10.314	5.862	8.138	-
two Al-sites	$Al_2Li_4Si_2B_{56}$-2	6.066	10.332	5.855	8.151	-
two Al-sites	$Al_2Li_4Si_2B_{56}$-3	6.061	10.314	5.860	8.138	-
two Al-sites	$Al_2Li_4Si_2B_{56}$-4	6.061	10.314	5.860	8.138	-
four Al-sites	$Al_0Li_4Si_4B_{56}$	6.034	10.295	5.854	8.163	191
four Li-sites	$Al_4Li_0Si_4B_{56}$	6.028	10.467	5.920	8.139	179
eight B2-sites	$Al_4Li_4Si_8B_{48}$	5.630	10.785	6.528	8.298	144

Table 2. Cohesive energy, lattice constants, and the bulk modulus, in the case of M = Mg

number and site of Si substitution	configuration	cohesive energy (eV/atom)	a (Å)	b (Å)	c (Å)	B_M (GPa)
0	$Al_4Mg_4B_{56}$ [8]	6.006	10.360	5.910	8.120	211
one Al-site	$Al_3Mg_4Si_1B_{56}$-1	5.993	10.346	5.911	8.137	197
one Al-site	$Al_3Mg_4Si_1B_{56}$-2	5.993	10.346	5.911	8.137	197
one Mg-site	$Al_4Mg_3Si_1B_{56}$	6.001	10.370	5.906	8.150	196
one B1-site	$Al_4Mg_4Si_1B_{55}$	5.930	10.430	5.935	8.179	194
one B2-site	$Al_4Mg_4Si_1B_{55}$	5.936	10.461	5.973	8.173	191
one B3-site	$Al_4Mg_4Si_1B_{55}$	5.936	10.441	5.951	8.159	194
eight B1-sites	$Al_4Mg_4Si_8B_{48}$	5.554	10.863	6.322	8.766	150
eight B2-sites	$Al_4Mg_4Si_8B_{48}$	5.554	10.862	6.322	8.764	150
eight B3-sites	$Al_4Mg_4Si_8B_{48}$	5.548	10.949	6.127	8.621	171
sixteen B3-sites	$Al_4Mg_4Si_16B_{40}$	5.232	11.034	6.254	9.602	143
one interstitial Si	$Al_4Mg_4Si_1B_{56}$	5.921	10.545	5.922	8.204	181

Si substitution. Note that the substitution on boron sites leads to a significant lattice expansion.

The bulk modulus, B_M, decreases to 191 GPa by changing one Si atom on Li site. Such a decrease was also found by Lowther for Si substitution in $AlMgB_4$. However, in this study Si was substituted on an Al site. Our calculations show that Si substitution is energetically more favorable on Mg site, though there is a small decrease in the cohesive energy as compared to the undoped case. The elastic properties also show a decrease and there is a small expansion in the unit cell volume. Furthermore we find that in general, an increase in the number of Si atoms decreases B_M more significantly. When adding one interstitial Si atom at (0.5 0.5 0.5) site of $Al_4Mg_4B_{56}$, the bulk modulus and the cohesive energy per atom again decrease compared with the case of no interstitial Si addition. For example, the B4-B4, B5-B5, and B4-B5 bond lengths increase from 1.758, 1.722, and 1.789 to 1.799, 1.829, and 1.830, respectively. An increase in the B-B bond lengths affects the bonding between boron atoms and decreases the bulk modulus.

Table 3. Cohesive energy, lattice constants, and bulk modulus, in the case of M = Na

number and site of Si substitution	configuration	cohesive energy (eV/atom)	a (Å)	b (Å)	c (Å)	B_M (GPa)
0	$Al_4Na_4B_{56}$ [8]	6.045	10.470	5.850	8.230	209
one Al-site	$Al_3Na_4Si_1B_{56}$-1	6.025	10.455	5.851	8.235	-
one Al-site	$Al_3Na_4Si_1B_{56}$-2	6.025	10.455	5.851	8.235	-
one Na-site	$Al_3Na_4Si_1B_{56}$	6.022	10.449	5.856	8.229	189
one B1-site	$Al_4Na_4Si_1B_{55}$	5.965	10.527	5.893	8.272	-
one B2-site	$Al_4Na_4Si_1B_{55}$	5.975	10.549	5.930	8.271	-
one B3-site	$Al_4Na_4Si_1B_{55}$	5.976	10.528	5.909	8.251	-
eight B2sites	$Al_4Na_4Si_8B_{48}$	5.561	10.950	6.477	8.530	144

In the row Li, Na, or Mg to Si, the $3p$ orbitals start playing an increasingly important role. Therefore, it is useful to analyze the electronic structure and the charge density distribution. Considering substitution on boron sites, we find that for all M, replacing eight bridging B2 atoms in the unit cell by Si leads to a significant decrease in the cohesive energy; the lattice constants increase and the bulk modulus B_M decreases. In order to further check particularly the case of M = Mg, we replaced eight boron atoms on B1 or B3 sites with Si. For substitution on boron sites, the highest B_M is realized when Si is substituted on B3 sites. However, the absolute value of B_M decreases compared with the undoped case. The decreasing tendency of B_M with increasing x became more significant when all the sixteen boron atoms on B3 sites are replaced by Si. In this case with M = Mg, B_M becomes 143 GPa as compared to the value of 211 GPa for the undoped case.

In order to understand the effect of Si doping on the elastic properties, we have given in table 4 the elastic constants, the bulk, shear and Young's moduli, and the Poisson ratio. The bulk modulus obtained using the Murnaghan equation of state is given in table 1. In order to obtain shear modulus, we considered a hypothetical case where all M or all Al atoms are replaced by Si in order to keep the same symmetry. That is, we calculated the elastic constants in the cases of $Al_4Si_4B_{56}$ and $Si_4Li_4B_{56}$. In general, our results show that all the elastic constants, shear, bulk, and Young's moduli decrease by replacing M or Al by Si compared to the values of Al$M$$B_{14}$, which can be seen in tables 1-3 of this paper and table 2 of ref. [8]. However, the values for $Si_4Li_4B_{56}$ are significantly higher than in case of $Al_4Si_4B_{56}$. Therefore, our results suggest that the effect of Si doping is more than a substitutional effect from the point of view of observed increase in hardness.

In order to understand the difference in the bonding character when Si is substituted, we have plotted in figure 2 the partial and the total densities of states (DOS) for the case of one Si atom doped on a Li site, namely for the compound $Al_4Li_3SiB_{56}$. The main contribution to the total DOS comes from $2s$ and $2p$ orbitals of boron atoms, which hybridize strongly on the nearest neighbor sites and have three center bonding on boron icosahedra as in undoped $AlLiB_{14}$ [8]. However, in contrast to $AlLiB_{14}$, for which there is little charge around the Li sites, as one can see in figure 2, there is significant density of states on Si in the occupied region for the Si doped system. The Fermi energy lies in a band and therefore, similar to the case of M = Mg, an increase in the number of electrons by substitution of Si makes the $AlLiB_{14}$ compound metallic. That is, substitution of Si could result in different transport properties of these compounds leading to new applications.

Figure 3 shows the isocontours of the total charge density and the difference in the charge densities of $Al_4Li_3SiB_{56}$ and the sum of the charge densities of a hypothetical case of $Al_4Li_3B_{56}$ (Si atom removed leaving the positions of the remaining atoms unchanged) and an isolated Si

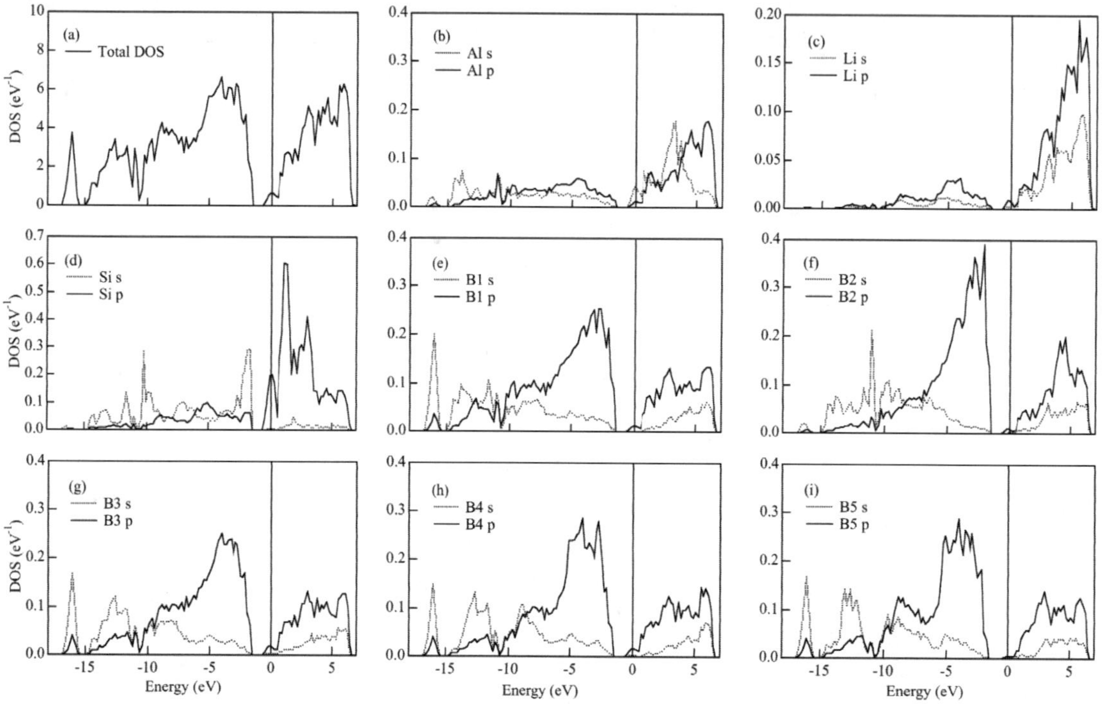

Figure 2. (a) Total DOS and (b-i) site-projected partial DOS on Al, Li, Si, B1, B2, B3, B4, and B5 atoms for the Si doped case ($Al_4Li_3SiB_{56}$), respectively. The Fermi energy is at the zero of energy.

Table 4. Elastic properties obtained by the present calculations in the unit of GPa

	$Al_4Si_4B_{56}$	$Si_4Li_4B_{56}$		$Al_4Si_4B_{56}$	$Si_4Li_4B_{56}$
C_{11}	412	447	B_v	179	193
C_{22}	474	517	B_R	177	191
C_{33}	394	439	B	178	192
C_{44}	180	197	G_v	173	181
C_{55}	164	135	G_R	171	175
C_{66}	48	162	G	172	175
C_{12}	44	27	Y	391	408
C_{13}	48	42	ν	0.13	0.15
C_{23}	72	98			

atom on the same position. The total charge density (figure 3a) has overall similar behavior as in $AlLiB_{14}$. There is a 3-D network of strongly bonded boron icosahedra and, as one could expect from the partial DOS of Si, we find electronic charge around Si ions. There is excess charge between neighboring icosahedra (figure 3b) primarily along a direction resulting from a contraction of charge (see figure 3 for the depletion of charge) between faces of two neighboring icosahedra that have B1, B3, and B5 atoms. Therefore, our results suggest that Si doping enhances inter-cluster covalent bonding in one direction.

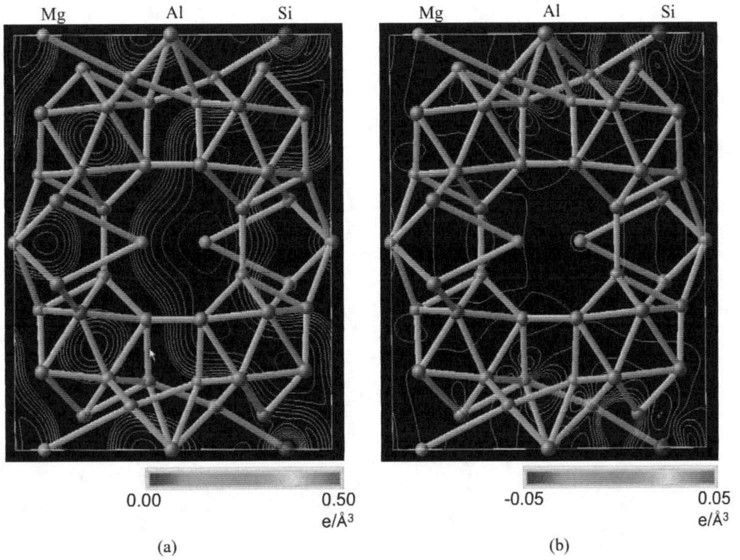

Figure 3. (a) Isocontours (values from 0 to 0.05 e/Å^3 with an increment of 0.005 e/Å^3) of the total electronic charge density for Si doped case ($Al_4Li_3SiB_{56}$), (b) isocontours (values from -0.05 to +0.05 e/Å^3 with an increment of 0.005 e/Å^3) of the difference between the charge density of $Al_4Li_3SiB_{56}$ and the such of the charge densities of a hypothetical case of $Al_4Li_3B_{56}$ and an isolated Si atom (positive and negative value show excess and depletion, respectively).

4. Summary

In summary, we have studied the effects of Si doping in Al$M$$B_{14}$ compounds having B_{12} icosahedral clusters with M = Li, Mg and Na by means of first principles calculations. By replacing Si atoms systematically on Al, M, B1, B2, or B3 sites and estimating the cohesive energy and the elastic properties, we find that Si substitution leads to a decrease in the cohesive energy as well as in the values of the bulk, shear, and Young's moduli.

Acknowledgements

We are grateful to the staff of the Center for Computational Materials Science of IMR, Tohoku University for allowing the use of the Hitachi SR- 11000 51/K2 supercomputing facility and their kind support. VK thankfully acknowledges the support of M.W. Chen and the kind hospitality at the International Frontier Center for Advanced Materials (IFCAM) of IMR as well as the support from Asian Office of Aerospace Research and Development.

References

[1] Higashi I and Ito T 1983 *J. Less-Common Metals* **92** 239
[2] Kudou K, Okada S, Mori T, Iizumi K, Shishido T, Tanaka T, Higashi I, Nakajima K, Rogl P, Andersson Y B and Lundström T 2002 Jpn. J. Appl. Phys. **41** L928
[3] S. Okada, T. Tanaka, A. Sato, T. Shishido, K. Kudou, K. Nakajima, and T. Ludstrom 2005 *J. Alloys Compd.* **395** 231
[4] Zhdanov G S and Sevastyanov N G 1941 *Compt. Rend. Acad. Sci. USSR* **32** 432
[5] Lee Y and Harmon B N 2002 *J. Alloys Compd.* **338** 242
[6] Lowther J E 2002 *Physica* B **322** 173

[7] Cook B A, Harringa J L, Lewis T L and Russell A M 2000 *Scr. Mater.* **42** 597
[8] Kojima H, Sahara R, Shishido T, Nomura A, Kudou K, Okada S, Kumar V, Nakajima K and Kawazoe Y *in submission*
[9] Blöchl P E 1994 *Phys. Rev.* B **50** 17953
[10] Kresse G and Joubert D 1999 *Phys. Rev.* B **59** 1758
[11] Vanderbilt D 1990 *Phys. Rev.* B **41** 7892
[12] Kresse G and Hafner J 1993 *Phys. Rev.* B **47** 558
[13] J.P. Perdew J P 1991 *Electronic Structure of Solids'91* eds P Ziesche and H Eschrig (Akademie Verlag, Berlin)
[14] Murnaghan F D 1944 *Proc. Natl. Acad. Sci. U.S.A.* **30** 244
[15] Mattesini M and Mater S F 2002 *Phys. Rev.* B **65** 075110
[16] Voigt W 1928 *Lehrbuch der Kristallpysik* (Taubner, Leipzig)
[17] Reuss A and Angew Z 1929 *Math. Meth.* **9** 55
[18] Hill R 1952 *Phys. Soc. London* **65** 350

Journal of Physics: Conference Series **176** (2009) 012019 doi:10.1088/1742-6596/176/1/012019

Present knowledge of electronic properties and charge transport of icosahedral boron-rich solids

Helmut Werheit

Experimental Physics, University Duisburg-Essen, D-47048 Duisburg, Germany

e-mail: helmut.werheit@uni-duisburg-essen.de and helmut.werheit@koeln.de

Abstract. B_{12} icosahedra or related structure elements determine the different modifications of elementary boron and numerous boron-rich compounds from α-rhombohedral boron with 12 to YB_{66} type with about 1584 atoms per unit cell. Typical are well-defined high density intrinsic defects: Jahn-Teller distorted icosahedra, vacancies, incomplete occupancies, statistical occupancies and antisite defects. The correlation between intrinsic point defects and electron deficiencies solves the discrepancy between theoretically predicted metal and experimentally proved semiconducting character. The electron deficiencies generate split-off valence states, which are decisive for the electronic transport, a superposition of band-type and hopping-type conduction. Their share depends on actual conditions like temperature or pre-excitation. The theoretical model of bipolaron hopping is incompatible with numerous experiments. Technical application of the typically p-type icosahedral boron-rich solids requires suitable n-type counterparts; doping and other possibilities are discussed.

1. Introduction

B_{12} icosahedra or related components determine the crystal structures of the different modifications of elementary boron and of numerous boron-rich compounds extending from α-rhombohedral boron with 12 to YB_{66} type with about 1608 atoms per unit cell. From careful X-ray fine structure investigations (Slack *et al* [1-3], Lundström (see [4] and references therein), Higashi *et al* [5]) some of these structures are known in detail. At boron carbide these methods failed because the scattering cross section of boron and carbon is very similar, and moreover the carbon atoms are statistically distributed on specific sites. However, this problem was solved by analyzing the phonon spectra (see [6], [7-10]).

Except α-rhombohedral boron, the icosahedral boron-rich solids are characterized by high concentrations of well-defined intrinsic structural defects. Apart from the Jahn-Teller distortion of the icosahedra, such defects are missing or incomplete occupation of specific atomic sites, statistical occupation of equivalent sites or antisite defects.

Such high defect concentrations in the icosahedral boron-rich structures exceed negligibly weak disorder by far. This was disregarded in the electronic band structure calculations performed until now. Consequently, qualitative discrepancies arose in contradiction to experimental results. Theoretical calculations predicted metallic character in consequence of high valence electron deficiencies, while experimental results proved semiconducting behavior. This contradiction has been clarified. At α-rhombohedral boron, β-rhombohedral boron and boron carbide (homogeneity range $B_{4.3}C$ to $B_{\sim 11}C$), the best-investigated icosahedral boron-rich solids, there is a quantitative correlation between intrinsic point defects and electron deficiencies [11] (table 1).

Journal of Physics: Conference Series **176** (2009) 012019 doi:10.1088/1742-6596/176/1/012019

Table 1. Icosahedral boron-rich structures: Comparison between theoretical electronic properties, experimental characterization and intrinsic point defects experimentally determined

Idealized structure						**Real structure**		
structure	Valence states [(unit cell)$^{-1}$]	Ref.	Valence electrons [(unit cell)$^{-1}$]	Electron deficiency [(unit cell)$^{-1}$]	Electronic character (theoretical)	Electronic character (experimental)	Intrinsic point defects [(unit cell)$^{-1}$]	Ref.
α-rhombohedral boron								
B_{12}	36	[12]	36	0	semiconductor	semiconductor	0	[13]
β-rhombohedral boron								
$(B_{12})_4(B_{28})_2B$	320	[14]	315	5	metal	semiconductor	4.92(20)	[1-3]
boron carbide $B_{13}C_2$								
$B_{12}(CBC)$	48	[15]	47	1	metal	semiconductor	0.97(5)	[7,8]
boron carbide $B_{4.3}C$								
$B_{11}C(CBC)$	48	[15]	47,83	0.17	metal	semiconductor	0.19(1)	[7,8]
boron carbide B_4C (hypothetical)								
$B_{11}C(CBC)$	48	[15]	48	0	semiconductor	—	—	—
YB_{66}								
$((B_{12})_{13})_8(B_{80})_8$	—		—	—	metal	semiconductor	304*)	[5]

*) ~ 42 sites of the B_{80} units are occupied implying a defect concentration of 16.1 % concerning the idealized structure $((B_{12})_{13})_8(B_{80})_8$ with 1584 of 1888 sites occupied

.Point defects in semiconductors generate split-off valence states in the band gap. Accordingly, the agreement of the concentration of experimentally proved point defects with calculated electron deficiencies implies that this way the unoccupied valence states of the idealized structures are exactly compensated in the real solids. Hence the real valence bands are completely filled, and the solids are semiconductors in accordance with experiment (see [6,11,16]). This implies that the electron deficiency evokes the generation of compensating intrinsic defects, probably for energetical reasons.

Such high concentrations of gap states attached to the valence band affect the electronic charge transport essentially; in particular they are responsible for the p-type character and the very low electrical conductivity. Aside from the electric conductivity in extended band states, hopping-type conduction in localized gap states must be expected. Their particular share depends on the actual conditions.

2. α-rhombohedral boron

α-rhombohedral boron, the low-temperature modification of elementary crystalline boron can be prepared below about 1500 K. An irreversible transformation into β-rhombohedral boron, the stable modification of crystalline boron, takes place at about 1670, 1890 and 1940 K via three different metastable phases [17]. The preparation of crystals of sufficient size for optical and electrical investigations turned out to be very difficult. Therefore, only few experimental results on α-rhombohedral boron, the only icosahedral boron-rich solid without high concentration of structural defects, have been performed. Some selected results are listed in table 2. For more details, see [16].

Table 2. α-rhombohedral boron, transition energies

Theory			**Experiment**			
eV	type	Ref	eV	Method	Transition type	Ref.
			0.73(2)	opt.	deep level to band	[18]
1.427	indirect	[19]				
~1.6	indirect	[20]	1.63(2)	opt.	indirect allowed interband	[18]
2.4	indirect	[19]	2.055(2)	opt.	indirect allowed interband	[21]
			2.0(2)	el.	conductivity	[22,23]

The theoretical and experimental interband transitions agree fairly well. The nature of the level in the gap is not yet explained. The effective masses of the charge carriers are estimated to 0.031(2) m_0, the mobility to 100 – 120 $cm^2V^{-1}s^{-1}$ [23,24], and the Seebeck coefficient 400 – 600 μVK^{-1} [22,23,25]. These data indicate electronic properties comparable to those of classical semiconductors, and confirm that they are not strongly influenced by defects.

These data make α-rhombohedral boron a promising candidate for technical application. Certainly, suitable preparation methods for bigger crystals are required.

3. β-rhombohedral boron

The unit cell of β-rhombohedral boron contains 106.5 B atoms, while the idealized structure $(B_{12})_4(B_{28})_2B$ describes 15 independent B sites with 100 % occupancy, 105 atoms in total. There are four icosahedra, one at the vertex, three on the edge centres of the unit cell, and two B_{28} units composed of three condensed icosahedra, each, arranged symmetrically to a centred single atom on the main diagonal of the unit cell. In particular group theoretical descriptions the alternative structure formula $B_{84}(B_{10})_2B$ is preferred with a B_{84} unit consisting of the B_{12} icosahedron at the vertex of the unit cell radially surrounded by 12 half-icosahedra completing each other to the B_{12} icosahedra on the edge centres. Two B_{10} units consisting of three condensed caps of icosahedra and the single atom on the diagonal complete this structure description of again the 105 atoms (see [16] and references therein).

In the real structure, 25.5 % of the site B(13) are unoccupied hence forming point defects of the idealized structure. Additional point defects are the sites B(16) to B(20) with 27.2 to 3.7 % occupancy [1,2]. In total their concentration agrees with the electron deficiency calculated for the idealized structure (see table 1).

The great number of atoms per unit cell in β-rhombohedral boron has prevented the calculation of the complete energy band structure. Only a density of states distribution has been reported [26] exhibiting a band gap of about 2 eV close to the experimentally determined value of about 1.5 eV. Theoretical calculations suggested assigning this gap to the Jahn-Teller distortion of the icosahedra [27]. The gap energies are listed in table 3

Table 3. Experimentally determined energy gaps of β-rhombohedral boron (extrapolated to T = 0 K)

	Natural isotope distribution			**^{10}B enriched**		**transition type**
	E ∥ c	E ⊥ c	Ref.		Ref.	
ΔE_1	1.32(1)	1.29(1)	[28]	1.27(2)	[29,11]	non-direct
ΔE_2	1.50	1.46	[28]	1.45(2)	[29]	indirect interband

Numerous experimental investigations yielded the consistent band scheme displayed in figure 1 (see [16] and references therein) [30-34].

Unfortunately, theoretical calculations on defect levels in icosahedral boron-rich solids are missing (apart from bipolaron states in boron carbide, which will be discussed below). Therefore in this respect the actual band schemes depend on experimental results only.

There are six equidistant electron traps attributed to distortions of the icosahedra evoked by electron-phonon interaction [35]. The defect states are attributed to the states separated from the valence band compensating the electron deficiency of the idealized structure. Occupied and unoccupied levels are energetically separated. The recombination of electrons excited from the valence into the conduction band essentially takes place via a cascade-like transition between the different electron traps with the electrons thermally excited from the occupied trap n to the unoccupied trap n+1. This activation energy of about 0.12 eV is the same for all traps.

Electrical transport takes place by free holes generated by thermal excitation of valence electrons into unoccupied gap states and by hopping processes between occupied and unoccupied gap states. The share of both processes depends on the actual conditions like temperature, pre-excitation and antecedent. At low temperatures hopping is prevailing, while at high temperatures the essential electrical conductivity is thermally activated.

Drift experiments with optically generated untrapped electron-hole pairs yielded the ambipolar carrier mobility $\mu_{ambipolar} = 565(120)\ cm^2V^{-1}s^{-1}$ [34]. Long-term drift experiments are dominated by the high probability trapping processes and show a soliton-type propagation of the carriers [36].

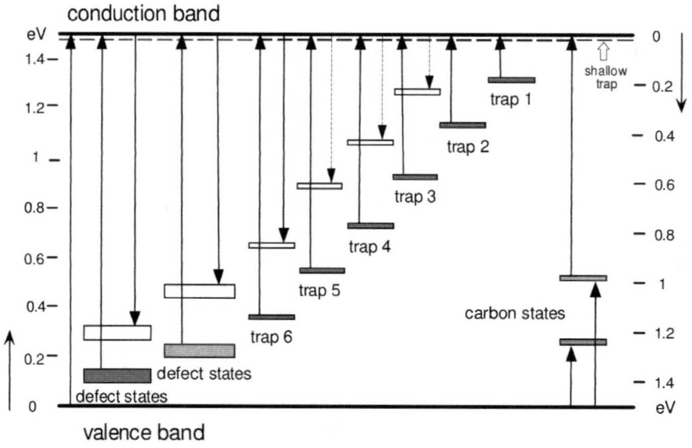

Figure 1. Actual Band scheme of β-rhombohedral boron. Full arrows indicate experimentally determined transitions; broken lines those, whose existence is indirectly concluded.

4. Boron carbide

In boron carbide, there is no unit cell representing the whole structure. This is composed of nearly isomorphous elementary cells, whose conformity is merely the 12-atomic, by the Jahn-Teller effect slightly distorted icosahedra at each vertex and the mostly three-atomic linear chains on the main diagonal parallel to the crystallographic *c*-axis. B_{12} and $B_{11}C$ icosahedra, C-B-C, C-B-B, B□B (□, vacancy), form differently composed elementary cells, statistically distributed over the whole structure (see [6] and references therein, [9,10]). The homogeneity range extends from $B_{4.3}C$ at the carbon-rich to $B_{\sim 11}C$ at the boron-rich limit. Since the carbon-rich limit $B_{4.3}C$ was proved by Schwetz and Karduck [37] for the first time, and is meanwhile often confirmed (see e.g., [6], figure 254, and assigned references, [39,40]), the previous designation "B_4C", based on the hypothetical structure formulas $(B_{12})C_3$ or $(B_{11}C)CBC$ with uniformly composed unit cells each, is no more acceptable. If this composition occurs, it consists of the compound $B_{4.3}C$ and precipitated graphitic free carbon. Therefore, in particular close to the carbon-rich limit of the homogeneity range, the separate analysis of bonded and free carbon in the structure is indispensable [4]. $B_{11.2}C$ synthesized by Gosset and Colin [42] seems to be the boron carbide with the lowest carbon content hitherto obtained.

Table 4. Concentration of structural elements in the elementary cell of boron carbide at some selected chemical compositions within the homogeneity range ([6-10,43])

	$B_{4.3}C$	$B_{13}C_2$	B_8C
B_{12} icosahedra	0	42	26
$B_{11}C$ icosahedra	100	58	74
C-B-C	81	62	7
C-B-B	19	19	77
B□B (□, vacancy)	0	19	16

The IR and Raman active phonons yielded the concentrations of the structural elements (see [6], [7-10,43]) leading to the general structure formula $(B_{12})_{1-x}(B_{11}C)_x(CBC)_{1-m-n}(CBB)_m(B\square B)_n$ (some examples in table 4; marginal deviations are discussed in [51]).

A strongly emphasized description of the electronic properties of boron carbide is the bipolaron hopping model developed by Emin and associates [44-50]. However, each theory requires the confirmation by relevant experiments. As shown below, numerous experimentally proved properties of boron carbide are incompatible with (bi)polarons. Therefore, this hypothesis must be taken as definitely disproved [51].

The bipolaron model for boron carbide is essentially based on the discrepancy between a hypothetically assumed high concentration of carriers (~ 10^{21} cm^{-3}) and the numerous experimental results of different authors (ESR, transport) consistently proving carrier concentrations of the order of ~ 10^{19} cm^{-3}. As shown above, this high carrier concentration was substantiated by inappropriate assumptions on the microstructure for band structure calculations and does not exist in reality. Accordingly, this basic assumption for the (bi)polaron model is not valid.

The lattice polarizability, an essential prerequisite for the formation of (bi)polarons, is responsible for the intensity of Raman scattering as well. At icosahedral boron-rich solids the Raman effect is very weak compared with other covalent solids; for boron carbide about three orders of magnitude smaller (see, e.g., references in [6]). Such extremely small polarizabilities make the formation of bipolarons very improbable. This was confirmed by Calandra *et al* [52] showing that the moderate electron–phonon coupling $\lambda = 0.81$ is too small for a bipolaronic state in $B_{13}C_2$.

After Emin *et al* [44-49] the electronic transport in boron carbides proceeds via bipolaron hopping between $B_{11}C$ icosahedra. However, the densities of hopping states determined according to Mott's law for dc variable-range hopping, according to the theory of Butcher and Morris [53] for the dynamical conductivity, and corresponding to the Seebeck coefficient S, which is proportional to the reciprocal density of carriers, vary opposite to the concentration of the $B_{11}C$ icosahedra in the homogeneity range of boron carbide (figure 2). Hence this assumption in the bipolaron hypothesis is disproved as well.

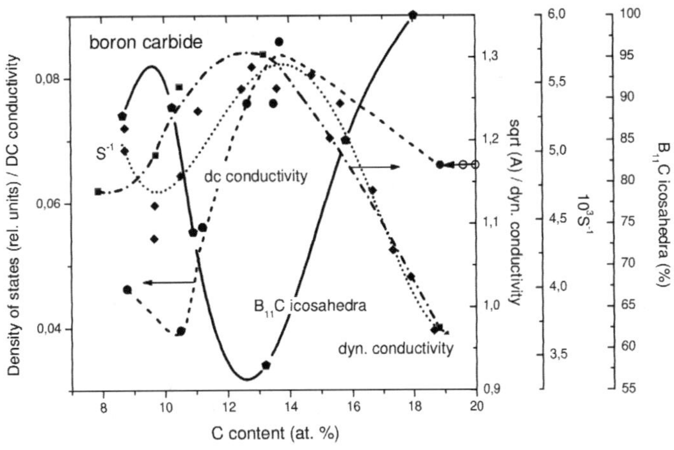

Figure 2. Boron carbide. Densities of hopping sites and B_{12} icosahedra (see [6,51,63] and references therein). •, dashed line, derived from DC conductivity after Mott's law for variable-range hopping; ○, DC results shifted to the carbon-rich limit of the homogeneity range; ■, dash-dotted line, derived from dynamical conductivity results (FIR); ♦, dotted line, 10^3S^{-1}, Seebeck coefficient S ~ 1/n (n, carrier density) (see [90]); ⬟, full line, density of B_{11} C icosahedra (data connected by a spline function).

The problem of the effective mass is a central one in (bi)polaron physics (see e.g., Devreese [54]). Dragging the polarized surrounding with them, the mass of polarons would be considerably higher than the band mass of free holes. Unfortunately, reliable experimental effective mass values as can be obtained by cyclotron resonance experiments are not available for boron carbide. Accordingly, the concerning discussion must be restricted to more or less rough estimations (see table 5).

Table 5. Estimations of hole masses in boron carbide

Band mass of free holes			Effective mass of mobile carriers		
	Method	Ref.		Method	Ref.
~ 10 m_0	curvature of valence band	[55]	> 3 m_0	DC conductivity and ESR	[56]
~ 10 m_0	luminescence	[57,58]	3…5 m_0	IR spectra	[59,60]
			1…10 m_0	Dynamical conductivity	[69,70]

Obviously, there is no indication that the effective mass of the mobile carriers in boron carbide is considerably larger than the band mass of free holes. Hence these results disprove polarons in boron carbide as well.

For supporting his bipolaron model Emin refers to the electrical conductivity of a "typical boron carbide" [44,48,49]. Figure 3 shows the discrepancy to the real conductivity in the high-temperature range consistently measured by different groups [61,62,63] (see also [6]).

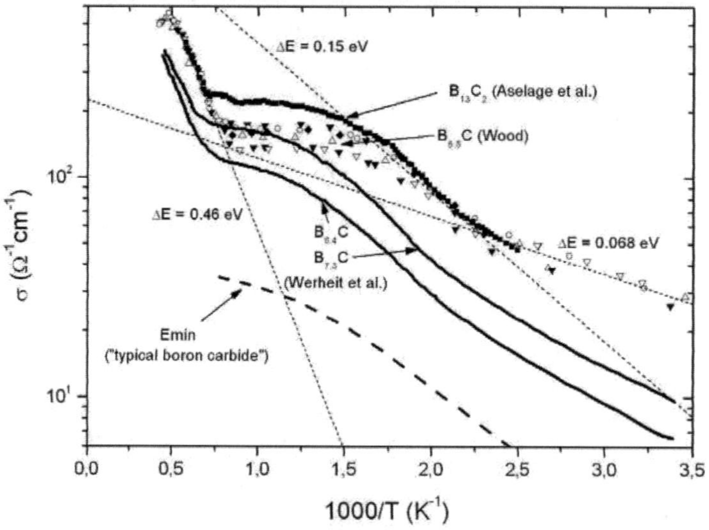

Figure 3. Boron carbide. High-temperature electrical conductivity. Dashed curve, "typical boron carbide", Emin [44,48,49]; ▽△▼◆, Wood [62]; ■, Aselage *et al* [63]; Solid curves, Werheit *et al* [64,65]; short-dashed curves, fits for determining the activation energies.

It is obvious that at high temperatures the electrical conductivity of boron carbide exhibits well-defined thermal activation energies in different ranges. This interpretation is strongly supported by corresponding gap states determined from the optical absorption edge [66,67,58]. Hence there is no sensible reason for a desperate interpretation within the bipolaron model.

After Emin [68] the coherent motion of bipolarons yields Drude-like optical spectra at low temperatures decreasing with increasing temperatures. The experimental finding for boron carbide and other icosahedral boron-rich solids is opposite to this prediction [69-72], constituting another inconsistency between polaron hypothesis and reality.

The existence of excitons in boron carbide associated to the central B atoms in the three-atomic chains was proved in two independent studies [57,73,74]. Excitons, which are formed by the interaction of free carriers, are incompatible with the existence of polarons.

As shown, there are many discrepancies and even incompatibilities between the (bi)polaron model and the real properties of boron carbide so that this hypothesis must be taken as definitely disproved. This conclusion cannot be put in question by apparently successful descriptions of properties of boron carbide like that of the very low electrical conductivity [44-50], and of the enhanced Seebeck coefficient evoked by carrier-induced vibrational softening (Emin [75] and Aselage *et al* [76]), because bipolarons, being the fundamental basis of these theories, obviously don't exist in boron carbide.

The alternative for the interpretation of the electronic properties of boron carbide is the band scheme displayed in figure 4. It allows the largely consistent description of all reliable experimental results, which have become known, and therefore this band scheme seems to be a serious basis for further work.

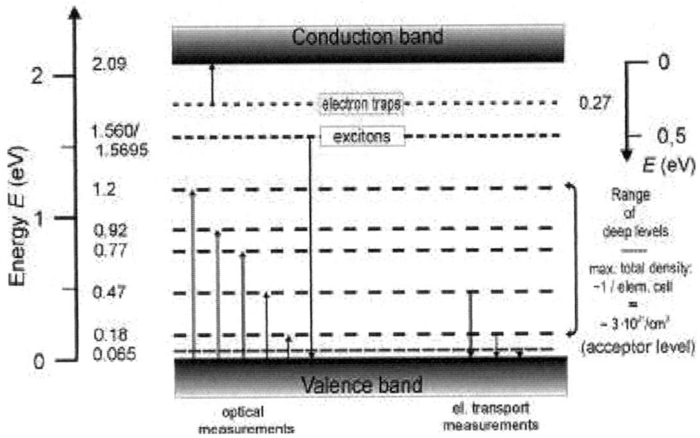

Figure 4. Boron carbide. Energy band scheme based on optical absorption, luminescence, XRS, and transport properties. Left ordinate, energies related to the valence band edge; right ordinate, energies related to the conduction band edge. Arrows indicate the direction of the measured transitions.

5. Required improvements of the energy band schemes

The band schemes presented are not intended to be exhaustive. They are essentially based on experimental results obtained more or less close to ambient conditions. However, for β-rhombohedral boron numerous experiments, in particular those on photo-effects show that the electronic properties discontinuously vary at certain temperatures (see [16],[31,77,78]) suggesting that the energetical positions of the gap states depend on temperature. Moreover the photo-effects considerably depend on the antecedent, for example, optical excitation and frozen-in carriers. In particular at lower temperatures, β-rhombohedral boron exhibits long-term memory effects after optical or thermal excitation.

On the other hand, temperature dependent structural modifications are indicated by internal friction maxima at about 150, 250, 340 530, and 570 K (see [16])[79,80]. The last two maxima are correlated with a hysteresis in the thermal expansion curves [81-83,28]. At least some of the internal friction maxima are obviously correlated with critical points of photo-effects. Considerable differences of the electrical conductivity for compression and decompression in high-pressure experiments indicate irreversible or slowly reversible modifications of the structure as well [83].

The correlation between the gap states and the atomic sites B(13), B(16)-B(20) suggests that the energetically most favourable stable positions of these sites in the structure depend on temperature, possibly caused by the anisotropy of the thermal expansion coefficient. The strong influence of

Journal of Physics: Conference Series **176** (2009) 012019 doi:10.1088/1742-6596/176/1/012019

interstitial doping on temperature and height of the internal friction maxima suggests interdependence with the partly occupied B sites. At a pressure of 160 GPa β-rhombohedral boron becomes superconducting [84]. All these aspects require further investigation.

For boron carbide the situation is probably similar. Indeed, there are no corresponding results on electronic properties, however internal friction maxima indicate temperature dependent structural variations for boron carbide as well [85], and their influence on the electronic properties seems natural.

6. Doping of icosahedral boron-rich structures

Doping of semiconductors is important for tailoring their electronic properties for application. The above-described icosahedral boron-rich solids are p-type because of their high concentration of largely unoccupied gap states close to the valence band. Therefore overcompensation to n-type requires high concentrations of doping elements.

Besides doping by the substitution of foreign atoms for boron, the rather open structures allow interstitial doping by the accommodation of foreign atoms in the voids of the structure. Because of its strong affinity to boron, carbon is a hardly avoidable doping element. C atoms typically substitute for B atoms at specific sites in the structures, for example, preferably in polar sites of the icosahedra. However, the expected n-doping resulting from the excess electron of C atom compared with B does not take place. The reason might be that the carbon levels in the gap are rather close to the valence band.

β-rhombohedral boron can be doped by the interstitial accommodation of metal atoms (for details and references see [16]). For Fe, V, Cu, Cr, Li, and Ni n-type behavior was achieved, however in a rather low degree and in limited temperature ranges only. This kind of doping is limited because a destabilization of the structure takes place. In how far the interstitial accommodation of foreign elements affects the occupancy of the sites B(13) and B(16) – B(20) (for details of these sites see ref. [1,2]) has not yet been investigated.

Systematical doping of boron carbide has become known for few elements only (see [16] and references therein). However, n-type conductivity was not realized; but sometimes the Seebeck coefficient became considerably higher. Because of the high concentrations of unoccupied gap states and the limited doping possibility compared with β-rhombohedral boron it is questionable, whether n-doping of boron carbide is possible at all. Systematic investigations on enhancing the p-type character of icosahedral boron-rich structures, e.g., by the substitution of regular B atoms by Be, are missing so far.

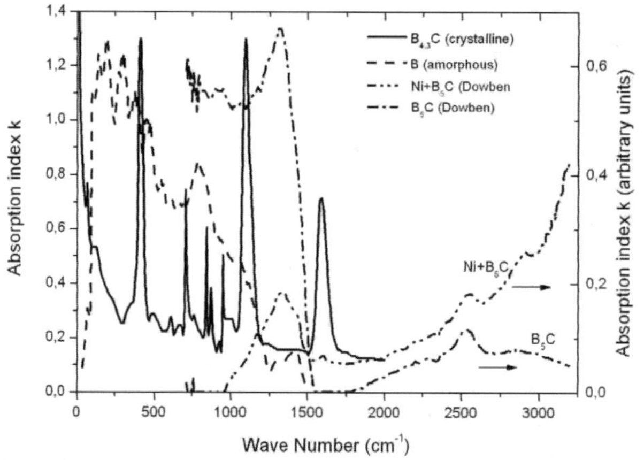

Figure 5. Comparison of the phonon spectra of B_5C and B_5C + Ni (Dowben *et al* [88] with those of boron carbide (see [6]) and amorphous boron (see [16])

Journal of Physics: Conference Series **176** (2009) 012019 doi:10.1088/1742-6596/176/1/012019

7. Chances for technical application.

Electronic application of the boron-rich solids requires highlighting properties compared with classical semiconductors; such as the high Seebeck coefficient of boron carbide monotonously increasing up to more than 2000 K. Unfortunately, as already mentioned, the high concentrations of unoccupied gap states prevent compensation to n-type in boron carbide.

Therefore, n-type counterparts with alternative structures but otherwise similar properties are required. Recently developed candidates are n-type alkaline hexaborides (Takeda *et al* [86]) or n-type RE boron carbonitrides (Mori *et al* [87]). A further very promising candidate is a new icosahedral boron-carbon compound (Dowben *et al* [88-90]). In pure form this compound is p-type, but becomes n-type by addition of Ni. This was proved by the fabrication of homojunction and heterojunction diodes combining both materials [91]. The structure of this compound is different from the boron carbide discussed above. It contains icosahedra as well, but these contain two carbon atoms, while in the icosahedra of boron carbide only one C atom substitute for B in a polar site. Moreover the three-atomic chain characteristic for boron carbide is missing. This becomes obvious by comparing the IR spectra in figure 5, showing that the 1500 cm^{-1} phonon band representing the stretching mode of the three-atomic chain is absent in this solid.

References

[1] Slack G A, Hejna C I, Garbauskas M F, Kasper J S 1987 *Proc. 9th Int. Symp. Boron, borides and Rel. Comp.*, ed H Werheit (University of Duisburg, Germany) p 132
[2] Slack G A, Hejna C I, Garbauskas M F and Kasper J S 1988 *J Solid State Chem.* **76** 52
[3] Slack G A, Hejna C I, Garbauskas M F and Kasper J S 1988 *J Solid State Chem.* **76** 64
[4] Lundström T 1986 *AIP Conf. Proc.* **140** 186
[5] Higashi I, Kobayashi K, Tanaka T and Ishizawa Y 1997 *J. Solid State Chem.* **133** 16
[6] Werheit H 2000 *Landolt-Börnstein, Numerical Data and Functional Relationships in Science and Technology, New Series, Group III Vol. 41D*, ed O Madelung (Berlin: Springer) p 1
[7] Kuhlmann U and Werheit H 1992 *Solid State Comm.* **83** 849
[8] Kuhlmann U, Werheit H and Schwetz K A 1992 *J. Alloys Compd.* **189** 249
[9] Werheit H, Au T, Schmechel R, Shalamberidze S O, Kalandadze G I and Eristavi A M 2000 *J. Solid State Chem.* **154** 79
[10] Werheit H, Rotter H W, Meyer F D, Hillebrecht H, Shalamberidze S O, Abzianidze T G and Esadze E G 2004 *J. Solid State Chem.* **177** 569
[11] Schmechel R and Werheit H 1999 *J. Phys.: Condens. Matter* **11** 6803
[12] Perrot F 1981 *Phys. Rev.* B **23** 2004
[13] Morosin B, Mullendore A W, Emin D and Slack G A 1986 *AIP Conf. Proc.* **140** 70
[14] Bullett D W 1982 *J. Phys. C: Solid State Phys.* **15** 415
[15] Armstrong D R, Bolland J, Perkins P G,Will G and Kirfel A 1983 *Acta Crystallogr.* B **39** 324
[16] Werheit H and Schmechel R 1998 *Boron, Landolt-Börnstein, Numerical Data and functional relationships in Science and Technology Group III, Vol. 41C* (ed) O Madelung (Springer, Berlin) p 3
[17] Runow P 1972 *J. Mater. Sci.* **7** 499
[18] Werheit H 1990 *The Physics and Chemistry of Carbides, Nitrides and Borides; NATO ASI Series E: Applied Sciences* Vol. 185 (ed) R Freer (Kluwer Academic Publishers, Dordrecht) p 691
[19] Lee S, Bylander D M and Kleinman L 1990 *Phys. Rev.* B **42** 1316
[20] Hatakeyama T 1989 *Electronic Structures of Icosahedral Boron Solids, Thesis* (University of Tokyo, Japan)
[21] Werheit H, Kuhlmann U, Solov'ev N E, Tsiskarishvili G P and Tsagareishvili G 1991 *AIP Conf. Proc.* **231** 350
[22] Golikova O A, Solov'ev N E, Ugai Ya A, Feigel'man V A 1979 *Fiz. Tekh. Poluprovodn.* **13** 825; *Sov. Phys. Semicond. (English Transl.)* **13** 486

[23] Golikova O A, Solov'ev, N E, Ugai, Ya A, Feigelman, V A 1981 in: *Proc. 7th Int. Symp. Boron, Borides and Related Compounds, Uppsala, Sweden*, p. 362.
[24] Golikova O A 1984 *phys. status solidi* (a) **86** K51
[25] Golikova O A1987 *phys. status solidi* (a) **101** 277
[26] Bullett D W 1990 *The physics and chemistry of carbides, nitrides and borides, NASA ASI series E: Applied sciences* (ed) R Freer Vol 185 (Kluwer, Dordrecht) p 513
[27] Fujimori M and Kimura K 1997 *J Solid State Chem.* **133** 178
[28] Werheit H, Laux M and Kuhlmann U 1993 *phys. status solidi* (b) **176** 415
[29] Kuhlmann, U, Werheit, H, Fahy, B and Perkins, P 1991 *AIP Conf. Proc.* **231** 148
[30] Schmechel R and Werheit H 2000 *J. Solid State Chem.* **154**, 68
[31] Werheit H, Schmidt M, Schmechel R and Lundström T 2000 *J. Solid State Chem.* **154** 93
[32] Werheit H and Wehmöller B 2004 *J. Solid State Chem.* **177** 556
[33] Werheit H and Moldenhauer A 2004 *J. Solid State Chem.* **177** 586
[34] Werheit H and Moldenhauer A 2006 *J. Solid State Chem.* **179** 2775
[35] Werheit H and Kuhlmann U 1993 *Solid State Commun.* **88** 421
[36] Werheit H and Kummer F 1995 *J. Phys: Condensed Matter* **40** 7851
[37] Schwetz K A and Karduck P 1990 *AIP Conf. Proc.* **231** 405
[38] Werheit H 2000 *Boron compounds, Landolt-Börnstein, Numerical Data and Functional Relationships in Science and Technology, New Series, Group III* Vol 41D (ed) O Madelung (Berlin, Springer) p 1
[39] Leithe-Jasper A and Tanaka T 1999 *13th Int. Symp. Boron, Borides and Rel. Comp.* (Dinard, France) O 22
[40] Werheit H, Leithe-Jasper A, Tanaka T, Rotter HW and Schwetz KA 2004 *J. Solid State Chem.* **177** 575
[41] Schwetz K A and Hassler J 1986 *J. Less-Common Met.* **117** 7
[42] Gosset D and Colin M 1991 *J. Nucl. Mater.* **183** 161
[43] Schmechel R, Werheit H, Kueffel V and Lundström T 1997 *Proc 16th Int Conf Thermoelectrics, Dresden, Germany* (eds) A Heinrich and J Schumann (IEEE 97TH8291) p 219
[44] Emin D 1986 *AIP Conf. Proc.* **140** 189
[45] Howard I A, Beckel C L and Emin D 1987 *MRS Symp. Proc.* **97** 177
[46] Howard I A, Beckel C L and Emin D 1987 *Phys. Rev.* B **35** 2929
[47] Howard I A, Beckel C L and Emin D 1987 *Phys. Rev.* B **35** 9265
[48] Emin D 1990 *AIP Conf. Proc.* **231** 65
[49] Emin D 1989 *The Physics and chemistry of carbides, nitrides and borides* (ed) R Freer (Dordrecht, Kluwer Acad. Publ.) p 691
[50] Emin D 2006 *J. Solid State Chem.* **179** 2791
[51] H Werheit 2007 *J. Phys.: Condens. Matter* **19** 186207
[52] Calandra M, Vast N and Mauri F 2004 *Phys. Rev.* B **69** 224505
[53] Butcher P N and Morris P L 1973 *J. Phys. C: Solid State Phys.* **6** 2147
[54] J T Devreese 2005 *Encyclopedia of Physics* Vol 2 (ed) R G Lerner and G L Trigg (Weinheim, Wiley-VCH) p 2004
[55] Armstrong R, Bolland J, Perkins P G, Will G and Kirfel A 1983 *Acta Crystallogr.* B **39** 324
[56] Geist D, Meyer J and Peussner H 1970 *Boron Vol 3* (ed) T Niemyski (Warszawa, PWN-Polish Scientific Publ.) p 207
[57] Schmechel R, Werheit H, Kampen T U and Mönch W 2004 *J. Solid State Chem.* **177** 566
[58] H Werheit 2006 *J. Phys.: Condens. Matter* **18** 10655
[59] Werheit H, Binnenbruck H and Hausen A 1971 *phys. status solidi* (b) **47** 153
[60] Binnenbruck H and Werheit H 1979 *Z. Naturforsch.* **34a** 787
[61] Werheit H, Kuhlmann U, Franz R, Winkelbauer W, Herstell B, Fister D and Neisius H 1991 *AIP Conf. Proc.* **231** 104
[62] Wood C 1985 *AIP Conf. Proc.* **140** 206

[63] Aselage T L, Emin D and McCready S S 2001 *Phys. Rev.* B **64** 054302
[64] Werheit H, Kuhlmann U, Franz R, Winkelbauer W, Herstell B, Fister D and Neisius H 1990 *AIP Conf. Proc.* **231** 104
[65] Werheit H, Herstell B and Winkelbauer W *unpublished results*
[66] Werheit H, Laux M, Kuhlmann U and Telle R 1992 *phys. status solidi* (b) **172** K81
[67] Werheit H, Leithe-Jasper A, Tanaka T, Rotter H W and Schwetz K A 2004 *J. Solid State Chem.* **177** 575
[68] Emin D 1993 *Phys. Rev.* B **48** 13691
[69] Schmechel R and Werheit H 1997 *J. Solid State Chem.* **133** 335
[70] Schmechel R and Werheit H 1998 *J. Mater. Process. Manufact. Sci.* **6** 329
[71] Schmechel R, Werheit H, Kueffel V and Lundström T 1997 *Proc. 16th Int. Conf. Thermoelectrics, Dresden, Germany* (eds) A. Heinrich, J. Schumann (IEEE 97TH8291) p 219
[72] Schmechel R and Werheit H 1996 *J. Phys.: Condens. Matter* **8** 7263
[73] Feng Y, Seidler G T, Cross J O, Macrander A T and Rehr J J 2004 *Phys. Rev.* B **69** 125402
[74] Werheit H 2006 *J. Phys.: Condens. Matter.* **18** 10655
[75] Emin D 1999 *Phys. Rev.* B **59** 6205
[76] Aselage T L, Emin D, McCready S S and Duncan R V 1998 *Phys. Rev. Lett.* **81** 2316
[77] Werheit H and Wehmöller B 2004 *J. Solid State Chem.* **177** 556
[78] Werheit H and Kummer F 1995 *J. Phys.: Condens. Matter* **7** 7851
[79] Werheit H and Kummer F 2008 *16th Int. Symp. Boron, Borides Rel. Mater.*
[80] Darsavelidze G Sh, Tsagareishvili G V, Tsagareishvili O A, Antadze M E, Zoidze N A, Dzigrashvili T V and Kutelia E K 1987 *Proc. 9th Int. Symp. Boron, Borides and Rel. Comp.* (ed) H Werheit (University of Duisburg, Duisburg) p 341
[81] Tsagareishvili G 1994 *Jpn. J. Appl. Phys.* **Ser. 10** 21
[82] Tsagareishvili G V, Tavadze F N, Darsavelidze G Sh and Metreveli V Sh 1970 *Boron, Vol 3* (ed) Niemyski T (Warsaw, Polish Sci. Publ.) p 281
[83] Zhang D, Gao C, Ma Y, He C, Huang X, Hao A, Yu C, Li Y, Liu J, Peng G, Li D, Liu H and Zou G 2007 *J. Phys.: Condens. Matter* **19** 425216
[84] Eremets M I, Struzhkin V V, Mao H and Hemley R J 2001 *Science* **293** 272
[85] Lezhava D, Darsavelidze G, Tsagareishvili O, Antadze M and Gabunia V 2006 *J. Solid State Chem.* **179** 2934
[86] Takeda M, Terui M, Takahashi N and Ueda N 2006 *J. Solid State Chem.* **179** 2823
[87] Mori T and Nishimura T 2006 *Proc. 25. Int. Conf. Thermoelectr.* (ed) P Rogl (IEEE Cat. No.: 06TH8931C) p 164
[88] Hwang S D, Yang K, Dowben P A, Ahmad A A, Ianno N J, Li J Z, Lin J Y, Jiang H X and McIlroy D N 1997 *Appl. Phys. Lett.* **70** 1028
[89] Lunca-Popa P, Brand J I, Balaz S, Rosa L G, Boag N M, Bai M, Robertson B W and Dowben P A 2005 *J. Phys. D: Appl. Phys.* **38** 1248
[90] Ahmad A A, Ianno N J, Snyder P G, Welipitiya D, Byun D and Dowben P A 1996 *J. Appl. Phys.* **79** 8643
[91] Caruso A N, Billa R B, Balaz S, Brand J I and Dowben P A 2004 *J. Phys.: Condens. Matter* **16** L139
[92] Werheit H 2006 *Proc. 25th Int. Conf. on Thermoelectr.* (ed) P Rogl (Piscataway, NJ: IEEE, Catalog Nr. 06TH8931) p 159

Fractional photoconductivity glow curves of β-rhombohedral boron

Helmut Werheit and Frank Kummer

Experimental Physics, University Duisburg-Essen, D-47048 Duisburg, Germany

e-mail: helmut.werheit@uni-duisburg-essen.de and helmut.werheit@koeln.de

Abstract. The energy band structure of β-rhombohedral boron is characterized by multiple electron and hole traps in the band gap. Fractional photoconductivity glow experiments were performed between 170 and 360 K for obtaining further information on the various traps in this semiconductor. The results can easily be interpreted within the actual band scheme and confirm the previous assumption of a cascade-like recombination of optically excited electrons with an activation energy of about 0.12 eV. The result support the assumption of optically induced structural changes regarding the partly occupied sites B(13) and B(16) to B(20).

1. Introduction

Valuable information on trap parameters of semiconductors can be obtained from photoabsorption and photoconductivity glow experiments. In such experiments, at low temperatures free electron-hole pairs are generated with excitation energies exceeding the band gap. In the traps, which are localized states in the band gap, the excited free carriers are entrapped depending on the relation between capture and recombination cross section. Thermal reexcitation into the adjacent energy band is prevented by the low temperature. However, in this situation the trap depth could be determined by optical absorption experiments ("photoabsorption") and optically excited conductivity ("photoconductivity"). Afterwards, reexcitation is systematically stimulated by continuously increasing the temperature. Depending on the number of released carriers and the process of subsequent interband recombination, the conductivity increases ("thermally stimulated current"). Simultaneously, the photoabsorption of trapped carriers decreases because of the decreasing occupation density of the traps. Sometimes, luminescence radiation from recombining electron-hole pairs can be obeyed as well. Such temperature dependent thermally stimulated photoabsorption, photoconductivity and luminescence experiments are called "glow curves".

If there is one trap only, the typical photoconductivity glow curve is bell-shaped. The ascending branch at increasing temperatures is determined by the increasing number of mobile carriers thermally released from the trap, followed by a decreasing branch in consequence of the progressing exhaustion of the initially filled trap. The individual shape of the glow curve depends on activation energy, capture cross section, transition probability, retrapping and recombination factor. The parameters concerned can be more or less well estimated from series of measurements under modified conditions.

In more complex cases, where various traps occur, the associated glow curves overlap, and hence qualifying of individual trap parameters is complicated or even impossible. This problem is surmounted or at least reduced by fractional glow experiments (see [1-3]). In such methods the optically excited semiconductor is heated to a temperature, where largely emptying of the shallowest trap is expected; this can be approximately assumed, when the fist glow peak is reached. In the next step the sample is cooled down to the starting temperature again and the heating procedure is repeated until the next trap is emptied. This way, the different traps are emptied step by step, and the trap parameters can be derived largely separately.

In β-rhombohedral boron electron and hole traps affect the electronic properties decisively (see [4-9]). For additional and possibly particular information on the different traps and gap states respectively, fractional photoconductivity glow experiments were performed between 175 and 360 K.

2. Sample material and experimental setup

A prismatic sample (1.6×1.6×10.5 mm) was cut from a high-purity β-rhombohedral boron single crystal (Wacker Chemistry, Munich; claimed purity 99.9999% except carbon (typically 60 ppm)) roughly parallel to the crystallographic *c*-axis. The sample was etched to remove Beilby layers and surface contamination (for the method see [10]). On opposite side faces of the prismatic sample, a pair of ohmic contacts (0.1 mm platinum wires) was fixed by capacitor discharge in about 1.5 mm distance from the front surface, illuminated by a Xenon arc-lamp for optical excitation. The side faces were carefully shielded from light.

For realizing well-defined starting conditions, the sample was heated to about 350 K, and remained there for one hour for setting thermal equilibrium. After cooling down to the starting temperature 170 K in complete darkness, it was kept there for another hour. Hereafter the excitation by a two hours illumination with a Xenon arc-lamp took place for achieving completely filled electron traps. Then, in the first cycle of measurement a normal conductivity glow curve was recorded heating the sample to 350 K (heating rate, 5 K/min). In additional sequences of measurements, fractional glow curves were obtained after heating the sample from the starting point at 170 K to the temperature, at which the maximum of the preceding curve had occurred. After re-cooling again to 170 K the respective fractional glow curve was recorded.

3. Experimental results

In figure 1 the normal and three fractional glow curves are displayed compared with the dark current obtained at otherwise same conditions as described. For clarity, the curves of intermediate heating to 186, 198 and 213.4 K, respectively, each well coinciding with the red curve, are not displayed. After re-cooling to the starting point, the next cycle of fractional electrical conductivity was measured up to 350 K, and yielded the glow curves with afore partly emptied electron traps.

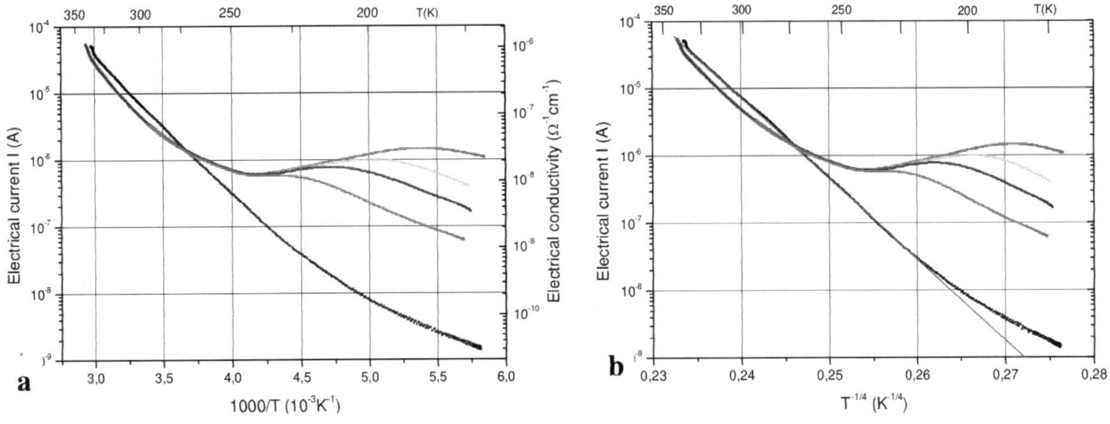

Figure 1. Normal dark conductivity (red) and fractional photoconduction (after optical excitation at 175 K and intermediate heating to 186 K (green), 198 K (blue), and 213.4 K (pink), respectively). Dark current (black) for reference (a, vs. T^{-1}; b, vs. $T^{-1/4}$)

Former investigations [11,12] showed that the electrical conductivity of β-rhombohedral boron follows Mott's law of variable-range hopping at lower temperatures. Therefore, in figure 1b the data are accordingly re-plotted vs. $T^{-1/4}$. The validity of Mott's law is confirmed by the present data in a wide range of temperature. However, towards low temperatures a certain deviation occurs, apparently indicating a different conductivity mechanism. It cannot be excluded that this deviation is not real, indeed it is marginal for the evaluation of the fractional photoconduction as difference to the dark current.

4. Discussion

For visualizing the fractional photoconduction glow curves, the measured data after subtracting dark current are plotted in figure 2. Since thermal activation processes of carriers in the involved traps are expected, plotting vs. reciprocal T is chosen.

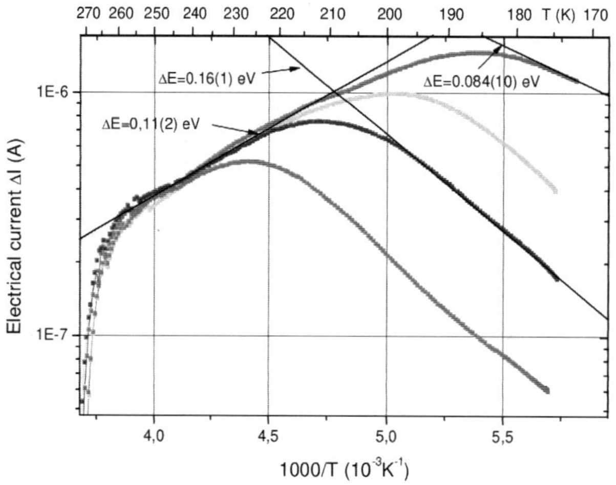

Figure 2. Fractional photoconductivity glow curves obtained as difference between measured data and dark current

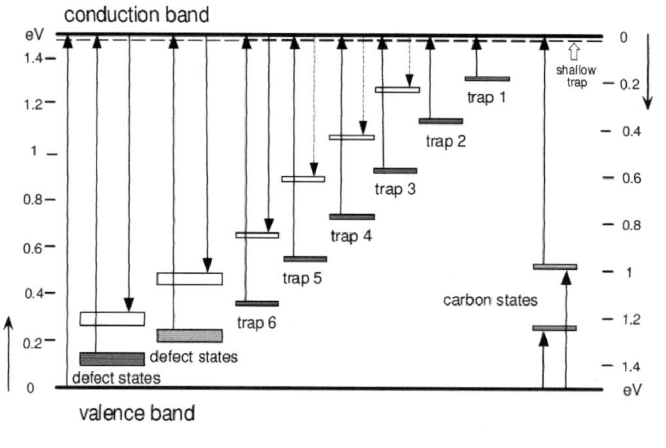

Figure 3. Actual energy band scheme of β-rhombohedral boron (reproduced from [6,8]).

Above ~260 K the curves abruptly decrease. This is caused by the crossing between dark current and glow curves indicating that thermal equilibrium at higher temperatures depends on the antecedent at low temperatures, here with and without strong optical excitation at low temperatures as well. The difference between dark current and glow curves at higher temperatures supports previous assumptions that the optical excitation of β-rhombohedral boron is correlated with structural relaxations. It seems natural attributing these interdependence with the incompletely occupied sites B(13) and B(16) to B(20). In this context we point to the internal friction maxima at about 140 K and 230 K by Tsagareishvili *et al* (see [4] and references therein) indicating structural modifications in β-rhombohedral boron at such low temperatures even in thermal equilibrium as well.

The thermal activation energies within the fractional glow curves indicated in figure 2 are to be interpreted within the band scheme of β-rhombohedral boron (figure 3).

The activation energy of the glow curve for full optical excitation (red curve) seems to correspond to the distance between the valence band edge and the nearest defect state. This would indicate that the current is essentially represented by free holes generated by thermal excitation of valence electrons into adjacent localized gap states. Indeed must be conceded that beneath the range of measurement the slope could possibly be steeper. For all fractional glow curves representing the conductivity under the condition of partly emptied electron traps, the situation is uniform: The activation energy of the electrical conductivity represents the distance between occupied and unoccupied defect states beyond the valence band edge.

The consistent decrease of the glow curves towards higher temperatures is obviously determined by the recombination of trapped electrons. This process in the fractionated glow curves is thermally activated with the energy 0.11(2) eV, which well agrees with the activation energy 0.12 eV of cascade-like recombination of trapped electrons via the series of intrinsic electron trapping levels [5].

5. Conclusion

The fractional glow curves of β-rhombohedral boron can be consistently described within its actual energy band scheme (figure 3) and confirm the cascade-like recombination of electrons trapped in different levels below the conduction band.

Concordantly to previous investigations on photoabsorption and photoconduction, the present results confirm that the electronic thermal equilibrium at low temperatures is remarkably altered by optical excitation. We suppose a correlation with the partly occupied sites B(13) or B(16) to B(20) proved to be relevant for the gap states in β-rhombohedral boron. This could be in the way that the positions or occupancies of these atomic sites depend on specific conditions of the electronic surrounding in the structure, which is changed by the optical excitation. We take the internal friction maxima at low temperatures for supporting our thesis of easily realisable structural changes in β-rhombohedral boron, and suppose that in this respect the mentioned sites are very probable aspirants.

References:

[1] Gobrecht H and Hofmann D 1966 *J. Phys. Chem. Solids* **27** 509
[2] Lang D V 1974 *J. Appl. Phys.* **45** 3023
[3] Fasbender R, Hirt G, Thoms M and Winnacker A 1996 *Semicond. Sci. Technol.* **11** 935
[4] Werheit H and Schmechel R 1998 *Boron, Landolt-Börnstein, Numersical data and functional relationships in science and technology, Group III* Vol. 41C (ed) O Madelung (Springer-Verlag, Berlin) p 3
[5] Schmechel R and Werheit H 2000 *J. Solid State Chem.* **154** 68
[6] Werheit H, Schmidt M, Schmechel R and Lundström T 2000 *J. Solid State Chem.* **154** 93
[7] Hori A, Tada T and Kimura K 1998 *J. Phys. Soc. Jpn.* **67** 4279
[8] Werheit H and Wehmöller B 2004 *J. Sold State Chem.* **177** 556
[9] Werheit H and Moldenhauer A 2006 *J. Solid State Chem.* **179** 2775
[10] Binnenbruck H, Hausen A, Runow P and Werheit H 1970 *Z. Naturforsch* **25a** 1431
[11] Werheit H and Leis H G 1970 *phys. status solidi* **41** 247
[12] Szadkowski A 1979 *J. Less-Common Met.* **67** 551

Anomalously high capacitance of β-rhombohedral boron induced by structural defects

Otar Tsagareishvili[1], Domenti Gabunia[1] and Levan Chkhartishvili[1,2]

[1] Laboratory for Boron and Its Compounds, F Tavadze Institute of Metallurgy and Materials Science, 15 A Kazbegi Avenue, Tbilisi, 0160, Georgia

[2] Department of Physics, Georgian Technical University, 77 M Kostava Avenue, Tbilisi, 0175, Georgia

E-mail: chkharti2003@yahoo.com

Abstract. Capacitance and dielectric loss tangent measurements have been carried out for β-rhombohedral boron within the frequency range $10^2 - 10^4$ Hz and temperature interval 77 – 373 K. Different specimens were studied: high-purity macrocrystals, vapor-liquid-grown faceted crystals, zone-melted single crystals, and crucible-melted polycrystals. Within the whole ranges of measurements, the capacitance of all samples was 10-60 times larger than their geometric capacitance. Temperature dependences of the capacitance reveal two-level step-like behavior. Location of levels is independent from the applied electric field frequency, but at higher frequencies, the capacitance steps shift to lower temperatures. For all frequencies, an abrupt rise in capacitance takes place at one and same temperature ~ 290 K. At low temperature, dielectric losses are negligible, but they increase towards the room temperature. At ~ 290 K (i.e., simultaneously with the abrupt rise in capacitance) their value abruptly falls down. Dielectric properties of β-rhombohedral boron are discussed on the basis of generalized barrier model of the heterogeneous semiconductor within its three-layer version. Planar defects like twins and stacking faults characteristic for real crystals are assumed to produce local elastic stresses sufficient for the local lowering in conductivity due to the piezoresistive effect. Consequently, layers adjacent to the planar defects should represent low-conducting inclusions in the relatively high-conducting matrix. In addition, the lesser-conducting barrier layers should appear at their boundaries. Effect at ~ 290 K seems to be related with a symmetry-restoring phase transition driven by the occupation patterns of certain atomic sites in β-rhombohedral crystalline boron lattice.

1. Introduction

Semiconductor can be considered as a dielectric with variable conductivity. When the changes in conductivity are heterogeneous, material reveals unusual dielectric properties such as "giant" capacitance. They may be useful in electronics, e.g., for making electrical elements with high capacitance. In recent years, perovskites and some related complex materials have attracted much attention due to their extremely high dielectric constants at low frequencies, which are unchanged over the wide temperature ranges [1-10]. Similar effect is known for a long time in elemental boron. Here we return to this interesting problem aiming to present new experimental results on β-rhombohedral boron (below, β-B), namely "giant" capacitance, and discuss them on the basis of a model of the heterogeneous semiconductor.

β-B is semiconductor, real samples of which quite often are characterized by the "giant" values of capacitance C: they noticeably exceed the geometric value C_0 calculated from the specimen form and static dielectric constant ε: $C_0 = \varepsilon_0 \varepsilon S / d$. Here S is the area of electrodes and d is the distance between them. Effective (apparent) dielectric constant ε_{eff} can be recalculated from the measured low-frequency capacitance using similar formula $C = \varepsilon_0 \varepsilon_{eff} S / d$. Thus, the ratio of effective dielectric constant and real one is defined by the ratio of measured and geometric capacitances of a sample, $\varepsilon_{eff} / \varepsilon = C / C_0$. To the best our knowledge, there are no recent studies devoted to the low-frequency dielectric properties of boron. As for the early data, summarizing them one can mark two main features: (1) polarizing dielectric losses in this material are small, and hence its dielectric constant weakly depends on the frequency of applied electric field; in any case, it is of the order of 10; (2) as a rule, a measured low-frequency capacitance is "giant", exceeding the value calculated from the sample geometry.

2. Experiment

Four types of samples were studied in the present work:

No. 1 – pure macrocrystalline material obtained by vacuum deposition of boron on boron filament with subsequent zone melting. It contains ≤ 30 ppm carbon and has room-temperature resistivity $\sim 2 \cdot 10^7\ \Omega \cdot$ cm.

No. 2 – faceted single crystals grown by vapor–liquid crystallization inside the voids nucleated in boron melt. In such a material foreign atoms are manly segregated at the planar defects, and their single-crystalline matrix contains significantly less impurities than crucible-melted polycrystals (see below No. 4). This assumption, based on the structural features, is supported by the room-temperature resistivity value $\sim 8 \cdot 10^6\ \Omega \cdot$ cm, which significantly exceeds that for the crucible-melted samples.

No. 3 – zone-melted single-crystal specimens with purity 99.99 % and room-temperature resistivity $\sim 2 \cdot 10^6\ \Omega \cdot$ cm.

No. 4 – polycrystals melted in crucible with purity 99.95 % and room-temperature resistivity $\sim 4 \cdot 10^5\ \Omega \cdot$ cm.

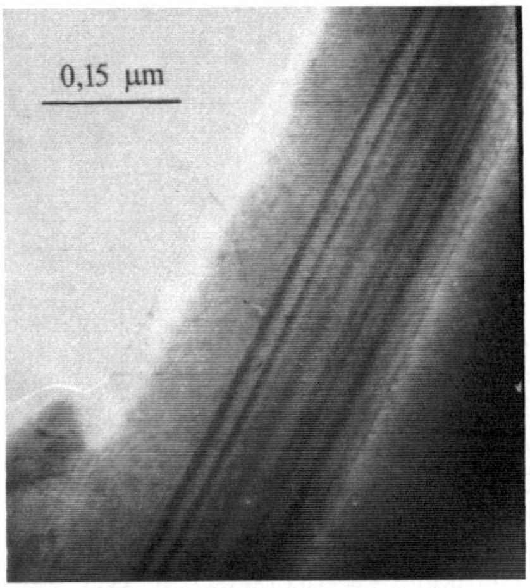

Figure 1. Electron-microscopic image of polysynthetic twin packet in β-rhombohedral boron single crystal.

X-ray diffraction confirms that all samples have the β-rhombohedral structure. Stacking faults and twinned layers are observed by electron diffraction. The image of the polysynthetic twin packet in β-rhombohedral boron single crystal presented in figure 1 is typical for all investigated samples. It is not easy to measure the planar defect concentrations. Even if they are assumed equal in specimens of various origins, they should be distinguished by orientations of the defect planes. Planar defects in pure β-B single crystals prefer $\{101\}$ planes of the elemental rhombohedron faces. However, impurities accumulated at the planar defects may randomize their orientations to some extent.

Prismatic specimens were cut out from rods by a diamond saw. Their surfaces were mechanically polished using boron carbide powders and then the formed amorphous surface layers were etched in boiling nitric acid solution. Tail-end surface cleaning was made using solution containing alkaline potassium before flushing with distilled water. Such treatment excludes any interface capacitance. The pairs of symmetrical layer-electrodes were formed by the vacuum sputtering of silver onto the opposite flat surfaces of the tested specimens.

Dielectric parameters were measured within the frequency range $\omega / 2\pi = 10^2 - 10^4\,\text{Hz}$ (ω is the cyclic frequency) in the temperature range between boiling points of nitrogen and water $T = 77 - 373\,\text{K}$. Electrical oscillations were excited applying a low-frequency generator. Temperature of a tested sample was estimated by a thermocouple, soldered end of which was placed inside a similar sample placed nearby in the same chamber.

The capacitance parameters are summarized in table 1. Geometric capacitances are specified in the second column. They were calculated taking the β-B static dielectric constant as $\varepsilon = 8.0$. In the whole frequency and temperature ranges, the measured values of capacitance for all samples were "giant", i.e., significantly exceeding corresponding geometric capacitances, $C / C_0 \sim 10 - 60$.

Table 1. Measured capacitance parameters of β-rhombohedral boron samples.

Sample	C_0, pF	C_{1+2} / C_0	C_1 / C_0	C_1 / C_{1+2}	p_2 / p_1
No. 1	1.0	15	43	2.9	1.9
No. 2	1.5	16	34	2.1	1.1
No. 3	2.4	10	23	2.3	1.3
No. 4	1.1	20	58	2.9	1.9

Frequency and temperature dependences of the dielectric properties of purer crystals (samples No. 1 and No. 2) are shown in figures 2 and 3. In the low-temperature region, capacitance practically does not depend on the temperature and has certain constant value C_1. With increasing temperature, it rapidly decreases to the new value C_{1+2}, and further, right up to ~ 290 K, again does not change. At this point, capacitance suddenly comes back to the former value C_1, and at higher temperatures also remains constant. As for the dielectric loss tangent $\tan\delta$, it is small in the whole low-temperature region and does not depend on temperature. Losses start to increase only near the room temperature, and the sudden increase in C at ~ 290 K is accompanied by the same-type reduction in $\tan\delta$ value. In the temperature ranges where the dielectric parameters of the samples are constant, they are frequency independent as well. Locations of the features on $C - T$ and $\tan\delta - T$ curves also do not depend on ω: at any frequency they are observed at ~ 290 K. On the contrary, with increase in frequency the regions of decrease in capacitance are displaced towards low temperatures (they can even fall outside the measurements range), and dielectric losses, which depend on temperature only near the room temperature and above, weaken.

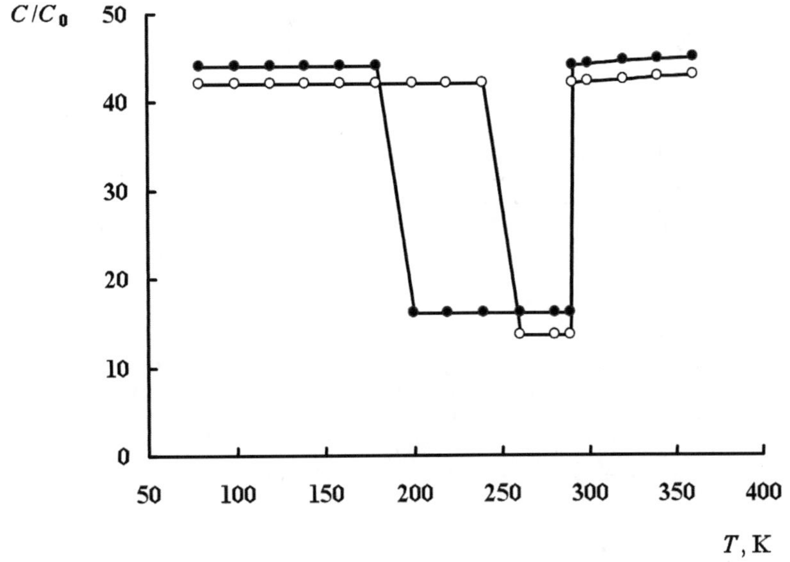

(1)

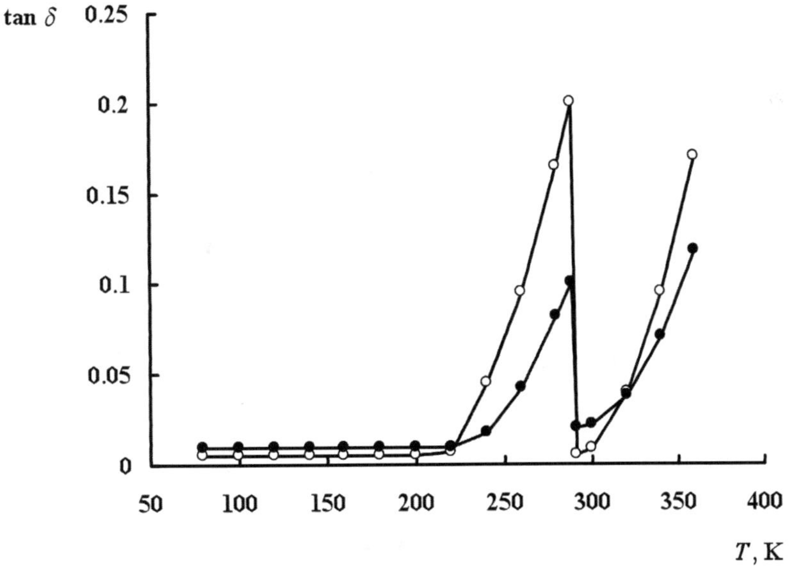

(2)

Figure 2. Temperature dependences of capacitance (1) and dielectric loss tangent (2) for pure macrocrystalline β-rhombohedral boron measured at frequencies $2 \cdot 10^2$ (white circles) and $5 \cdot 10^3$ Hz (black circles).

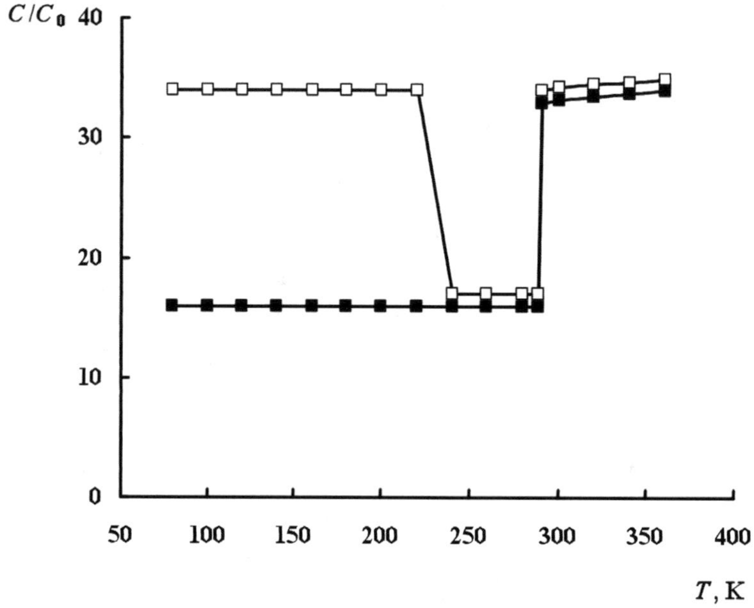

(1)

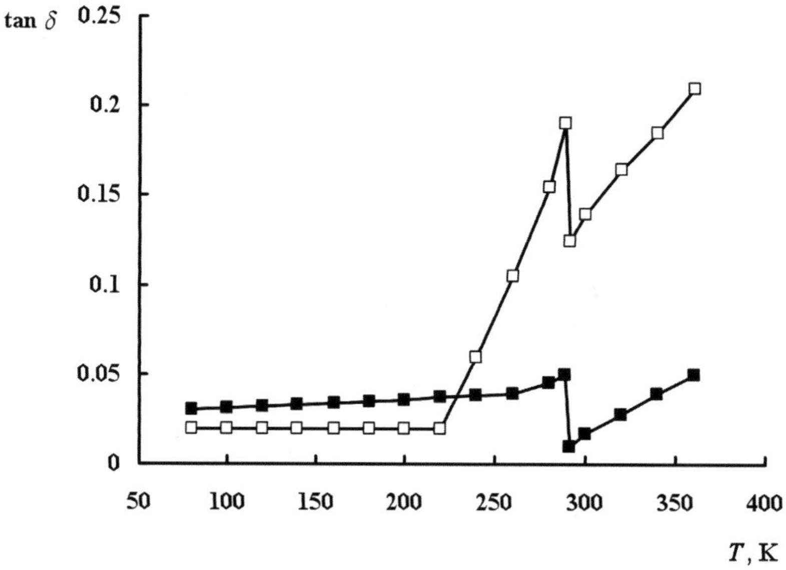

(2)

Figure 3. Temperature dependences of capacitance (1) and dielectric loss tangent (2) for faceted single-crystalline β-rhombohedral boron measured at frequencies $1 \cdot 10^3$ (white squares) and $1 \cdot 10^4$ Hz (black squares).

In samples with lesser purity (No. 3 and No. 4), dielectric losses are stronger. In general, their temperature characteristics are similar to the described above. However, curves for the given frequency were only partially reproducible. Within the measurement errors, capacitance values C_1 and C_{1+2}, and temperature of jumps ~ 290 K are the same. However, neither measured values of $\tan\delta$ nor regions of decrease in capacitance are reproducible.

3. Discussion

In general, heterogeneity is known to be a rationale for any abnormal change in measured values of dielectric or other electrophysical parameters of a material [11].

Results of the present measurements, together with earlier empirical data on the dielectric properties of β-B real crystals, can be explained within a model of the crystalline semiconductor heterogeneous by the conductivity. Indeed, temperature dependence of the capacitance of such semiconductor should be a step-like function with levels determined by the sum (in the specimen thickness) of layers with relatively low conductivity within given temperature region, because they are "short-circuited" by the relatively high-conductive layers; and these levels may significantly exceed the geometric capacitance. As for the temperature dependence of dielectric losses in heterogeneous semiconductor, it should follow that of the specimen conductivity averaged in some way (for efficient numerical simulation of the alternating current conduction in heterogeneous materials see, for example, [12]). All semiconductors, including β-B, reveal monotonous growth in their ohmic conductivity with temperature. Correspondingly, dielectric losses should reveal a general growth trend. However, inside the inhomogeneous specimen, between two adjacent layers with different ohmic conductivities, a barrier-layer with non-ohmic conductivity can be formed with an unusual temperature dependence of the conductivity.

There are many different physical mechanisms causing heterogeneous conductivity related with chemical impurities and structural defects in semiconductors, like the heterogeneous doping, impurities segregation at the grain boundaries in polycrystalline specimens, formation of complexes of impurity atoms with radiation defects in irradiated material (see [13]), etc. There is a strong evidence for heterogeneity effects in some single-crystalline and chemically pure semiconductors. Apparently, in such cases the layers with different conductivities can be formed only due to piezoresistance effect induced by the local elastic stresses, which are related with twins and stacking faults existing in real single crystals. Indeed, elastic textures, such as twin and anti-phase boundaries, can generate electronic heterogeneity in materials with strong electron-phonon coupling [14]. Note that such coupling is characteristic for β-B.

To answer a question, what kind of heterogeneity is responsible for unusual dielectric properties of β-B, we should pay attention to the fact that the high capacitances are measured not only for contaminated polycrystals, but also for pure single crystals. Therefore, the models, based on significant concentration of impurities and/or grain boundaries between fine crystallites, have to be excluded. On the one hand, even pure single crystals of β-B can include twinned layers and stacking faults, which are capable to combine into micro-twins; and metallographic analysis of the fracture faces in zone-melted boron reveals accumulation of high residual stresses in twinning planes. On the other hand, β-B exhibits piezoresistance, which can lead to the local changes in conductivity of the stressed layers adjacent to planar structural defects and to the increase in the electrical resistance of β-B single crystal due to the internal stresses during the plastic deformations. Correlation was found [15] between heat-treatment modifications in mechanical and electric properties of samples, which reveals influence of stresses on the β-B conductivity.

For estimation of the internal pressure value p in stressed layers adjacent to twinning planes in β-B crystals, we can use its elastic parameters, e.g., Young modulus $E = 4.58 \cdot 10^{11}$ Pa [16]: $p \sim 10^{12}$ Pa. Piezoresistance coefficient Π of β-B single crystal for uniaxial deformation is known to be $\Pi \sim 10^{-9}/\,\text{Pa}$ [17]. Then, we obtain $\Pi p \sim 10^3 >> 1$. Thus, the conductivity within the stressed layers will be smaller by few orders of magnitude than in the rest of crystalline β-B. Note that piezoresistance effect in β-B can reflect two possible effects causing heterogeneity by the conductivity: changes in hopping mobility of the localized charge carriers and/or in concentration of the localization centers controlling hopping conductivity. As for the thickness d of the stressed layers including twinning planes in β-B crystals, it can be estimated using the lattice constant value $a = 10.15\,\text{Å}$ [18]. For example, for the sample No. 4, presence of the low-conductive layers with thickness $d \sim 10\,\text{Å}$ increases capacitance by $C_1 / C_0 \sim 60$ times. Then, at least one planar defect should exist approximately on every 60 × 10 Å = 600 Å of the specimen thickness. The obtained value is reasonable for linear density of such defects in β-B.

Presence in our $C - T$ characteristics of only two different horizontal levels C_1 and C_{1+2} placed noticeably higher than C_0 level shows that for describing the low-frequency dielectric properties of the tested β-B samples it is sufficient to consider only two kinds of low-conductivity layers. They can be treated within a simple barrier model of the heterogeneous semiconductor, according to which in a crystal there are ohmic inclusions with relatively lower conductivity together with adjacent barrier layers. At low temperatures, probably, only low-conductive stressed ohmic layers contribute to the capacitance. If their sum in the sample thickness is p_1 then we get $C_0 / C_1 = p_1$. Upon temperature increase, when conductivity of the barrier layers decreases, those two start to affect the measured capacitance. If the sum of the barrier layers is p_2, we get $C_0 / C_{1+2} = p_1 + p_2$. Suggested sequence of "actuations" of ohmic and barrier layers follows from the fact that averaged ohmic conductivity of β-B always increases with temperature. The fact that values C_1 / C_0 and C_{1+2} / C_0 for samples of various origins vary noticeably is not unexpected, because these ratios are determined by the concentration and orientation of planar defects. Meanwhile, the ratio $C_1 / C_{1+2} = 1 + p_2 / p_1$ of two levels of capacitance expressed by the ratio p_2 / p_1 is constant to some extent. The given result is one more argument in favor of the barrier model accepted for β-B, which suggests correlation between sums of the low-conductive layers of two kinds.

Analysing obtained differences in C / C_0 ratios between tested samples, we should take into account that maximal effect of the relatively low-conductive thin planar inclusions on the sample capacitance takes place when inclusions are perpendicular to the applied bias, while they have almost no effect being parallel to the field. In case of randomly oriented inclusions, only layer projection on plane perpendicular to the applied electric field acts effectively. On the other hand, measured ratio C / C_0 increases as the sum of low-conductive layers decreases. Thus, higher values are expected in polycrystalline samples with random orientations of the planar defects regarding the applied bias. Such conclusion of the proposed theory agrees well with experimental data showing general trend that ratios C / C_0 of the single-crystal samples No. 2 and 3 are smaller compared with those of macro- and polycrystalline samples No. 1 and 4.

As it was mentioned, there are no recent studies devoted to the "giant" capacitance of the real β-B crystals. There are few new experimental studies [19-21] of other related properties, like the time-dependent photoconductivity of pure and carbon-doped β-B, as well as the correlation between electronic properties and structural defects in β-B, which could be helpful in explaining our results. According to them, in β-B the influence of structural defects on the electronic properties is much stronger than in common semiconductors. Conductivity of the pure β-B is controlled by the centres of carrier localization related to the point structural defects, preferably certain vacancies, high concentrations of which are characteristic for the real structure of β-B. At ~ 380 K and probably at lower temperatures slow structural relaxations, e.g., diffusion of certain interstitial atoms into energetically more favourable vacant sites, should take place. It influences electrical conductivity causing statistical fluctuations (stochastic precipitous jumps) and is responsible for some structural hysteresis effects in β-B crystals. Besides, carbon donor-states in the β-B band gap were identified, which can affect time-dependent conductivity at the sufficient concentrations of carbon. Consequently, irreproducibility of some dielectric parameters of the unrefined β-B crystals No. 3 and 4 can be considered as a result of the transient conductivity characteristic for β-B.

In the proposed model, abrupt capacitance recovery at temperature ~ 290 K, which is independent of the applied electric field frequency and is accompanied by the jump in dielectric losses, specifies elimination of conditions for existence of the low-conductive barrier layers. This effect can not be connected with displacements of the trace impurities segregated on planar defects as it is found not only in contaminated polycrystals, but also in purest macrocrystalline material with conductivity controlled by the intrinsic point defects. The reason seems to be related to the above-mentioned structural relaxations in β-B. The most reliable explanation can be based on a moderate-temperature structural transition recently predicted [22] in β-phase of boron. Using density functional theory methods and an extensive search of the configuration space, a unique, energy-minimizing pattern of occupied and vacant sites was found that could be stable at low temperatures and break β-rhombohedral symmetry. However, alternative configurations are close in energy, allowing the entropy of partial occupancy to stabilize the β-rhombohedral structure through a phase transition. The strong heat capacity peak around ~ 300 K revealing intrinsic phase transition is only a crude estimate (because of several approximations made) and difficult to observe experimentally (because of slow atomic diffusion at this temperature). Nevertheless, the closeness of the transition temperature with ~ 290 K measured for abrupt changes in dielectric properties does not seem accidental. However, without detailed comparative analysis of the fine microstructures characteristic for β-B matrix and stressed inclusions, it is impossible to imagine how such transition promotes sample homogeneity by the concentration of hopping carriers.

4. Conclusions

Summarizing investigations carried out on temperature and frequency dependences of the dielectric properties of β-rhombohedral boron, we come to conclusion that the principal cause of the measured "giant" low-frequency capacitance of this semiconductor is propensity of its crystalline structure to form planar defects, namely twins. In the adjacent regions, twinning boundaries generate stresses causing piezoresistance, which is sufficient to form inclusions in form of thin layers with noticeably reduced conductivity. Besides, barrier layers arise at the borders of such layers with the matrix; those layers could have sufficiently low conductivity. Such heterogeneity of the conductivity in real β-rhombohedral boron crystals means an opportunity of electric charge accumulation not only on the sample interfaces with electrodes, but also in the bulk – at borders between areas with relatively low and relatively high conductivities. As a result, effective thickness of the equivalent capacitor decreases and the measured value of capacitance increases in comparison with the geometric one.

References

[1] Ramirez A P, Subramanian M A, Gardel M, Blumberg G, Li D, Vogt T and Shapiro S M 2000 *Solid State Commun.* **115** 217
[2] Subramanian M A, Li D, Duan N, Reisner B A and Sleight A W 2000 *J. Solid State Chem.* **151** 3232
[3] Homes C C, Vogt T, Shapiro S M, Wakimoto S and Ramirez A P 2001 *Science* **293** 676
[4] Sinclair D C, Adams T B, Morison F D and West A R 2002 *Appl. Phys. Lett.* **80** 2153
[5] Si W, Cruz E M, Johnson P D, Barnes P W, Woodward P and Ramirez A P 2002 *Appl. Phys. Lett.* **81** 2056
[6] Homes C C, Vogt T, Shapiro S M, Wakimoto S, Subramanian M A and Ramirez A P 2003 *Phys. Rev.* B **67** 092106
[7] Cohen M H, Neaton J B, He L and Vanderbilt D 2003 *J. Appl. Phys.* **94** 3299
[8] Wu L, Zhu Y, Park S, Shapiro S, Shirane G and Tafto J 2005 *Phys. Rev.* B **71** 014118
[9] Cohn J L, Peterca M and Neumeier J J 2005 *J. Appl. Phys.* **97** 034102

[10] Shi N and Ramprasad R. 2007 *Phys. Rev.* B **75** 155429
[11] Kuchis E V 1990 *Galvano-Magnetic Effects and Methods of Their Investigation* Moscow (Radio i svyaz')
[12] Sahimi M, Naderian M and Ebrahimi F 2005 *Phys. Rev.* B **71** 0942208
[13] Pagava T A and Chkhartishvili L S 2004 *Ukr. J. Phys.* **49** 1006
[14] Ahn K H, Lookman T, Saxena A and Bishop A R 2005 *Phys. Rev.* B **71** 212102
[15] Dugladze G M, Khvedelidze A G, Lominadze J V and Chkhartishvili L S 1984 *Abs. 8th Int. Symp. Boron, Borides, Carbides, Nitrides & Rel. Comp.* Tsagareishvili G V (Ed) Tbilisi (Metsniereba) 140
[16] Tsagareishvili G V and Tsagareishvili D Sh 1990 *Thermal and Elastic Properties of Boron* Tbilisi (Metsniereba)
[17] Umarov B V and Tashmirsaev M A 1986 *J. Less-Comm. Met.* **117** 199
[18] Tsagareishvili G V Antadze M E and Tavadze F N 1991 *Synthesizing and Structure of Boron* Tbilisi (Metsniereba)
[19] Schmechel R and Werheit H 2000 *J. Solid State Chem.* **154** 68
[20] Werheit H, Schmidt M, Schmechel R and Londström T 2000 *J. Solid State Chem.* **154** 93
[21] Werheit H and Wehmöller B 2003 *J. Solid State Chem.* **177** 556
[22] Widom M and Mihalkovič M 2008 *Phys. Rev.* B **77** 064113

Journal of Physics: Conference Series **176** (2009) 012022 doi:10.1088/1742-6596/176/1/012022

Isotopic composition dependences of lattice constant and thermal expansion of β-rhombohedral boron

Domenti Gabunia[1], Otar Tsagareishvili[1], Levan Chkhartishvili[1,2] and Levan Gabunia[1]

[1] Laboratory for Boron and Its Compounds, F Tavadze Institute of Metallurgy and Materials Science, 15 A Kazbegi Avenue, Tbilisi, 0160, Georgia

[2] Department of Physics, Georgian Technical University, 77 M Kostava Avenue, Tbilisi, 0175, Georgia

E-mail: chkharti2003@yahoo.com

Abstract. Boron isotope substitution in boron-containing crystalline structures is important because many properties of these materials are structure sensitive. In the present work, we studied the unit cell expansion at the fixed temperature and temperature dependence of thermal expansion for the β-rhombohedral modification of boron, when ^{11}B is replaced by ^{10}B isotope. Theoretical analysis of experimental results associates the observed changes in structures of the β-rhombohedral boron samples of various isotope compositions with isotope effects rather than with technologically concomitant impurities.

1. Introduction

Boron, together with boron-rich alloys and compounds, forms the class of materials suitable for fabrication of devices and products, which can be used in extreme conditions of high temperatures, chemically aggressive environments, strong electric fields, irradiation, etc. Since in the Nature boron can be found in form of two stable isotopes ^{10}B and ^{11}B with masses differing by ~ 10 %, the physical properties of boron modifications and boron-rich solids are expected to be significantly affected by their isotopic compositions.

Indeed, there is a number of boron isotopic effects: fluctuations of the equilibrium positions of atoms and of the electron-phonon coupling constant can lead to the intrinsic electron localization in semiconducting boron characterized by the narrow electron bands [1]; changes in the isotopic composition shift the phonon frequencies in boron carbide [2]; the heat flow increases in isotopically pure β-rhombohedral boron crystals, because of decrease in the phonon scattering [3]; replacement of one of the boron isotope by another changes the critical temperature of the high-temperature superconductor magnesium diboride [4], as well as melting temperature [5] and melting enthalpy [6] of pure boron itself.

As it is known, variations in isotopic composition affect main characteristics of the crystalline semiconductors not only by the renormalization of the phonon spectrum, but also through the changes in lattice parameters [7-9]. The purpose of present work is experimental and theoretical estimation of the isotopic effect on the structure of β-rhombohedral boron (β-B).

2. Experimental details

Macrocrystalline β-B samples were produced by floating-zone refining of the ^{10}B (enrichment 97.2 %, purity 98.5 %) and ^{11}B (enrichment 99.8 %, purity 98.3 %) isotopically modified powders' mixed at certain compositions. With the purpose of relaxation of the strains arisen in the production process, elimination of pores, and healing of microdefects in crystals, before testing samples were exposed to annealing in argon atmosphere at temperatures up to ~ 1300 K for 10 h. Contents of main metallic impurities in zone-refined specimens are following: Al – about 0.001, Ca – 0.004 ÷ 0.008, Cu – 0.001 ÷ 0.003, Fe – less than 0.001, Mg – 0.001 ÷ 0.003, Mn – less than 0.001, and Ni – 0.001 ÷ 0.002 wt %. Besides, they contained 0.010 ÷ 0.014 wt % Si. As for the carbon C, the major impurity in β-B, its content in initial powders was ~ 1 wt %, while that in refined samples was less than 0.06 wt %. All these β-B specimens were produced using the same initial material (high-pure boron powders) in the same technological conditions; measurements were also performed using the same equipment and methods.

X-ray diffraction patterns were recorded at room temperature using automatic X-ray diffractometer HZG – 4A. Software allowed scanning of the most reliable maxima for the purpose of measuring the β-B unit cell parameters. Continuous diffractograms from the rotating samples were obtained using the nickel-filtered radiation, anode voltage 20 ÷ 30 kV, anode current 15 ÷ 20 mA, counting rate 10^3 pps, time constant 3 s^{-1}, plotter sensitivity 100 mV /cm^2. Averaged errors of measurements of the hexagonal unit cell parameters a_{hex} and c_{hex} were 0.0007 and 0.0009 Å, respectively. This translates into averaged error of 0.0005 Å for the unit rhombohedron edge a values, which are calculated from a_{hex} and c_{hex}.

X-ray structural analysis of the investigated samples reveals the general trend of growth of the unit rhombohedron edge a in dependence on the ^{10}B isotope content x in the range 0.2 – 97.2 %. The results are presented in table 1 and figure 1.

Table 1. Lattice constant values of β-rhombohedral boron samples with various isotopic compositions at room temperature.

No	^{10}B content, %	Lattice parameter a, Å
1	0.2	10.1233
2	1.8	10.1279
3	10.0	10.1335
4	19.0	10.1366
5	60.0	10.1436
6	87.8	10.1543
7	97.2	10.1560

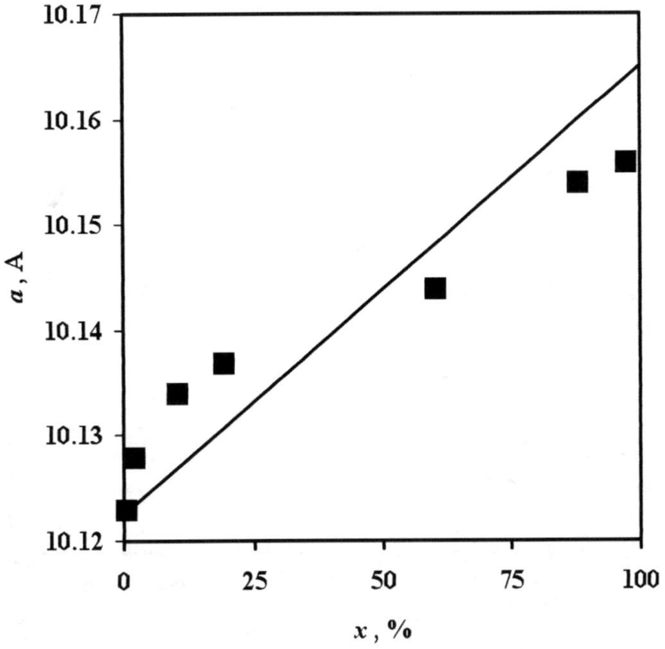

Figure 1. β-rhombohedral boron lattice constant in dependence on ^{10}B isotope content; black squares – experimental values, solid line – trend theoretically expected from the quasi-classical B–B interatomic potential.

High-temperature thermal expansion $\Delta \equiv \Delta l / l$ of the β-B specimens with sizes of $3 \times 3 \times (10 \div 12)$ mm^3 has been investigated in the temperature range 400 – 1100 K using vacuum quartz dilatometer equipped with inductive sensor. Heating rate was 3 K / min. Errors of relative elongation and temperature measurements were 3 % and 0.5 K, respectively. The elongations temperature-curves for samples with relatively low and relatively high composition of ^{10}B isotope are shown in figures 2 and 3.

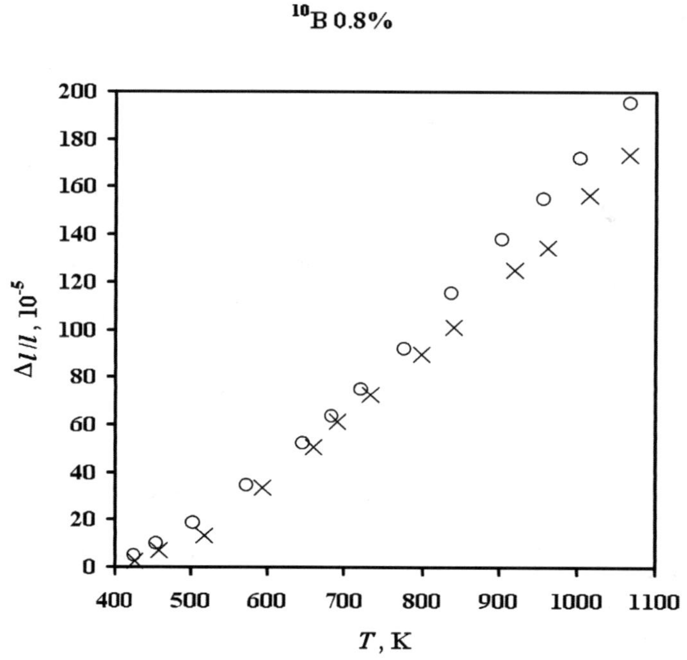

(1)

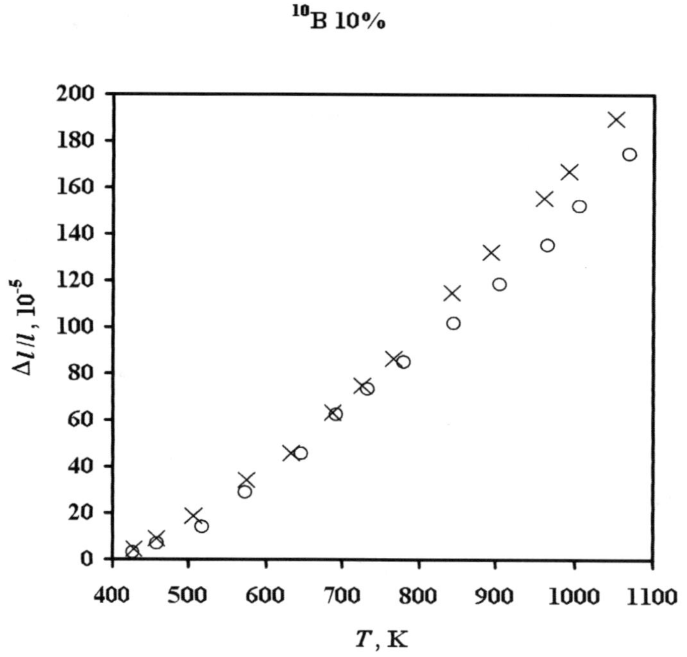

(2)

Figure 2. Thermal expansion of β-rhombohedral boron crystals containing 0.8 (1) and 10 % ^{10}B (2); circles – experimental points, crosses – data calculated from the experimental results using universal theoretical relation.

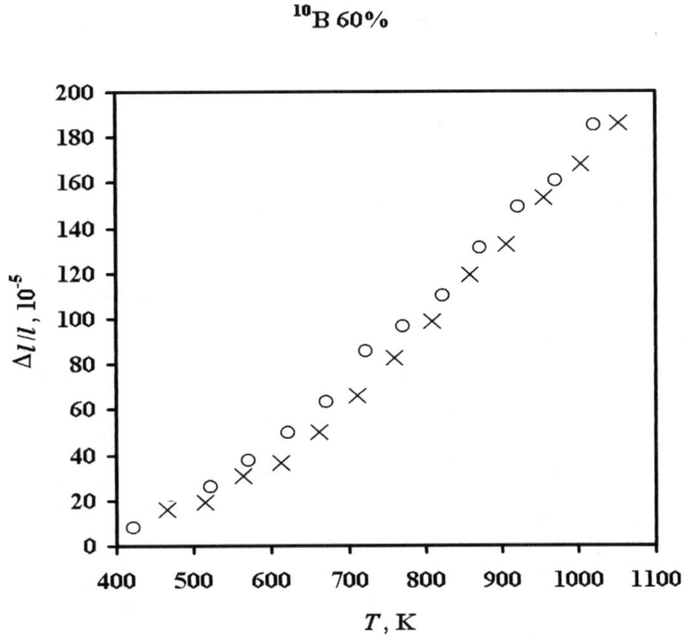

(1)

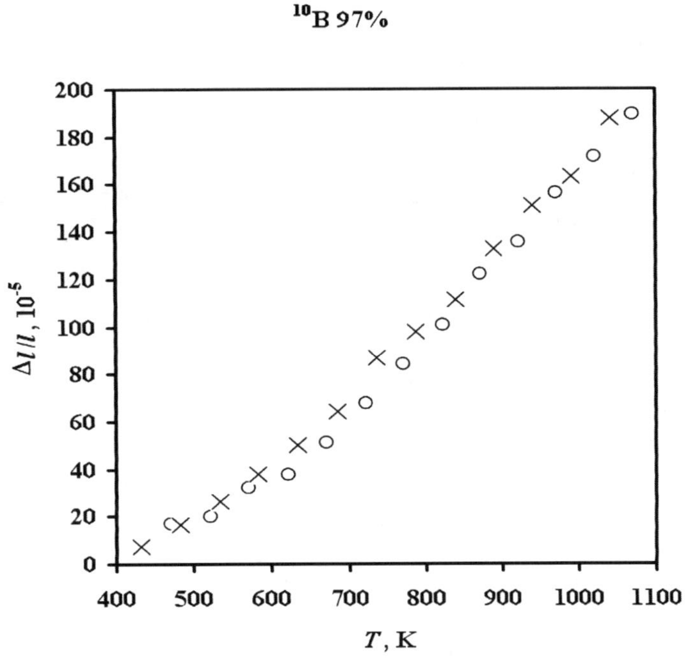

(2)

Figure 3. Thermal expansion of β-rhombohedral boron crystals containing 60 (1) and 97 % ^{10}B (2); circles – experimental points, crosses – data calculated from the experimental results using universal theoretical relation.

3. Results and discussion

Isotopic composition effect on the crystalline structure built up from the light atoms theoretically can be considered using some first-principles approaches [10-12] or special semi-empirical pair interatomic potentials (for instance, parameterized in form of Mie–Lennard–Jones [13,14]). However, such calculations for semiconductors with simple crystalline structures are practically impossible in case of β-B structure with 105 regular atomic sites in the unit cell. Therefore, our analysis will be carried out within the frames of the diatomic model of a crystal [15].

In this model, the pair B–B interatomic interaction potential energy U can be written as

$$U(a) = U(a_0) + c(a - a_0)^2 - g(a - a_0)^3, \tag{1}$$

where a is the interatomic distance; a_0 is its equilibrium value corresponding to the potential energy minimum $U(a_0)$; $c > 0$ and g are the constants at harmonic and anharmonic terms, respectively.

Constant g is positive or negative depending on the sign of thermal expansion coefficient. There are boron-rich solids (some rare-earth-metal dodecaborides) with negative thermal expansion coefficient in certain temperature ranges [16]. This fact is considered as a consequence of nearly non-interacting free oscillations of the metal ions in cavities of a simple cubic lattice formed by B_{12} cage units. However, β-B structure has only covalent bonds (some deficit of the bonding electrons characteristic for ideal structure in real β-B crystals is exactly compensated by the point defects stabilizing the structure [17]). Then, it is not surprising that thermal expansion coefficient for all crystalline modifications of boron is positive [18-21]. This is why a boron crystal can be modelled by the assembly of diatomic bonds with $g > 0$.

According to the model, the averaged relative change in equilibrium interatomic distance $\overline{\Delta a_0} / a_0$ caused by anharmonic correction to the harmonic potential equals to

$$\frac{\overline{\Delta a_0}}{a_0} = \frac{3g\overline{E}}{4a_0c^2} = \frac{3g}{4a_0c^2}\left(\frac{\hbar\omega}{2} + \frac{\hbar\omega}{\exp(\hbar\omega/kT) - 1}\right). \tag{2}$$

Here $\overline{E}$ is the averaged energy of relative interatomic vibrations in diatomic system within the harmonic approximation. It was demonstrated [22] that same simple relation can be obtained using rigorous statistical method, while macroscopic theory of thermal expansion based on corresponding diatomic model allows to estimate [23] thermal expansion coefficient values for number of crystalline substances. The diatomic model is successfully used in calculations of the anharmonicity effects in solids [24].

In general, the averaged energy of relative interatomic vibrations in diatomic system within the harmonic approximation is a sum of two terms. One of them, $\hbar\omega/2$, is temperature independent and related to the zero-point vibrations with cyclic frequency ω, while another is temperature dependent and related to the thermal excitations. If the lattice temperature T is low, $\hbar\omega/kT >> 1$, the main contribution in $\overline{E}$ is zero-point vibrations and thus it can be approximated as $\overline{E} \approx \hbar\omega/2$. At high temperatures, $\hbar\omega/kT << 1$, the vibration energy of atoms is mainly determined by the crystal temperature, $\overline{E} \approx kT$, and not by their masses. This results in no isotopic composition influence on the interatomic distances mediated by the thermally excited phonons. Thus, at any temperature (for example, at room temperature), isotopic composition influence on the lattice parameter is quantum effect exclusively related to the zero-point vibrations and, consequently, in the formula (2) one can substitute $\overline{E}$ for $\hbar\omega/2$ even in the high-temperature range.

Let us firstly consider the case when the masses of all constituent atoms have one and same value and denote reduced mass of any atomic pair by μ. In this case $\omega = \sqrt{2c/\mu}$ and

$$\frac{\overline{\Delta a_0}}{a_0} = \frac{3\hbar g}{4a_0c\sqrt{2c\mu}}. \tag{3}$$

The right part of the above formula would describe anharmonic relative changes in any structure parameter of a crystal if the crystal contains only chemically identical atoms with equal masses. It is convenient to analyze the isotopic composition effect on the structure parameters by considering elemental boron crystal with a random distribution of two isotopes in its otherwise perfect lattice. Namely, aiming to separate isotopic contribution, we suppose that crystal consists of two isotopes with masses M_1 and M_2 at $M_1 < M_2$ and with relative concentrations x and $1-x$ randomly distributed in the lattice sites ($0 \le x \le 1$). There exist three types of diatomic pairs with reduced masses of $M_1/2$, $M_2/2$, $1/(1/M_1+1/M_2)$, and probabilities of x^2, $(1-x)^2$, $2x(1-x)$, respectively. Then, generalized equation (3) leads to

$$\frac{\overline{\Delta a_0(x)}}{a_0} = \frac{3\hbar g}{4a_0 c\sqrt{c}}\left(\frac{x^2}{\sqrt{M_1}} + \frac{(1-x)^2}{\sqrt{M_2}} + x(1-x)\sqrt{2\left(\frac{1}{M_1}+\frac{1}{M_2}\right)}\right). \quad (4)$$

Here term $3\hbar g/4a_0 c\sqrt{cM_2}$ expresses the relative change in the structural parameter of monoisotopic crystal (built up only from atoms of the heavier isotope) caused by the thermal effect. As for the isotopic effect, it is expressed by the reminder

$$\delta(x) = \frac{3\hbar g x}{4a_0 c\sqrt{c}}\left(\left(\sqrt{2\left(\frac{1}{M_1}+\frac{1}{M_2}\right)} - \frac{2}{\sqrt{M_2}}\right) - x\left(\sqrt{2\left(\frac{1}{M_1}+\frac{1}{M_2}\right)} - \frac{1}{\sqrt{M_1}} - \frac{1}{\sqrt{M_2}}\right)\right). \quad (5)$$

In general case, function $\delta = \delta(x)$ exhibits the maximum at the concentration of

$$x_{\max} = \frac{\sqrt{2(1+M_2/M_1)} - 2}{2(\sqrt{2(1+M_2/M_1)} - \sqrt{M_2/M_1} - 1)}. \quad (6)$$

However, condition $M_1 < M_2$ yields inequality $x_{\max} > 1$ contradicting to the restriction $0 \le x \le 1$.

Equation (5) determines a relative change in a structural parameter value arisen as a result of substitution of the x part atoms of isotope with mass M_2 by the atoms of isotope with mass M_1. Note that both coefficients of the linear form in parenthesis are positive, and first coefficient exceeds second one if $M_1 < M_2$; besides, $g > 0$. Consequently, substituting the heavier isotopes by the lighter ones one can expect the quadric (in respect to concentration) expansion of crystal structure parameters: $\delta(x) > 0$. However, for most chemical elements, including boron, $M_1 \approx M_2$ (only exception is hydrogen) and the linear form in parenthesis is almost constant. It means that function $\delta = \delta(x)$ is nearly linear:

$$\delta(x) \approx \delta_{\max} x, \quad (7)$$

where

$$\delta_{\max} = \frac{3\hbar g}{4a_0 c\sqrt{c}}\left(\frac{1}{\sqrt{M_1}} - \frac{1}{\sqrt{M_2}}\right) \quad (8)$$

denotes maximal expansion when all heavier isotopes are substituted by the lighter ones ($x = 1$).

The numerical estimations of isotopic composition effect on the boron crystal structure parameters given below are based on B–B interatomic interaction potential function obtained previously [25,26] within the quasi-classical approach. Corresponding dependence of the binding energy $E_{Binding} = E_{Binding}(a)$ of this diatomic system from the interatomic distance a is presented in figure 4.

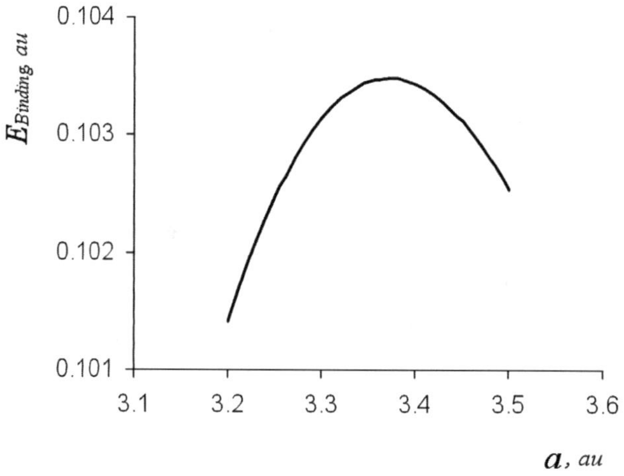

Figure 4. Boron–boron binding energy in dependence on interatomic distance.

Note that a parameter of the corresponding potential curve like the core charge of $Z^* \approx 3.151$, which is near the boron atom valence of 3, already was with success applied [6] to interpret isotopic composition effect on boron melting characteristics. This curve leads to equilibrium values of interatomic distance and binding energy of $a_0 = 3.37\,\text{au}$ and $E_{Binding}(a_0) = 0.103\,\text{au}$. If the atomic mass of boron is taken as the mass weighted according to its natural isotopic composition, the relative vibration frequency ω in B–B diatomic system is $\omega = 4.80 \cdot 10^{-3}\,\text{au}$. These quasi-classical values deviate by only by few percents from experimental bond lengths in the main structural unit of boron and boron-rich compounds – icosahedron with boron atoms at vertexes – 3.40 au [27], dissociation energy $0.099\,\text{au}$ [28], and oscillatory quantum $4.79 \cdot 10^{-3}\,\text{au}$ [29] of diboron molecule B_2, respectively. Thus, the quasi-classical B–B interaction potential curve together with equation (5) seems to be useful for estimation of the isotopic composition effect on the boron lattice parameters.

Trend analysis of the quasi-classical potential curve in the vicinity of equilibrium point yields the factors at harmonic and anharmonic terms: $c \approx 0.12\,\text{au}$ and $g \approx 4.0\,\text{au}$. Taking into account masses of isotopes ^{10}B and ^{11}B ($M_1 \approx 10.01$ and $M_2 \approx 11.01$, respectively) we get the constant standing in equation (8) for the relative change in structure parameter under the boron atoms isotopic replacement in the lattice sites: $\delta_{\max} \approx 0.0042$. Thereby, the maximal isotopic effect (it occurs at $x = 1$ meaning that ^{11}B is completely substituted by ^{10}B) is estimated as 0.42 %.

A line described by the approximated equation (7) (i.e. with slope of ~ 0.0042) is shown in figure 1 together with the experimental points. It is evident that the theory explains general trend of the lattice constant growth with lighter isotope concentration. The sign and order of magnitude are in agreement with the theoretical estimation based on quasi-classical B–B interatomic potential, from which follows the approximately linear concentration-dependence of the boron-rich crystal lattice constant.

The temperature dependence of thermal expansion $\alpha = \alpha(T)$ has been analyzed based on the Zhernov's universal relation between the thermal expansion coefficient for two crystals with different averaged masses of constituent atoms M' and M'' [8]:

$$\alpha'(T) = \alpha''(T\sqrt{M'/M''}). \tag{9}$$

In case of two isotopes, the averaged mass M is determined as

$$M = xM_1 + (1-x)M_2 . \tag{10}$$

It is easy to modify equation (9) for elongation Δ:

$$\Delta'(T) = \sqrt{M''/M'}\ \Delta''(T\sqrt{M'/M''}) . \tag{11}$$

Figures 2 and 3, together with experimental data, show points calculated from the elongation-temperature curves using universal relation (11). There is a good agreement between observed and calculated points. Thus, in temperature range 400 – 1100 K, the experimental curves of thermal expansion of the β-B samples with various isotope compositions with satisfactory accuracy follow the required universal relations [8].

Note that figures 2 and 3 represent data of thermal expansion for β-B samples with relatively low (0.8 and 10 %) and relatively high (60 and 97 %) contents of ^{10}B isotope, respectively. A curve is made up from the experimental points for one (labelled) isotopic composition and theoretical points evaluated for this composition from the experimental ones for another (counterpart) isotopic composition using universal relation between two temperature-curves of the samples with different isotopic compositions. On the one hand, errors in elongation measurements are about 3 %. It is evident that relative difference between curves obtained for β-B samples with different isotopic composition should be less than maximal relative difference between masses of two boron stable isotopes ~ 10 %. Therefore, if the points shown in figures 2 and 3 are combined, the difference between curves will be barely distinguishable. In spite of this, one may believe that these experimental curves demonstrate universal theoretical relation valid for isotopic effect.

The isotopic changes in β-B lattice parameter measured at room temperature (figure 1) are larger in ^{10}B enriched crystals, in agreement with quantum correlation formula (8) determining isotopic effect at any fixed temperature. However, the same order does not to work for the thermal elongation temperature-curves presented in figures 2 and 3. The explanation may be related with the losses of linearity: according to general formula (2), at sufficiently high temperatures, $\hbar\omega / kT << 1$, averaged relative elongation of a crystal is a linear function of temperature almost independent from its isotopic composition, whereas at relatively lower temperatures it should reveal slight, but complicated temperature and averaged isotopic mass dependences.

Regardless of the fact that β-B specimens with different isotopic composition have different impurity contents too, a satisfactory agreement obtained between experiment and theory shows that detected changes in β-rhombohedral micro- and macro-structural parameters to a greater extent are of isotopic origin.

References

[1] Berezin A A 1984 *J. Chem. Phys.* **80** 1241
[2] Werheit H, Au T, Schmechel R, Shalamberidze S O, Kalandadze G I and Eristavi A M 2000 *J. Solid State Chem.* **154** 792
[3] Nogi N, Noda T and Tanaka S 2000 *J. Solid State Chem.* **154** 296
[4] Bud'ko S L, Lapertot G, Petrovic C, Cunnigham E C, Anderson N and Canfield P C 2001 *Phys. Rev. Lett.* **86** 1877
[5] Chkhartishvili L, Gabunia D, Tsagareishvili O and Metreveli V 2004 *Bull. Georg. Acad. Sci.* **170** 530
[6] Chkhartishvili L S, Gabunia D L and Tsagareishvili O A 2007 *Inorg. Mater.* **43** 594
[7] Zhernov A P 1998 *JETP* **114** 2153
[8] Zhernov A P 1998 *Phys. Solid State* **40** 1829
[9] Zhernov A P 2002 *Phys. Solid State* **44** 992
[10] Pavone P and Baron S 1994 *Solid State Commun.* **90** 295
[11] Cardona M 2000 *phys. status solidi* b **220** 5
[12] Zhernov A P and Inyushkin A V 2001 *Usp. Fiz. Nauk* **171** 827
[13] Magomedov M N 2005 *Tech. Phys. Lett.* **31** 381
[14] Magomedov M N 2006 *High Temp.* **44** 8612

Journal of Physics: Conference Series **176** (2009) 012022 doi:10.1088/1742-6596/176/1/012022

[15] Fermi E 1938 *Moleküle und Kristalle*
[16] Czopnik A, Shitsevalova N, Pluzhnikov V, Krivchikov A, Paderno Yu and Onuki Y 2005 *J. Phys.: Condens. Matter* **17** 5971
[17] Jemmis E D and Prasad D L V K 2006 *J. Solid State Chem.* **179** 2768
[18] Tsagareishvili G V, Nakashidze T G, Jobava S Sh, Lomidze G P, Khulelidze D E, Tsagareishvili D Sh and Tsagareishvili O A 1986 *J. Less-Comm. Met.* **117** 159
[19] Lundström T, Lönnberg B and Bauer J 1998 *J. Alloys & Comp.* **267** 54
[20] Patokin A P, Khrebtov V L and Shirokov B M 2005 *Modern Materials Science – Progress and Problems* Skorokhod V V (Ed) Kiev (IPMS) 463
[21] Gabunia D, Tsagareishvili O, Lezhava D, Gabunia L, Antadze M, Darsavelidze G and Tanaka T 2006 J. *Solid State Chem.* **179** 2944
[22] Frenkel' Ya I *Introduction in Theory of Metals* 1950 Moscow – Leningrad (GITTL)
[23] Novikova S I 1974 *Thermal Expansion of Solids* Moscow (Nauka)
[24] Slutsker A I, Gilyarov V L and Luk'yanenko A S 2006 *Phys. Solid State* **48** 1832
[25] Chkhartishvili L, Lezhava D, Tsagareishvili O and Gulua D 1999 *Trans. AMIAG* **1** 295
[26] Chkhartishvili L, Lezhava D and Tsagareishvili O 2000 *J. Solid State Chem.* **154** 148
[27] Tsagareishvili G V and Tavadze F N 1978 *Semiconducting Boron* Moscow (Nauka)
[28] Vedeneev V I, Gurvich L V, Kondrat'ev V N, Medvedev V L and Frankevich E L 1962 *Chemical Bonds Energy – Ionization Potentials and Electron Affinity. Reference Book* Moscow (Acad. Sci. USSR)
[29] Huber K-P and Herzberg G 1979 *Molecular spectra and molecular structure IV. Constants of diatomic molecules* New York (van Nostrand Reinhold Co)

Journal of Physics: Conference Series **176** (2009) 012023 doi:10.1088/1742-6596/176/1/012023

Thermal switching effect in β-rhombohedral boron

Domenti Gabunia[1], Levan Chkhartishvili[1,2] and Otar Tsagareishvili[1]

[1] Laboratory for Boron and Its Compounds, F Tavadze Institute of Metallurgy and Materials Science, 15 A Kazbegi Avenue, Tbilisi, 0160, Georgia

[2] Department of Physics, Georgian Technical University, 77 M Kostava Avenue, Tbilisi, 0175, Georgia

E-mail: chkharti2003@yahoo.com

Abstract. Considering heat-balance equation for the sample of solid β-rhombohedral boron overheated through the Joule effect, its normalized current–voltage characteristic has been obtained. On this basis, a universal correlation is established between the ambient-temperature resistance and the voltage of thermal switching: when resistance decreases by a factor of ~ 2 the current switches on. According to the criterion, the switching effect observed in boron near the room temperature reveals mainly thermal nature.

1. Introduction

Electric current switching phenomena in the bulk of solids or at the solid-state barrier junctions are known for a long time, but still are a subject of intensive studies (e.g., see [1-4]). This is not only because of the wide field of their technical applications in controlling, measuring, cutting out, rectifying, memory, protective, etc., systems, but also for the diversity of switching mechanisms that can take place in solids. Among them, the thermal switching is one of the most interesting. This kind of the electric current instability is explained by the heat generated due to the current passage in the material with a negative resistivity-temperature coefficient. However, the effect in pure form is difficult to find.

On the one hand, thermal switching is almost impossible in metals, as they commonly have a positive resistivity-temperature coefficient. Because of the electron-phonon scattering, the mobility of the quasi-free electrons is a decreasing function of temperature, whereas their concentration remains constant. On the other hand, the ionic mechanism of conduction, peculiar to dielectrics, can lead only to the irreproducible thermal breakdown accompanied by the irreversible structural changes in the material.

As for current switching observed in semiconductors, it is termed as a thermo-electrical or electro-thermal effect. In the former case, the thermal switching assists the avalanche breakdown, which occurs at voltages exceeding a certain threshold value and means the steep increase of current with voltage. The avalanche current is related with impact ionization of the localized or band electron states otherwise not participating in the electrical conduction at given temperature. In the latter case, the current carriers in semiconductor may be multiplied by their injection from the electrodes or due to the electric field effect. So, the almost pure thermal effect can appear only in semiconductors and only under the special conditions favourable to such kind of switching.

Both crystalline (β-rhombohedral) and amorphous boron can be considered as most probable candidates for realization of the current thermal switching. There are numerous studies on electrical current instabilities in boron (see [5-14] and others). They had shown that at low temperatures, switching reveals an electric character, but above ~ 250 K contribution of the heat generated by the current flow becomes important [11]. In [15], new measurements of the electric field dependence of electrical conductivity in β-rhombohedral boron were presented and discussed according to the various theories (the classical model of hot electrons, the Poole–Frenkel model, models of non-thermally activated hopping by Mott and Shklovskii, the model of the field dependence of small polaron mobility, space charge limited currents) together with a critical review of the previous measurements. In general, three ranges of the $I - E$ dependence are distinguished (I is the electric current and E is the mean electric field strength in the sample): (i) Ohmic behaviour $I \sim E$; (ii) non-Ohmic behaviour and temperature-dependent field dependence; and (iii) non-Ohmic behaviour $I \sim E^2$ independent of temperature, which reveals a change in the transport mechanism rather than an "electrical breakdown". These results have confirmed hopping as the prevailing transport mechanism in high-purity β-rhombohedral boron at room temperature, at least up to fields of about 8 kV / cm. Resistance of the samples of lesser purity usually had been assumed high enough to exclude the Joule heating below the room temperature. On the contrary, in [15] it was found that in high-purity β-rhombohedral boron, already at room temperature and at fields of about 1.5 kV / cm, the samples are significantly heated. In [15], Joule heating was considered problematic and was avoided by means of accommodation of a sample in the evacuated chamber of a close-cycle refrigerator equipped with an additional heater.

In the present work, we consider the different case, when the sample temperature is not fixed and the thermal switching can freely develop. Earlier we mapped [16-17] the detailed temperature field within the sample overheated through the electric current and found that theoretical current–voltage characteristic provided a good fit to the measurements. However, the suggested analytical formulas were too complicated, and it was not clear whether or not the thermal switching takes place under given conditions.

It seems that for a long time this interesting problem was left behind. Here we return to same task aiming to consider simplified, but more general model of the overheated sample at the certain averaged temperature and analyze current–voltage characteristics available for boron.

2. Criterion of thermal switching

Below the static current vs. voltage characteristic will be constructed under the assumption of a pure temperature effect in semiconductor sample, resistance of which uniformly reduces due to Joule heating. In such case, a good approximation of the current dependence of the sample's steady resistance near the fixed ambient temperature T is given by

$$R(I) = R_0\,(1 - \alpha\,\Delta T(I)) \tag{1}$$

Here R_0 and $\alpha > 0$, respectively, are the resistance of the sample and absolute value of its temperature coefficient at the ambient temperature, and $\Delta T(I)$ is the rise in temperature at current I flowing through the sample. According to the Ohm's law, $U = R(I)I$ is the voltage across the sample. Consequently, the Joule power input UI equals to

$$W_{Input} = R_0\,(1 - \alpha\,\Delta T(I))I^2 . \tag{2}$$

If $\Delta T(I) << T$ the power loss is proportional to $\Delta T(I)$,

$$W_{Output} = \lambda\,\Delta T(I) . \tag{3}$$

Here λ is the effective heat-transfer coefficient, which in general includes not only thermal conduction through the electrodes, but also the thermal radiation from the sample's lateral surfaces.

In steady state (i.e., for a stable operating point) the condition $W_{Input} = W_{Output}$ must be held meaning the heat-balance equation

$$R_0(1-\alpha\Delta T(I))I^2 = \lambda\Delta T(I). \tag{4}$$

With $I_S^2 = \lambda/\alpha R_0$ and equations (1) and (4) we obtain

$$\Delta T(I) = \frac{1}{\alpha}\frac{I^2/I_S^2}{1+I^2/I_S^2} \tag{5}$$

and

$$\frac{R(I)}{R_0} = \frac{1}{1+I^2/I_S^2}. \tag{6}$$

Then, with $U_S = R_0 I_S/2$ the equation (6) yields relation between the normalized voltage U/U_S and normalized current I/I_S:

$$\frac{1}{2}\frac{U}{U_S} = \frac{I/I_S}{1+I^2/I_S^2}. \tag{7}$$

For the relaxation point, where (normalized) differential resistance of sample

$$\frac{1}{R_0}\frac{dU}{dI} = \frac{1-I^2/I_S^2}{(1+I^2/I_S^2)^2} \tag{8}$$

equals to zero, we found $I = I_S$ and $U = U_S$ (see equation (7)). Hence, temporary application of the voltage higher than U_S to the sample transforms a state of high resistance to a state of poor resistance, which then is retained even at a lower voltage. So, the arrangement under the consideration operates as a current switch and, according to equation (6), the switching point has to be attained by a current through the sample, which decreases its resistance by a factor of $R_0/R(I_S) = R_0 I_S/U_S = 2$.

The equation (7) can be solved for the current value. There exist two roots. The first branch

$$\frac{I}{I_S} = \frac{U_S}{U} - \sqrt{\frac{U_S^2}{U^2} - 1} \tag{9}$$

expresses current as a function of voltage below the switching point. If the voltage vanishes $U \to 0$ the current also $I \to 0$ and, in accordance with Ohm's law, $I \to U/R_0$. The first branch of the current–voltage characteristic from the point (U_S, I_S) is followed by the second one

$$\frac{I}{I_S} = \frac{U_S}{U} + \sqrt{\frac{U_S^2}{U^2} - 1}, \tag{10}$$

where the differential resistance is negative and increases in absolute value. Now at $U \to 0$ the current depends inversely of the voltage $I \to R_0 I_S^2/U$ and, consequently, unrestrictedly large $I \to \infty$.

The normalized theoretical curve calculated from equations (9) and (10) is shown in the figure 1.

In the proposed generalized model of the solid sample overheated by electric current, the theorizing is based upon two approximations: (i) the rise in temperature is relatively small in comparison with ambient temperature; (ii) there are negligible temperature variations across the sample. Let us consider their influence on the obtained criterion of thermal switching.

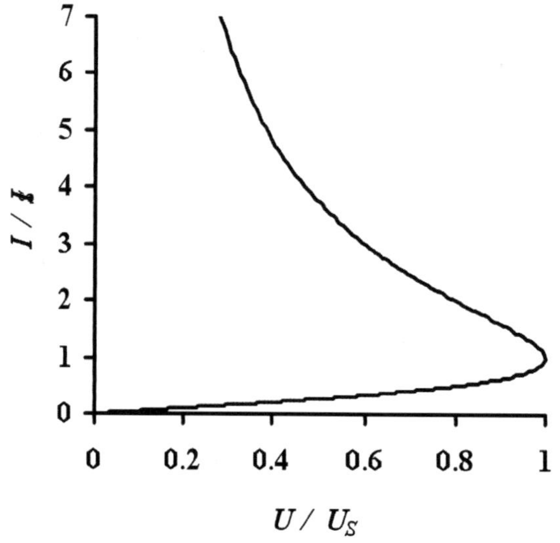

Figure 1. Normalized current–voltage characteristic of the thermal switch.

If the averaged rise in temperature of the sample is not small in comparison with ambient temperature then the linear approximations for resistance and output power are no longer valid. For instance, the radiation part of the power should be presented as $\sim ((T + \Delta T(I))^4 - T^4)$. However, necessity of overheating at $\Delta T(I) \geq T$ to reach the thermal switching means that absolute value of resistivity-temperature coefficient of the material is very low, which obviously tends to thermal or some kind of electrical breakdown masking thermal switching phenomenon.

If the temperature field within the sample at a non-zero current is not uniform then Thomson and Peltier heat contributions are generated in the bulk and at the electrodes, respectively. These are first-order effects vs. the current ($\sim I$), while the Joule heat is a second order effect ($\sim I^2$). Thus, their heat densities are small in comparison with Joule heat density generated at any given point. Moreover, if the sample and electrodes are symmetrically shaped then the temperature distribution is also symmetrical with respect to midpoint and the temperatures of both electrodes are the same. In such case, Thomson heat release inside one half of a sample is exactly compensated by Thomson heat absorption inside another one. Analogously, Peltier heat release and absorption at the electrodes are compensated by each other.

Thus, if the current switching in solid has a pure thermal nature, it should satisfy the obtained criterion: resistivity of material at the switching current makes up almost the half of its Ohmic resistivity, i.e., the low current limit.

3. Testing of boron samples

For the measurements of the current–voltage characteristic of β-rhombohedral boron we used two specimens of the almost identical geometry which were cut out from pure (99.9999 wt. % B) vacuum-deposited and zone-refined macro-crystalline rod. Their surfaces were mechanically polished and then the amorphous layers formed at the surface were etched in boiling nitric acid solution. Tail-end surface cleaning was performed using solution containing alkaline potassium before flushing with distilled water. Such treatment excludes any kind of surface conductivity or surface breakdown frequently observed in high-resistivity materials.

The pairs of symmetrical electrodes were formed by the vacuum sputtering of silver onto the opposite flat surfaces of the tested specimens. In addition to the electrical connection they also provided good heat dissipation. These devices were clamped between flawless steel probes the contact pressure of which was sufficiently high to guarantee the negligible series resistances.

The $I-U$ characteristics were displayed by the semiconductor device curve tracer. The switching effect has been investigated using static (DC) and so-called normal (sinusoidal AC of line frequency 50 Hz) regimes.

When very slowly increasing voltage was applied to the specimen, at first, there was only a slight increase in current; but, above a certain voltage, the current rose sharply while the voltage drop at the specimen decreased asymptotically to zero. The obtained characteristics are symmetric with respect to the current direction. If the time for the measurement of the individual points is sufficiently long (to reach the steady state), the return curves closely approach the initial ones for increasing current. The differences between the characteristics determined with DC and AC are only slightly marked below the switching point. Therefore, the switching time is quite large (much greater than period of AC oscillations 0.02 s). The detected characteristics were fully reproducible as switching did not cause any irreversible changes in the specimens' structure or any kind of electrostatic memory. It should be mentioned that all these features are inherent to the thermal effect.

Table 1. Switching ratios for tested boron specimens.

Specimen	$R_0 I_S / U_S$	References
Pure macrocrystalline β-B (1)	1.97	This work
Pure macrocrystalline β-B (2)	2.03	This work
Polycrystalline β-B	1.87	[7]
Amorphous B (1)	2.18	[9]
Amorphous B (2)	2.00	[10]
Amorphous B (3)	2.04	[10]

Table 1 presents the values of switching ratio $R_0 I_S / U_S$, i.e. ratio of resistances at zero and switching currents measured in static regime for two tested β-rhombohedral boron specimens. The values of the mean electric field strength at the switching points are 6.1 and 6.4 kV / cm, respectively, less than the suggested [15] upper limit 8 kV / cm of the hopping conductivity region. Thus, up to switching, these samples indeed are characterized with a negative resistivity-temperature coefficient. One can see that in both cases, the switching factors 1.97 and 2.03 almost equal 2, which is also true for some earlier results listed in the table. The thermal switching ratio can deviate from 2 because of varying contact resistances and also possibly due to variations in heat dissipation. Thus there is a good agreement between experiment and proposed theory for current–voltage characteristic. We can confidently explain the current instability effect observed in boron at room temperature as a process of thermal switching.

4. Conclusions

Finally, let us consider the probable reasons of thermal nature of current switching in β-rhombohedral boron. At room temperature ($T \approx 300$ K) chemically pure material is characterized by the high resistivity (~ 10^7 Ωcm) and the large and negative resistivity-temperature coefficient ($\alpha \approx 0.0587$ /K). The nature of hopping conductivity realized in boron seems suitable for observation of the thermal switching: (i) as there are practically no free current carriers, the breakdown via impact ionization is excluded; (ii) due to the high sample resistance, the overheating sufficient for the thermal switching is reached at relatively low voltage, which can not induce an electric field ionization of the low-lying hopping levels or inject free carriers from the electrodes before the switching point; (iii) because of the large absolute value of resistivity-temperature coefficient, the rise in temperature should to be small for the thermal excitation of the trapped carriers. Therefore, the switching in β-rhombohedral boron near the room temperature is predominantly a heat effect.

References

[1] Chen T P, Tse M S, Sun Ch Q and Fung S 2002 *Jpn. J. Appl. Phys.* **41** 3047
[2] Liu Q and Seabaugh A 2005 *IEEE Trans. Circ. Syst.* **II-EB** 572
[3] Tretiakov O A and Matveev K A 2006 *Phys. Rev.* B **71** 165326
[4] Tretiakov O A and Matveev K A 2006 *Phys. Rev.* B **73** 115302
[5] Gaulé G K, Ross R L and Bloom J L 1965 *Boron – Preparation, Properties, and Applications* Gaulé G K (Ed) New York (Plenum Press) 317
[6] Feldman Ch 1968 *Mat. Res. Bull.* **3** 95
[7] Klein W 1968 *J. Appl. Phys.* **39** 5797
[8] Babaev R M, Iglitsyn M I, Kiskachi Yu A and Tikhonov V I 1969 *Dev. & Contr. Syst.* **8** 49
[9] Feldman Ch and Moorjani K 1970 *J. Non-Cryst. Solids* **2** 82
[10] Feldman Ch and Moorjani K 1970 *Thin Solid Films* **5** R1
[11] Prudenziati M, Lanzi A, Majni G and Malavasi G 1973 *Phys. Status Solidi* A **18** 651
[12] Prudenziati M, Majni G and Alberigi Quaranta A 1973 *Solid State Commun.* **13** 1927
[13] Feldman Ch and Charles Jr H K 1974 *Solid State Commun.* **15** 551
[14] Gulyamov K B, Muminov A and Sultanov M 1974 *Problems of Dielectric Electronics* Gulyamov K B (Ed) Tashkent (Fan) 32
[15] Weheit H and Moldenhauer A 2004 *J. Solid State Chem.* **177** 586
[16] Chkhartishvili L S and Tsagareishvili G V 1987 *Bull. Georg. Acad. Sci.* **126** 305
[17] Chkhartishvili L S, Tsiskarishvili G P, Tsagareishvili O A, Gvelesiani G G, and Tsikaridze J N 1987 *Proc. 9th Int. Symp. Boron, Borides Rel. Compd.* Werheit H (Ed) Duisburg (UDG) 360

Journal of Physics: Conference Series **176** (2009) 012024 doi:10.1088/1742-6596/176/1/012024

Relative stability of α and β boron

Michael Widom[1] **and Marek Mihalkovič**[2]

[1] Department of Physics, Carnegie Mellon University, Pittsburgh PA 15213, USA

[2] Department of Physics, Slovakian Academy of Sciences, Bratislava Slovakia

E-mail: `widom@andrew.cmu.edu`

Abstract. α and β rhombohedral boron are closely related structures with nearly identical stability. Unlike α, in which every site is fully occupied, β is characterized by partial occupancy and intrinsic disorder. Optimizing the site occupation yields a superstructure of β that we call β'. Its energy is lower than α but its symmetry lower than β. The configurational entropy of partial occupation drives a $\beta' \to \beta$ transition around room temperature. Comparing the vibrational density of states, we find that β' has an excess of low frequency modes that are strongly localized around the partially occupied sites. The vibrational entropy of these low frequency modes adds to the configurational entropy of partial occupation to further stabilize the β' structure compared to α.

1. Introduction

The relative stability of α and β rhombohedral boron has long puzzled researchers [1, 2, 3]. While β is observed experimentally to be the stable phase and remain so down to low temperatures, initial calculations of total energies [4, 5, 6] found $E_\alpha < E_\beta$. The β structure was taken as Pearson type hR105, which is rhombohedral with 105 atoms in the 105-site primitive cell [7]. However, improved crystallographic studies of the β phase reveal partially occupied sites [8, 9], leading to Pearson type hR141 with approximately 107 atoms distributed among 141 sites. By optimizing the placement of atoms on these sites a variant of β (henceforth named β') is found with $E_{\beta'} < E_\alpha$, indicating that the true ground state is likely a superstructure of β with a specific symmetry-broken assignment of atoms among the partially occupied sites [10]. We propose that the symmetric β phase is restored above a moderate temperature and stabilized relative to β' by the entropy of partial site occupation.

The hR12 structure of α-boron shown in figure 1 is relatively simple with just 2 Wyckoff positions (i.e., 2 separate groups of symmetry-equivalent atomic sites) and a total of 12 atomic sites per rhombohedral cell. Every atomic site is fully occupied. In contrast, the hR141 structure [9] of β comprises 20 Wyckoff positions labeled B1-B20. Wyckoff positions B1-B15 coincide with those of the earlier hR105 structure [7]. Site B15 occupies the rhombohedral cell center. The sites reported to be partially occupied are B13 and B16-20, and are color-coded according to occupancy in figure 2.

The following section of this paper reviews the proposed optimized β' structure and its discrete low-lying excitations related to changes in site occupation. However, the complete configuration space is continuous, owing to the phonon degrees of freedom and their associated vibrational free energy. To rigorously model the free energy of a phase we should evaluate both the full partition function which integrates the Boltzmann factor $\exp -E/k_BT$ over all positions of all atoms. We

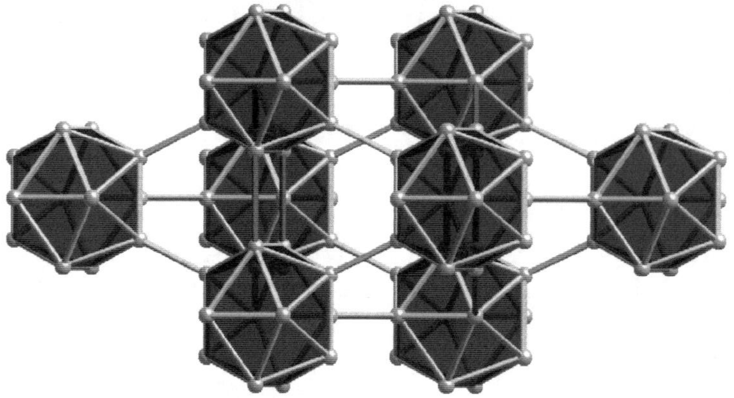

Figure 1. Structure of α-B (Pearson hR12).

Figure 2. Structure of β-B (Pearson hR141) viewed from the body center (B15 site) to cell vertex along the (111) direction. Non-bonded atoms are partially occupied sites B16-B20. Partially occupied sites are shown in color: B13 (74.5% average occupancy cyan); B16 (27.2% pink); B17 (8.5% yellow); B18 (6.6% indigo); B19 (6.8% blue); B20 (3.7% orange).

Journal of Physics: Conference Series **176** (2009) 012024 doi:10.1088/1742-6596/176/1/012024

Name	Pearson	Atoms	V	$E - E_\alpha$	ΔR	Comments
β-R	hR105	105	7.72	25.87	0.023	full occupation
α-R	hR12	12	7.18	0.00	0.002	full occupation
β'	hR111	105	7.69	13.02	0.013	B13bcdefB16a
β'	hR111	106	7.64	0.15	0.012	B13bcdefB16bd
β'	hR141	107	7.57	-0.86	0.005	B13bcefB16bdB17aB18a
β'	aP282	214	7.57	-1.75	NA	optimized supercell

Table 1. Structural data [10] including atoms per primitive cell, volume (Å^3/atom) and energy relative to α-R (meV/atom). ΔR (Å) measures the deviation of the symmetry-averaged relaxed positions from the crystallographically reported positions, averaged over the fully occupied Wyckoff positions. Comments list occupied sites using notation discussed in text. Structures labeled β' exploit partially occupied sites of hR111 or hR141 and break rhombohedral symmetry.

will approximate this by integrating over small displacements (phonons) around nearly optimal discrete configurations C, as

$$e^{-F_{tot}/k_BT} = \int \Pi_{i=1}^{N} d\mathbf{r}_i e^{-E/k_BT} \approx \sum_C e^{-(E(C)+F_{vib}(C,T))/k_BT} \tag{1}$$

where $E_0(C)$ is the cohesive energy of configuration C and the vibrational free energy $F_vib(C,T)$ depends on the phonon spectrum for each configuration C. The vibrational contributions will be approximated as being *independent* of configuration, yielding

$$F_{tot} \approx F_{vib}(C_{opt}, T) - k_BT \log Z_{conf} \tag{2}$$

where

$$Z_{conf} = \sum_C e^{-E(C)/k_BT}. \tag{3}$$

2. Configurational free energy

The crucial step in determining the low temperature state of elemental boron is identifying the lowest energy configuration. Using electronic density functional theory we can accurately evaluate the energy of any specific configuration. Our precise methods, based on the VASP code [11] are described in Ref. [10]. As can be seen in in table 1, the energy of α lies below the energy of the original (hR105) β structure. To explain the observation of β as the stable structure, we must evaluate the energies of specific configurations of the more highly refined hR141 structure. Choosing the proper configuration to evaluate is a challenge. Within a single unit cell of hR141 there are 42 partially occupied sites, among which we must distribute 8 atoms, yielding more than 10^8 configurations that must be tested, in principle. Since each calculation takes approximately one day, we clearly need a way to simplify the search.

Our strategy is to first optimize a (hR111) structure [8] that contains only 16 Wyckoff positions, only two of which (B13 and B16) are partially occupied. Based on the reported 74.5% site occupancy of B13, and the 27.2% occupancy of B16, we consider structures with one or two vacant B13 sites (out of 6 per primitive cell)and one or two occupied B16 sites. From this we learn that it is favorable to place a single atom in each equilateral triangle of B16 atoms (i.e., one among B16abc and one among B16def, see figure 3). We also learn that when two B13 are vacant, the vacant sites prefer to be adjacent to each other.

Now consider the remaining sites, B17-20. Based on the reported 8.5% occupancy of B17 we attempt to occupy just a single site (out of six per primitive cell). The bond length of B17 from the nearest B13 site is only 1.57 Å, slightly too short for simultaneous occupancy. Hence we conjecture, and verify by total energy calculation, that the occupied B17 sites should be close to a vacant B13 site. Since the 6.8% reported occupancy of B18 is similar to B17, and the B17B18 bond is a favorable 1.62 Å length, we conjecture, and verify, the presence of a B17B18 bonded pair.

There are six available B17B18 pairs per primitive cell. One of the B13 sites collinear with this pair must be vacant, and one of the two nearby B13 sites must also be. Together with the three possible sites for each B16 atom, this yields $6 \times 2 \times 3^2 = 108$ configurations to test. One of these (see figure 3) turns out to be the best structure we found. In particular, despite intensive efforts we found no lower energy structure utilizing either a B19 or a B20 site. The energy of our optimal structure turns out to be 0.86 meV per atom (0.092 eV per primitive cell) *below* E_α. Among the 108 structures that obey the rules just stated, the energy differences were generally within about 0.1 eV per primitive cell. At a temperature $T = T_m/2$, where T_m is the melting temperature, fluctuations between these configurations should be active, yielding an entropy of $k_B \log 108$ and reducing the free energy of β by a further $k_B T \log 108 = 0.48$ eV per primitive cell.

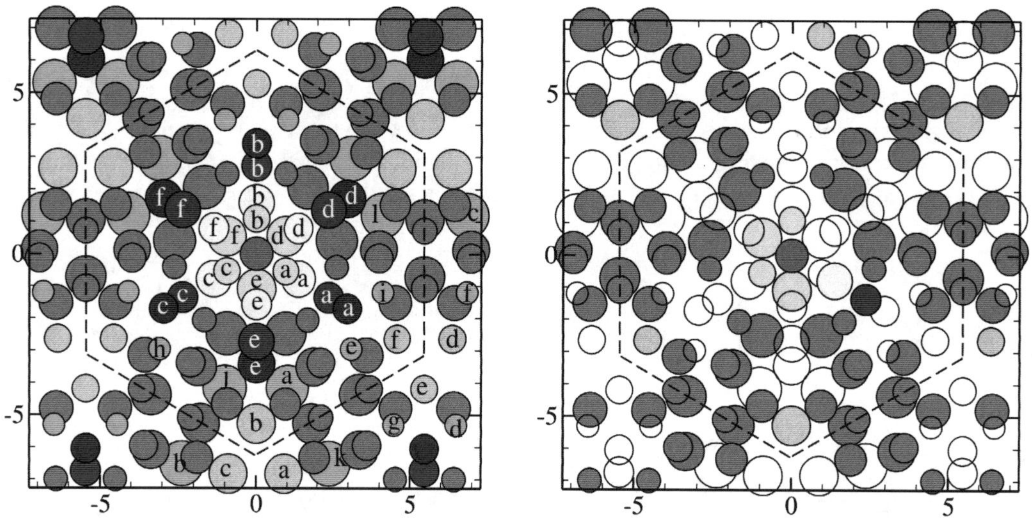

Figure 3. View of body center along (111) axis. Coloring as in figure 2: B13 (cyan); B16 (pink); B17 (yellow); B18 (indigo); B19 (blue); B20 (orange). (left) site labeling scheme; (right) optimal structure places atoms on sites "B13bdefB16bdB17aB18a".

The electronic density of states offers potential insight into relative stability and optimal numbers of atoms. Figure 4 illustrates the DOS of hR12 and hR105. Evidently α is semiconducting (the Fermi energy lies at a gap) while β is predicted to be metallic (the Fermi energy lies in a band) in contradiction to experiment which finds β to be insulating. Placement of the Fermi energy in a gap can provide a means of lowering the total energy, potentially explaining $E_\alpha < E_\beta$ while simultaneously indicating the possibility of modifying β to reduce its energy. An additional 5 electrons per primitive cell would place E_F^β in its band gap, indicating an additional 1.67 boron atoms are needed. However, the additional atoms in turn create new states, so in our hR141 structure $E_F^{\beta'}$ remains below the gap by 1 electron per primitive cell.

While we believe we have found the optimal arrangements of atoms within a single primitive cell of β, we now test for a possible further reduction in energy by superlattice ordering. To do

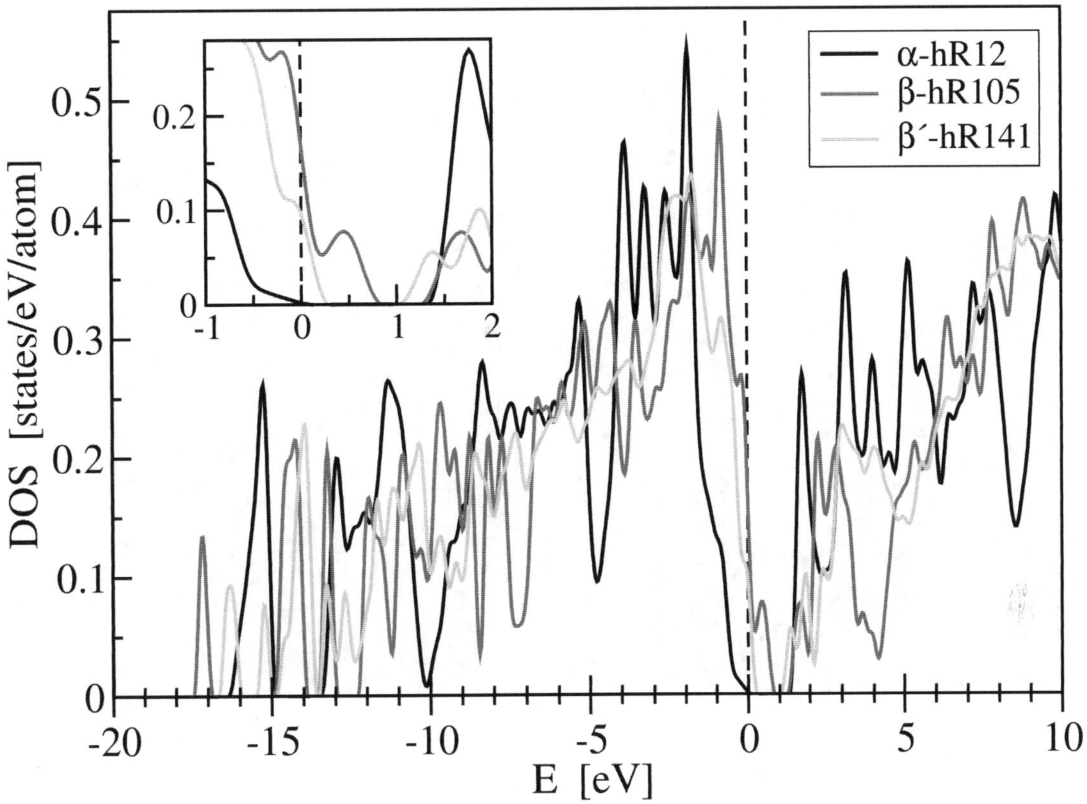

Figure 4. Single electron density of states for α (hR12) and β (hR105). The Fermi energies (E_F^{α}=6.15eV, E_F^{β}=4.47eV, $E_F^{\beta'}$=4.86eV) have been shifted to zero to aid comparison. Note the DFT single electron band gap is not expected to match the experimentally observed electronic excitation band gap. Inset: enlargement of gap region.

so, we combine two independent cells into a $2\times1\times1$ supercell, and consider differing assignments of partially occupied sites in the two independent cells. To simplify our study, we only consider structures built out of our fundamental motif. Thus in each cell we replace a B13 pair with a B17/18 pair, but we vary the choices of specific pairs in the two cells. Since there are $6\times2=12$ ways to move these pairs in a single cell, there are $12^2=144$ ways to do it in our supercell, of which 22 are symmetry unrelated. For each of these 22 we obtain the optimal positions of the B16 atoms, resulting in 22 energies E_k, each of which has a multiplicity Ω_k due to symmetry.

Several of our supercells achieve energy lower than our optimal primitive cell, with the best one (see figure 5) reaching energy 1.75 meV/atom below E_α. Since superlattice ordering is favorable, we conclude that the optimal boron structure is not β, but some other structure β' that breaks the translational symmetry of β, leading to a larger lattice constant. Our particular supercell structure also breaks rhombohedral rotational symmetry (it has Pearson type aP214). Conceivably some other structure with superlattice ordering along the other two directions could restore rhombohedral symmetry, but still the translational symmetry of β would be broken.

According to the Landau theory of phase transitions, structures of differing symmetry must be separated by a thermodynamic phase transition. In the present case that means that a phase transition must separate the familiar β from the true low temperature phase, β'. To test this

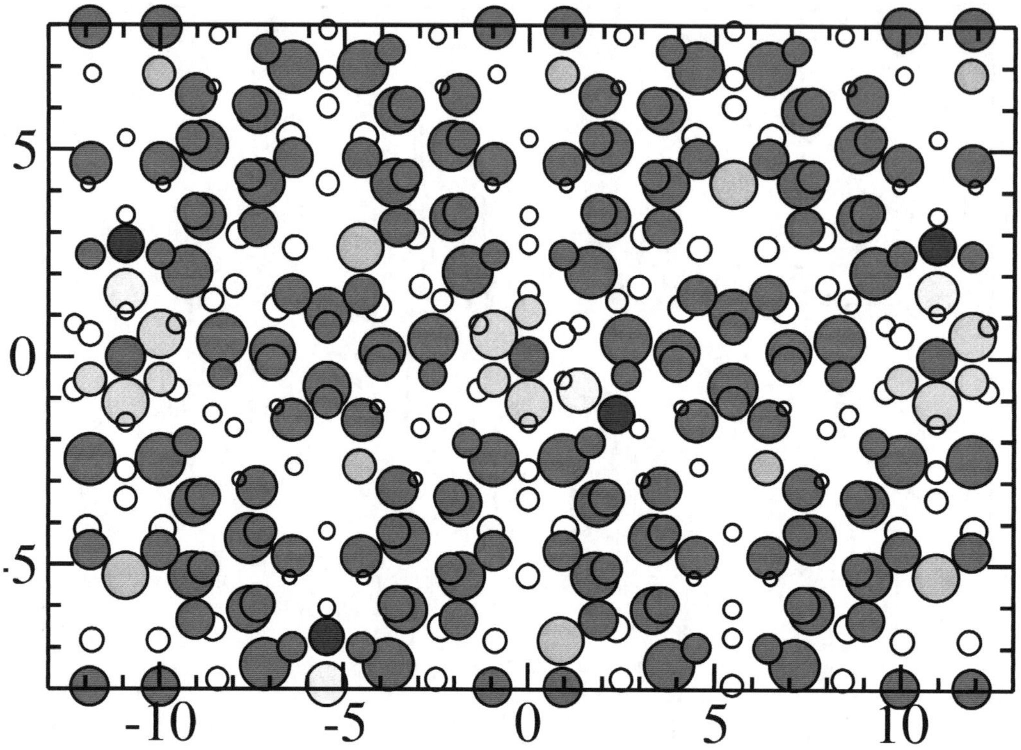

Figure 5. Optimal supercell as proposed in [10].

hypothesis, we evaluate the partition function

$$Z_{conf} = \sum_{k=1}^{22} \Omega_k e^{-E_k/k_B T} \tag{4}$$

using the energies and multiplicities obtained for the complete set of supercell models. Conventionally the canonical ensemble takes (N, V, T) as independent variables. Here consider only the dependence on T. N is fixed at the optimal value (214) for two primitive cells. The volume V is determined by relaxation of each specific configuration k, and is nearly a constant. Alternatively, this calculation can be considered in the isothermal-isobaric (N, P, T) ensemble with $P = 0$. Since the energies E_k are relaxed values, they hold only at $T = 0$K, and thus we neglect coupling of vibrations to the configurational free energy and we also neglect thermal expansion.

By thermodynamic differentiation we obtain internal energy, entropy and heat capacity, as illustrated in figure 6. Notice the entropic reduction in free energy reaches $k_B T \log 12^2/214$=4 meV/atom, indicating complete freedom to sample all 12^2 differing configurations within the two cells by the time the temperature reaches T=2000 K. evidently, around room temperature a phase transition occurs as fluctuations of atoms among different partially occupied sites stabilize the highly symmetric β phase over the energetically preferred but lower symmetry β' phase.

We only include a small subset of all low energy configurations in our sum (equation 4) because we select only the optimal B16 arrangement for each of our 22 supercell models. Hence the actual configurational entropy will be greater than the 4 meV/atom quoted above. Likewise, inclusion of a larger structural ensemble might shift the transition temperature somewhat. The transition is broadened due to finite size effects, which can also shift the position of the heat capacity peak. So our entropy and transition temperatures are just crude estimates.

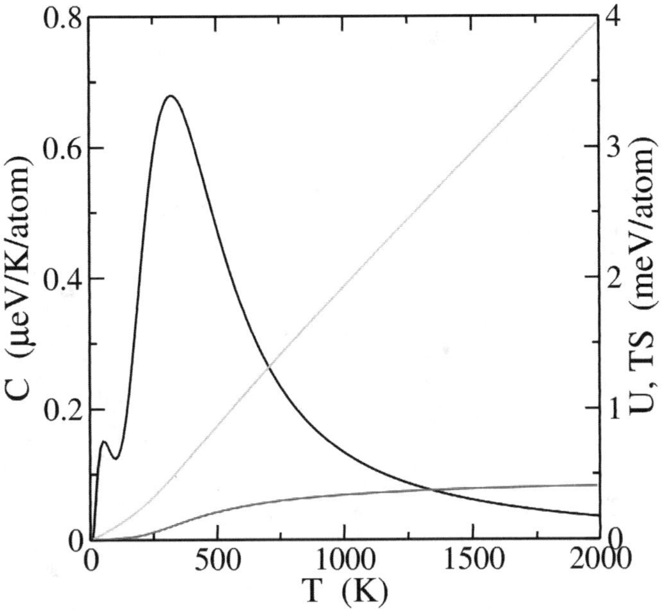

Figure 6. Thermodynamic data for supercell model [10]. Black line shows heat capacity, red line shows internal energy and green line shows TS, with S the entropy.

3. Vibrational free energy

Vibrational free energy enhances the stability of both β and β' relative to α. Owing to the complexity of the β structure, one previous calculation of vibrational free energy used only a single phonon k-point (the Γ point) to estimate the density of states [5]. Another [6] carried out both Γ-point and full Brillouin zone calculations, but only published the spectrum at the Γ-point. Phonon dispersion curves for α have been previously calculated [12, 3], but again the full density of states was not published. Here we carry out true phonon calculations integrating over the full Brillouin zone.

Given vibrational density of states $g(\omega)$, the vibrational free energy is obtained by integrating the single mode free energy, resulting in

$$f_{vib}(T) = k_B T \int g(\omega) \log 2 \sinh \hbar\omega / 2k_B T d\omega \tag{5}$$

Low frequencies, with $\hbar\omega << k_B T$, make negative contributions to the vibrational free energy. Structures with larger $g(\omega)$ at low frequencies generally have lower vibrational free energy, lending greater thermodynamic stability. At sufficiently low temperature, f_{vib} approaches the quantum zero point energy

$$E_0 = \int g(\omega) \frac{\hbar\omega}{2k_B T} \tag{6}$$

Again, lower frequency modes lead to lower zero point energy and greater stability.

To calculate $g(\omega)$ we evaluate the interatomic force constant matrix consisting of the second derivatives of total energy E

$$\frac{\partial^2 E}{\partial \mathbf{R}_i \partial \mathbf{R}_j} \tag{7}$$

for atom positions $\mathbf{R}_{i,j}$ within an interaction range that we took as 5.1 Å. Specifically, we extract these values from linear response, by displacing atoms from their equilibrium positions and calculating the resulting forces using VASP. Data is collected using our in-house phonon code. This matrix is then Fourier transformed at wavevector $\mathbf{k}$ and then diagonalized yielding the dispersion relation $\omega(\mathbf{k})$. We evaluate ω on a dense mesh in k-space, smear the values with gaussians of width 0.3 meV, then superpose the results.

All calculations are done within the harmonic approximation, holding the lattice parameters fixed at their optimal values. Some results including thermal expansion obtained within the quasiharmonic approximation are presented by Shang, *et. al.* [3].

The resulting $g(\omega)$ for α and for our optimized 107-atom β' structure are shown in figure 7. The DOS are similar over the low frequency (note energy $E = \hbar\omega$) "acoustic" portion of the spectrum. Although α exhibits some sharper features than β', the contribution of these modes to thermodynamic stability is similar. The crucial difference in the spectra lies around 60 meV, where α exhibits a gap between acoustic and optical modes, while β' possesses a high density of localized modes. This excess of low frequency modes leads to a reduction of f_{vib} for β' compared to α that ranges from 3 meV/atom at low temperature (i.e., due to quantum zero point motion) to 15 meV/atom at T =2000 K.

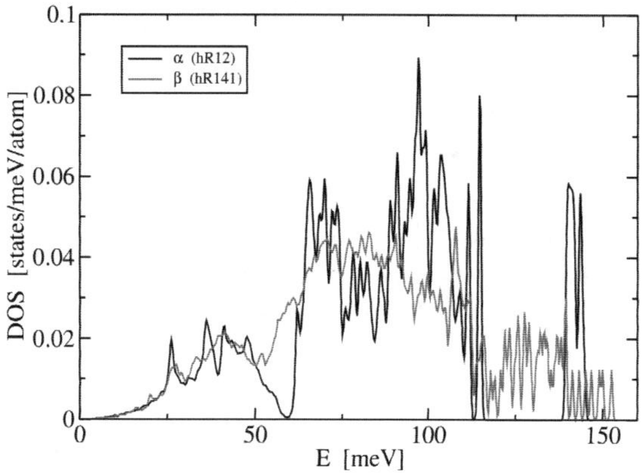

Figure 7. Vibrational densities of states for α (black) and β' (red). Horizontal axis is phonon energy $E = \hbar\omega$. Vertical axis is density of states.

Localization of the modes is judged by their participation ratio,

$$PR = \frac{(\sum_k |u_k|^2)^2}{N \sum_k |u_k|^4} \tag{8}$$

where $PR \sim 1$ indicates extended modes in which all atoms participate equally, and low PR indicates a localized mode in which only a few atoms are active.

For localized modes, the atom-projected density of states reveals which specific atoms are mobile at a given frequency. We find that the localized modes of β' in the energy range 55-65 meV lie predominantly on the B13 and B16 sites. The localized modes around 25 meV are predominantly associated with the B17 and B18 sites. Since these modes occur at low energies, these atoms are expected to be mobile at relatively low energies ($\hbar\omega$=60 meV corresponds to k_BT at T=700K). Indeed, in solid state molecular dynamics simulations [10] at T = 2000K precisely these atoms exhibited high mobility while the remaining atoms remained near their ideal positions.

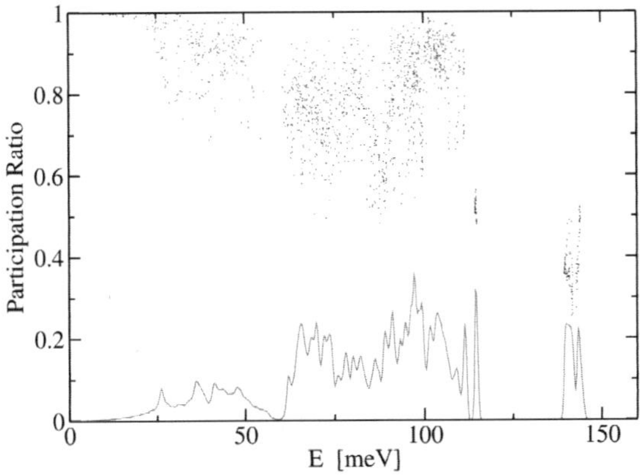

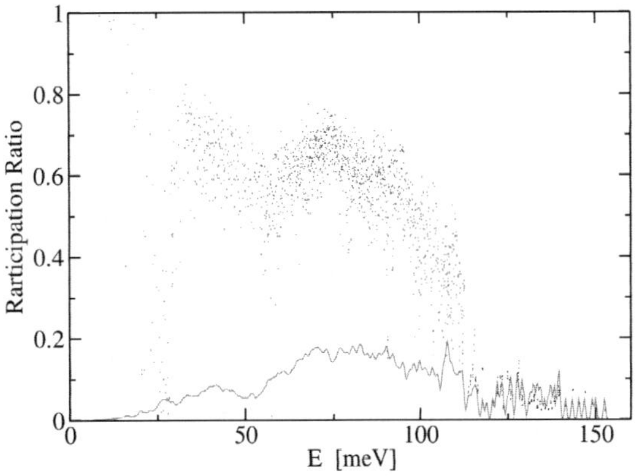

Figure 8. Participation ratios (see equation 8) of α (left) and β' (right). PR is shown only for every $50th$ mode, and the densities of states are multiplied by 4, for visual clarity.

4. Conclusions

In summary, we find that a specific arrangement of atoms among the partially occupied sites of β-rhombohedral boron achieve a lower total energy than α. The optimal configuration within the primitive cell of β breaks rhombohedral symmetry, and further reduction in energy is possible by means of superlattice ordering. Thus we conclude that the true low temperature phase is not β-rhombohedral boron, but a related structure that we name β'. Careful experimental study of the X-ray diffraction pattern would be appropriate, seeking Bragg peaks or diffuse scattering (indicating partial ordering) in the vicinity of reciprocal space points where diffraction is forbidden by the symmetry of β. Selected area electron diffraction could also be helpful because the domain size of the ordered superlattice might be small, owing to the low temperature of the predicted ordering transition.

As temperature grows, β' transforms into β through a phase transition in the vicinity of T =300 K. A variety of circumstantial evidence supports the existence of a structural transition in β around this temperature. These are summarized in Ref. [13, 14] and include changes in photoabsorption spectra [15], internal friction [16] and dielectric properties [17, 18].

Within the β phase, the disorder of partial occupancy causes the internal energy to grow and to exceed the energy of α by approximately 3 meV/atom by the time the temperature reaches T =2000 K. However, the entropy of β is sufficiently large that β has lower free energy than α at all temperatures, up to melting. At T=2000 K the components of stability are the partial occupancy entropy of at least 4 meV/atom, and vibrational free energy of 15 meV/atom.

Acknowledgments

We wish to acknowledge useful discussions with K. Shirai, N. Vast and H. Werheit. MM acknowledges support from grants VEGA-2/0157/08 and APVV-0413-06.

References

[1] Hoard J L and Hughes R E 1967 *The chemistry of boron and its compounds* ed Muetterties E L (Wiley) chap 2, p 25
[2] Shirai K, Masago A and Katayama-Yoshida H 2006 *Phys. Stat. Sol. B* **244** 303
[3] Shang S, Arroyave R and Liu Z K 2007 *Phys. Rev. B* **75** 092101
[4] Mihalkovic M and Widom M 2004 *Phys. Rev. B* **70** 144107
[5] Masago A, Shirai K and Katayama-Yoshida H 2006 *Phys. Rev. B* **73** 104102
[6] van Setten M J, Uijttewaal M A, de Wijs G A and de Groot R A 2007 *J. Am. Chem. Soc.* **129** 2458

[7] Hughes R E, Kennard C H L, Sullenger D B, an d D E Sands H A W and Hoard J L 1963 *J. Am. Chem. Soc.* **85** 361
[8] Callmer B 1977 *Acta Crystallogr. B* **33** 1951
[9] Slack G A, Hejna C I and Kasper J S 1988 *J. Solid State Chem.* **76** 52
[10] Widom M and Mihalkovič M 2008 *Phys. Rev. B* **77** 064113
[11] Kresse G and furthmuller J 1996 *Phys. Rev. B*
[12] Beckel C L, Yousaf M, Fuka M, Raja S Y and Lu N 1991 *Phys. Rev. B* **44** 2535
[13] Wehrheit H and Franz R 1984 *Phys. Stat. Sol. B* **125** 779
[14] Werheit H and Schmechel R 1998 *Landolt-Bornstein, Numerical data and functional relationships in science and technology Group III* vol 41C (Springer, Berlin) p 3
[15] Werheit H 1970 *Phys. Stat. Sol.* **39** 109
[16] Tsagareishvili G V, Tavadze F N, Darsavelidze G S and Metreveli V S 1970 *Electron Technol.* **3** 281
[17] Tsagareishvili O, Gabunia D and Chkhartishvili L 2008 *These Procedings*
[18] Tsagareishvili O A, Chkartishvili L S and Gabunia D L 2009 *Semiconductors* **43** 18

Soft X-ray emission spectroscopy study of the valence electron states of α-rhombohedral boron

Masami Terauchi[1], Yohei Sato[1], Hiroshi Hyodo[2] and Kaoru Kimura[2]

[1] Institute of Multidisciplinary Research for Advanced Materials, Tohoku University, 2-1-1 Katahira, Aoba-ku, Sendai 980-8577, Japan

[2] Department of Advanced Materials Science, University of Tokyo 5-1-5 Kashiwanoha, Kashiwa-shi, 277-8561 Chiba, Japan

E-mail: terauchi@tagen.tohoku.ac.jp

Abstract. High energy-resolution boron K-emission spectra of single-crystalline α-rhombohedral-boron (α-r-B) were measured with a newly developed soft X-ray emission spectroscopy (SXES) instrument attached to a conventional transmission electron microscope. The intensity profiles of the spectra, which correspond to the density of states of the valence bands with *p*-symmetry, were compared with those of amorphous boron (am-B), single-crystalline β-rhombohedral-boron (β-r-B) and B_4C. A characteristic shoulder structure appears in the spectrum of α-r-B but not of am-B, β-r-B or B_4C. This structure should be due to the inter-cluster three-center bonding states among B_{12} clusters in α-r-B.

1. Introduction

Boron forms four allotropes: α-rhombohedral boron (α-r-B), α-tetragonal boron, β-rhombohedral boron (β-r-B) and β-tetragonal boron. Those allotropes and amorphous boron (am-B) are constructed by B_{12} icosahedral clusters [1-3]. The crystal structure of α-r-B is the simplest among them, in which B_{12} clusters are placed at the corners of the rhombohedral unit cell [4,5]. Each B_{12} cluster in α-r-B is covalently bonded to twelve nearest-neighbor B_{12} clusters. The B_{12} clusters in the (111) plane are connected by the three-center bonding and those between two (111) planes by the two-center bonding.

Horn measured the band gap energy of α-r-B single crystals as about 2eV from optical absorption experiments [6]. Theoretical calculations of the electronic structure indicate that α-r-B is a semiconductor with an indirect energy gap [7-9]. A theoretical calculation reported peaks in the density of states (DOS) of α-r-B near the Fermi level [10]. The peaks originate from small deviation of α-r-B from an ideal B_{12} cluster with icosahedral symmetry. The peak structures of the DOS of the conduction band (CB) were confirmed by a high energy-resolution electron energy loss spectroscopy (EELS) of single-crystalline α-r-B [11]. After that, metal-doped α-r-B and β-r-B have been investigated by high energy-resolution EELS based on transmission electron microscopy [12-14].

Domashevskaya *et al* investigated the DOS of the valence band (VB) of α-r-B, β-r-B and am-B using X-ray emission spectroscopy, but could not find a significant difference among them [15]. It may be due to insufficient energy resolution and specimen quality. Fujimori *et al* derived the spatial distribution of valence electrons of α-r-B from powder X-ray diffraction experiment using maximum-entropy analysis combined with Rietveld structure refinement [16]. They visualized the spatial distribution of the three-centered bonding of intra- and inter-clusters. The inter-cluster three-centered bonding is unique for α-r-B. To detect the electronic state of the inter-cluster three-center bonding, high energy-resolution measurements are required on high-quality single crystalline α-r-B.

In this report, high energy-resolution soft X-ray emission spectroscopy measurements are reported on single-crystalline α-r-B and compared with those of am-B, β-r-B and B_4C. From the comparison of boron K-emission spectra, the characteristic structure in the DOS of VB of α-r-B is ascribed to the inter-cluster three-center bonding states of the crystal.

2. Experimental details

Specimens of α-r-B were prepared by annealing am-B powder, which was deposited by electron-beam evaporation, at 1200 °C for 10 h in a BN crucible. The crucible was held in an Ar-sealed quartz tube, in which tantalum foil was placed to prevent oxidation of the specimen. After annealing, the specimens were composed of small single crystals of α-r-B (20-40 μm) and β-r-B (50-100 μm). The crystals were crushed in a crucible and put on a grid for transmission electron microscopy experiments. Electron diffraction patterns showed that the specimens were high quality single crystals of α-r-B and β-r-B.

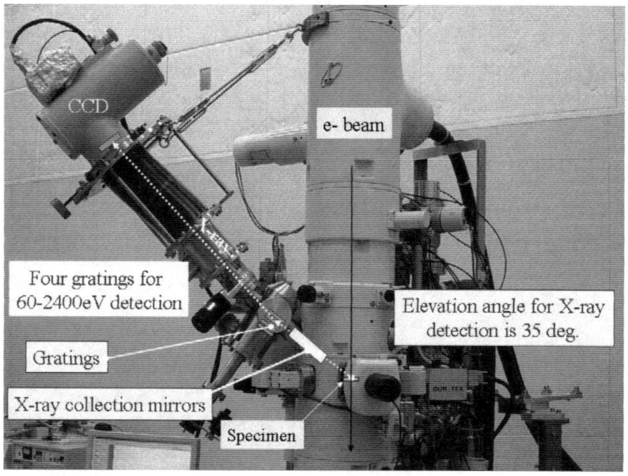

Figure 1. X-ray emission spectrometer attached to a conventional transmission electron microscope JEM2010. A Ni-coated varied-line-spaced grating was used in the experiment.

Figure 1 shows a soft X-ray emission spectroscopy (SXES) instrument, which is attached to a conventional transmission electron microscope JEM2010 [17]. Soft X-ray emission was induced by 100 kV electrons. It was dispersed by a grating and detected by a charge-coupled-device (CCD) camera. We used a Ni-coated varied-line spaced grating with an average groove density of 1600 lines/mm and a CCD camera without an anti-reflection coating. The calculated energy-dispersion of the grating for boron K-emission energy of 180 eV was about 0.07 eV/channel (1 channel: 13.5 μm) on the CCD surface. Probe current, probe diameter and acquisition time were 5-10 nA, 500 nm and 30-60 min, respectively.

3. Results and discussions

Figure 2 shows boron K-emission spectrum of α-r-B obtained from a single-crystalline area of about 500 nm in diameter. K-emission spectra of β-r-B and am-B are also shown for comparison. The energy scale was determined by the energy dispersion of the grating, and the origin was taken arbitrarily. The spectra are due to electronic transitions from the VB states to a K-shell core-hole state created by electron-beam illumination. A dipole transition is associated with soft-X-ray emission. Therefore, the intensity distribution of each spectrum corresponds to the partial DOS of the VB with *p*-symmetry for each material. The intensity maximum of each spectrum, indicated by a dotted vertical line, occurs at almost the same energy. The top (the dotted vertical line on the right-hand side) and the bottom (the dotted vertical line on the left-hand side) of the VB are also almost the same for all materials. This may be due to that all those materials have common structural unit of B_{12} cluster. It

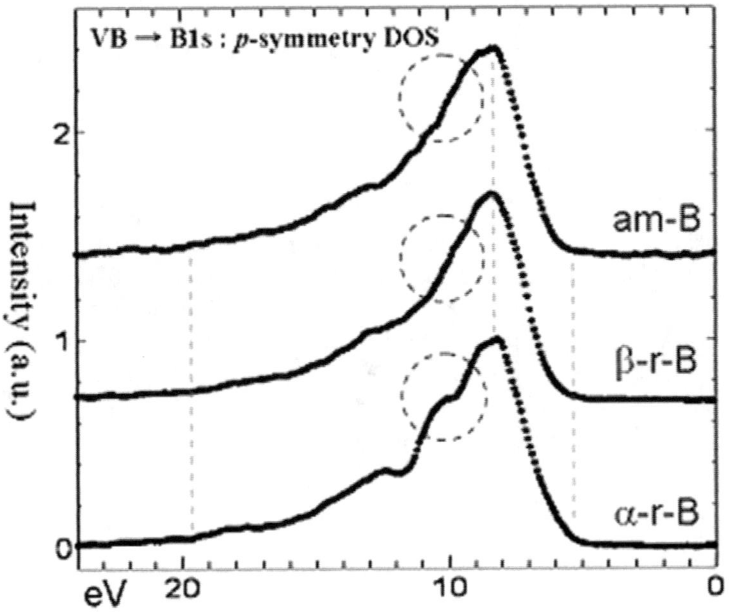

Figure 2. Boron K-emission spectra of am-B, β-r-B and α-r-B. The spectrum of α-r-B shows an additional shoulder structure compared with those of β-r-B and am-B.

should be noted that a circled area of α-r-B shows a characteristic additional shoulder structure compared with those of β-r-B and am-B. Furthermore, the intensity profile of the top part of the main peak of β-r-B is the sharpest among the three spectra. Those differences in the DOS of the VB of the three boron allotropes were successfully identified for the first time using a developed SXES-TEM instrument [17].

Figure 3 shows the DOS with *p*-symmetry of the VB calculated for α-r-B using the Wien2k code. A characteristic DOS peak is indicated by an arrow. The DOS is attributed to boron atoms, which form three-centered bonding between B_{12} clusters. Three-centered bonding state between B_{12} clusters is known as a characteristic of the crystalline α-r-B. The energy position of the peak is about 2 eV

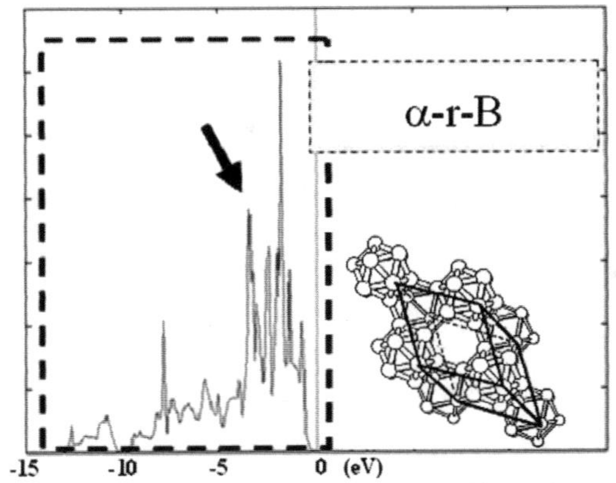

Figure 3. DOS of the VB for α-r-B calculated using the Wien2k code. A characteristic DOS peak indicated by an arrow, which corresponds to the shoulder structure in figure 2.

lower than the top of the DOS peak. The energy difference between the experimentally observed shoulder structure and the DOS peak in figure 2 is almost the same with this calculated result. Thus, the shoulder structure observed in figure 2 can be due to the three-centered bonding state. To confirm this, more detailed DOS calculations for β-r-B and am-B are desired, but have not performed those due to the structural complexity of the materials.

When the shoulder structure of α-r-B is due to the three-centered bonding, anisotropic X-ray

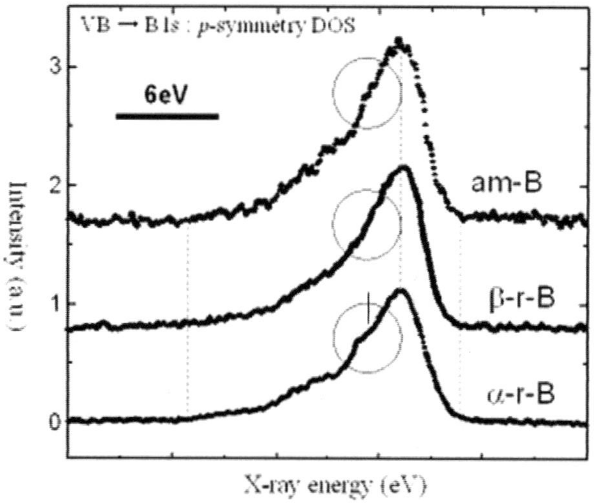

Figure 4. Boron K-emission spectra of α-r-B with a different crystal setting from that of figure 2. The shoulder structure is different from that of figure 2.

emission intensity is expected due to the anisotropic distribution of the three-centered bonding. Figure 4 shows boron K-emission from crystalline α-r-B, measured with a different crystal orientation from that of figure 2. The spectra of β-r-B and am-B are also obtained form different specimens. The shoulder structure of α-r-B is also seen as indicated by a vertical line. However the intensity distribution is different from that in figure 2. This different intensity distribution for different crystal settings can be due to the anisotropic distribution of the three-centered bonding.

The structure of B_4C is similar to that of α-r-B except that carbon atoms are placed on the body

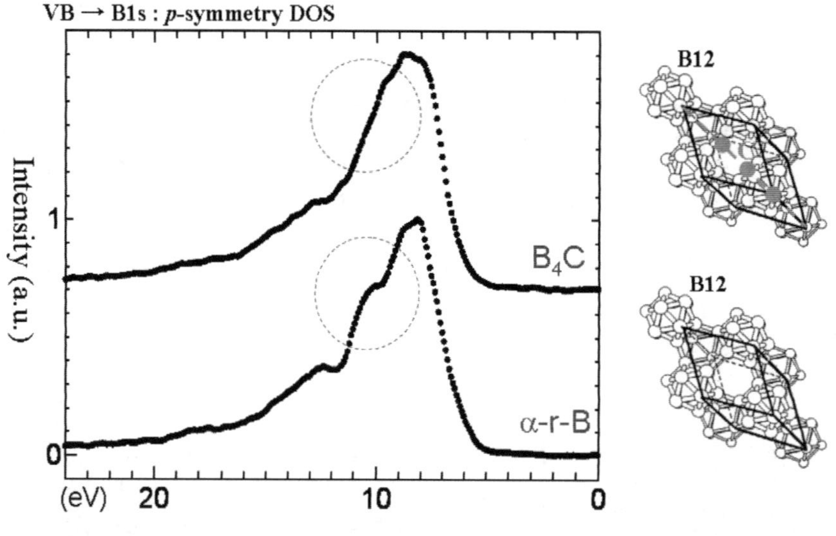

Figure 5. Boron K-emission spectra of B_4C and α-r-B. The shoulder structure is clearly seen in α-r-B but not in B_4C.

diagonal line, [111]. Although the stoichiometry of this system is not well defined and is considered between $B_{13}C_2$ and B_4C, this material may not have the inter-cluster three-center bonding between B_{12} clusters as in α-r-B [18,19]. Figure 5 shows a boron K-emission spectrum of a single-crystalline B_4C. The α-r-B spectrum of figure 2 is also shown for comparison. Contrary to α-r-B, the spectrum of B_4C does not show a shoulder structure. This again suggests that that the origin of the shoulder structure of the DOS of α-r-B is mainly due to the inter-cluster three-center bonding, which is a characteristic electronic state of α-r-B.

4. Concluding remarks

Newly developed SXES instrument based on transmission electron microscope has been applied for measuring boron K-emission spectra from identified single-crystalline areas of α-r-B, β-r-B and B_4C and also from amorphous boron. It is clearly revealed that the DOS of α-r-B has a characteristic shoulder structure, which may be due to the inter-cluster three-center bonding. For discussing the DOS in terms of the inter-cluster three-center bonding and for comparison with the result of MEM/Rietveld analysis [16] on the three-centered bonding in real space, an angle-resolved SXES measurement is necessary on single-crystalline α-r-B by changing the take-off angle of X-ray emission.

Acknowledgements

The authors thank Mr. F. Sato for his skillful technical assistance. The development of the XES-TEM instrument was conducted under a name of "Development of an EELS/XES electron microscope for electronic structure analyses", which is one of the leading projects of Ministry of Education, Culture, Sports, Science and Technology, Japan. The present work on boron materials was partly supported by Grant-in-Aid for Scientific Research on Priority Areas 'New Materials Science Using Regulated Nano Spaces–Strategy in Ubiquitous Elements' by the Ministry of Education, Culture, Sports, Science and Technology of Japan.

References

[1] Golicova O A 1987 *Phys. Status Solidi* **101** 277
[2] Katada K 1966 *Jpn. J. Appl. Phys.* **5** 582
[3] Kobayashi M 1988 *J. Mater. Sci.* **23** 4392
[4] McCartyL V, Kasper J S, Horn F H, Decker B F and Newkirk A E 1958 *J. Am. Chem. Soc.* **80** 4507
[5] Decker B F and Kasper J F 1959 *Acta Crystallogr.***12** 503
[6] Horn F H 1959 *J. Appl. Phys.* **30** 1611
[7] Bullet D W 1982 *J. Phys. C* **15** 415
[8] Lee S, Bylander D M and Kleinman L 1990 *Phys. Rev.* B **42** 1316
[9] Li D, Xu Y N and Ching W Y 1992 *Phys. Rev.* B **45** 5896
[10] Gunji S and Kamimura H 1996 *Phys. Rev.* B **54** 13665
[11] Terauchi M, Kawamata Y, Tanaka M, Takeda M and Kimura K 1997 *J. Solid State Chem.* **133** 156
[12] Terauchi M, Kawamata Y, Tanaka M, Matsuda H and Kimura K 1997 *J. Solid State Chem.* **133** 152
[13] Soga K, Oguri A, Araake S, Terauchi M, Fujiwara A and Kimura K 2004 *J. Solid State Chem.* **177** 498
[14] Terauchi M, Sato Y, Nakayama T, Oguri A and Kimura K 2004 *J. Solid State Chem.* **177** 2916
[15] Domashevskaya E P, Solovjev N E, Terekhov V A and Ugai Ya A 1976 *J. Less-Common Metals* **47** 189
[16] Fujimori M, Nakata T, Nakayama T, Nishibori E, Kimura K, Takata M and Sakata M 1999 *Phys. Rev. Lett.* **82** 4452
[17] Terauchi M, Koike M, Fukushima M and Kimura J 2006 *Microsc. Microanal.* **12** (suppl.2) 866
[18] Tallant D R, Aselage T L, Campbell A N and Emin D 1989 *Phys. Rev.* B **40** 5649

[19] Kleinman L and Bylander D M 1991 *Phys. Rev.* B **43** 1487

Journal of Physics: Conference Series **176** (2009) 012026 doi:10.1088/1742-6596/176/1/012026

Electron density distribution of β-rhombohedral boron constructed from icosahedral cluster

Shizuka Hosoi[1,5], Kenji Tanabe[1], Hongki Kim[1], Kazuhiro Kirihara[2], Kohei Soga[3], Kaoru Kimura[1], Kenichi Kato[4] and Masaki Takata[4]

[1]Department of Advanced Material Science, The University of Tokyo, Chiba 277-8561, Japan
[2]National Institute of Advanced Industrial Science and Technology, Ibaraki 305-8568, Japan
[3]Department of Materials Science and Technology, Tokyo University of Science, Chiba 278-8510, Japan
[4]RIKEN SPring-8 Center/Japan Synchrotron Radiation Research Institute/CREST/JST, Hyogo 679-5198, Japan

E-mail: bkimura@phys.mm.t.u-tokyo.ac.jp

Abstract. Boron forms unusual structures and chemical bonds. Pure boron and many boron-rich borides have local icosahedra in their crystalline structures. The electron density distribution (EDD) in β-rhombohedral boron (β-B_{105}) is visualized by the maximum entropy method (MEM)/Rietveld analysis of powder X-ray diffraction (XRD) data in order to discuss the peculiarity of the chemical bonds. In α-rhombohedral boron (α-B_{12}) and its derivative crystals, peculiar covalent bonds, i.e., intracluster three-center bonds and intercluster bent bonds, have been observed. In β-B_{105}, the intracluster three-center bonds are observed, but the intercluster bent bonds are not. The origin of the bent bond can be explained by the mismatch in the directions of the bond and the lattice vector.

1. Introduction

Boron and some boron-rich borides have local icosahedral clusters in their crystalline structures. The pure boron elemental solid has mainly two rhombohedral allotropes, α-rhombohedral boron (α-B_{12}) and β-rhombohedral boron (β-B_{105}), whose structures are shown in figure 1. β-B_{105} is the most common allotrope of the elemental boron at room temperature and atmospheric pressure. β-B_{105} has a very complex structure that consists of B_{12} clusters (icosahedra), B_{28} clusters (three face-sharing icosahedra), a single atom site connecting two B_{28} clusters and several partially occupied interstitial sites [1]. In practice, there are 106.5 atoms per rhombohedral unit cell in β-B_{105}. However, in this study, we adopt the notation β-B_{105}, for simplicity, disregarding the partial occupation. Much research has been carried out to elucidate the physical properties of metal-doped β-B_{105} [2]. However, no understanding of the chemical bonds of this phase has been established. In the past studies, we discussed the chemical bonds in α-B_{12} and in its derivative crystals, boron carbide and boron

[5]Present affiliation: Materials Analysis Center, Sony Corporation, Kanagawa 243-0021, Japan

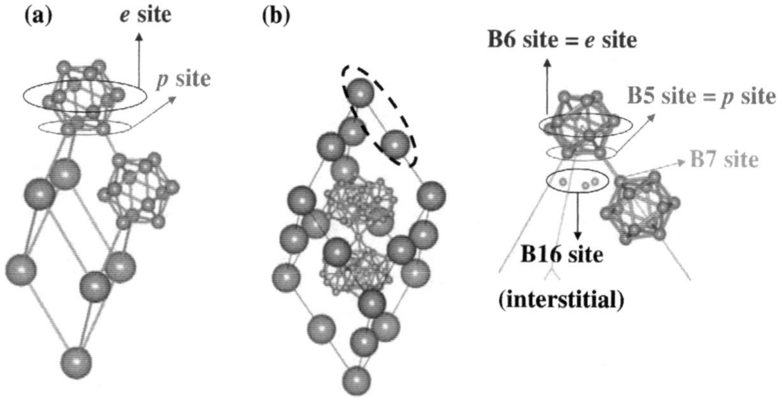

Figure 1. Structures of (a) α-B_{12} and (b) β-B_{105}. The largest balls represent the B_{12} icosahedra.

phosphide [3, 4]. Their chemical bonds were evaluated using the electron density distribution (EDD) obtained from experimental data and the maximum entropy method (MEM) analysis. The experimental EDD of the local B_{12} cluster agreed well with the results of calculation [5]. α-B_{12} has peculiar chemical bonds such as the three-center covalent bond and the bent two-center bond. The former is also considered to be a multiple-center bond of all the 12 atoms in the B_{12} cluster. Since β-B_{105} has a huge unit cell and a complex structure, few calculational studies have been carried out. Hence, in order to compare the form of the chemical bonds in β-B_{105} with that in α-B_{12}, we obtained the EDD of β-B_{105} by MEM/Rietveld analysis using the powder X-ray diffraction (XRD) data acquired at the synchrotron radiation facility. In this paper, we report the EDD of β-B_{105} and discuss the origin of the peculiar bonds by comparison with the EDD of α-B_{12} and its derivative crystals.

2. Experiment

We crushed chunks of β-B_{105}, (3N purity, Furuuchi Chem. Corp.), then collected grains smaller than 1 μm. We confirmed that they do not have specific crystal orientation and do not include huge grains by measuring the powder XRD patterns in the ω-scanning mode. They were enclosed in a Lindemann glass capillary of 0.3 mm diameter. The 2θ/θ powder XRD patterns were measured at room temperature using a large Debye-Scherrer camera with an imaging plate (IP) at the BL02B2 beamline of SPring-8, a synchrotron radiation facility in Japan. By setting the wavelength to 0.8 Å, an XRD pattern for 2θ from 0 to 75° in 0.01° steps was obtained for each sample. These patterns were analyzed by the MEM/Rietveld method. The numbers of crystal structure factors used for MEM analysis as "observed structure factors" were 1094.

After applying the Rietveld refinement, the MEM analysis was carried out using the computer code, ENIGMA [6], to obtain the EDD. To obtain the same resolution as that of the results of α-B_{12} [4], we used 180×180×240 pixels in a hexagonal lattice in the MEM analysis. A rhombohedral lattice is reconsidered to be a hexagonal one. To reduce the dependence on the initial structure model used in the Rietveld analysis, we adopted a "G-type MEM" [7]. The details of the experiment and analysis are described in the previous report [4], and in the following, we reproduce the results of α-B_{12} and its derivative crystals for comparison.

3. Results and discussion

The structures of α-B_{12} and β-B_{105} are shown in figure 1 (the crystalline structure was visualized using the program, "VESTA" [8]). α-B_{12} has only two crystalline sites in the rhombohedral lattice: "polar" (p) and "equatorial" (e). The polar site corresponds to the atoms that have intercluster two-center bonds, and the other boron atoms are at the equatorial sites. In contrast, the structure of β-B_{105} is not so simple. In addition, the linkage between two B_{12} clusters is different between α-B_{12} and β-B_{105}. In β-

B_{105}, two B_{12} clusters are connected with ~π/5 rotation around the pseudo-fivefold axes. However, for the icosahedra at the vertex of the rhombohedral cell of β-B_{105}, the p and e sites in α-B_{12} correspond to the B5 and B6 sites in β-B_{105}, respectively. The bent p-p bonds in α-B_{12} correspond to B5-B7 bonds in β-B_{105}.

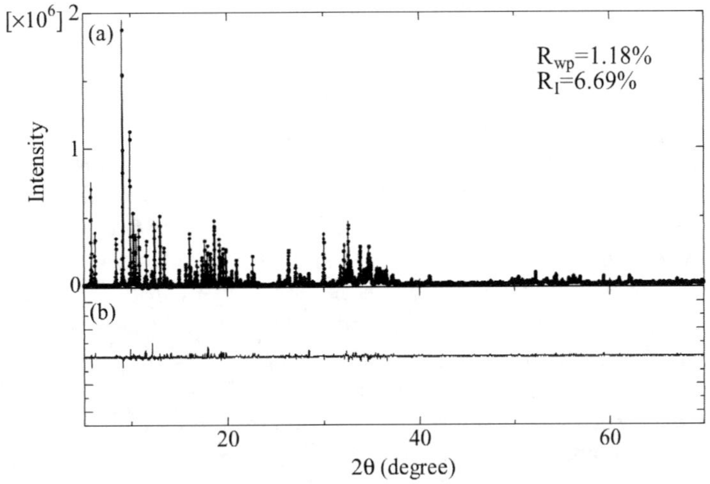

Figure 2. XRD pattern of β-B_{105}. The background from the capillary is removed. (a) The data (black dots) and the results of the Rietveld refinement (solid line). (b) Residual.

Table 1. Results of Rietveld refinement of β-B_{105}. R_{wp} and R_I are reliability factors of the analysis. R_{wp} is based on the weighted XRD pattern and R_I is based on the Bragg intensities. Lattice constants and coordinates are described in the hexagonal lattice (space group $R\bar{3}m$, No.166). g is the occupancy at each atomic site. x, y, and z are fractional coordinates. B is the isotropic atomic displacement parameter.

		β-B_{105}			
	R_{wp} = 1.18%			R_I = 6.69%	
	lattice constants		a=10.92206(3) Å	c=23.80674(13) Å	
site	g	x	y	z	B [Å^2]
B1	1.0066 (17)	0.17422 (16)	0.17420 (15)	0.17626 (10)	0.492 (12)
B2	0.982 (2)	0.3167 (2)	0.29601 (17)	0.12897 (9)	0.520 (15)
B3	1.0198 (17)	0.2640 (2)	0.21996 (19)	0.41989 (9)	0.510 (8)
B4	1.0000 (19)	0.23258 (18)	0.24947 (18)	0.34773 (10)	0.517 (17)
B5	1.0105 (6)	0.05110 (6)	-x	0.94376 (5)	0.510 (8)
B6	1.0000 (7)	0.08905 (9)	-x	0.01270 (5)	0.504 (9)
B7	1.0118 (10)	0.11083 (5)	-x	0.88527 (8)	0.508 (7)
B8	0.9862 (10)	0.17096 (14)	-x	0.02629 (7)	0.534 (10)
B9	1.0223 (13)	0.12770 (14)	-x	0.76772 (9)	0.505 (13)
B10	1.0245 (14)	0.10158 (13)	-x	0.69855 (10)	0.498 (8)
B11	0.9739 (13)	0.0586 (3)	-x	0.32676 (10)	0.498 (12)
B12	0.9877 (16)	0.0873 (3)	-x	0.39832 (10)	0.524 (16)
B13	0.6835 (14)	0.0595 (4)	-x	0.55371 (15)	0.493 (18)
B14	1.0892 (8)	0	0	0.385640 (7)	0.495 (16)
B15	1.0360 (7)	0	0	1/2	0.50 (3)
B16	0.2082 (15)	0.0504 (10)	-x	0.1215 (5)	0.50
B17	0.1384 (19)	0.075 (3)	-x	0.4646 (12)	0.50
B18	0.0846 (18)	0.156 (3)	-x	0.4994 (14)	0.50
B19	0.1038 (18)	0.1623 (19)	-x	0.5471 (11)	0.50
B20	0.0553 (19)	0.181 (5)	0.128 (7)	0.079 (3)	0.50

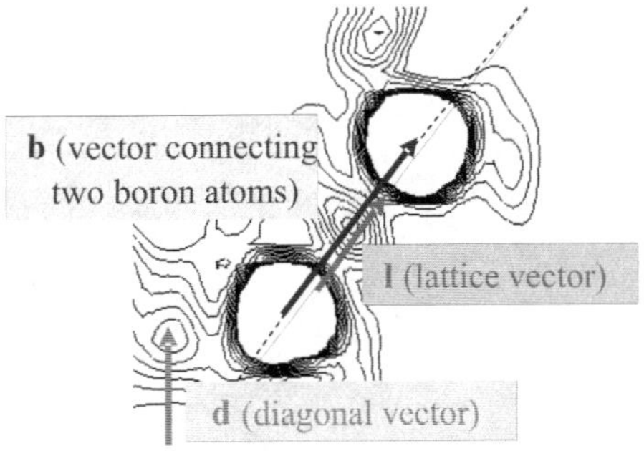

Figure 3. Definition of vectors.

Figure 2 shows the XRD data of β-B_{105}, and in table 1, the results of the Rietveld refinement are summarized. The estimated standard deviations in the table are calculated by a standard method [9]. In the Rietveld analysis, in addition to the lattice constants (a, c), coordinates (x, y, z) and isotropic atomic displacement parameters (B), we refined the site occupancies (g). Since β-B_{105} has many partially occupied sites and MEM requires the total number of electrons in the unit cell as a constraint, we must confirm the site occupancies and estimate the total number of electrons by the refinement. The reliability factors of the Rietveld refinement (R_{wp}, R_I) are low enough to apply the results to MEM analysis.

Table 2 shows the distortion of the icosahedra calculated from the results of the Rietveld refinement. In our past paper, we reported that the bonds in α-B_{12} connecting two B_{12} clusters are bent. The reason why the bonds are bent is the mismatch between the direction of the bond along the pseudo-fivefold axes and the lattice vector [4]. In this study, we confirmed whether the bonds between two icosahedra are bent or not in β-B_{105} by using the EDD. The reliability factors of MEM analysis calculated from nonweighted (R) and weighted structure factors (R_w) were R=3.10% and R_w=2.23%, respectively. The EDD around p-p (B5-B7) bonds is shown in figure 4. Although the bonds (p-p) connecting the two B_{12} clusters are bent in α-B_{12} and its derivative crystals (a), (b) and (c), the bond (B5-B7) connecting the vertex and edge-centered B_{12} clusters is not bent in β-B_{105} (d). As shown in table 2 and figures 3 and 4, the angle between the direction of the B5-B7 bond (**b**) and the lattice vector (**l**) of the rhombohedron is less than in other solids. Vector **b** (connecting two boron atoms) and vector **l** (lattice vector of rhombohedra) are almost parallel to each other. In this case, the positions of

Table 2. Distortion of icosahedra in the four solids. O stands for the center of B_{12} icosahedra. In the case of β-B_{105}, p corresponds to a B5 site, and e corresponds to a B6 site. The distortion ratio is shown in parenthesis. The angular distortion ratio is the ratio of the measured angle (p-O-p, p-O-e, and e-O-e) to the regular central angle (63.43°) of icosahedra. The distance distortion is measured using the ratio of each measured distance to their average distance. The definitions of vectors **b**, **d**, and **l** are shown in figure 3.

	α-B_{12}	$B_{12}C_3$	$B_{12}P_2$	β-B_{105}(vertex)
		Intracluster		
Angle [deg] (distortion ratio)				
p-O-p[a]	61.77(-2.62%)	64.01(0.91%)	65.29(2.94%)	60.92(-3.96%)
p-O-e	64.07(1.01%)	62.46(-1.53%)	62.40(-1.63%)	64.35(1.45%)
e-O-e[b]	63.20(-0.36%)	63.06(-0.58%)	63.99(0.88%)	63.05(-0.60%)
Distance [Å] (distortion ratio)				
O-p	1.696(-0.04%)	1.739(1.42%)	1.775(3.47%)	1.651(-3.64%)
O-e	1.697(0.04%)	1.690(-1.42%)	1.656(-3.47%)	1.712(3.64%)
Avr.	1.697	1.714	1.716	1.681
p-p[a]	1.741(-2.19%)	1.843(2.36%)	1.916(5.67%)	1.674(-5.37%)
p-e	1.800(1.12%)	1.778(-1.24%)	1.781(-1.77%)	1.791(1.24%)
p-e[b]	1.800(1.13%)	1.813(0.70%)	1.800(-0.71%)	1.794(1.41%)
e-e[b]	1.779(-0.07%)	1.767(-1.82%)	1.755(-3.18%)	1.79(1.19%)
Avr.	1.780	1.800	1.813	1.769
		Intercluster		
Distance [Å]				
p-p[c] (vector **b**)	1.640	1.715	1.712	1.793
Angle [deg]				
Rhomb. lattice	58.05	65.68	69.52	65.20
Between two vectors **d b**	29.54	40.88	47.13	38.18
Between two vectors **d l**	34.07	38.77	41.17	38.48

[a] Symmetry codes: -x/2, y, z.
[b] Symmetry codes: y, x, -z.
[c] Neighbor boron atom along the rhombohedral lattice vector.

the two-center bonds are shifted from the edge of rhombohedra, but there is no clear bending because the bonds and lattice vector cannot cross each other. (No symmetry center exists at the center of the bonds in the case of β-B_{105}.) The reasons behind the shift of the position are considered to be the mismatch between the direction of the pseudo-fivefold axis and the lattice vector, and the existence of B16 sites. Since the dangling bonds of a local B_{12} cluster are directed towards the pseudo-fivefold axes, it is natural for the two-center bonds to follow this direction. On the other hand, there are some interstitial sites of boron (B16) [1] near these bonds. It is possible to make three-center bonds among the B5-B7-B16 triangle. The distribution of the electrons toward the B16 site shown in figure 4 (d) can correspond to this three-center bond.

Figure 5 shows the intracluster three-center bonds of the two elemental phases. The three-center bonds have been observed in the MEM EDD of α–B_{12} (a) and its derivative crystals, and can also be observed in β-B_{105} (b, c, d). This finding indicates that the three-center bonds are a universal feature of the boron-rich icosahedral cluster solids.

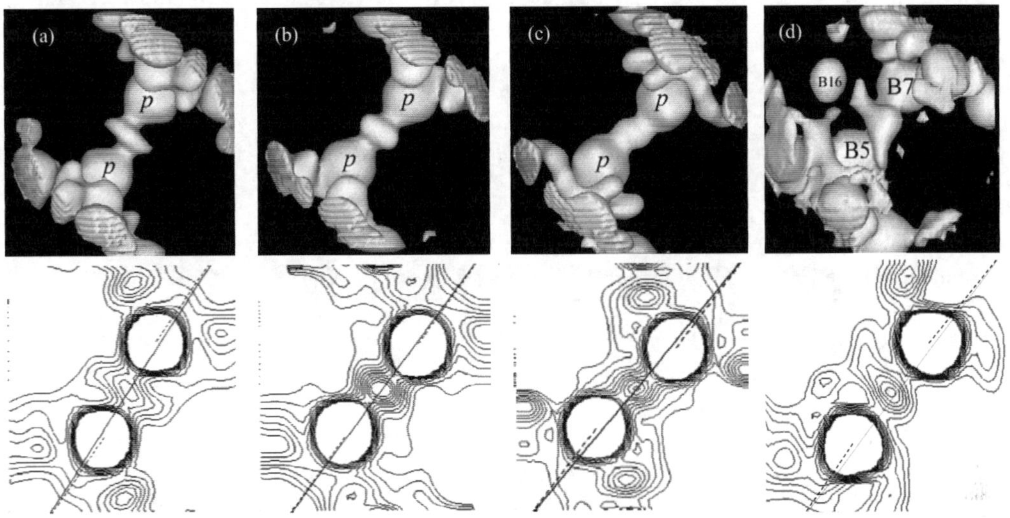

Figure 4. Equidensity surfaces (0.69e/Å^3) and contour maps (hexagonal lattice 110 plane) of electron density around the two-center bonds: (a) α-B_{12}, (b) $B_{12}C_3$, (c) $B_{12}P_2$ and (d) β-B_{105}. Solid lines are edges of rhombohedral lattice and dashed lines are pseudo-fivefold axes of the icosahedra.

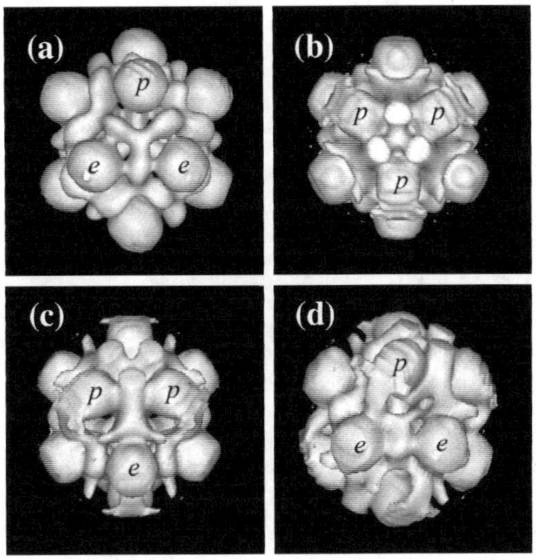

Figure 5. Equidensity surfaces (0.62e/Å^3) of three-center bonds in (a) α-B_{12} (p-e-e), (b) β-B_{105} (p-p-p), (c) β-B_{105} (p-p-e) and (d) β-B_{105} (p-e-e).

4. Conclusion

We succeeded in the observation of the EDD of β-B_{105} experimentally for the first time. There are no bent bonds in this compound, unlike α-B_{12} and its derivative crystals, because the bond direction is parallel to the lattice vector. However, β-B_{105} also has the peculiar three-center bonds, similarly to α-B_{12} and its derivative crystals.

Acknowledgement

The synchrotron radiation experiments were performed at the BL02B2 station in SPring-8 with the approval of the Japan Synchrotron Radiation Research Institute (Proposal No. 2001A0373-ND-np, 2005B0557). Two of the authors (S. H. and K. T.) acknowledge the support for this work provided by the 21st Century COE Program "Applied Physics on Strong Correlation" and the Global COE Program "The Physical Sciences Frontier", MEXT, Japan, respectively. This work was partly supported by Scientific Research on Priority Areas of New Materials Science Using Regulated Nano Spaces, MEXT, Japan, Grant-in-Aid for No. 19051005 and by the Sasakawa Scientific Research Grant from the Japan Science Society.

References

[1] Slack G A, Hejna C I, Garbauskas M F and Kasper J S 1988 *J. Solid State Phys.* **76** 52
[2] e.g., Matsuda H, Tanaka N, Nakayama T, Kimura K, Murakami Y, Suematsu H, Kobayashi M and Higashi I 1996 *J. Phys. Chem. Solids* **57** 1167
[3] Fujimori M, Nakata T, Nakayama T, Nishibori E, Kimura K, Takata M and Sakata M 1999 *Phys. Rev. Lett.* **82** 4452
[4] Hosoi S, Kim H, Nagata T, Kirihara K, Soga K, Kimura K, Kato K and Takata M 2007 *J. Phys. Soc. Jpn.* **76** 044602
[5] Li D and Ching W Y 1995 *Phys. Rev.* B **52** 17073
[6] Tanaka H, Takata M, Nishibori E, Kato K, Ishii T and Sakata M 2002 *J. Appl. Crystallogr.* **35** 282
[7] Sakata M, Mori R, Kumazawa S, Takata M and Toraya H 1990 *J. Appl. Crystallogr.* **23** 526
[8] Momma K and Izumi F 2006 *Commission on Crystallogr. Comput., IUCr Newslett.* No. **7** 106
[9] Young R A 1993 *The Rietveld Method. IUCr Monographs on Crystallography* No. **5**, ed R A Young (Oxford University Press) pp 1-38

Metallic-covalent bonding conversion in boron icosahedral cluster solids studied using electron localizability indicator

Michitaka Yamaguchi[1], Yuji Ohishi[2], Sizuka Hosoi[2], Kohei Soga[1] and Kaoru Kimura[2]

1 Department of Material Science and Technology, Tokyo University of Science, 2641 Yamazaki, Noda, Chiba 278-8510, Japan
2 Department of Advanced Materials Science, The University of Tokyo, Kiban-toh 502, 5-1-5 Kashiwanoha, Kashiwa, Chiba 277-8561, Japan

E-mail: mail@ksoga.com

Abstract. The metallic-covalent bonding conversion of clusters in boron icosahedral cluster solids was studied using the electron density and electron localizability indicator (ELI). The comparison of B_{12} and B_{13} clusters revealed that the bonding conversion from covalent to metallic one involved a decrease in both the electron density and ELI between two boron atoms in the clusters. On the other hand, in the case of vanadium doping into the B_{48} cluster surrounding the vanadium doping site in β-rhombohedral boron, the electron density increased in some parts of the cluster and decreased in other parts. This uncertainty regarding the bonding conversion was clarified by the use of the ELI. The ELI analysis clearly revealed that bonding conversion occurs in the cluster surrounding the vanadium doping site.

1. Introduction

Some of the solids based on group III elements are known to form icosahedral cluster solids (ICSs) consisting of clusters with icosahedral symmetry such as icosahedral clusters with 12 boron atoms (B_{12}) or aluminum atoms (Al_{12}) [1]. The icosahedral clusters as structural units impart peculiar characteristics to ICSs. β-rhombohedral boron (β-B) has been investigated extensively as the most typical ICS consisting of boron [2-4]. It has been revealed that metal doping into β-B can control the bonding nature in and between the clusters, which results in the variation of physical properties such as the electrical and thermal conductivities and the Seebeck coefficient [5-10]. Among the dopant metals, vanadium (V) has attracted much interest because doping of only about 1 at.% into β-B changes the temperature dependence of the electrical conductivity of β-B to an almost metallic dependence [5,7]. The origin of this sensitivity to doping has been considered to be the metallic-covalent bonding conversion (MCBC) of the clusters [1,11]. The MCBC may also be referred to as a multicenter – two-center bonding conversion.

The MCBC in ICSs has been considered as a conversion of the bonding nature of the bonds in and around the cluster that occurs as a result of the occupation of the cluster center by an atom. The MCBC has also been discussed in comparison with the metal-insulator transition (MIT) [1, 7, 8, 12-14]. The MCBC is commonly observed both in boron-rich ICSs (B-ICSs) and aluminum-rich ICSs (Al-ICSs) and can be clearly distinguished from the MIT. α-AlReSi and $Al_{12}Re$ are typical Al-ICSs and have an Al_{12} icosahedral cluster in the structure without and with an atom at the center of the cluster, respectively. It has been suggested that the MCBC occurs in these ICSs on the basis of electron density analysis [12,13]. The above-described change in the electrical conductivity by doping a small amount of V in β-B as one of the B-ICS may also originate from the MCBC around the V doping site. The V atom is known to occupy the A_1 site which is located at the center of the B_{48} cluster. The B_{48} cluster is located at the sharpest vertex among the rhombohedron and the tetrahedron made of four B_{12} clusters. Hosoi *et al* analyzed the electron density distribution in the B_{48} cluster upon V-doping [14] and discussed the MCBC on the basis of the decrease in the electron density at the midpoint of B-B bonds. Recent works on the MCBC in B- and Al-ICSs have focused on the comparison of electron density between clusters without and with an atom occupying the cluster center.

Recently, the electron localization function (ELF) and the electron localizability indicator (ELI) have been proposed as methods for analyzing the localization in chemical bonding. The ELF is a function proposed by Becke and Edgecombe in 1990 for analyzing the nature of chemical bonding [15] and has been used to analyze a wide range of materials from organic molecules to solids such as Li and V metals [16,17]. The ELF was also applied to analyze boron clusters to clarify the existence of bonding among more than two atoms [18-20]. The quantitative analysis of metallic and covalent bonding natures has also been performed on carbon-based solids with a diamond structure, as well as on silicon and β-tin [21], and on intermetallic compounds such as $CaAl_2$ and $BaAl_4$ [22]. Although the ELF has contributed to the investigation of the bonding nature, Kouhout has suggested that the use of the energy of a free electron gas causes the discussion of the localization of the electron to be unclear. He proposed the ELI by investigating the spatial distribution of same-spin electron pairs [23,24]. The ELI consists of the density function of the same-spin electrons.

The ELI in direct space is defined by the following three equations (1)-(3) [23-25]. Its major difference from the ELF is the exclusion of the free electron gas energy. The ELI takes a value greater than or equal to zero but is not limited to less than or equal to one as in the case of the ELF.

$$\gamma(r) = \rho(r)\tilde{V}(r) \tag{1}$$

$$\tilde{V}(r) = \left[\frac{12}{G(r)}\right]^{3/8} \tag{2}$$

$$G(r) = \sum_{i<j}^{\mathrm{occ},\sigma} \left|\phi_i \nabla \phi_j - \phi_j \nabla \phi_i\right|^2 \tag{3}$$

The ELI was defined to monitor the number of σ-spin electrons necessary to locally build a fixed fraction of a σ-spin electron pair [24,26], which is given in the continuous representation as the position where the Fermi hole curvature G for a time-independent single-determinant wave function from orbitals ϕ leads (equations 2 and 3). Thus, the ELI distribution in position space can be represented as a product of two scalar fields in the position space (equation 1): the well-known σ-spin electron density ρ (a one-particle property), and the so-called pair-volume function $\tilde{V}$ (a two-particle property), because it is proportional to the volume needed to locally encompass a fixed fraction of a same-spin electron pair of σ-spin electrons [27].

The larger the ELI value is, the more the electrons are localized. The ELI has also been applied to molecules and solids [24-26,28]. The objective of this paper is to discuss the MCBC of boron clusters in B-ICSs, on the basis of ELI analysis. Electronic-state calculations on the clusters extracted from B-ICSs were performed without and with the occupation of an atom at the center of the clusters,

2. Calculation method

Figure 1 shows clusters used for the analysis. The ELI analysis was first applied to quantitatively evaluate the bonding conversion using two clusters having typical covalent bonding (Si_{17}) and typical metallic bonding (Al_{19}) to clarify how the metallic and covalent bonding natures are revealed in the ELI distribution. Accordingly, we compared the ELI distributions of pairs of clusters without and with an atom at the cluster centers. The cluster pairs are boron clusters (B_{12}^{2+} and B_{13}^{-}) and boron clusters terminated by hydrogen atoms ($B_{12}H_{12}^{2-}$ and $B_{13}H_{12}^{+}$; and $B_{48}H_{36}$ and $VB_{48}H_{36}$). The $B_{48}H_{36}$ cluster corresponds to the structure surrounding the A_1 site in β-B.

Electronic-state calculations were performed on all clusters using GAUSSIAN98. The structures of Si_{17}, Al_{19}, B_{48} and VB_{48} were extracted from the structures determined by X-ray powder diffraction. The electronic-state calculations of Si_{17} and Al_{19} were performed at the DFT-B3LYP theory level using the 6-31G(d) basis set. The dangling bonds of B_{48} and VB_{48} were terminated by hydrogen atoms and the electronic-state calculations were performed at the HF theory level using the 3-21G basis set. The optimized structures of B_{12}^{2+}, $B_{12}H_{12}^{2-}$ and $B_{13}H_{12}^{+}$ were refined at the DFT-B3LYP theory level using the 6-31G(d) basis set. The structure of B_{13}^{-} was determined by the radius between the atom at the cluster center and B_{12} at the DFT-B3LYP theory level using the 6-31G(d) basis set because B_{13}^{-} did not have a stable structure. The ELI was calculated using DGrid [29] in direct space.

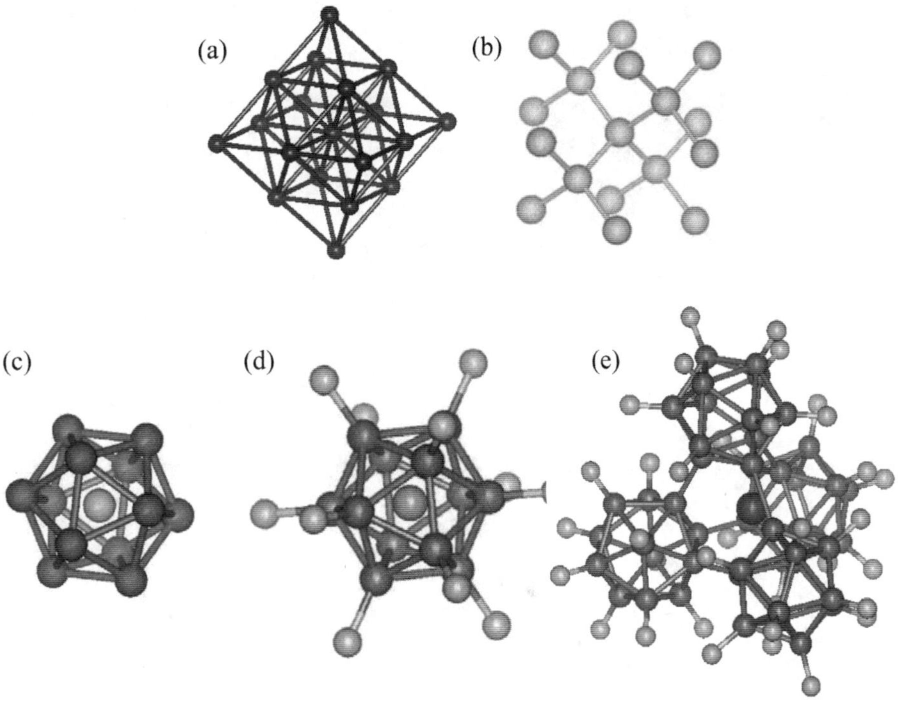

Figure 1. Clusters used for the analysis:, (a) Si_{17} cluster, (b) Al_{19} cluster, (c) B_{13} cluster, (d) $B_{13}H_{12}$ cluster and (e) $VB_{48}H_{36}$ cluster.

3. Results and discussion

To evaluate how the metallic and covalent bonding natures are reflected in the ELI distribution, both the electron densities and the ELI values of the electrons in the Si_{17} cluster with typical covalent bonding and in the Al_{19} cluster with typical metallic bonding were compared. As shown in figure 2(a), the electrons in the Si_{17} cluster are concentrated between pairs of Si atoms in the cluster, while figure 2(b) shows that there is no concentration of electrons between Al atoms in the Al_{19} cluster, as expected from the nature of metallic bonding. According to the spatial distribution of ELI in figures 3(a) and (b), a clear two-center covalent bond between Si-Si pairs in the Si_{17} cluster was observed and the maximum value of ELI between the cluster center and the nearest-neighbor atoms is 2.39, while ELI in the Al_{19} cluster is 1.57, indicating as more metallic bonding than that in Si_{17}.

Journal of Physics: Conference Series **176** (2009) 012027 doi:10.1088/1742-6596/176/1/012027

The above discussion of metallic and covalent bonding appears to be reasonable. However, in the case of a smaller difference in the bonding nature, it is difficult to observe a difference between equidensity surfaces, and a more quantitative analysis is necessary. As preparation for a more difficult case, the following analysis is performed for the typical case of Si_{17} and Al_{19} clusters. Figure 4 shows ELI distribution histograms of the electrons in the Si_{17} cluster (a), the Al_{19} cluster (b) and the difference between them (c). The ELI values were obtained by integration from the position of the cluster center atom up to the distance corresponding to that between the center atom and the nearest neighbor atom. The histogram for the Si_{17} cluster (figure 4(a)) indicates a relatively high ELI of approximately 2.0 with no concentration at a specific ELI level. This represents the covalent nature of the bonding electrons illustrated in figure 3(a). On the other hand, the histogram for the Al_{19} cluster (figure 4(b)) is concentrated at an intermediate level of approximately 1.5, which corresponds to the metallic nature of the bonding electrons illustrated in figure 3(b). Accordingly, as shown in figure 4(c), the difference between the ELI for the Si_{17} and Al_{19} clusters is positive at the intermediate level and negative elsewhere, which represents the change in ELI due to the change in the bonding nature from covalent to metallic. Following this discussion, the change in the bonding nature of the cluster from covalent to metallic will be discussed on the basis of the increase in the frequency of the ELI at the intermediate level (approximately 1.5).

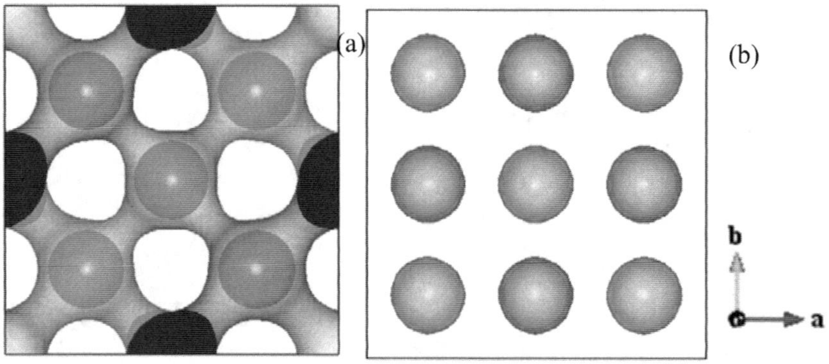

Figure 2. Equidensity surface (isosurface) at an electron density of 0.47 e/Å^3 of in (a) Si_{17} and (b) Al_{19} clusters. The spheres in (a) and (b) denote the Si and Al atoms, respectively.

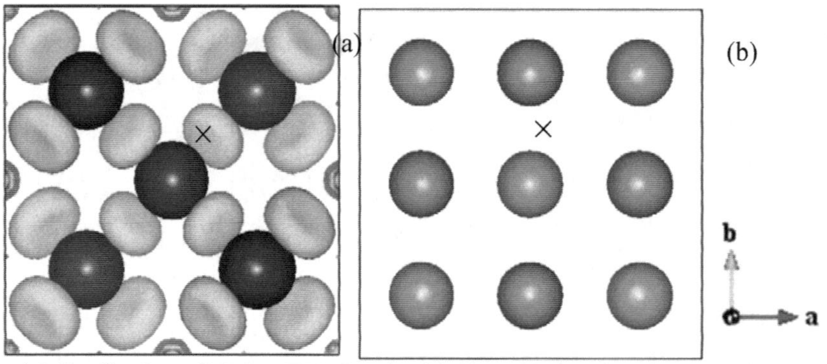

Figure 3. Spatial distribution of ELI at ELI=2.0 for (a) Si_{17} and (b) Al_{19} clusters. The spheres in (a) and (b) denote the Si and Al atoms, respectively. ×: The position of maximun ELI between the atom at the cluster center and the nearest neighbor atom. The ELI value for Si_{17} is 2.39and that for Al_{19} is 1.57.

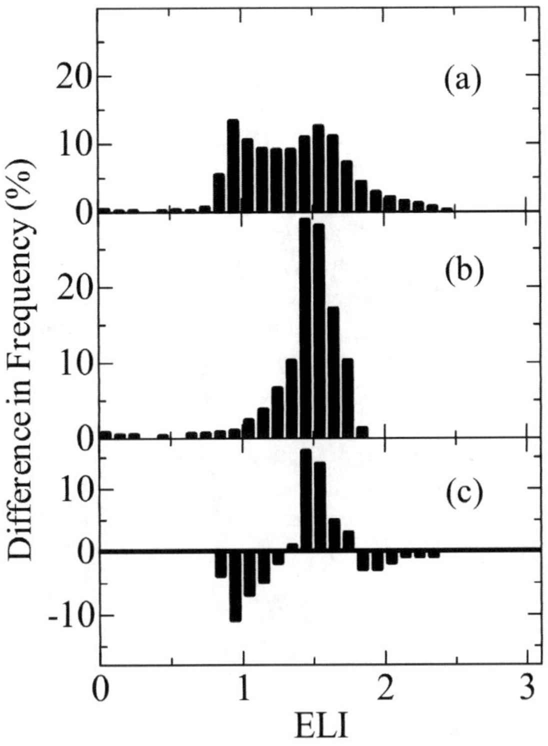

Figure 4. Histogram of ELI values for (a) Si_{17} and (b) Al_{19} clusters. (c) shows the difference between the frequences in (a) and (b).

Figure 5 shows a comparison of the spatial distribution of the electron density of the B_{12}^{2+} (a) and B_{13}^{-} (b) clusters. The electron distribution in the B_{12}^{2+}cluster along the line connecting neighboring boron atoms suggests the existence of B-B covalent bonds. The equidensity surface at an electron density of 0.94e/Å^3 disappears at the midpoint between the boron atoms in the B_{13}^{-} cluster, implying weaker covalent bonding than that in the B_{12}^{2+}cluster. As shown in figures 5(c) and (d), the electron distribution corresponding to B-B covalent bonds in the $B_{12}H_{12}^{2-}$ cluster also disappears in the $B_{13}H_{12}^{+}$ one, though the value of electron density is 0.78 e/Å^3 and lower than that for the B_{12}^{2+} and B_{13}^{-} clusters. However, the comparison between $B_{48}H_{36}$ (figure 5(e)) and $VB_{48}H_{36}$ (figure 5(f)) suggests that the electron density decreases in the bottom plane of the B_{48} tetrahedron of B_{12} clusters and increases on the ridge line of the tetrahedron (the rhombohedron of β-B) owing to the occupation of the center of the tetrahedron (the A_1 site) by the V atom. In contrast to the case of B_{12} clusters, no overall decrease in the bonding electron densities among the B-B bonds surrounding the cluster center was observed in the B_{48} clusters.

Figure 6(a) shows the ELI spatial distribution of the B_{12}^{2+} cluster. It reveals the existence of a dodecahedral bonding frame in the cluster. On the other hand, no localized region is observed in the B_{13}^{-} cluster (figure 6(b)) and the bonding nature in the cluster was found to be metallic. The $B_{12}H_{12}^{2-}$ and $B_{13}H_{12}$ clusters, shown in figures 6(c) and (d), respectively, had the same spatial distributions as the clusters without hydrogen termination in figures 6(a) and (b). The comparison of the $B_{48}H_{36}$ and $VB_{48}H_{36}$ clusters in figures 6(e) and (f), respectively, reveals that the region with localized electrons shrinks upon the occupation of the cluster center by a V atom. A difference of the change in the electron density from those in the previous two cluster pairs is that the more highly localized region does not grow along the ridge line of the tetrahedron (the rhombohedron of β-B).

Similarly to the quantitative discussion of the bonding nature of Al_{19} and Si_{17} clusters using the ELI histogram, the difference between the ELI histograms of the B-B bonds surrounding the cluster center for clusters without and with a cluster-center atom was calculated as shown in figure 7. For all cluster pairs -$B_{12}H_{12}^{2-}$ and $B_{13}H_{12}^{+}$ clusters, $B_{12}H_{12}^{2-}$and $B_{13}H_{12}^{+}$ clusters, and $B_{48}H_{36}$ and $VB_{48}H_{36}$ clusters- the frequency of electrons with an intermediate ELI level of approximately 1.5 increased upon the occupation of the cluster center by an atom. This can be interpreted as a change in the bonding nature surrounding the cluster center from covalent to metallic upon the occupation of the cluster center. In particular, in the case of the bonding conversion from $B_{48}H_{36}$ to $VB_{48}H_{36}$ clusters, the change in the bonding nature is not clear from the spatial distribution of the electron density because electron density increased at some of the bonds and decreased at others. The reason for this change is the increase in the total electron density due to the supply of electrons from the V ion. The increased electron density of some of the bonds may be due to the electron supply, while the decrease in the electron density of others may be due to the bonding conversion. The ELI is not directly affected by the total electron density; thus, it is unaffected by the electron supply indicating an MCBC. In general the ELI analysis revealed that the bonding nature in the B-ICS changes from covalent to metallic owing to the occupation of the center of the clusters by an atom.

Journal of Physics: Conference Series **176** (2009) 012027 doi:10.1088/1742-6596/176/1/012027

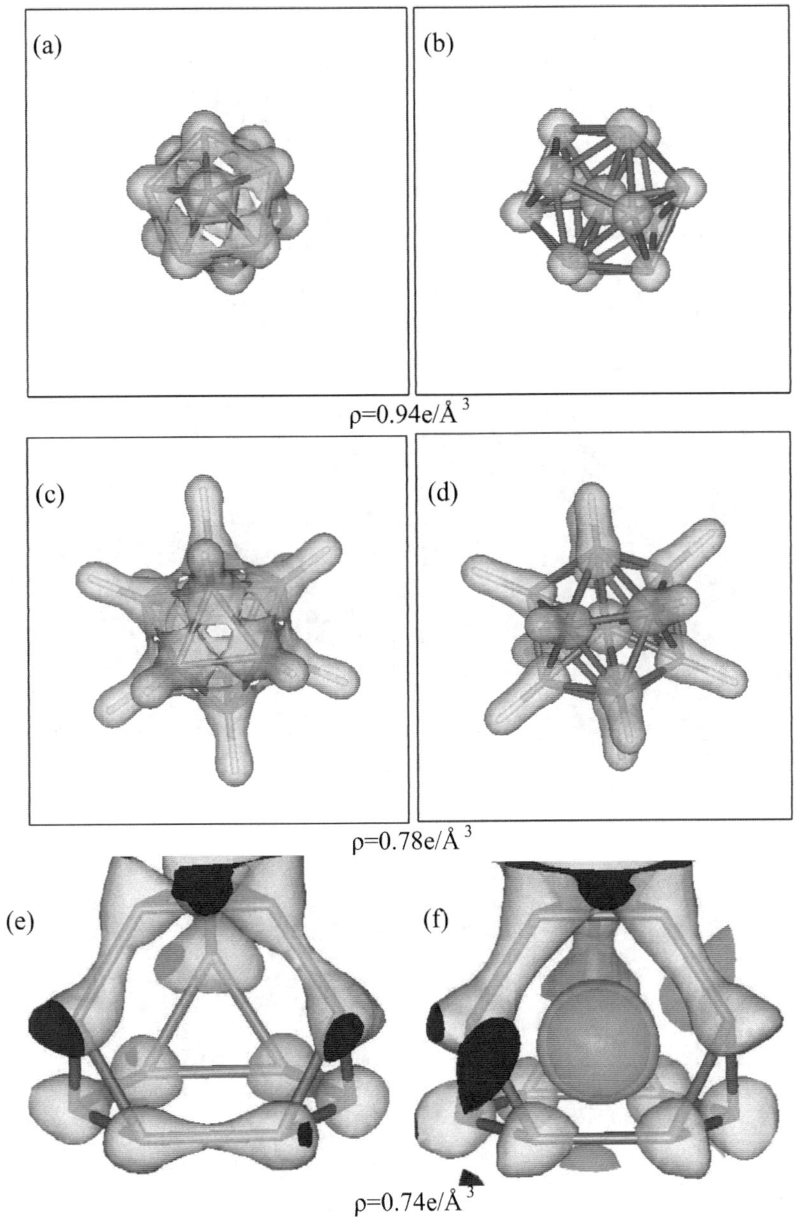

Figure 5. Equidensity surface of electron density for clusters (a) B_{12}^{2+}, (b) B_{13}^{-}, (c) $B_{12}H_{12}^{2-}$, (d) $B_{13}H_{12}^{+}$, (e) $B_{48}H_{36}$ and (f) $VB_{48}H_{36}$. The sphere in (f) denotes the V atom.

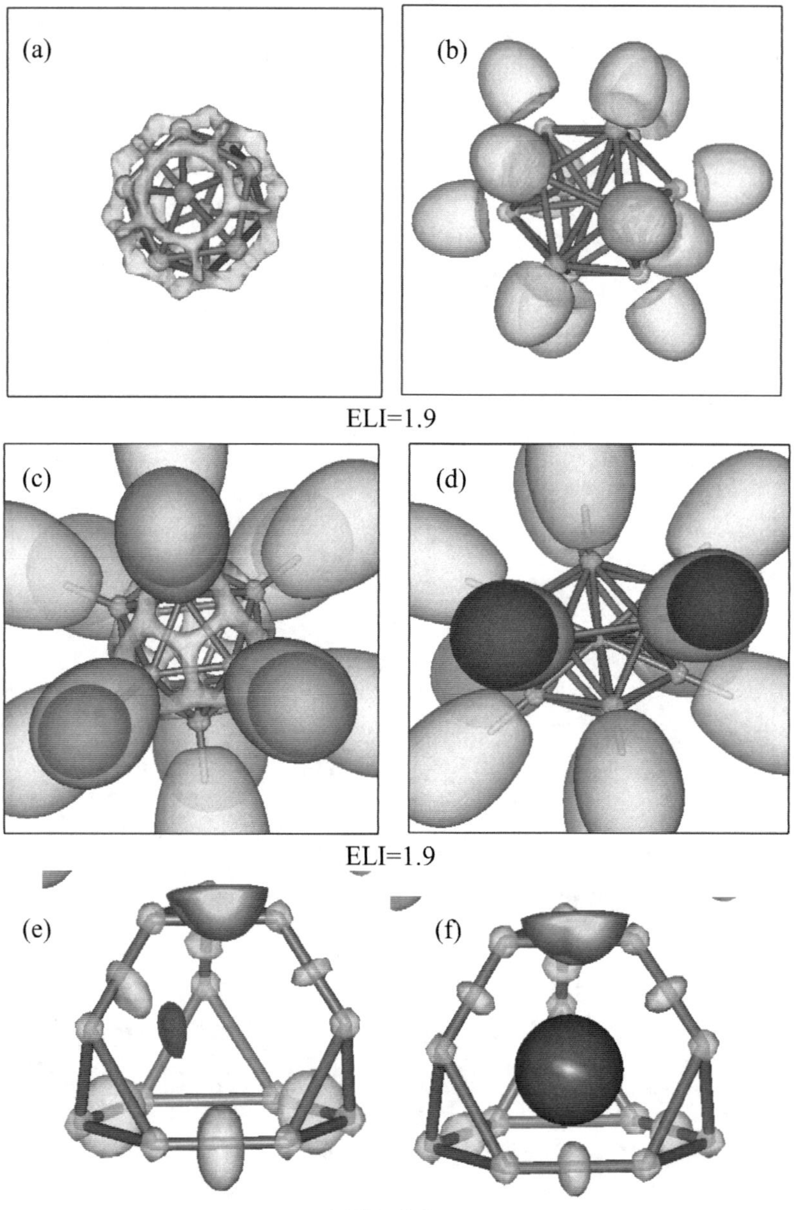

ELI=2.2

Figure 6. Spatial distribution of ELI for clusters (a) B_{12}^{2+}, (b) B_{13}^{-}, (c) $B_{12}H_{12}^{2-}$, (d) $B_{13}H_{12}^{+}$, (e) $B_{48}H_{36}$ and (f) $VB_{48}H_{36}$. The sphere in (h) shows the V atom.

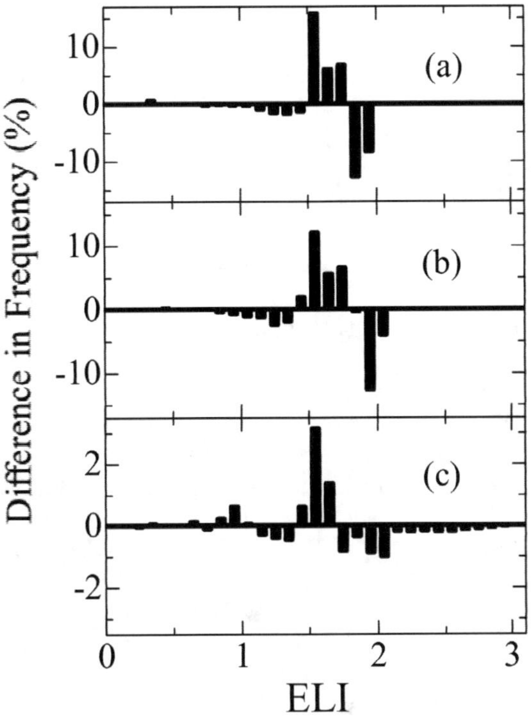

Figure 7. Histograms of difference in ELI between clusters without a center atom and clusters with a center atom for cluster pairs (a) B_{13}^{-} and B_{12}^{2+}, (b) $B_{13}H_{12}^{+}$ and $B_{12}H_{12}^{2-}$ and (c) $VB_{48}H_{36}$ and $B_{48}H_{36}$.

4. Conclusion

ELI distribution analysis was applied to clarify the MCBC in B-ICS. According to the ELI distribution analysis of an Al_{19} cluster as a typical metallic bonding cluster and a Si_{17} cluster as a typical cluster containing covalent bonding, the metallic bonding nature is reflected in the ELI distribution as a relative increase in the frequency of the intermediate level (an ELI of approximately 1.5). The occupation of the central site of the B_{12} cluster by a boron atom resulted in both a decrease in the electron density and an increase in the frequency of the intermediate level in the ELI distribution. On the other hand, in the case of the occupation of the A_1 site in a B_{48} cluster by a V atom, the electron density does not decrease as in the case of the B_{12} cluster. On the basis of the change in electron density, the change in the bonding nature of the B_{48} cluster is unclear. However, the frequency of the ELI distribution of the B-B bonds around the A_1 site atom clearly increased in the intermediate level upon the occupation of a V atom, which implies that the change in the bonding is reasonably isotropic. The histogram of the ELI revealed that its distribution changed from that of more covalent bonding to that of more metallic bonding upon the occupation of the A_1 site in the B_{48} cluster. In conclusion, the ELI analysis revealed that the bonding nature in the B-ICS changes from more covalent to more metallic owing to the occupation of the center of the clusters by an atom.

References

[1] Kimura K, Takeda M, Fujimori M, Tamura R, Matsuda H, Schmechel R and Werheit H 1997 *J. Solid State Chem.* **133** 302
[2] Werheit H, Laux M and Kuhlmann U 1993 *Phys. Status Solidi (b)* **176** 415
[3] Schmechel R and Werheit H 1996 *J. Phys.: Condens. Matter* **8** 7263
[4] Hori A, Tada T and Kimura K 1998 *J. Phys. Soc. Jpn.* **67** 4279
[5] Slack G A, Rosolowski J H, Hejna C, Garbauskas M and Kasper J S 1987 *Proc. 9th Int. Symp. on Boron, Borides and Related Comp.,* ed H. Werheit (University of Duisburg, Duisburg, Germany) p 132
[6] Matsuda H, Nakayama T, Kimura K, Murakami Y, Suematsu H, Kobayashi M and Higashi I 1995 *Phys. Rev.* B **52** 6102
[7] Matsuda H, N. Tanaka, Nakayama T, Kimura K, Murakami Y, Suematsu H, Kobayashi M and Higashi I 1996 *J. Phys. Chem. Solids* **57** 1167
[8] Nakayama T, Shimizu J and Kimura K 2000 *J. Solid State Chem.* **154** 13
[9] Soga K, Oguri A, Araake S, Terauchi M, Fujiwara A and Kimura K 2004 *J. Solid State Chem.* **177** 498
[10] Hyodo H, Araake S, Hosoi S, Soga K, Sato Y, Tarauchi M and Kimura K 2008 *Phys. Rev.* B **77** 024515
[11] Fujimori M and Kimura K 1997 *J. Solid State Chem.* **133** 310
[12] Kirihara K, Nagata T, Takata M, Kubota Y, Nishibori E, Kimura K and Sakata M 2000 *Phys. Rev. Lett.* **85** 3468
[13] Kirihara K, Nagata T, Kimura K, Kato K, Takata M, Nishibori E and Sakata M 2003 *Phys. Rev.* B **68** 014205
[14] Hosoi S, H. Hyodo, Nezu A, Kim H, Kirihara K, Soga K, Kimura K, Kato K and Takata M 2008 *J. Phys. Soc. Jpn.* submitted
[15] Becke A D and Edgecombe K E 1990 *J. Chem. Phys.* **92** 5397
[16] Savin A, Silivi B and Colonna F 1996 *Can. J. Chem.* **74** 1088
[17] Silvi B and Gatti C 2000 *J. Phys. Chem.* A **104** 947
[18] Honle W, Grin Y, Burkhardt A, Wedig U, Schultheiss M, Schnering H G, Kellner R and Binder H 1997 *J. Solid State Chem.* **133** 59
[19] Kellner R, Berger N, Stockle R, Vaas K, Soylemez S, Pfitzner A, Binder H, Honle W, Grin Y, Burkhardt A, Schultheiss M, Wedig U, J. Nub, Schnering H G, H. Borrmann and A. Simon 1998 *Advances in Boron Chemistry* ed W. Siebert (The Royal Society of Chemistry) p 38
[20] Binder H, Kellner R, Vaas K, Hein M, Baumann F, Wanner M, Winter R, Kaim W, Honle W, Grin Y, Wedig U, Schultheiss M, Kremer R K, Schnering H G, Groeger O and Engelhardt G 1999 *Z. Anorg. Allg. Chem.* **625** 1059
[21] Savin A, Jepsen O, Flad J, Andersen O K, Preuss H and Schnering H G 1992 *Angew. Chem. Int. Ed. Engl.* **31** 187
[22] Haussermann U, Wengert S, Hofmann P, Savin A, Jepsen O and Nesper O 1994 *Angew. Chem. Int. Ed. Engl.,* **33** 2069
[23] Kouhout M 2004 *Int. J. Quant. Chem.* **97** 651
[24] Kouhout M 2007 *Faraday Discuss.* **135** 43
[25] Wagner F R, Kouhout M and Grin Y 2008 *J. Phys. Chem. A* **112** 9814
[26] Wagner F R V. Bezugly, Kouhout M and Grin Y 2007 *Chem. Eur. J.* **13** 5724
[27] Kouhout M , to be published
[28] Liang Y, Borrmann H, Baenitz M, Schnelle W, Budnyk S, Zhao J T and Grin Y 2008 *Inorg. Chem.* **47** 9489
[29] Kouhout M 2007 Program DGrid, version 4.2. Max-Planck-Institut fur Chemische Physik fester Stoffe, Dresden

Novel spherical boron clusters and structural transition from 2D quasi-planar structures to 3D double-rings

Saikat Mukhopadhyay[1], Haying He[1], Ravindra Pandey[1], Yoke Khin Yap[1] and Ihsan Boustani[2]

[1] Michigan Technological University, Department of Physics, Houghton, Michigan, 49931 USA
[2] Bergische Universität Wuppertal, FB C - Mathematik und Naturwissenschaften, Gaußstraße 20, D-42097 Wuppertal, Germany

E-mail: boustani@uni-wuppertal.de

Abstract. Based on *ab initio* quantum-chemical and density functional methods we determined the geometry, electronic and structural properties of three cluster-families: boron spheres, double-rings and quasi-planars up to a cluster size of 122 atoms. The most stable structure is the B_{100} sphere showing similar shape but more stability than the B_{80} cage recently proposed by Yakobson *et al* [PRL 98, 166804 (2007)]. In addition we compared the stability of the three cluster families to each other, and reported the structural transition from 2D quasi-planar clusters to 3D double-rings. This transition occurs between the B_{16} and B_{19} clusters.

1. Introduction

The ultimate goal of this research field is to develop, simulate and predict new arrays of nanoclusters and nanotubes with desired, pre-selected and specific uniform properties. The incorporation of nanoclusters, nanotubes and nanocrystals into nanostructured materials can be expressed in form of miniaturization of these new materials and could lead to nanoscaled electronic devices. For instance, possible application fields are in oil-industry, medicine and materials science. During the last decade many researchers focused on a systematic search for new materials mainly consisting of pure or mixed boron, carbon, nitrogen, boron-hydrogen and metal-boron types of systems [1, 2].

Though some variety of possible structures in nanostructured materials like quasi-crystals [3], nanowires [4], nanoribbons [5], was identified, many forms however remain to be discovered. In order to achieve this, there are several theoretical methods based on a variety of concepts, ranging from the most accurate ab initio first-principle methods suited for small or medium sized systems to semi-empirical methods for very large systems. Those methods are reliable tools to detect, to analyze and optimize most of materials, and in many cases they already turn out to be some sort of inexpensive, fast and reliable alternative to standard experimental methods in materials science. These methods provide an opportunity for 'materials engineering', a systematic understanding and development of new nanoscale materials with desired properties. Ultimately, an understanding of "self-assembling" of boron-based materials could lead to manufacturing of useful structures [6].

The physics and chemistry of boron resembles in its ability to configure and form molecular networks. Unlike carbon, bulk boron cannot be found in nature and all known boron allotropes were obtained in the laboratory. All of them are based on different arrangements of B_{12} icosahedra. It is very natural to believe that - like carbon - boron also can form molecular allotropes (fullerenes and nanotubes). Experimental and theoretical research on the chemistry of boron nanomaterials is developing rapidly. The existence of quasi-planar [7] and tubular [8] boron clusters was predicted theoretically and confirmed more recently experimentally [9, 10]. In addition, double-walled boron nanotubes were predicted to be highly stable [11]. These authors found that the interaction between the walls in form of σ-bonds increases the stability of the system in opposite to the carbon nanotubes where the walls are connected by Van-der-Waals forces only, in which no chemical bonds are formed among the neighboring walls.

Lipscomb and co-workers proposed boron spheres as large closo boron hydrides in an analogon to carbon fullerenes using the Descartes-Euler formula $P + F = E + 2$ (P for vertices, F for faces, and E for edges) [12]. They also predicted "conjuncto closo boranes" and their relationships to the dual structures [13]. Based on quantum chemical methods they extended their studies and proposed multicage boron fullerenes B_nH_n [14, 15]. A new concept of conjugate polyhedra was introduced by Tang *et al* [16]. Considering the above Descartes-Euler formula and by a given polyhedron of carbon (boron) one can obtain the corresponding conjugate polyhedron of carbon (boron) by interchange of P and F, with E being kept constantly. These conjugate polyhedra share the same symmetry. For example, for the icosahedral symmetry group I_h, B_n: $n= 30\ k^2 + 2$ is the conjugated polyhedron to carbon C_n: $n = 60\ k^2$. In other words, one can determine for carbon fullerenes C_{60}, C_{240}, the corresponding boron spheres to be B_{32}, B_{122}, for $k = 1$ and 2, respectively.

Small pure boron spheres were predicted by Boustani [17]. Besides the icosahedron B_{12} he constructed small boron B_{22}, B_{32} and B_{42} cages, based on the "Aufbau Principle" [7] as a combination of pentagonal and hexagonal pyramids. In addition, spherical boron hydrides were the subject of a further study by Boustani *et al* [18]. They attached hydrogen atoms to the boron spheres and concluded that the idea of Lipscomb and Massa for geometrical mapping between fullerenes and boron is consistent with the ab initio proved proposals. Besides their equality in the number of carbon faces and boron vertices and vice versa we may add another equality to generate new boron spheres in the way that the number of vertices and faces of carbon cab be replaced by boron vertices.

A new strategy to define reliable geometries for spherical boron cages without resorting too complicated and heavy ab initio calculations was presented by Amovilli and March [19]. Their method reflects the "Aufbau Principle" proposed for boron clusters but it is not restricted to a series of magic numbers. It is always possible, in fact, to recover a sphere with distorted pentagons and hexagons not necessarily displaced in an ordered manner. This situation coincides with the projection - over the surface of the sphere - of the faces of a polyhedron generated by a system of equal charges, with the motion constrained over the sphere surface itself, in their more stable configuration. Such geometries can be easily generated for a given number n of boron atoms by a simple Monte Carlo sampling in a few minutes. The radius R_e of the sphere, however, is finally found by imposing a constant numeral surface density $n/4\pi R_e^2$ [19].

Recently boron fullerenes were presented by Yakobson *et al* [20]. Based on ab initio first-principle methods they generated spherical boron cages as a combination of several staggered double-rings crossing together in a rhombus. The energies of the spheres were compared to the corresponding reference energies of the double-rings. They found that the B_{80} cage, with a binding energy per atom of 5.76 eV, was the most stable sphere symmetrically similar to the C_{60} structure. The only difference between B_{80} and B_{60} is the presence of an additional atom at the center of each hexagon. These facets of the B_{80} follow the "Aufbau Principle" although the hexagonal pyramidal units are rather planar here.

Larger fullerenes with a cluster size of up to B_{300} were proposed by Szwacki [21], using ab initio DFT methods. These cages have similar structures consisting of six interwoven boron double-rings. He predected that the most stable fullerene is made up of 180 atoms and has almost a perfect spherical shape and is more stable than the B_{80} fullerene by 0.01 eV. He also showed that each fullerene of B_{80}, B_{180}, and B_{300} is accompanied by its precursor sheet. The sheets have atomic motifs with a C_{3v} symmetry, similar to the quasi-planar B_7, B_{12} and B_{18} clusters. The precursor sheets of B_{80}, B_{180}, and B_{300} exhibit holes characterized respectively by 6, 9 and 12 surrounding atoms. The existence of holes in very stable α-sheets was already proposed by Tang and Ismail-Beigi [22].

Furthermore, the spherical B_{84} cluster, known as a subunit of β-boron, as well as the B_{156} cluster were experimentally observed by Oku [23]. Both clusters were directly detected by high-resolution electron microscopy and crystallographic image processing, embedded in $B_{105}Al_{2.6}Cu_{1.8}$ and $B_{56}Y$ crystals. The detection technique, developed by Oku [23], seems to be useful for evaluation of the arrangement of light elements like boron clusters. However, a recent ab initio DFT study presented by Prasad and Jemmis [24] claims that stuffing of fullerenelike boron clusters leads to improving their stability. The authors showed that boron clusters such as B_{98}, B_{99}, B_{100}, B_{101}, B_{102} based on icosahedral B_{12} stuffed fullerenes are more stable than the hollow clusters. The most stable fullerenelike cluster was the B_{101} cage, which is around 0.081 eV more stable than the Yakobson's B_{80} fullerene. The basic idea arises from the spherical B_{84} cluster which was extracted from β-rhombohedral boron. The B_{84} is a stellation of B_{12} icosahedron and of 12 B_6 pentagonal "umbrellas". The authors said that the B_{84} is deficieint by 50 electrons according to the Wades n+1 and n+2 skeletal electron rules [25]. The 50 electron devicient B_{84} cluster can be stabilized by adding 16.66 atoms to be made electron sufficient. Therefore, the clusters surrounding $B_{100.66}$ should be the right candidates.

Further study on a bundle of B_{80} fullerenes condensed to a faced-centered-cubic solid was carried out by Yan, Zheng and Su [26]. They showed that when forming a crystal, the icosahedral structure of the B_{80} cages is destorted leading to lowering the symmetry from I_h to T_h. They also found that there are chemical bonds between the nearest neighboring five B_{80} clusters in (001) plane. The interaction between these cages improves the stabilty so that the cohesive energy of the fcc B_{80} solid is about 0.23 eV/atom with respect to the isolated B_{80} fullerenes. They also proposed a different elctron counting rule which could might explain the stability of the B_{80} fullerene. They considered that each triangle would consume two electrons in a three-center two-electrons bond (3c2e) while no electrons are considered by the 12 pentagonal holes. For 120 triangles (6 triangles for each of 20 hexagons) there are 240 electrons required for bonds. This number is eqal to the number of the valence electrons of the B_{80}. In other words, this rule says that the number of the total electrons demanded for bonding should be equal to the total number of the valence electrons. Another interesting result of their calculations is that the electronic structure reveals the fcc B_{80} solid to be metallic.

In this paper we will report on the structures of three cluster families: i) of nearly round boron cages built from pentagons and hexagons or from pentagonal and hexagonal pyramids, ii) of double-rings (DR), and iii) of quasi-planar (QP) clusters, up to a cluster size of 122, and of the structural transition from 2D quasi-planar clusters to 3D double-rings in comparison to an infinite boron strip. The results will be finally discussed and the main conclusions will be given.

2. Theoretical Details

The calculations for the current work were performed using ab initio first-principle methods for solving the many-electron problems in atoms, clusters, and solids, in the framework of the Hartree-Fock (HF) self-consistent-field (SCF) and the density functional theory [27, 28]. These important concepts are the kernels of a variety of program packages that we use for our simulations, depending on the size of the system to be examined. Due to the fact that the size of the investigated systems is relatively large, it was advisable to use the standard minimal STO-3G basis set to calculate large systems containing over 100 atoms and to consider symmetry constraints in the form of a point group symmetry. The optimization process was carried out at the nonlocal corrections to exchange correlation B3LYP [29, 30], using the program packages GAMESS_UK [31] and Gaussian 03 [32].

In the present contribution, the geometrical and electronic structures of the boron clusters B_n for $n = 12-122$ were investigated. Only the final geometries of the lowest energy selected cluster are presented. The optimization procedure, based on the analytical gradient method, was carried out for the ground state energies using the restricted Hartree-Fock theory, followed by B3LYP throughout. In order to compare the stability of all clusters we calculated the binding energy per atom which is defined as ($E_b = E_1$ - E_n/n), where n is the cluster-size, E_1 is the energy of atomic boron, and E_n is the cluster energy. Since the geometrical structures of the clusters influence their electronic properties we compared their highest occupied molecular orbital (HOMO) and the lowest unoccupied molecular orbital (LUMO) in relationship with the cluster size.

3. Results and discussion

3.1. Boron Spheres

Inspired by Yakobson *et al* [20], our investigations on larger boron spheres were carried out using the ab initio methods mentioned above. We considered spherical boron clusters B_n for n = 12, 20, 32, 42, 60, 72, 80, 92, 100, 110, and 122, most of them with icosahedral symmetry [figure 1]. The cage B_{12} is the well known icosahedron of the α-rhombohedral crystalline boron.

Table 1. The considered spherical boron clusters [figure 1] and the corresponding binding energy per atom E_b, obtained at the B3LYP-STO3G level of theory

Cluster	Symmetry	E_b(eV)	Number of		
			Triangles	Pentagons	Hexagons
B_{12}	I_h	5.68	20	12	-
B_{20}	I_h	6.07	-	12	-
B_{32}	I_h	6.12	60	12	12
B_{42}	D_{5d}	6.27	70	12	20
B_{60}	I_h	6.42	-	12	20
B_{72}	I_h	6.63	60	12	20
B_{80}^h	I_h	6.65	-	12	30
B_{80}^f	I_h	6.72	120	12	20
B_{92}	I_h	6.70	180	12	20
B_{100}	D_{5d}	6.81	120	12	30
B_{110}	I_h	6.53	180	12	30
B_{122}	I_h	6.19	240	12	20
$^aB_{12}$	C_{3v}	6.29			
$^b Strip$	C_1	6.78			
$^cB_{81}$	D_{2h}	6.39			
$^d\alpha-Boron$	I_h	7.66			

[a] The quasi-planar structure [figure 2], [b] The infinite strip [figure 5], [c] Precursor Sheet of B_{100} [figure 6],[d] Crystalline rhombohedral boron.

The structure of B_{12} - I_h has 12 vertices and 20 faces. It has a binding energy per atom E_b of 5.68 eV, calculated at the STO3G-B3LYP level of theory. However, since the five-fold degenerate HOMOs of B_{12} are not fully occupied the Jahn-Teller-effect follows and distorts the icosahedral

Journal of Physics: Conference Series **176** (2009) 012028 doi:10.1088/1742-6596/176/1/012028

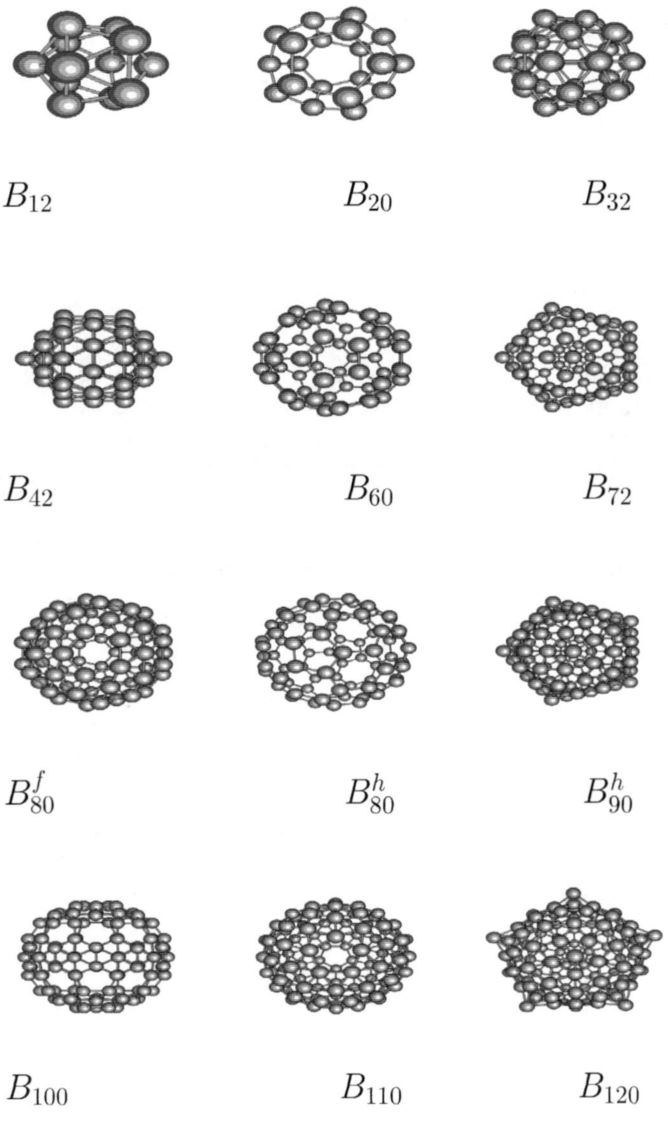

Figure 1. Spherical structures of the B3LYP-optimized boron clusters B_n, obtained with the STO3G basis set. The symmetry point group of the clusters is I_h except of B_{42} and B_{100} clusters which have D_{5d} symmetry.

symmetry to a lower one so that the icosahedron flats out into a quasi-planar structure of C_{3v} symmetry [figure 2] approaching a high stability of 6.29 eV [table 1]. The dodekahedron B_{20}, with a E_b of about 6.07 eV, has 20 vertices but 12 faces, in opposite to the icosahedron. Both structures, icosahedron and dodekahedron have 30 edges regarding the Descartes-Euler formula mentioned above.

Journal of Physics: Conference Series **176** (2009) 012028 doi:10.1088/1742-6596/176/1/012028

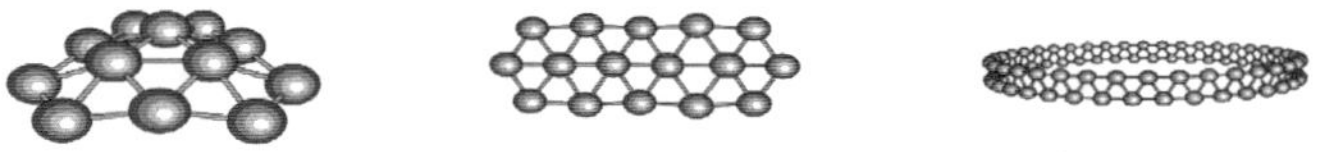

$B_{12} - C_{3v}$ $B_{16} - C_{2v}$ $B_{80} - D_{40d}$

Figure 2. Examples of quasi-planar and double-rings structures of obtaiend at the B3LYP-STO3G level of theory.

Fusing both the icosahedral and dodekahedral cages we receive the B_{32} - I_h which has 30 faces and 60 edges. It can be seen as dovetailed 12 pentagons and 12 hexagons. The E_b of the B_{32} cage is around 6.12 eV. The B_{42} - D_{5d} sphere can be generated by adding an atomic ring of 10 boron atoms perpendicular to an arbitrary axis going through two polar atoms of the B_{32} - I_h cage. The B_{42} structure has 70 faces and 110 edges. The binding energy per atom of the B_{42} increases to 6.27 eV. The B_{60} - I_h cage can be achieved by replacing each icosahedral atom in B_{12} - I_h by a pentagonal B_5 cluster keeping the icosahedral symmetry. Its geometry is simply that one of the C_{60} fullerene. The related E_b is about 6.43 eV. The structure B_{72} - I_h arises from B_{60} - I_h by setting one boron atom in each center of the 12 pentagons. The binding energy per atom of the B_{72} cage is around 6.63 eV.

There are two structures of B_{80} both with I_h symmetry. The first but less stable one called hexagonal B_{80} (B^h_{80}) can be generated by rotating the pentagons in B_{60}, so that the edges of neighboring pentagons are parallel, then adding a dodecahedron B_{20} having the same radius. This B^h_{80} - I_h cage has the same form as the C_{80} fullerene. The B^h_{80} structure has 42 faces (12 pentagons and 30 hexagons). The number of hexagons n can be calculated after (n = a/2 - 10), where a is the number of boron atoms of the sphere. This B^h_{80} structure is the starting geometry for the most stable cage B_{100}. The other and more stable structure B_{80} fullerene (B^f_{80}) was proposed by Yakobson et. al. [20]. It can be generated by adding one boron atom to each center of the 20 hexagons of the B_{60} fullerene. The binding energy per atom of B^f_{80} is nearly 6.72 eV. The average radius (AR) of the cage is about 4.15 Å.

The structure B_{92} - I_h is arising from B_{60} - I_h by setting one boron atom to each center of the 12 pentagons and 20 hexagons of the B_{60} fullerene. The corresponding E_b is about 6.70 eV with an AR of 4.7 Å. The cages B_{100} - D_{5d}, and B_{110} - I_h can be obtained by setting 20 and 30 boron atoms into the centers of the hexagons of the B^h_{80} cage, respectively. The calculated binding energy per atom for B_{100} is 6.81 eV and for B_{110} is about 6.53 eV and their AR is about 5 Å. Finally, the cage B_{122} can be generated by setting boron atoms to each center of the 12 pentagons of the B_{110} cage. B_{122} seems to be less stable than the previous one and has a E_b of about 6.28 eV and an AR of 5.19 Å. For comparison, the cohesive energy of the periodic rhombohedral α-boron, calculated at B3LYP-STO3G, is about 7.66 eV/atom.

In order to assess the idea of Tang and Ismail-Beigi [22] which says that the hexagonal holes (defects) enhance more stability in boron sheets and nanotubes, we determined the cages between B_{80} and B_{110} by gradually adding two atoms to two centers of these 30 hexagons of B^h_{80}. We found out that filling only 20 hexagons of a maximum of 30 leads to the highest stability of all structures [figure 4]. In other words, the spherical B_{100} cage seems to be the most stable cluster between all 2D and 3D clusters and the idea of holes is working well. The B_{100} sphere is consisting of a ring of 10 centered hexagons bicapped by five connected centered hexagons.

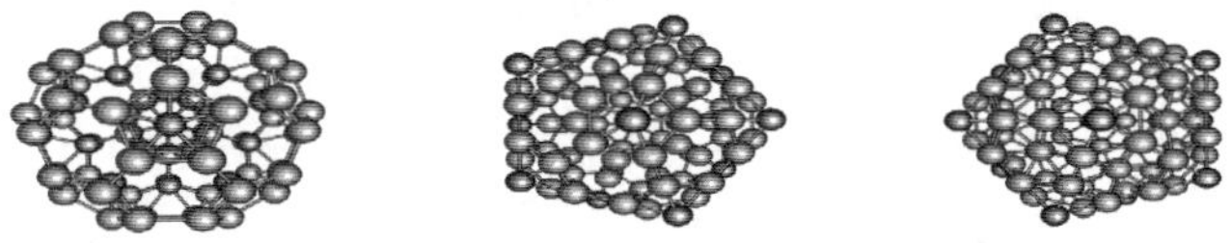

B_{84} $B_{92}-a$ $B_{92}-b$

Figure 3. The B_{84} cage is composed of three shells: B_{12} (red), B_{12} (blue) and B_{60} (green). The B_{92} cages give an example for stretching the inner shell depicted by the nonbonding dodecahedron (red balls) in $B_{92}-a$ outwards to the external shell $B_{92}-b$ forming new chemcial bonds.

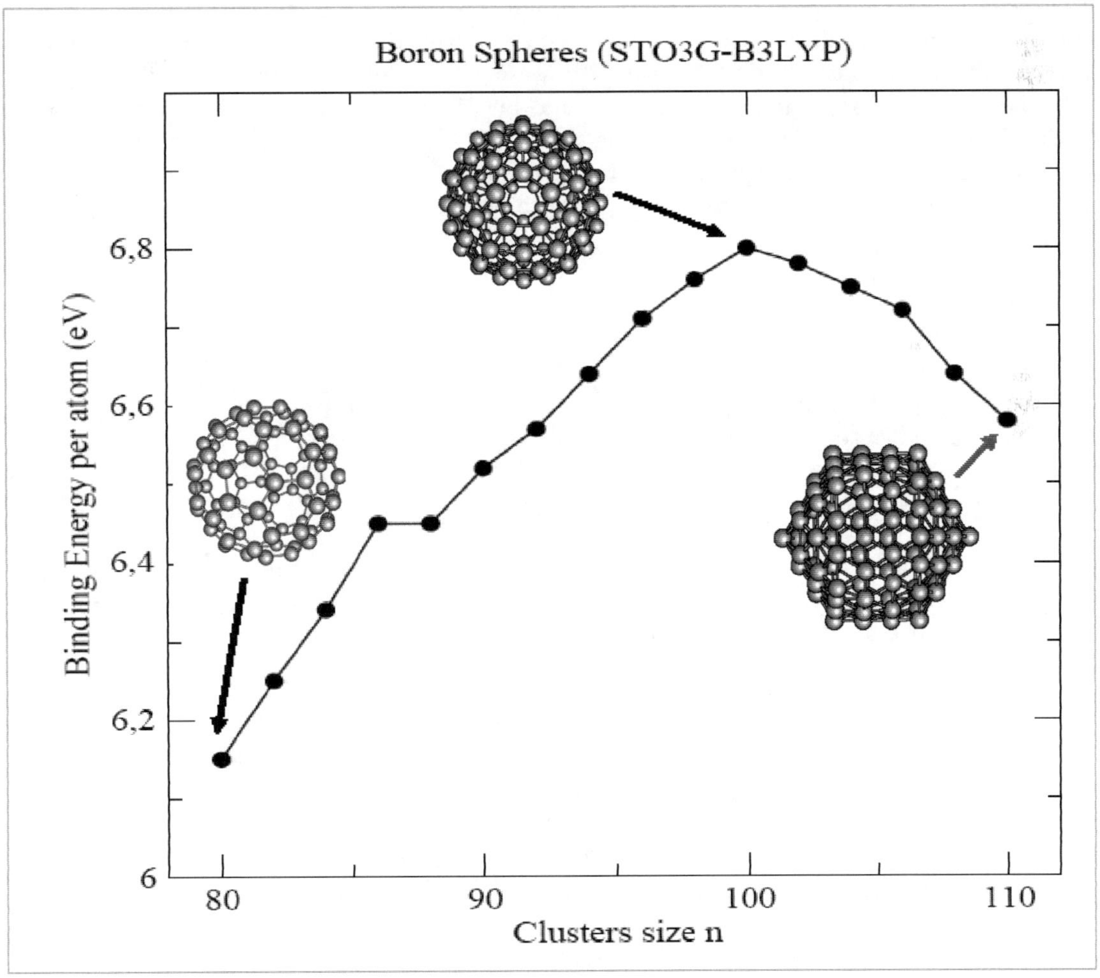

Figure 4. Binding energy per atom as a function of the cluster size n of B_n for n = 80 to 120, via adding two by two boron atoms in each of the 30 hexagonal centers of the spherical B_{80}^{h} cluster.

This B_{100} sphere is accompanied by a segment of its precursor sheet [figure 6.]. This sheet (B_{81} with a D_{2h} symmetry) is by 0.42 eV less stable than the B_{100} sphere.

Furthermore, we studied few spherical clusters composed of different shells [see figure 3]. The first one, a B_{84} sphere, is consisting of two shells: $B_{12}@(B_6)_{12}$. The inner shell B_{12} is an icosahedron connected to 12 pentagonal pyramids in a form of open umbrellas B_6, considered as the outer shell. The calculated binding energy per atom is about 6.43 eV and lies below the corresponding E_b of B_{80}^f which is 6.72 eV. Further examples for stuffing are the B_{92} isomers, illustrated in [figure 3]. The sphere $B_{92} - a$ is consisting of an inner shell B_{20} which is not connected to the outer shell of B_{72}. The inner shell stretches outward and leans forward to build up new chemical bonds with the outer shell, as can be seen by the $B_{92} - b$ cluster.

It was determined experimentally that the QP B_{12} has a relatively large HOMO-LUMO gap of 2.0 eV [9]. Theoretically, the HOMO-LUMO gap of the QP B_{12} according to our calculations is about 2.47 eV which is the largest gap of all clusters. The HOMOs and the LUMOs of the spherical cluster B_{80} are both fivefold degenerate exhibiting a gap of 2.27 eV, being the next largest value. It is comparable to the value of 1.94 eV calculated by Gopakumar [33] for the buckyball B_{80}^f. The estimated gap of the B_{80}^f cage, obtained by Yakobson [20], amounts to 1.01 eV, nearly to less than the half of the previous values. In contrast, the gap of the B_{100} sphere is of a relatively small value 0.14 eV. That can be connected with the symmetry of the cluster. When the HOMOs are degenerate and fully occupied the cluster exhibits a large HOMO-LUMO gap, as in the case of the B_{80}^f cage. But when these orbitals are partially occupied, the degeneracy finishes and the gap is small, as in the case of the B_{100} sphere. In general the HOMO-LUMO gaps of all clusters are alternating as a function of the cluster size in a decreasing manner approaching small values for larger cluster sizes converging to be metallic as in the case of the infinite boron strip.

3.2. Boron Double-Rings

Regarding our previous studies on segments of single-walled boron nanotubes with different diameters and sizes, we found that staggered double-rings of the following boron clusters B_{24} [34], B_{32} [35], B_{36} [36], and B_{96} [37] have the highest stability compared to all other multi-ring structures. In order to study the stability of the boron double-rings family in relationship to the cluster-size, we carried out new ab initio calculations starting from the smallest DR B_{12} up to the largest DR of B_{122}. An example of these structures is given in [figure 2]. We optimized these structures at the B3LYP and STO3G level of theory. The stability of these structures, expressed as the binding energy per atom, increases with increasing cluster-size, which is shown in [figure. 5]. We also calculated the energy of an infinite strip of double-rings as a reference for all structures. The binding energy per atom of this infinite structure is 6.78 eV.

In comparison of the double-rings and spherical structures of B_{80} and B_{100} clusters we found that the binding energy per atom obtanied for the B_{80} DR, calculated at the HF-SCF and B3LYP level of theory, are 5.39 eV and 6.75 eV, respectivley. They are higher than the corresponding ones 5.22 eV and 6.74 eV, determined for the spherical B_{80}^f. This means that the DR of B_{80} is more stable than the spherical one at both levels of theory. Concerning the B_{100} structure, the E_b of B_{100} DR obtained at the HF-SCF level is about 5.39 eV, while the corresponding E_b of the spherical B_{100} cluster is 5.27 eV. At the more reliable B3LYP level of theory we have a reverse picture. The E_b of B_{100} DR is about 6.77 eV while the B_{100} sphere is 6.81 eV. In other words, the B_{100} DR is more stable at the HF-SCF level than the B_{100} sphere, while the sphere is more stable at the B3LYP level of theory due to its high correlation energy.

3.3. Quasi-planar Clusters and Structural Transition

It was established that different boron cluster topologies in form of quasi-planars, sheets or nanotubes possess highly stable structures. It was also found that the most stable boron clusters are small and have quasi-planar structures. But what happens with growing the cluster size? Therefore the simple question rises: Up to which size do the 2D quasi-planar structures remain as the most stable ones? This item was the subject of further studies exploring the transition from 2D to three-dimensional (3D) structures. Kiran *et al* [38] proposed B_{20} as the embryo of single-walled boron nanotubes for a planar-to-tubular structural transition in boron clusters. They investigated experimental and computational simulation methods revealing that boron clusters, which favour planar 2D structures up to 18 atoms, prefer 3D structures beginning at 20 atoms.

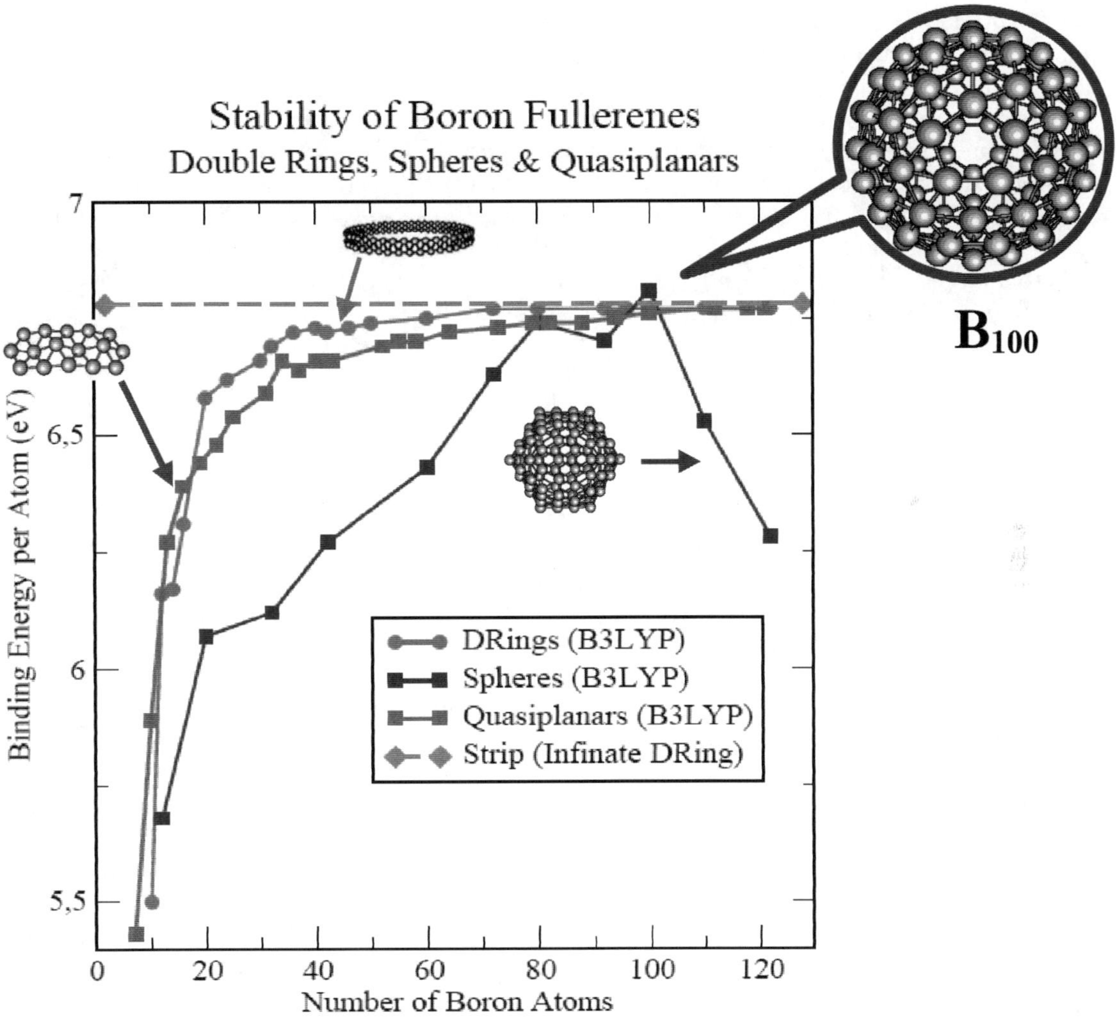

Figure 5. Binding energy per atom as a function of the uster size for spherical, quasi-planar, and double-ring boron clusters showing other perspective of the spherical B_{100} cluster namely along the 5-fold symmetry axis.

Marques and Botti [39] examined the structural transition question from the optical point of view. They applied the time-dependent density functional theory using the real-time and space to solve the time-dependent Kohn-Sham equations. They also showed that the optical spectroscopy can be applied to distinguish without ambiguity between the different low-energy members of the B_{20} family. The most stable neutral B_{20} isomer is the tubular cluster which can be unequivocally identified due to the presence of a very sharp resonance at about 4.8 eV.

An *et al* [40] studied the relative stability among four low-lying isomers of neutral and anionic B_{20} clusters. They carried out highly accurate MP4(SDQ) and CCSD(T) calculations yielding the same energy ordering for the neutral B_{20} isomers. Both calculations show that the double-ring is the lowest-energy structure, has a large negative nucleus-independent chemical shift and therefore is strongly aromatic. Oger *et al* [41] asserted that boron cluster cations were found to undergo a transition between quasi-planar and cylindrical molecular structures at B_{16}^+. The experimentally determined collision cross sections are consistent with those calculated for the global minimum structures as obtained from the theory. For the cations B_{17}^+ and larger, the cylindrical geometries dominate the low-energy structures and for neutral clusters, the transition from 2D to double-ring structures occurs for B_{20}.

As mentioned above, several authors [38] to [41] showed theoretically and experimentally that a structural transition from 2D into 3D structures occurs between B_{16} and B_{20}. Therefore, we investigated different configurations of quasi-planar clusters starting from B_{12} in gradual steps of three atoms up to B_{122} quasi-planar clusters using the same theoretical methods mentioned in section II. An example of quasi-planar clusters is shown in [figure 2]. The calculations of the quasi-planar clusters expressed in form of binding energy per atom are shown in [figure 5]. One can recognize that the binding energies of the 2D quasi-planar clusters are crossing those of the 3D structures of double-rings confirming that the structural transition from 2D into 3D occurs between the B_{16} and B_{19} clusters.

4. Discussion and Conclusions

We developed novel structures in the form of spheres, double-rings and quasi-planar structures. These nanostructures were optimized at the B3LYP level of theory using the STO3G basis set. Generally, the optimized boron spheres range from the smallest B_{12} icosahedron to B_{20}, B_{32}, B_{42}, B_{60}, B_{72}, B_{80}, B_{92}, B_{100}, B_{110} and up to B_{122}. The average radius of the spheres are between 4.15 and 5.19 Å and the interatomic distances are about 1.67 to 1.94 Å. We also optimized double-rings as segments of nanotubes from a DR of B_{12} (2xB_6) up to a DR of B_{122} (2xB_{61}). In addition we studied two-dimensional structures in the form of quasi-planar clusters from 12 up to 122 cluster size. The smallest QP cluster is the B_{12} - C_{3v}, the conjugate of the icosahedron. The next QP cluster is the B_{16} - C_{2v} strip which is composed of three parallel staggered chains [figure 2]. Further QP clusters are generated in gradual steps by adding three atoms to each cluster up to the largest one B_{122}.

We conclude that the most stable clusters is the B_{100} sphere whose binding energy per atoms goes behind that of the infinite strip of the DR clusters. The structure of B_{100} contains triangle and hexagonal motives. These motives contribute essentially to the cluster stability as recently found in boron sheets [43]. Our B_{100} sphere is about 0.09 eV more stable than the B_{80}^f porposed by Yakobson. Similar value of around 0.081 eV obtained by Jemmis for the stuffed B_{101} structur in relationship to Yakobson B_{80}^f cage. "Stuffing improves the stability" of boron clusters proposed by Jemmis could not be confiremd by our the B_{84} and B_{92} clusters. However our B_{100} is one of the favourites determined by the counting rules and fulfills the stability requirement to stabilize B_{84} by adding exact 16 atoms. Furthermore, our B_{100} sphere does not fulfill the counting rules proposed by Su *et al* [26]. The B_{100} sphere has 120 triangles 3c2c bonds consuming 240 electrons and 20 two-center two-electron 2c2e bonds counting two 2c2e for each of the 10 hexagonal holes, consuming 40 electrons. The sume of the required electrons 280 is not equal to the valence

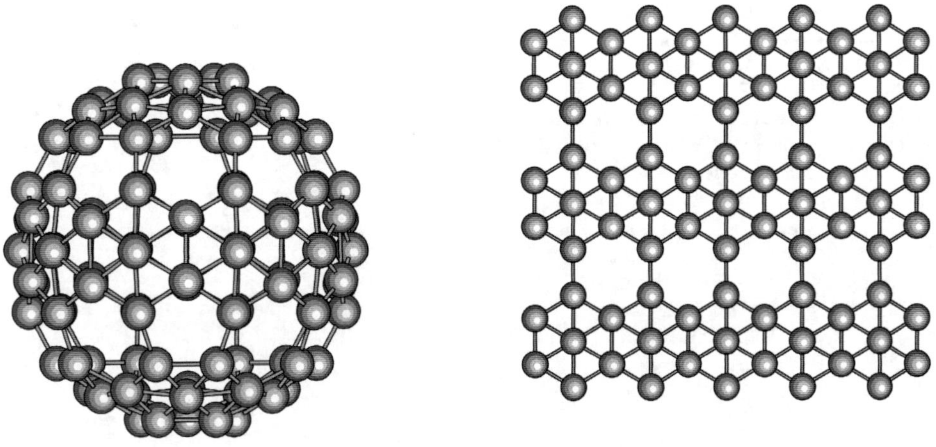

B_{100} **Precursor Boron Sheet**

Figure 6. The spherical B_{100} cluster and the related precursor boron sheet. The structure of B_{100} is composed of ten dovetailed equatorial centered hexagons consisting 50 atoms and two polar cups each is consisting of five dovetailed centered hexagons with 25 atoms.

electrons of B_{100}.

The structural transition from 2D quasi-planar clusters into 3D double-rings occurs beyond the B_{16} cluster and is in excellent agreement with the experimental results obtained by Oger *et al* [41], detemined by photoelectron specsroscopy. Furthermore, the quasi-planar strcuture of B_{16} is identical to that one found experimenatlly by Sergeeva *et al* [42]. Finally, the most stable sphere B_{100} has a small HOMO-LUMO gap and less stable precursor sheet in opposite to the less stable B_{80} fullerene with a relatively larger HOMO-LUMO gap and very stable precursor α-sheet. In other words, the most stable structure of the spheres does not necessarily have a very stable precursor sheet or to exhibit a large HOMO-LUMO gap. The both precursors of B_{80}^{f} and B_{100} spheres are only two of many pattern which could be generated by different combinations of hexagonal holes of the precursor sheets. Since there exists at least more than one precursor for the stable B_{100} sphere, we believe that the energy of this cage corresponds to one of the local minima on the potential energy hypersurface of these clusters. Therefore, further investigations are required to search for the global minimum, electronic, and thermodynamic properties.

References

[1] Terrones M 2004 *Int. Mater. Rev.* **49** 325
[2] Ivanovskii A L 2002 *Russ. Chem. Rev.* **71** 175
[3] Boustani I, Quandt A, and Kramer P 1996 *Europhys. Lett.* **36** 583
[4] Zhang Y, Ago H, Yumura M, Ohshima S, Uchida K, Komatsu T and Iijima S 2004 *Chem. Phys. Lett.* **385** 177

[5] Ding Y, Yang X and Ni J 2008 *Appl. Phys. Lett.* **93** 043107
[6] Quandt A and Boustani I, and the references their in 2005 *Minireviews, ChemPhysChem* **6** 1
[7] Boustani I 1997 *Phys. Rev. B* **55** 16426
[8] Boustani I, Quandt A, Hernandez E and A. Rubio 1999 *J. Chem. Phys.* **110** 3176
[9] Zhai H J, Kiran B, Li J and Wang L-S 2003 *Nature Mat.* **2** 827
[10] Ciuparu D, Klie R F, Zhu Y and Pfefferle L 2004 *J. Phys. Chem.B* **108** 3967
[11] Sebetci A, Mete E and Boustani I 2008 *J. Phys. Chem. Solid* **69** 2004
[12] Lipscomb W N and Massa L 1992 *Inorg. Chem.* **31** 2299
[13] Lipscomb W N and Massa L 1994 *Inorg. Chem.* **33** 5155
[14] Derecskei-Kovacs A, Dunlap B I, Lipscomb W N, Lowrey A, Marynick D S and Massa L 1994 *Inorg. Chem.* **33** 5617
[15] Gindulyte A, Krishnamachari N, Lipscomb W N and Massa L 1998 *Inorg. Chem.* **37** 6546
[16] Tang A-C, Li Q-S, Liu C-W and Li J 1993 *Chem. Phys. Lett.* **201** 465
[17] Boustani I 1997 *J. Solid State Chem.* **133** 182
[18] Boustani I, Rubio A and Alonso J A 2000 *Contemporary Boron Chemistry* (eds) M G Davidson, A K Hughes, T B Marder and K Wade (the Royal Society of Chemistry, Cambridge) p 493
[19] Amovilli C and March N H 2001 *Chem. Phys. Lett.* **347** 459
[20] Szwacki N G, Sadrzadeh A and Yakobson B 2007 *Phys. Rev. Lett.* **98** 166804
[21] Szwacki N G 2008 *Nanoscale Res. Lett.* **3** 49
[22] Tang H and Ismail-Beigi S 2007 *Phys. Rev. Lett.* **99** 115501
[23] Oku T 2003 *Solid State Commun.* **127** 689
[24] Prased D L V K and Jemmis E D 2008 *Phys. Rev. Lett.* **100** 165504
[25] Wade K 1971 *J. Chem. Soc. D* **15** 792
[26] Yan Q-B, Zheng Q-R and Su G 2008 *Phys. Rev. B* **77** 224106
[27] Hohenberg P and Kohn W 1964 *Phys. Rev. B* **136** 864
[28] Kohn W and Sham L J 1965 *Phys. Rev. A* **140** 1133
[29] Becke A D 1988 *Phys. Rev. A* **38** 3098
[30] Lee C, Yang W and Parr R G 1988 *Phys. Rev. B***37** 785
[31] Guest M F, Bush I J, van Dam H J J, Sherwood P, Thomas J M H, van Lenthe J H, Havenith R W A and Kendrick J, Gamess_UK, *http://www.cfs.dl.ac.uk/*
[32] Frisch M J, Trucks G W, Schlegel H B, Scuseria G E, Robb M A, Cheeseman J R, Montgomery J A, Jr, Vreven T, Kudin K N, Burant J C, Millam j M, Iyengar S S, Tomasi J, Barone V, Mennucci B, Cossi M, Scalmani G, Rega N, Petersson G A, Nakatsuji H, Hada M, Ehara M, Toyota K, Fukuda R, Hasegawa J, Ishida M, Nakajima T, Honda Y, Kitao O, Nakai H, Klene M, Li X, Knox J E, Hratchian H P, Cross J B, Bakken V, Adamo C, Jaramillo J, Gomperts R, Stratmann R E, Yazyev O, Austin A J, Cammi R, Pomelli C, Ochterski J W, Ayala P Y, Morokuma K, Voth G A, Salvador P, Dannenberg J J, Zakrzewski V G, Dapprich S, Daniels A D, Strain M C, Farkas O, Malick D K, Rabuck A D, Raghavachari K, Foresman J B, Ortiz J V, Cui Q, Baboul A G, Clifford S, Cioslowski J, Stefanov B B, Liu G, Liashenko A, Piskorz P, Komaromi I, Martin R L, Fox D J, Keith T, Al-Laham M A, Peng C Y, Nanayakkara A, Challacombe M, Gill P M W, Johnson B, Chen W, Wong M W, Gonzalez C and Pople J A 2004 *Computer code GAUSSIAN 03, revision B.03* (Gaussian, Inc., Wallingford, CT)
[33] Gopakumar G, Nguyen M T and Ceulemans A 2008 *Chem. Phys. Lett.* **450** 175
[34] Chacko S, Kanhere D G and Boustani I 2003 *Phys. Rev. B* **68** 035414
[35] Boustani I, Rubio A and Alonso J A 1999 *Chem. Phys. Lett.* **311** 21
[36] Boustani I and Quandt A 1997 *Europhys. Lett.* **39** 527
[37] Boustani I, Quando A and Rubio A 2000 *J. Solid State Chem.* **154** 269
[38] Kiran B, Bulusu S, Zhai H-J, Yoo S, Zeng X C and Wang L-S 2005 *Proc. Natl. Acad. Sci. U.S.A.* **102** 961
[39] Marques M A L and Botti S 2005 *J. Chem. Phys.* **123** 014310
[40] An W, Bulusu S, Gao Y and Zeng X C 2006 *J. Chem. Phys.* **124** 154310
[41] Oger E, Crawford N R M, Kelting R, Weis P, Kappes M M and Ahlrichs R 2007 *Angew. Chem. Int. Ed.* **46** 8503
[42] Sergeeva A P, Zubarev D Y, Zhai H-J, Bodyrev A I and Wang L-S 2008 *J. Am. Chem. Soc.* **130** 7244
[43] Lau K C and Pandey R 2008 *J. Phys. Chem. B* **112** 10217

Electron energy-loss and soft X-ray emission study of boron nanobelts

Yohei Sato[1], Masami Terauchi[1], Kazuhiro Kirihara[2], Takeshi Sasaki[2], Kenji Kawaguchi[2], Naoto Koshizaki[2] and Kaoru Kimura[3]

[1]Institute of Multidisciplinary Research for Advanced Materials, Tohoku University, Sendai 980-8577, Japan.
[2]Nanoarchitectonics Research Center, National Institute of Advanced Industrial Science and Technology, Tsukuba 305-8565, Japan
[3]Depertment of Advanced Materials Science, The University of Tokyo, Chiba 277-8561, Japan

E-mail: y-sato@tagen.tohoku.ac.jp

Abstract. Electronic structure of individual boron nanobelts, which supposedly have α-tetragonal boron structure, was studied with electron energy-loss spectroscopy (EELS) and soft X-ray emission spectroscopy (SXES) in a transmission electron microscope. A boron K-shell excitation spectrum by EELS and boron K-emission spectra by SXES correspond to partial density of states of the conduction and the valence bands, respectively. Experimental results reveal that boron nanobelts are either a semimetal or narrow-gap semiconductor.

1. Introduction

Crystal structures of boron allotropes, for example, α- and β-rhombohedral boron (α-r-B and β-r-B) [1, 2] and α-tetragonal boron (α-t-B) [3], are composed of B_{12} clusters. High symmetry of a B_{12} cluster should increase the degeneracy and thus the value of the density of electronic states (DOS). A theoretical study of the α-r-B reported DOS peaks near the Fermi level [4]. The DOS peaks of the conduction band in single-crystalline α-r-B were confirmed by high-energy-resolution electron energy-loss spectroscopy (EELS) [5]. A theoretical study indicated that Li-doped α-r-B could be a superconductor when the Fermi level coincides with the DOS peaks [4]. Electronic structures of Li- and Mg-doped α-r-B were characterized by EELS [6]. Chemical shifts of the boron K-edge and a step-like onset, which indicates presence of the Fermi edge due to electron doping, were reported. Electrical transport measurements could not be conducted due to a small sample size. Transport measurements were reported, however, for the carrier doped β-r-B [7-9]. Although the electrical conductivity of β-r-B increased upon doping, metallization was not attained. Metallization of β-r-B was achieved by applying high pressure of 175 GPa [10]. Compressed β-r-B also showed superconductivity, and the transition temperature could be as high as 11.2 K at 250 GPa. Theory predicts pure α-t-B to be metal [11, 12]. Crystalline boron materials $B_{50}C_2$ or $B_{50}N_2$ have been produced [13], but not pure single-crystalline α-t-B.

Boron nanowires composed of B_{12} clusters have been synthesized. Their structures could be amorphous [14], β-r-B [15], α-t-B [16, 17] or a yet unknown form [18]. In order to utilize those nanowires in nano-sized devices [19], it is necessary to characterize their physical properties.

Wang *et al* produced pure single-crystalline α-t-B nanowires with rectangular cross-section and boron nanobelts (BNB) by laser ablation without any catalysts [16]. Kirihara *et al* measured electrical transport of single BNBs [19, 20]. They reported that the BNB is a semiconductor whose electrical conductivity exceeds 1000 times that of bulk β-r-B. They also reported field effect transistor measurements of the BNBs and electrical conductivity of Mg-doped BNBs [20]. BNB was classed as *p*-type semiconductor and the activation energies for carrier concentration and mobility were 0.07 eV and 0.19 eV, respectively. The electrical conductance of Mg-doped BNBs was 100 times higher than that of the undoped BNB at room temperature. We are not aware of spectroscopic studies of this material.

Electron energy-loss spectroscopy (EELS) and soft-X-ray emission spectroscopy (SXES) are powerful techniques coupled to a transmission electron microscope (TEM). Sato *et al* reported high energy-resolution EELS on individual double-walled carbon nanotubes (DWCNTs) [21]. They confirmed that the observed peak structure originated from interband transitions intrinsic for a chiral structure of each DWCNT. Terauchi *et al* attached a wavelength-dispersive SXES device to a TEM [22]. They applied SXES to partial DOS of valence band and analyzed in detail the electronic structure of cubic BN, hexagonal BN, Al and an Al-Ni-Co alloy by combining EELS and SXES [23, 24].

In this paper, valence-electron and boron K-shell excitation spectra and a boron K-emission spectrum were measured by EELS and SXES, respectively, from single BNBs. The results were compared with those of β-r-B, and the band gap of BNB was estimated.

2. Experiment

BNBs were synthesized by laser ablation [16]. A single crystal of β-r-B (99.999 % purity) was studied for comparison. TEM images and electron diffraction patterns of BNBs were obtained using a JEM-2000FX TEM operated at 100 kV. Electron diffraction confirmed that the specimens were high quality single crystals of β-r-B and BNB. Individual BNBs were also characterized with a high energy resolution EELS microscope operated at 70 kV [25-27]. A collection semiangle was 5 mrad. The energy resolution was 0.17-0.25 eV at the full widths at half maximum of the zero-loss peak. SXES was carried out with a wavelength-dispersive soft X-ray spectrometer [22-24] attached to a JEM-2000FX TEM operated at 100 kV. The energy dispersion on a CCD detector was 0.2 eV/channel at the boron K-emission energy of 180 eV.

3. Result and discussion

Figure 1 shows a TEM image (a) and electron diffraction pattern (b) from a 40 nm wide BNB. The diffraction pattern was assigned to the $[0\ \bar{1}\ 0]$ zone axis of α-t-B.

Figure 2 shows a valence electron excitation spectrum of a single BNB in the range 1 - 40 eV. A spectrum of β-r-B is also shown for comparison. The steep intensity decrease around 1 eV is the tail of the zero-loss peak. Apparent onset energy of the BNB spectrum is observed at 1.7 eV marked by an arrow. This onset energy and small humps at 4.5, 7.0 and 9.5 eV should be due to interband transitions. Prominent peaks at 24 eV are due to the collective excitation of valence electrons (volume plasmon). The same energy for the two materials means that the density of valence electrons is the same in BNB and β-r-B.

Figure 3 shows boron K-shell excitation spectra (1s → conduction band) of the BNB and β-r-B. The spectral shape corresponds to the DOSs of the conduction band with *p*-symmetry because the spectra were obtained under the dipole transition condition. The onset energy at 186.4 eV of the BNB is 1.4 eV lower than that of β-r-B. This lower onset energy indicates that the band gap is narrower in BNB than in β-r-B. In the BNB spectrum, peaks at 187.5, 189 and 191.8 eV should be attributed to

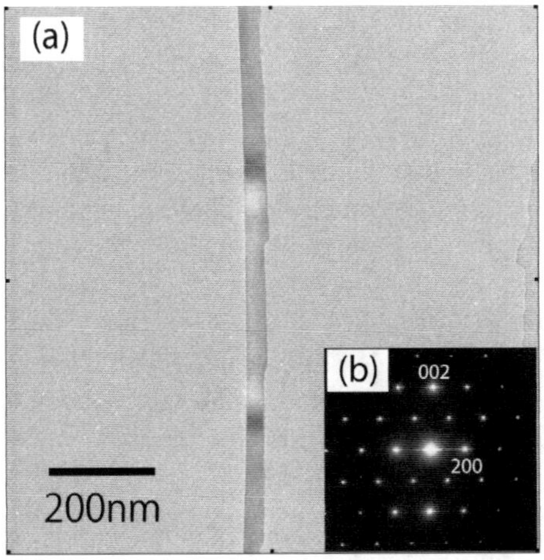

Figure 1. (a) TEM image and (b) diffraction pattern of BNB.

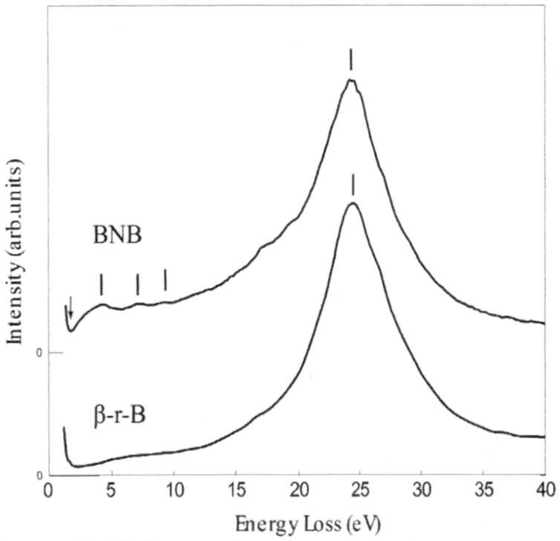

Figure 2. Valence electron excitation spectra of BNB and β-r-B.

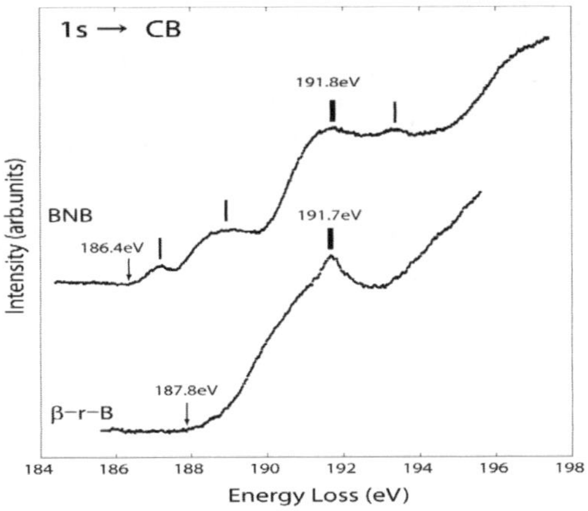

Figure 3. Inner-shell excitation spectra of BNB and β-r-B. CB refers to conduction band.

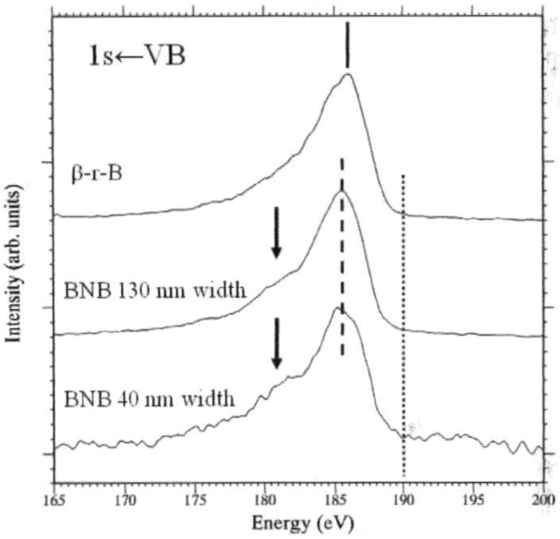

Figure 4. K emission (valence band→1s) spectra of BNB and β-r-B.

electronic states of B_{12} clusters of BNB. The difference of the number of the peaks between the two materials could reflect unequal distortions of the boron clusters [28].

Figure 4 shows boron K-emission spectra (valence band → 1s) of single BNBs having the widths of 130 and 40 nm. The spectrum of β-r-B is also shown for comparison. The spectral shape corresponds to the DOS of the valence band with *p*-symmetry. The BNBs peak at 185.5 eV has slightly lower energy than the β-r-B peak at 186 eV. The onsets indicated by the vertical dashed lines, whose positions correspond to the top of the valence band, are almost the same in BNB and β-r-B. Additional shoulders marked by arrows are observed at 182 eV in BNB. They may indicate different inter-icosahedral cluster bonding states in BNB and β-r-B.

The DOS of the valence and conduction bands of BNB were recorded for the first time in this study. K-shell excitation spectra reveal that the onset energy is 1.4 eV lower in BNB than in β-r-B. Considering that the band gap of the β-r-B is 1.6 eV [5] and the energy of the top of the valence band is almost the same in BNB and β-r-B, the band gap of the BNB can be estimated as 0.2 eV ± 0.2 eV. We therefore conclude that BNB is a semimetal or a narrow-gap semiconductor. This result is inconsistent with the metallic behavior predicted by theory [12], which might be explained by deviation of the real structure of BNB from the ideal α-t-B. A further detailed study of the atomic structure of BNB is required.

4. Conclusion

Density of states of the valence and conduction bands in BNB was studied for the first time by EELS and SXES. The structure of BNB was believed to be same as that of pure α-t-B, but the present results indicate it is not.

Acknowledgements

The authors thank Mr. F. Sato for his skillful technical assistance. The present work was partly supported by Grant-in-Aid for Scientific Research on Priority Areas 'New Materials Science Using Regulated Nano Spaces−Strategy in Ubiquitous Elements' by the Ministry of Education, Culture, Sports, Science and Technology of Japan.

References

[1] Decker B F and Kasper J S 1959 *Acta Crystallogr.* **12** 503
[2] Hoard J L, Sullenger D B, Kinnard C H L and Hughes R E 1970 *J. Solid State Chem*, **1** 268
[3] Hoard J L, Hughes R E and Sands D E 1958 *J. Am. Chem. Soc.* **80** 4507
[4] Gunji S and Kamimura H 1996 *Phys. Rev.* B **54** 13665
[5] Terauchi M, Kawamataq Y, Tanaka M, Takeda M and Kimura K 1997 *J. Solid State Chem.* **133** 156
[6] Soga K, Oguri A, Araake A, Terauchi M, Fujiwara A and Kimura K 2004 *J. Solid State Chem.* **177** 498
[7] Matsuda H, Tanaka N, Nakayama T, Kimura K, Murakami Y, Suematsu H, Kobayashi M and Higashi I 1996 *J. Phys. Chem. Solids* **57** 1167
[8] Nakayama T, Matsuda H, Kimura K and Ino H 1997 *J. Solid State Chem.* **133** 342
[9] Nakayama T, Shimizu J and Kimura K 2000 *J. Solid State Chem.* **154** 13
[10] Eremets M I, Struzhkin V V, Mao H K and Hemley R J 2001 *Science* **293** 272
[11] Bullett D W 1982 *J. Phys.* C **15** 415
[12] Li D, Xu Y N, Ching W Y 1992 *Phys Rev.* B **45** 5895
[13] Will G and Kossobutzki K H 1976 *J. Less-Common Met.* **47** 33
[14] Cao L, Zhang Z, Sun L, Gao C, He M, Wang Y, Li Y, Zhang X, Li Gong Zhang J and Wang W 2001 *Adv. Mater.* **13** 1701
[15] Wang Y Q and Duan X F 2003 *Appl. Phys. Lett.* **82** 272
[16] Wang Z, Shimizu Y, Sasaki T, Kawaguchi K, Kimura K and Koshizaki N 2003 *Chem. Phys. Lett.* **368** 663
[17] Xu T T, Zheng J G, Wu N, Nicholls A W, Roth J R, Dikin D A and Ruoff R S 2004 *Nano Lett.* **4** 963
[18] Otten C J, Lourie O R, Yu M F, Cowley J M, Dyer M J, Ruoff R S and Buhro W 2002 *J. Am. Chem. Soc.* **124** 4564
[19] Kirihara K, Wang Z, Kawaguchi K and Shimizu Y, Sasaki T, Koshizaki N, Soga K and Kimura K 2005 *Appl. Phys. Lett.* **86** 212101
[20] Kirihara K, Hyodo H, Fujihisa H, Wang Z, Kawaguchi K, Shimizu Y, Sasaki T, Koshizaki N, Soga K and Kimura K 2006 *J. Solid State Chem.* **179** 2799
[21] Satoh Y, Terauchi M, Saito R and Saito Y 2006 *J. Electro. Microsc.* **55** 137
[22] Terauchi M, Yamamoto H and Tanaka M 2001 *J. Electro. Microsc.* **50** 101
[23] Terauchi M and Kawana M 2006 *Ultramicrosc.* **106** 1069
[24] Terauchi M 2006 *Microsc. Res. Technol.* **69** 531
[25] Tsuno K, Terauchi M and Tanaka M 1989 *Optik.* **83** 77
[26] Terauchi M, Kuzuo R, Satoh F, Tanaka M, Tsuno K and Ohyama J 1991 *Microsc. Microanal. Microstruct.* **2** 351

Journal of Physics: Conference Series **176** (2009) 012029 doi:10.1088/1742-6596/176/1/012029

[27] Terauchi M, Tanaka M, Tsuno K and Ishida M 1999 *J. Microsc.* **194** 203
[28] Fujimoto M and Kimura K 1997 *J. Solid State Chem.* **133** 178

Energy barrier of structure transition from icosahedral $B_{12}H_6^+$ to planar $B_{12}H_5^+$ and $B_{12}H_4^+$ clusters

Yuji Ohishi[1], Kaoru Kimura[1], Masaaki Yamaguchi[1], Noriyuki Uchida[2] and Toshihiko Kanayama[1]

[1] Department of Advanced Material Science, University of Tokyo, 5-1-5 Kashiwanoha, Kashiwa, Chiba 277-8561, Japan

[2] Nanodevice Innovation Research Center, National Institute of Advanced Industrial Science and Technology (NIRC), 1-1-1 Higashi, Tsukuba, Ibaraki 305-8562, Japan

three-stones@phys.mm.t.u-tokyo.ac.jp

Abstract. Ab initio calculations indicated that icosahedral structure is more stable for $B_{12}H_n^+$ clusters with n = 6-12, but planar configuration is more stable for $B_{12}H_n^+$ with n = 0-5. Here, we report the energy barriers of structure transition by detaching H atoms from icosahedral $B_{12}H_6^+$ to planar $B_{12}H_5^+$, planar $B_{12}H_4^+$, icosahedral $B_{12}H_5^+$ or icosahedral $B_{12}H_4^+$ cluster. We performed density functional theory calculations to explore the energy barriers and the reaction path of the structure transition. The resultant energy barriers are almost the same, enabling the transition to proceed and suggesting that the structure of $B_{12}H_n^+$ clusters is controlled by the number of hydrogen atoms n.

1. Introduction

There is much interest in utilizing small atomic clusters to build well-controlled nanostructures. Boron has rich chemistry [1-2] and forms cluster solids, where structural units are icosahedra containing 12 boron atoms [3-6]. On the other hand, it has been proved theoretically and experimentally that all small boron clusters known so far have planar or quasiplanar structures [7-11]. These features make boron clusters an attractive unit for building nanostructures.

Hydrogen-terminated icosahedral (I_h) $B_{12}H_{12}^{2-}$ cluster is the most stable among the various polyhedral boranes synthesized thus far. The three-dimensional aromaticity of its highly symmetrical structure [1] and the perfect matching of its electron number with that required by the Wade's rule [12, 13] accounts for its stability. On the other hand, the B_{12} cluster (without hydrogen) has quasi-planar structure with six delocalized π-electrons, similar to C_6H_6 (benzene) [7-11]. Compared with clusters of carbon or the other elements neighboring boron in the periodic table, elemental boron clusters have received relatively little experimental and theoretical attention. In particular, few reports are known on the gas-phase reactions of boron clusters [14, 15].

In a recent article, we produced $B_{12}H_8^+$ cluster cations experimentally [16] using an external quadrupole static attraction ion trap (EQSIT) [17-20]. In that article [16], we reported the stability of $B_{12}H_n^+$ clusters calculated at B3LYP/6-31G(d). Comparison of the energies of icosahedral $B_{12}H_n^+$ and planar $B_{12}H_n^+$ clusters reveals that they are planar for n = 0-5 and icosahedral for n = 6-12 [16].

Calculations of the energies of formation reactions and detachment pathways of H_2 molecules and H atoms from the produced cluster ions indicate that the $B_{12}H_8^+$ cluster, produced in the EQSIT, is icosahedral. Assuming that the energy barrier of the structure transition is not high compared to that of hydrogen detachment, one can expect that planar $B_{12}H_n^+$ is produced by the detachment of hydrogen atoms. This suggests that the structure of $B_{12}H_n^+$ is controlled by the number of H atoms. In this study, we calculated the pathway by which hydrogen molecules and atoms are eliminated from icosahedral $B_{12}H_6^+$, thus forming icosahedral and planar $B_{12}H_n^+$ (n = 5,4) clusters. The objective was to predict the possibility of transition from icosahedral to planar structure through the calculation of the energy barriers.

2. Quantum chemical calculations

The geometries of low-lying minima and transition states for $B_{12}H_n^+$ are determined at the B3LYP/6-31G(d) level of the density functional theory. For all the species considered in this study, the natures of the calculated stationary points are determined by frequency calculation at the B3LYP/6-31G(d) level. The potential energy surfaces are calculated to investigate the reaction energy, following previous studies on the mechanism of gas-phase reactions [21-23]. All minima are characterized by the presence of all positive vibration frequencies. All transition states are characterized by the presence of one imaginary vibration frequency. The intrinsic reaction coordinate [24-26] calculations are further carried out at the B3LYP level to confirm that the transition states connect the right minima. Zero-point energies are calculated, scaled by a factor of 0.9804, and included in the calculation of relative energies. All calculations are carried out using the GAUSSIAN 03 program package [27].

3. Results and discussion

To find the transition state which connects icosahedral and planar structures, first, the most stable planar structure of $B_{12}H_6^+$ (figure 1(a)) is chosen as a starting geometry. As shown in figure 1(a), hydrogen atoms are bonded with peripheral boron atoms, and the structure is completely planar. Because the hydrogen-bonded boron atom is pulled toward the direction of hydrogen atom, spherical structure (figure 1(b)) is produced by moving hydrogen atoms to the inner boron atom. As the movement of one hydrogen atom to the inner boron atom is insufficient to obtain closed structure, one more hydrogen atom is moved to the inner boron atom. Three structures shown in figures 1(b), (c) and (d) are chosen as a starting geometry; (c) and (d) are obtained from (b) by moving the hydrogen atom bonded with the inner boron atom to the neighboring other inner boron atoms. We calculated all possible structural isomers of $B_{12}H_6^+$, which are obtained by moving one more peripheral hydrogen atom to the inner boron atom, starting from (b), (c) and (d). Among these trials, the optimization of (e), which is obtained from (c), results in the most spherical structure (f). The transition state between (c) and (e) is calculated, leading to the structure (g).

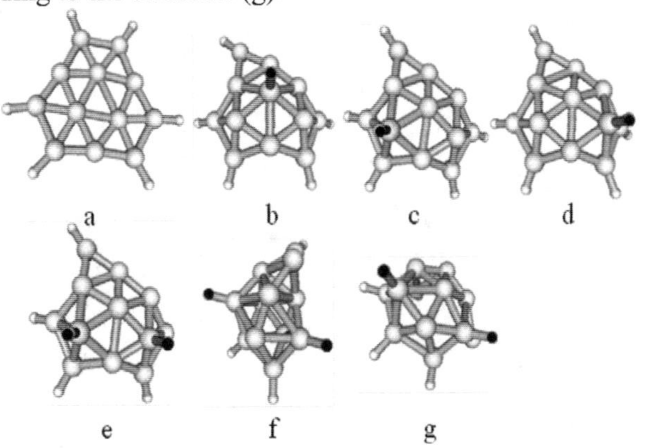

Figure 1. Procedure for determining the transition state between planar and icosahedral structures. The black filled circles indicate the moved hydrogen atoms.

To examine whether the structure (g) is the transition state which connects icosahedral and planar structures, we calculated the geometrical structures (see figure 2) and relative energies of each state in the reaction path which connect the most stable structure of icosahedral $B_{12}H_6^+$ and planar $B_{12}H_5^+$, $B_{12}H_4^+$ clusters. In figure 2, structure (6) is same as structure (g) in figure 1. At first, intrinsic reaction coordinate calculation is conducted for structure (6) to obtain the local minimum structures, which are connected by structure (6). This means that the structure transition is bidirectional. The calculation results in the closed structure (5) and planar structure (7) as shown in figure 2. Icosahedral structure (3) is obtained from (5) through the transition state (4). The most stable isomer of icosahedral $B_{12}H_6^+$ (1) is obtained from (3) by migration of a hydrogen atom through transition state (2). In a similar way, the most stable isomer of planar $B_{12}H_6^+$ (15) is obtained from (7) by migration of a hydrogen atom through (8) to (14) structures. The hydrogen atom detaches from planar $B_{12}H_6^+$ (15) without transition state, leading to the planar $B_{12}H_5^+$ configuration (16), which is the most stable isomer of $B_{12}H_5^+$. The hydrogen molecule detaches from the planar $B_{12}H_6^+$ (15) through transition state (17), leading to the planar $B_{12}H_4^+$ (18), which is the most stable isomer of $B_{12}H_4^+$. The relative energies of each state are shown in figure 3. The resultant relative energies indicate that the energy barriers of the structure transition and of the following detachment of hydrogen atom and molecule are both equal to 3.67 eV.

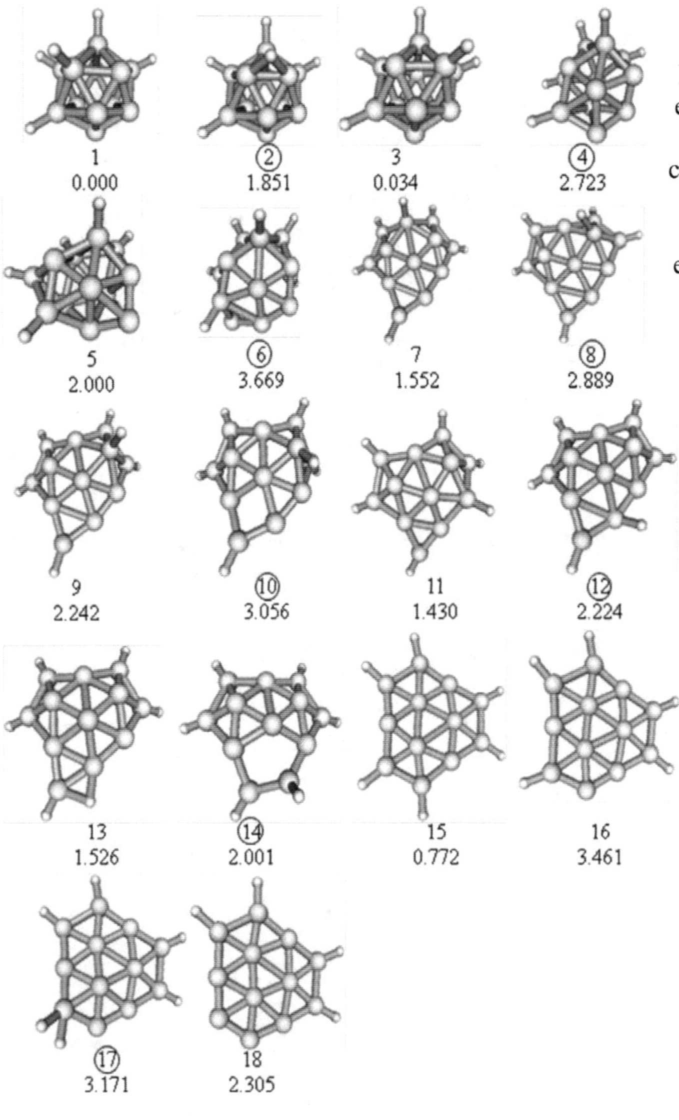

Figure 2. Geometrical structures of each state in the reaction path which connect the most stable configuration of icosahedral $B_{12}H_6^+$ (1) and planar $B_{12}H_5^+$ (16), $B_{12}H_4^+$ (18) clusters with their relative energies in eV. The structures (2-15, 17) correspond to the locally stable or transition states of $B_{12}H_6^+$ in the reaction path. The circled numbers indicate transition states.

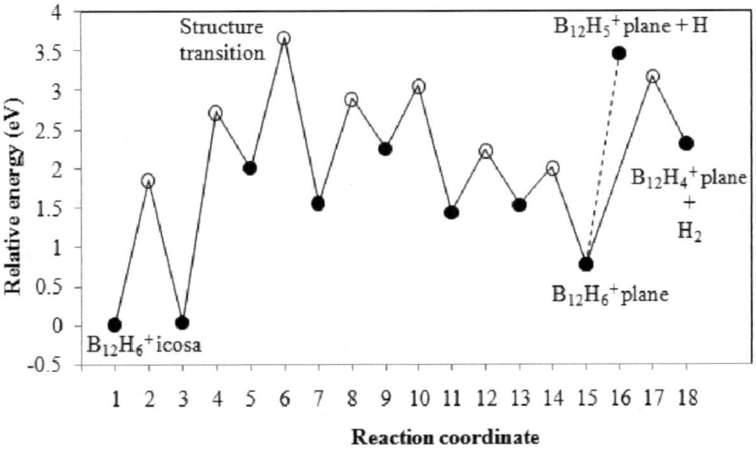

Figure 3. The relative energies of each state in figure 2. The dashed line shows the detachment of H atoms. The filled circle shows the local minima and open circle shows the transition state.

We have also calculated hydrogen detachment without changing the structure of the cluster. The geometrical structures and energies of each state are shown in figures 4 and 5, respectively. The energy of icosahedral $B_{12}H_6^+$ (1) is chosen as the reference, and it is the same as in figures 2 and 3. The structure of icosahedral $B_{12}H_6^+$ cluster (20) is obtained by moving the hydrogen atom of starting structure (1) to the neighboring bare boron atom through a transition form (19). A hydrogen atom is detached from (20) without a transition state, leading to the most stable isomer of $B_{12}H_5^+$ (21). In case of the detachment of a hydrogen molecule, a hydrogen atom approaches the neighboring hydrogen atom, forming a $B_{12}H_4^+ \cdot H_2$ complex (23) through a transition state (22). The hydrogen molecule detaches from the complex with zero activation energy, resulting in icosahedral $B_{12}H_4^+$ (24). The resultant relative energies indicate that the energy barrier for transition from icosahedral $B_{12}H_6^+$ to icosahedral $B_{12}H_5^+$ and H atom is 3.82 eV and the barrier for transition from icosahedral $B_{12}H_6^+$ to icosahedral $B_{12}H_4^+$ and H_2 molecule is 3.59 eV.

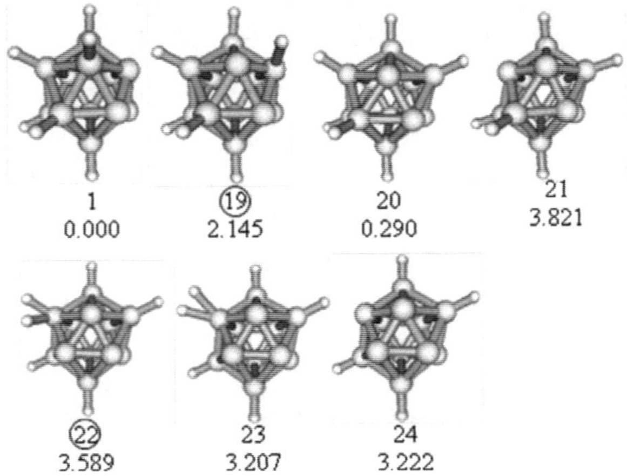

Figure 4. The geometrical structures of each state in the reaction path from icosahedral $B_{12}H_6^+$ (1) to icosahedral $B_{12}H_5^+$ (21) and icosahedral $B_{12}H_4^+$ (24) with their relative energies in eV. The structures 19, 20, 22 and 23 correspond to the local stable or transition states of $B_{12}H_6^+$ in the reaction path. The circled numbers indicate transition states.

The calculated reaction processes and the corresponding energy barriers are summarized in table 1. When one hydrogen atom is detached from icosahedral $B_{12}H_6^+$, the formation of planar $B_{12}H_5^+$ with H atom needs 0.152 eV less energy. On the other hand, when one hydrogen molecule is detached from icosahedral $B_{12}H_6^+$, the icosahedral $B_{12}H_4^+$ is favored by 0.08 eV. Because these differences are 4.14% and 2.23% of the energy barriers, energetically favorable structure should be produced by the detachment of hydrogen atoms or molecules.

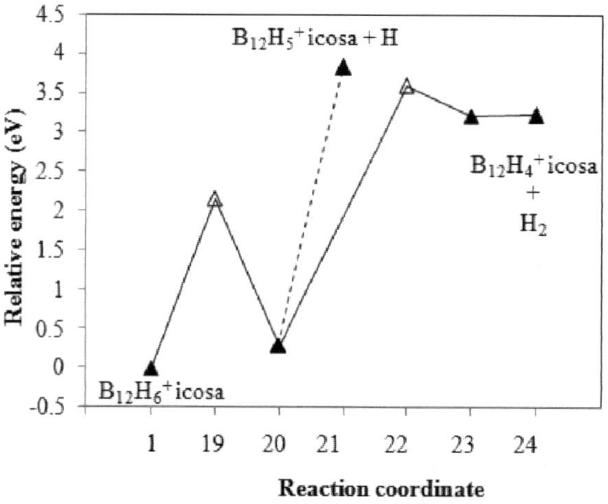

Figure 5. The relative energies of each state in figure 4. The dashed line shows the detachment of H atom. The filled triangle shows the local minima and open triangle shows the transition state.

Table 1. The reaction processes and the corresponding energy barriers.

Reaction process	Barrier (eV)
$B_{12}H_6^+$ (icosahedral) → $B_{12}H_5^+$ (planar) + H	3.669
$B_{12}H_6^+$ (icosahedral) → $B_{12}H_4^+$ (planar) + H_2	3.669
$B_{12}H_6^+$ (icosahedral) → $B_{12}H_5^+$ (icosahedral) + H	3.821
$B_{12}H_6^+$ (icosahedral) → $B_{12}H_4^+$ (icosahedral) + H_2	3.589

4. Conclusion

The energy of each state in the reaction path from icosahedral $B_{12}H_6^+$ to planar $B_{12}H_5^+$, planar $B_{12}H_4^+$, icosahedral $B_{12}H_5^+$ or icosahedral $B_{12}H_4^+$ cluster is calculated. The resultant energy barriers are 3.67, 3.67, 3.82 and 3.59 eV, respectively. Because their differences are within 5% of the energy barriers, energetically favorable structure should be produced by the detachment of hydrogen atoms or molecules. This implies that the structure of $B_{12}H_n^+$ clusters is controlled by the number of H atoms.

Acknowledgment

This work was partly supported by KAKENHI No. 19051005. One of the authors (Y. O.) acknowledges support by the Global COE Program “the Physical Sciences Frontier”, MEXT, Japan.

References

[1] King R B 2001 *Chem. Rev.* **101** 1119
[2] Jemmis E D, Balakrishnarajan M M and Pancharatna P D 2002 *Chem. Rev.* **102** 93
[3] Li D, Xu Y N and Ching W Y 1992 *Phys. Rev.* B **45** 5895
[4] Fujimori M, Nakata T, Nakayama T, Nishibori E, Kimura K, Takata M and Sakata M 1999

Phys. Rev. Lett. **82** 4452
[5] Prasad D L V K, Balakrishnarajan M M and Jemmis E D 2005 *Phys. Rev.* B **72** 195102
[6] Hosoi S, Kim H, Nagata T, Kirihara K, Soga K, Kimura K, Kato K and Takata M 2007 *J. Phys. Soc. Jpn.* **76** 044602
[7] Boustani I 1997 *Phys. Rev.* B 55 16426
[8] Zhai H J, B. Kiran B, Li J and Wang L S 2003 *Nat. Mater.* **2** 827
[9] Lau K C, Deshpande M and Pandey R 2005 *Int. J. Quantum Chem.* **102** 656
[10] Aihara J I, Kanno H and Ishida T 2005 *J. Am. Chem. Soc.* **127** 13324
[11] Alexandrova A N, Boldyrev A I, Zhai H J and Wang L S 2006 *Coord. Chem. Rev.* **250** 2811
[12] Wade K 1971 *J. Chem. Soc.* D 792
[13] Stone A J and Alderton M J 1982 *Inorg. Chem.* **21** 2297
[14] Hales D A, Kautzman K E, Williams N G, Haile P A and Barker M P 2007 *J. Phys. Chem.* A **111** 2266
[15] Hales D A, Haile P A, Barker M P and Hunt H L 1997 *J. Phys. Chem.* A **102** 8305
[16] Ohishi Y, Kimura K, Yamaguchi M, Uchida N and Kanayama T 2008 *J. Chem. Phys.* **128** 124304
[17] Kanayama T 1994 *Jpn. J. Appl. Phys.* **33** L1792
[18] Kanayama T and Murakami H 1997 *J. Vac. Sci. Technol.* B **15** 2882
[19] Hall S G, Murakami H, Palmer R E and Kanayama T 1997 *Jpn. J. Appl. Phys.,* Part2 **36** L639
[20] Hiura H and Kanayama T 2001 *Rev. Sci. Instrum.* **72** 2893
[21] Antoniotti P, Operti L, Rabezzana R, Splendore M, Tonachini G and Vaglio G A 1997 *J. Chem. Phys.* **107** 1491
[22] Qu Z W, Zhu H, Li Z S and Zhang Q Y 2002 *J. Comput. Chem.* **23** 414
[23] Raghavachari K 1991 *J. Chem. Phys.* **95** 7373
[24] Fukui K 1981 *Acc. Chem. Res.* **14** 363
[25] Gonzalez C and Schlegel H B 1989 *J. Chem. Phys.* **90** 2154
[26] Gonzalez C and Schlegel H B 1990 *J. Chem. Phys.* **94** 5523
[27] Frisch M J *et al* 2004 *Gaussian 03, Revision E.01* (Gaussian, Inc., Wallingford CT)

Journal of Physics: Conference Series **176** (2009) 012031 doi:10.1088/1742-6596/176/1/012031

Valence dependence of interatomic interactions in RB_6

Takumi Hasegawa, Norio Ogita and Masayuki Udagawa
Graduate School of Integrated Arts and Sciences, Hiroshima University, 1-7-1 Kagamiyama, Higashi-Hiroshima, Hiroshima 739-8521, Japan
E-mail: hase@hiroshima-u.ac.jp

Abstract. Previous Raman measurements of phonon in hexaborides have shown that bandwidth of low-frequency vibrations of R in divalent hexaborides is very broad in comparison with that in trivalent ones. To investigate this valence dependence, we have calculated interatomic force constants (IFC) from first principles for RB_6 (R=Ca, Sr, Ba, Y, and La). The calculated phonon dispersion curves reproduce the valence dependence. From the investigation of calculated IFC, we conclude that the different bandwidth is originated from the flatness of the dispersion curve of R ion vibration produced by IFC between R ions. Thus, the dispersion curve is highly dispersive for the divalent case. Since the IFC in divalent hexaborides are similar to those of dipole-dipole interactions, we suggest that an origin of the difference of IFC is the screening effect of the dipole field by conduction electrons in metallic trivalent hexaborides.

1. Introduction

Atomic vibration in a cage structure has attracted much interest because of its possibility to improve thermoelectric properties [1], or to realize superconductivity [2, 3] and heavy fermion state [4, 5] through interactions with electronic states. For the relatively large cage size compared with that of *guest* atom trapped in the cage, its vibration has large amplitude and low frequencies. Therefore, the dynamical properties of the guest atoms with low frequency are very important for the cage structural compounds.

Many compounds, constructed by boron and metallic atoms, reveal the cage structure, because very hard network structure is constructed by boron and metallic ions locate in the interspaces of the network. Therefore, we can recognize the metallic atoms as the guests in the context of cage compounds. However, in boron compounds, amplitude of the guest vibrations might not be so large to affect electronic states because of not so large interspaces for most metallic ions. Nevertheless, boron compounds will be useful to investigate the dynamics of the guest atoms, because their structures are relatively simple in comparison with other cage compounds. In this paper, we have focused on hexaborides, RB_6, which is one of the simplest compounds among the cage compounds.

Our previous Raman scattering investigation [6] of divalent and trivalent hexaborides have clarified two results with the valence dependence of R ion. One is the frequency of T_{2g} mode, which takes about 700 cm^{-1} or 800 cm^{-1} for trivalent or divalent hexaborides, respectively. The other is spectrum of second-order scattering from low-frequency vibrations of R (LFV). In trivalent hexaborides, this spectrum is sharp and its intensity largely depends on temperature. On the other hand, in divalent hexaborides, the spectrum is broad and weak. Its intensity is

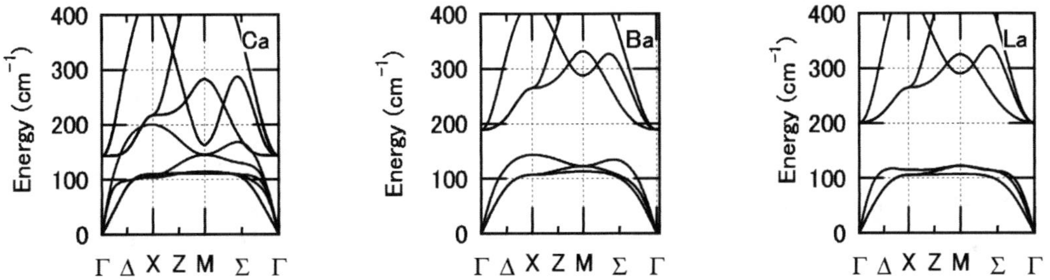

Figure 1. Calculated phonon dispersion curves at a low-energy region for CaB_6 (left), BaB_6 (center), and LaB_6 (right).

almost independent of temperature. These two facts indicate that the interatomic interactions are strongly affected by electronic state. The spectrum width of the second-order scattering reflects dispersion of LFV band. The observed feature means the small dispersion for trivalent R^{3+} and large one for divalent R^{2+}. This behavior of the dispersion has been also confirmed by inelastic neutron scattering [7]. In this paper, to quantitatively clarify the latter result, we have calculated the interatomic force constants (IFC) using first principles calculations.

2. Calculation method

Calculations have been performed with ABINIT package [8]. The package relies on density functional theory with local density approximation (LDA) and plane-wave basis together with norm-conserved pseudopotential. For exchange-correlation energy functional, the LDA potential presented by Perdew and Wang [9] was used. The Troullier-Martins type [10] pseudopotentials have been obtained using FPI98PP [11] and the pseudopotentials of R include semi-core states, except for Y. Cut-off energy for the plane-wave basis was from 34 Ry for Ba to 54 Ry for La, where 1 Ry = 13.6 eV. For an integration of the Brillouin zone, special points [12, 13] on 8 × 8 × 8 **k**-point grids with a smearing width 0.02 Ry were employed. Calculations of lattice dynamics were proceeded by density functional perturbation theory [14, 15]. To obtain IFC and phonon dispersion curves, the Fourier interpolation method was used on 4 × 4 × 4 grids. Lattice constants and parameter x of boron position have been optimized. Then, IFC was calculated for the obtained structure. As the results, the calculated lattice constants are smaller than the experimental ones by about 0.10 Å, and the calculated phonon frequencies are larger than the experimental results by 30 $\sim$ 50 cm^{-1} for three Raman active modes.

3. Result and discussion

Calculated results of phonon dispersion curves in a low-energy region are shown in figure 1 for CaB_6, BaB_6, and LaB_6. The calculated phonon dispersions are similar to the reported results of first principles calculations [16]. The relatively flat phonon bands around 100 cm^{-1} correspond to LFV. Their widths are determined by variation of force constant $k(q)$ at wavevector q, and frequency of the vibration is expressed by $\sqrt{k(q)/M}$, where M is the mass of R. From this expression, the width is narrow for heavy R and also for small R ion because of weak IFC. Therefore, the result in figure 1 is inevitable at first sight. However, as shown below, the IFC are different between divalent and trivalent hexaborides, especially between Ca^{2+} and La^{3+}. Since the ionic radii of Ca^{2+} and La^{3+} are close, the difference of force constants between them are originated from the different electronic states.

The IFC $D_{i\alpha j\beta}$ is defined by force $f_{i\alpha}$ acted on atom i, when atom j move by $u_{j\beta}$, as $f_{i\alpha} = \sum_\beta D_{i\alpha j\beta} u_{j\beta}$. Since it is very cumbersome to treat all components of $D_{i\alpha j\beta}$, we define a

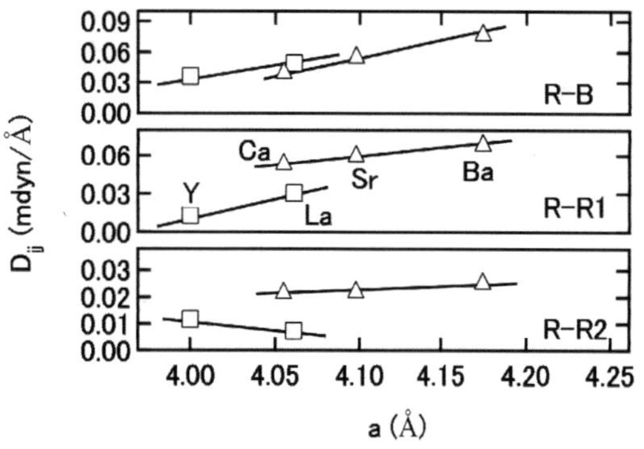

Figure 2. *Norm* D_{ij} of IFC from R at (0 0 0) to B at (x 0.5 0.5) (top), R at (1 0 0) (middle), and R at (1 1 0) (bottom), for divalent (triangle) and trivalent (square) hexaborides.

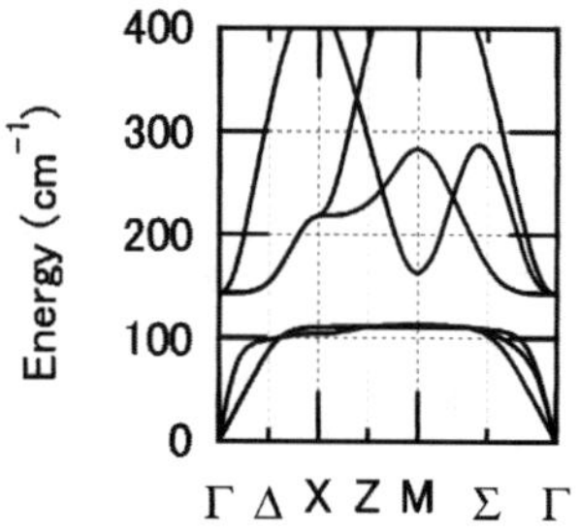

Figure 3. Phonon dispersion curve calculated using IFC, where IFC between Ca ions are set to 0 and the other IFC are same as those obtained for CaB_6.

norm D_{ij} by following formula.

$$D_{ij}^2 = \frac{1}{4\pi}\int_{-1}^{1} d\cos\theta \int_{0}^{2\pi} d\phi \sum_{\alpha} \left| \sum_{\beta} D_{i\alpha j\beta} e_{\beta}(\theta, \phi) \right|^2 = \frac{1}{3}\sum_{\alpha\beta} D_{i\alpha j\beta}^2, \qquad (1)$$

where $e_\beta(\theta, \phi)$ is a unit vector in polar coordinate. D_{ij} from R at (0 0 0) to B (x 0.5 0.5), R (1 0 0), and R (1 1 0) are shown in figure 2. As seen in figure 2, the IFC between R atoms are quite different between divalent and trivalent hexaborides, while the difference for IFC between R and B is small. From this result, we can conclude that the IFC between R atoms are important for the difference of bandwidth of LFV. This conclusion can be easily confirmed by calculating phonon dispersion using IFC, where IFC between R atoms are set to 0 and the others are same as the calculated IFC. Figure 3 shows the calculated results using such artificial IFC for CaB_6. The figure clearly demonstrates that the bandwidth is caused by IFC between R atoms and that the difference of the width is due to a difference of IFC between R atoms.

Roughly speaking, in the hexaboride system, the nearest neighbor IFC between R and B organizes the bottom of LFV band, and the IFC between R atoms determine the width of the band. This gives good contrast to the skutterudite system, where IFC between guest atoms are very small, and width of guest vibration band is determined by hardness of cage structure [17]. Since the guest atom motion causes a cage distortion with wavevector dependence, the variation of distortion produces the bandwidth in skutterudites. In the hexaboride case, the boron network is very hard and is not deformed by R ion motion. Therefore, the origin of the bandwidth of guest vibrations is quite different between hexaborides and skutterudites.

Now, we discuss an origin of the valence dependence of IFC. One possibility is difference of screening ability of dipole electric field by conduction electrons as discussed in the previous paper [6]. It must be examined from first principle results, but it is difficult to divide the IFC to dipole interaction part and the other part in metallic compounds. Unfortunately, our calculations produce metallic states even for divalent hexaborides. This fault is perhaps due to the use of LDA which tends to underestimate the band gap. Nevertheless, as shown in figure 2, different norm of the IFC gives us a hint to discuss the problem. If the large IFC in divalent hexaborides are dominated by dipole interactions between R^{2+}, directions of forces acting on R^{2+} are similar to those of dipole electronic field. Figure 4 shows forces acting on neighbor R ions in CaB_6 and LaB_6 due to the displacement of the center ion. As clearly seen, the directions

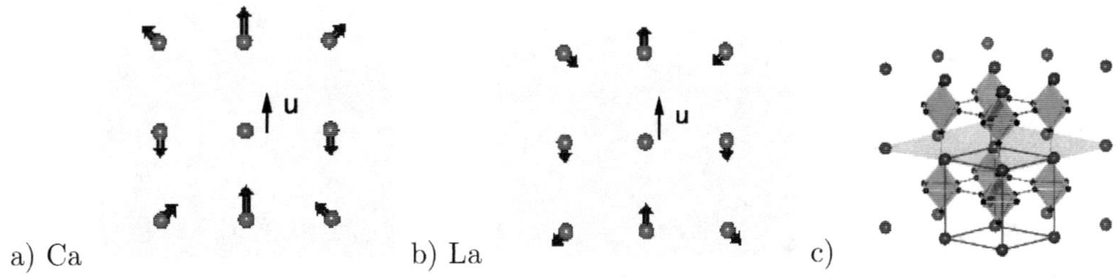

Figure 4. Acting forces on neighbor R ions due to the displacement u of center ion for a) CaB_6 and b) LaB_6. c) The crystal structure. The R ions on the gray plane are shown in a) and b).

of forces are quite similar to those of dipole field for CaB_6 and different for LaB_6. Therefore, we suggest that the large IFC between R are due to dipole-dipole interactions between R^{2+} in divalent hexaborides, and that the small IFC in trivalent hexaborides are due to the screening effect by conduction electrons.

4. Conclusion

We have calculated IFC using first principles method for CaB_6, SrB_6, BaB_6, YB_6, and LaB_6. The calculated result of phonon dispersion curves shows valence dependence of bandwidth of LFV as observed. From the calculated IFC, we conclude that the difference of the bandwidth is due to difference of IFC between R. In the hexaboride case, the bandwidth is determined by the IFC between R, and large bandwidth in divalent hexaborides is caused by large IFC between R. The large IFC in divalent hexaborides are similar to those of dipole-dipole interactions. Therefore, we suggest that the small bandwidth in trivalent hexaborides is due to the screening effect of the dipole field by conduction electrons.

Acknowledgement

This work was partially supported by a Grant-in-Aid Scientific Research on Priority Areas of "Strategy in Ubiquitous Elements" and "Skutterudite" from The Ministry of Education, Culture, Sports, Science, and Technology, Japan, and was also supported by N-BARD and IAMR at Hiroshima University.

References

[1] Slack G A 1995 *CRC Handbook of Thermoelectrics* (Boca Raton, FL: Chemical Rubber Co.)
[2] Hiroi Z, Yamaura J, Yonezawa S and Harima H 2007 *Physica* C **460-462** 20
[3] Kuneš J, Jeong T and Pickett W E 2004 *Phys. Rev.* B **70** 1745100
[4] Sanada S, Aoki Y, Aoki H, Tsuchiya A, Kikuchi D, Sugawara H and Sato H 2005 *J. Phys. Soc. Jpn.* **74** 246
[5] Hattori K, Hirayama Y and Miyake K 2005 *J. Phys. Soc. Jpn.* **74** 3306
[6] Ogita N, *et al.* 2003 *Phys. Rev.* B **68** 224305
[7] Kohgi M, Kuwahara K, Ogita N, Udagawa M and Iga F 2006 *J. Phys. Soc. Jpn.* **75** 086003
[8] Gonze X, *et al.* 2002 *Comput. Mater. Sci.* **25** 478
[9] Perdew J P and Wang Y 1992 *Phys. Rev.* B **45** 13244
[10] Toullier N and Martins J L 1991 *Phys. Rev.* B **43** 1993
[11] Fuchs M and Scheffler M 1999 *Comput. Phys. Commun.* **119** 67
[12] Baldereschi A 1973 *Phys. Rev.* B **7** 5212
[13] Chadi D J and Cohen M L 1973 *Phys. Rev.* B **8** 5747
[14] Gonze X 1997 *Phys. Rev.* B **55** 10337
[15] Gonze X and Lee C 1997 *Phys. Rev.* B **55** 10355
[16] Shang S, Wang Y and Liu Z 2007 *Phys. Rev.* B **75** 024302
[17] Hasegawa T, Takasu Y, Ogita N and Udagawa M 2008 *J. Phys. Soc. Jpn.* **77** Suppl. A 248

Raman scattering of rare earth hexaborides

Norio Ogita[1], Takumi Hasegawa[1], Masayuki Udagawa[1], Fumitoshi Iga[2] and Satoru Kunii[3]

[1] Graduate School of Arts and Sciences, Hiroshima University, 1-7-1 Kagamiyama, Higashi-hiorshima-shi, Hiroshima 739-8521, Japan
[2] Graduate School of Advanced Sciences of Matter, Hiroshima University, 1-3-1 Kagamiyama, Higashi-hiorshima-shi, Hiroshima 739-8530, Japan
[3] Department of Physics, Tohoku University, Sendai 980-8578, Japan

E-mail: nogita@hiroshima-u.ac.jp

Abstract. Raman scattering spectra were measured for the rare-earth hexaborides RB_6 (R = Ce, Gd, or Dy). All Raman-active phonons due to B_6 vibrations were observed in the range 600 - 1400 cm^{-1}. Anomalous peaks were detected below 200 cm^{-1}, which correspond to vibrations of rare-earth ion excited by second-order Raman scattering process. The intensity and energy of the rare-earth mode decrease with decreasing temperature. This suggests that the rare-earth ion vibrates in a shallow and anharmonic potential due to the boron cage. Using the reported values of mean square displacement of rare-earth ion, we estimated the anharmonic contribution for the rare-earth vibrations.

1. Introduction

Rare-earth hexaborides, RB_6, have a CsCl type structure with $Pm\bar{3}m$ symmetry, where rare-earth ion is located at Cs site and octahedral B_6 molecule occupies the Cl place. The isostructural compounds of RB_6 exhibit interesting physical properties, such as superconductivity in YB_6 [1], dense Kondo behavior in CeB_6 [2], and intermediate valence state in SmB_6 [3]. DyB_6 exhibits successive phase transitions accompanied by the ferroquadrupolar ordering at 30K and the antiferromagnetic transition at 25 K. GdB_6 shows magnetic structure change at two successive phase transitions (15 K and 8 K). Because the interactions between any adjacent borons are extremely strong, the boron network stabilizes the structure and can be regarded as a cage; the rare-earth ion weakly interacts with the cage. This lattice dynamical behavior is observed in the specific heat [4] and the phonon dispersion curve [5]. In our previous Raman scattering study [6], we found that the rare-earth motion is excited by a second-order Raman scattering process. The caged compounds are expected to be high-performance thermoelectric materials, because the cage structure has a possibility to reduce the phonon thermal conductivity by a rattling motion of the guest atoms. One of the key properties of the rattling motion is anharmonicity. In order to clarify the anharmonicity of rare-earth motion in the hexaborides, we measured Raman scattering spectra of RB_6 (R = Ce, Gd, or Dy).

2. Experimental setup

Single crystals of RB_6 were synthesized with a floating zone method [7]. Raman scattering was measured in a quasi-backscattering geometry. An Ar ion laser operated at 488.0 nm or a dye laser operated at 589 nm were employed as an excitation light with the output power of 10 mW at the specimen. The scattered light was analyzed by a triple-grating spectrometer and detected by a liquid nitrogen cooled CCD. Three phonons, $\Gamma_{\rm R} = A_{1g} + E_g + T_{2g}$, "Raman-active phonons", are allowed in the first-order Raman scattering. These phonons can be selected by their polarization dependence in the following geometries: A_{1g} appears in the (x,x), E_g in (x,x) and $(x+y,x-y)$, and T_{2g} in (x,y) geometry. In this (α,β) notation, α and β are the polarization directions of incident and scattered light, respectively.

3. Results and discussion

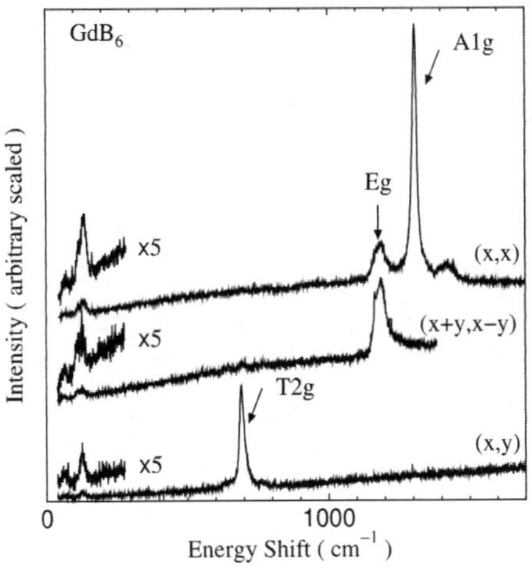

Figure 1. Polarization dependence of Raman spectra in GdB_6, excited by 488 nm light at room temperature.

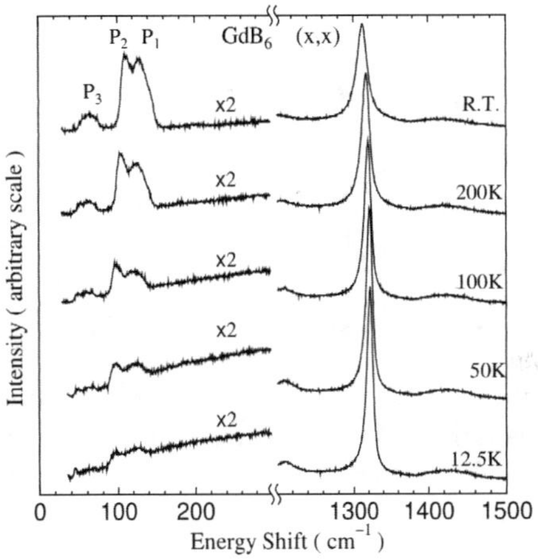

Figure 2. Temperature dependence of (x,x) Raman spectra in GdB_6 excited by 589 nm light. The intensities of the spectra at lower energy side are doubled.

Figure 1 shows representative polarization dependence of Raman scattering spectra in GdB_6, excited by 488 nm light at room temperature. Raman-active phonons are marked by arrows. These phonons satisfy the polarization selection rules mentioned above. This dependence is also observed in the other hexaborides. On the other hand, the small peaks at around 100 cm^{-1} have weak polarization dependence. These low-energy peaks correspond to rare-earth vibrations due to the second-order Raman scattering process [6]. Hereafter, we name this vibration as the rare-earth mode. The rare-earth modes are clearly observed in trivalent hexaborides $R^{3+}B_6$, and their energies and intensities are well scaled by a cage space $\delta r = a - 2r_R$, which is a available space for the rare earth ion in the boron cage; a and r_R are the lattice parameter and ionic radius of rare-earth ion, respectively. The intensities of rare-earth modes strongly depend on the wavelength (energy) of incident light, and increase for longer wavelength of incident light [6].

Figure 2 shows temperature dependence of (x,x) Raman spectra of GdB_6 excited by 589 nm light. Three peaks labeled $P_1 \sim P_3$ are clearly observed as the rare-earth modes in GdB_6. Three rare-earth modes are also present in DyB_6, but only one (P_1) in CeB_6. With decreasing temperature, the intensity of the rare-earth mode decreases monotonically, in spite of the increase of the intensity for Raman-active phonons. In the classical picture, the Raman scattering intensity is proportional to the mean square displacement (MSD) of the vibration. Therefore, the displacement of the rare-earth mode in the cage decreases with decreasing temperature. GdB_6 shows an antiferromagnetic transition at T_N=15.4 K, but figure 2 reveals little change in the phonon spectra at this temperature.

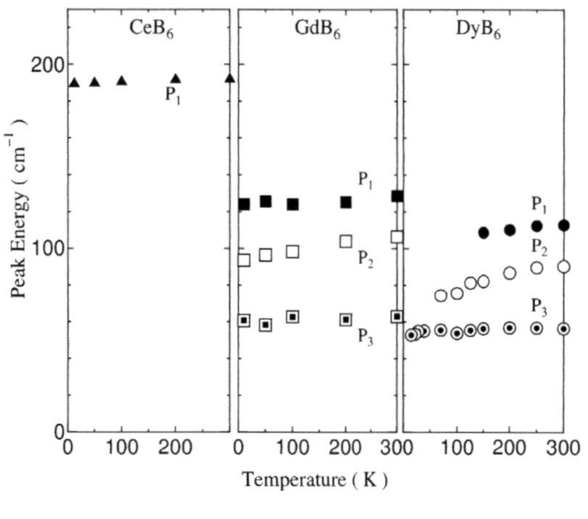

Figure 3. Temperature dependence of the position of the rare-earth mode.

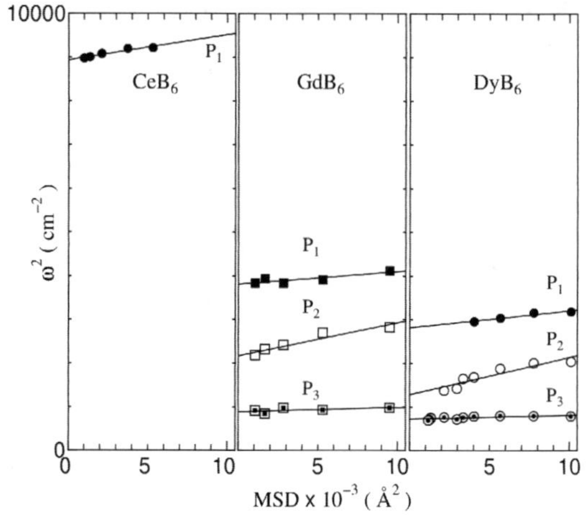

Figure 4. Square of the energy of rare-earth mode vs. mean square displacement.

Figure 3 shows the temperature dependence of the energy of the rare-earth modes $P_1 \sim P_3$. With decreasing temperature, the decrease of the energy is clearly observed for DyB_6. Since the cage space of these samples is $\delta r_{Dy} > \delta r_{Gd} > \delta r_{Ce}$, the larger cage space corresponds to the larger energy decrease, as shown in figure 3. Usually, the phonon energy increases due to the thermal shrinkage with decreasing temperature. This opposite to the ordinary phonon behavior is well understood as an anharmonic vibration [8].

To estimate the contribution of anharmonicity, we assume the simplified potential for the guest atom vibration in the cage

$$V(u) = \frac{m\omega_0^2}{2}u^2 + \frac{mk_4}{4}u^4,$$

where m, ω_0, and u are the atomic mass, the vibration energy for harmonic oscillation, and the atomic displacement from the center of the cage, respectively. The second term denotes the anharmonic potential. In this potential, the energy is derived as

$$\omega^2 = \omega_0^2 + k_4 < u^2 > .$$

$< u^2 >$ corresponds to MSD of the rare-earth mode. In this model, the third order effect of u is renormalized into ω_0. In the case of hexaboride, we assume ω_0 as constant, because the lattice constant of GdB_6 decreases by about 0.1% from 300 to 10 K [9], and the energies of the Raman-active phonons increase by 1% from 300 to 10 K. Thus, the energy increment is estimated as less than 0.5 cm^{-1} for the rare-earth mode. In the second equation, the anharmonic potential contributes to the phonon energy through MSD $< u^2 >$. Among many reports concerning the temperature dependence of the MSD for rare-earth ions in RB_6, we use the MSD data of ref. [10]. The square of ω as a function of MSD for each rare-earth ion is plotted in figure 4; ω is a half of the observed rare earth mode, because this mode corresponds to a two-phonon excitation. As shown in the figure, the fitted lines reproduce the measured data.

Figure 5 presents the temperature dependence of the ratio of anharmonic contribution $k_4 < u^2 > /\omega_o^2$. The middle peak P_2 shows the largest anharmonicity among three modes for GdB_6 and DyB_6. The three peaks correspond to the longitudinal or transverse phonons traveling to the X, M, or R point in the Brillouin zone [11]. Therefore, the obtained results

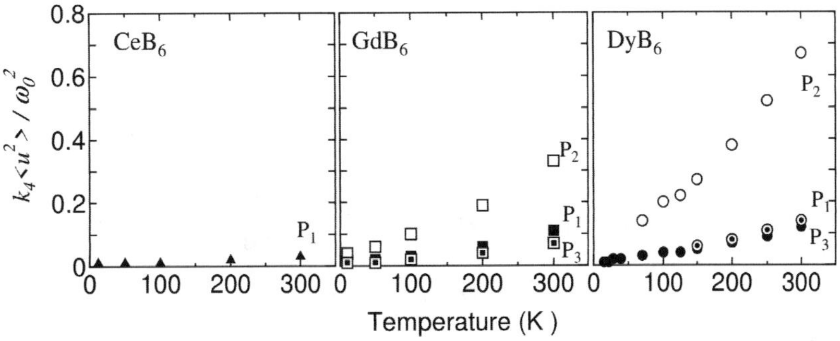

Figure 5. Temperature dependence of the ratio of anharmonic contribution to harmonic part $k_4 < u^2 > /\omega_o^2$ for the rare-earth mode.

suggest the anharmonicity of the rare-earth mode is anisotropic, that is, it depends on the vibrating and traveling directions of the rare-earth mode in DyB_6 and GdB_6. Because the energies for CeB_6 are degenerate, the anharmonic contribution in CeB_6 seems to be isotropic. Therefore, the anisotropy is enhanced for the larger cage space.

4. Concluding remark

Here we comment on the reported MSD values. Temperature dependence of the MSD for rare-earth ion is well described by a Debye model or an Einstein model. On the other hand, the present analysis gives us the important information on the anharmonicity. The anisotropic potential in GdB_6 and DyB_6 suggests us that the directional dependence of MSD is necessary for further understanding of the dynamical properties of rare-earth ion.

Acknowledgments

This work was partially supported by the Grant-in-Aid for Scientific Research Priority Area of "Strategy in Ubiquitous Elements" and "Skutterudite" from the Ministry of Education, Culture, Sports, Science and Technology, Japan, and by a research grant from the Mazda Foundation. The low-temperature experiments have been supported by N-BARD and IAMR of Hiroshima University.

[1] Kunii S, Kasuya T, Kadowaki K, Date M and Woods S B 1984 *Solid State Commun.* **52** 659
[2] Komatsubara T, Sato N, Kunii S, Oguro I, Furukawa Y, Ōnuki Y and Kasuya T 1983 *J. Mag. Magn. Mat.* **31-34** 368
[3] Kasuya T, Kasaya M, Takegahara K, Fujita T, Goto T, Tamaki A, Takigawa M, Yasuoka H *et al.*, *J. Magn. Magn. Mat.* **31-34**, 447 (1983).
[4] Mandrus D, Sales B C and Jin R 2001 *Phys. Rev.* B **64** 012302-1
[5] Smith H G, Dolling G, Kunii S, Kasaya M, Liu B, Takegahara K, Kasuya T and Goto T 1985 *Solid State Commun.* **53** 15
[6] Ogita N, Nagai S, Okamoto N, Udagawa M, Iga F, Sera M, Akimitsu J and Kunii S 2003 *Phys. Rev.* B **68** 224305
[7] Kunii S 1988 *J. Phys. Soc. Jpn.* **57** 361
[8] Takasu Y, Hasegawa T, Ogita N, Udagawa M, Avila M A, Suekuni K, Takabatake T and Takasu Y 2008 *Phys. Rev. Lett.* **100** 165503
[9] Takahashi Y, Fujimoto M, Tsuchiko M and Ohshima K 2001 *J. Appl. Crystallogr.* **34** 208
[10] Novikov V V 2003 *Phys. Solid State* **45** 1543
[11] Kohgi M, Kuwahara K, Ogita N, Udagawa M and Iga F 2006 *J. Phys. Soc. Jpn.* **75** 085003

Journal of Physics: Conference Series **176** (2009) 012033 doi:10.1088/1742-6596/176/1/012033

^{149}Sm nuclear resonant inelastic scattering of SmB_6

Satoshi Tsutsui [1], Takumi Hasegawa[2], Yuichi Takasu[2,3], Norio Ogita[2], Masayuki Udagawa [2], Yoshitaka Yoda[1,4] and Fumitoshi Iga[5]

[1] Japan Synchrotron Radiation Research Institute, SPring-8, Sayo, Hyogo 679-5198, Japan

[2] Graduate School of Integrated Arts and Sciences, Hiroshima University, Higashi-Hiroshima, Hiroshima 739-8521, Japan

[3] St. Marianna University School of Medicine, Kawasaki, Kanagawa 216-8511, Japan

[4] CREST, Japan Science and Technology Agency, Kawaguchi, Saitama 332-0012, Japan

[5] Graduate School of Advanced Science of Matter, Hiroshima University, Higashi-Hiroshima, Hiroshima 739-8530, Japan

E-mail: satoshi@spring8.or.jp

Abstract. ^{149}Sm nuclear resonant inelastic scattering was carried out in SmB_6 in order to investigate the lattice dynamics due to Sm atoms. A sharp phonon excitation associated with Sm atoms is observed around 10 meV, suggesting the presence of a Sm dispersionless mode. The Sm phonon energy shows significant temperature dependence, correlated with temperature dependence of Sm valence state. The nuclear resonant inelastic scattering spectra also indicate that the electronic configuration at the Sm site is $4f^5$ hybridized with the conduction electrons.

1. Introduction

SmB_6 is one of the attractive compounds in a series of hexaborides. This is a valence fluctuating system, as revealed by X-ray absorption spectroscopy, photoemission and ^{149}Sm Mössbauer spectroscopy [1-3]. Anomalous temperature dependence of the lattice constant also suggests the correlation with Sm valence fluctuation. The lattice constant shows a minimum around 150 K, and negative thermal expansion coefficient is observed at lower temperatures [4, 5]. Because the photoemission spectroscopy suggests that the insulator gap opens around 150 K [2], the opening of the insulator gap seems to be correlated with the lattice dynamics in this compound.

Correlation between the lattice dynamics and valence state of the guest atoms is expected by the previous Raman scattering investigation and recent *ab initio* calculation using the ABINIT package [6-8]. The most significant difference due to the valence state of the guest atoms is the phonon dispersion relation at the zone boundaries. The phonon energies of the guest atomic motion at the zone boundaries are split in divalent hexaborides, whereas they are degenerate in trivalent hexaborides [9]. This affects the phonon density of states (DOS) as shown in figure 1. The partial DOS of the

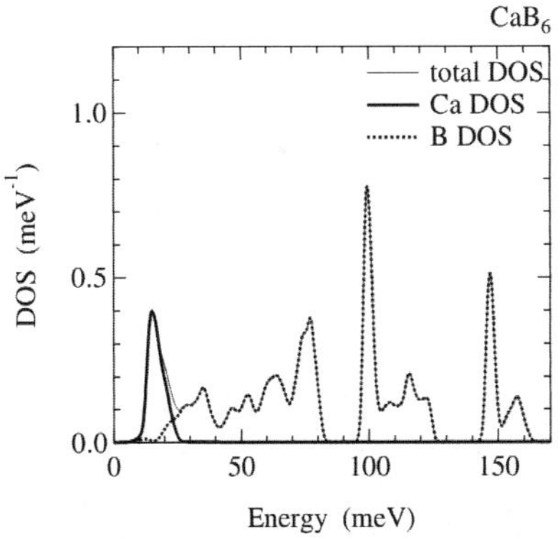

Figure 1 (a) Calculated phonon density of states in CaB_6 convolved with 1.5 meV Gaussian resolution function.

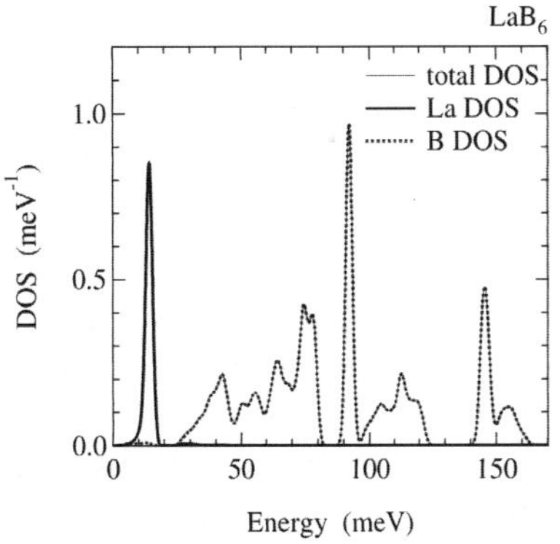

Figure 1 (b) Calculated phonon density of states in LaB_6 convolved with 1.5 meV Gaussian resolution function.

guest modes clearly depends on the valence state of the guest ions. The calculation suggests that the difference in the phonon energy between divalent and trivalent hexaborides is caused by the number of carriers. Because SmB_6 is recognized as a Kondo insulator and as a valence fluctuating system, a temperature dependence of phonon spectra is expected, correlated with the change of the number of carriers.

Nuclear resonant inelastic scattering (NRIS) is a powerful tool to investigate the element-specific partial DOS [10]. Here, nuclear resonance corresponds to the Mössbauer effect. Thus, Sm compounds are available for NRIS measurements, as Sm atom includes several Mössbauer isotopes. In addition, since the elastic scattering in Mössbauer resonance means the probability of recoil-free nuclear resonance [11], the inelastic part of the Mössbauer resonance corresponds to phonon excitation produced by nuclear recoil. Therefore, no excitations except for phonons can be observed by NRIS. As mentioned above, the calculation suggests that the partial DOS of the guest atoms is sensitive to the valence state of the guest ion. When the lattice dynamics in SmB_6 is strongly correlated with the Sm valence state and/or electronic configuration, temperature dependence of ^{149}Sm NRIS spectra is expected. Therefore, the lattice dynamical properties and electronic states of Sm atom have been investigated in this paper by ^{149}Sm NRIS.

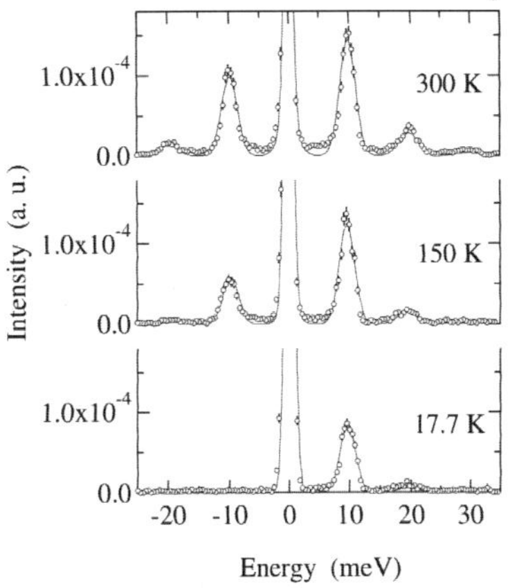

Figure 2. ^{149}Sm nuclear resonant inelastic scattering spectra of SmB_6 at various temperatures.

Journal of Physics: Conference Series **176** (2009) 012033 doi:10.1088/1742-6596/176/1/012033

2. Experimental setups

^{149}Sm NRIS measurements were carried out at the BL09XU beamline of SPring-8. The energy resolution was 1.5 meV at the ^{149}Sm Mössbauer resonance of 22.494 keV. The details of the X-ray optics are reported in [12]. The detector was a multi-element avalanche photodiode with the large solid angle. The chosen operation mode of SPring-8 was 203 bunches with the interval between the X-ray pulses of 23.6 ns, in order to distinguish nuclear scattering from electronic scattering.

3. Experimental results and discussion

^{149}Sm NRIS spectrum of SmB_6 at 300 K shows periodic and sharp excitations as shown in figure 2. The line width of the excitation is 3 meV at all temperatures. The calculation shown in figure 1 indicates that the width of the DOS associated with the La atoms is 3 meV in LaB_6 and that associated with Ca atoms is 6 meV in CaB_6. This demonstrates the presence of the dispersionless modes, as was indicated by the inelastic neutron scattering [13]. This is similar to Sm filled skutterudites [14, 15]. Assuming the Gaussian-shaped partial DOS, the obtained spectrum is well reproduced with multiple phonon contributions [16]. This result does not deny the contribution of Sm atomic motion to the acoustic modes in SmB_6, but demonstrates small contribution to the acoustic modes. Because the present result is very similar to the previous ^{161}Dy NRIS of DyB_6 [17], Sm valence expected from the Sm phonon DOS is a trivalent state. This also agrees with the contrast between the cases of the divalent and trivalent hexaborides as shown in figure 1. However, as reported previously [2, 3], the Sm valence state is an intermediate state between divalent and trivalent states. This suggests that the electronic configuration at the Sm site consists of a strong hybridization between 4f and conduction electrons.

Strong temperature dependence of the spectra shown in figure 2 indicates that the excitations at high energies come from multi-phonon contribution [16]. As mentioned above, because the cross section of the NRIS is the only contribution of phonon excitations, the other contributions are ruled out. In addition, the balance of NRIS spectra is caused by the Bose factor. Unless the observed temperature dependence of the spectra correlates with multi-phonon contributions, the intensity of the lower energy excitations is expected to show minor temperature dependence, and the peak of the higher energy excitations must remain even at low temperature. The energy observed at the second excitation around 20 meV agrees well with the unknown dispersionless excitation reported in the neutron scattering experiment [13]. However it might be hard to assert whether or not the unknown excitation comes from a two-phonon contribution in SmB_6, because temperature dependence of inelastic neutron scattering measurements has not been reported to our knowledge. Probably, the present result reveals the possible origin of the unknown excitation.

Figure 3 shows temperature dependence of the averaged phonon energy in SmB_6. The averaged energy is sensitive to the temperature dependence of the Sm valence state [18]. Above ~ 150 K, the

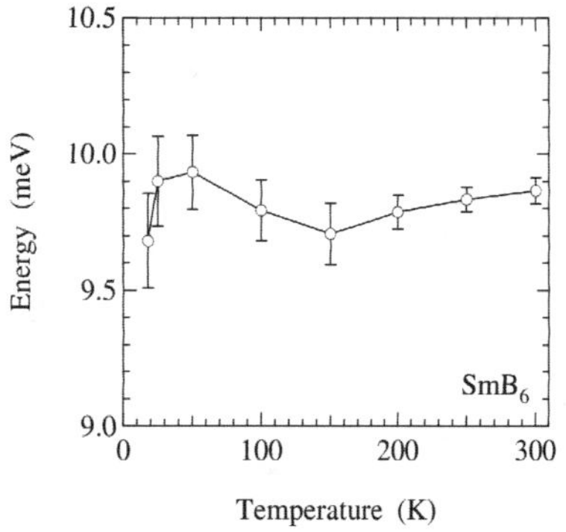

Figure 3. Temperature dependence of Sm phonon energy.

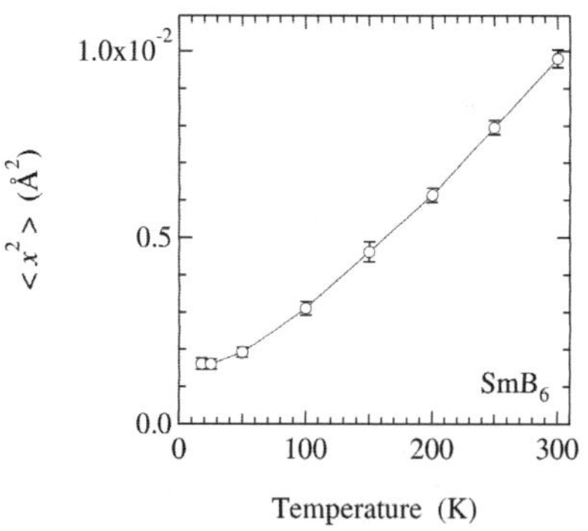

Figure 4. Temperature dependence of the atomic displacement factor obtained from the recoil-free fraction at the ^{149}Sm nuclear site.

energy decreases with the decrease of temperature, suggesting an anharmonic phonon mode. Between ~ 30 and ~ 150 K, the energy increases with decreasing temperature. In this temperature region, the temperature dependence of the averaged Sm valence state indicates an increase of Sm ionic radius towards 30 K. The increase of the phonon energy seems to be correlated with the free space of the Sm ions. In addition, the averaged phonon energy decreases again below ~ 30 K. This also suggests the correlation between Sm valence state and phonon energy associated with the guest free space at the Sm site.

In NRIS measurements, generally, a sum rule exists based on the Mössbauer effect [19]. The inelastic contribution provides the probability of the Mössbauer effect, recoil-free fraction. The relationship between the inelastic contribution of NRIS spectra and the recoil-free fraction is given by

$$f = 1 - E_R \int S_{inelastic}(\omega)d\omega \Big/ \int \omega\, S_{inelastic}(\omega)d\omega ,$$

where E_R = 1.81 meV is the recoil energy in ^{149}Sm Mössbauer effect and $S_{inelastic}(\omega)$ is an inelastic contribution of the NRIS spectra. The recoil-free fraction f is written as

$$f = \exp\left(-\frac{4\pi^2}{\lambda^2}\left\langle x^2 \right\rangle \right),$$

where λ is the wavelength of Mössbauer resonance and $\langle x^2 \rangle$ is the atomic displacement factor at the probe atomic site, the Sm site in the present case. The obtained atomic displacement factor is plotted against temperature in figure 4. This dependence agrees with the reported neutron diffraction data [4]. It also demonstrates one of the reasons not only for the adaptability of the sum rule [19] but also for denying any contribution of the observed excitations except for phonons. The temperature dependence of the atomic displacement factor shows no anomalies at about 25, 50 and 150 K where the slope of the averaged Sm valence changes with temperature.

4. Summary

We have succeeded in the direct measurement of the Sm phonon energy in SmB_6. The present results demonstrate the presence of the dispersionless modes around 10 meV at 300 K. This agrees with the calculation in LaB_6 [7] and with the ^{161}Dy NRIS of DyB_6 [17]. Therefore, the presence of the dispersionless modes infers that a possible electronic configuration at Sm site is a mixture between $4f^5$ and conduction electrons. The temperature dependence of the dispersionless mode energy indicates that the Sm mode is anharmonic in SmB_6. Besides, the temperature dependence of the Sm phonon energy shows significant temperature dependence associated with the averaged Sm valence state.

Acknowledgments

The author ST is grateful to Prof. H. Kobayashi and Dr. M. Mizumaki for their fruitful discussion. The NRIS experiment was carried out under the approval of JASRI (Proposal No. 2007A1233). The present work is partially supported by the Grant-in-Aid for Scientific Research Priority Area "Skutterudite" (No. 18027017) of the Ministry of Education, Culture, Science and Technology, Japan and the Grant-in-Aid for Young Scientists from the Japan Society of Promotion of Science (No. 20740181).

References

[1] Beaurepaire E , Kappler J P and Krill G 1990 *Phys. Rev.* B **41** 6768
[2] Souma S, Kumigashira H, Ito T, Takahashi T and Kunii S 2002 *Physica* B **312-313** 329
[3] Cohen R L, Eibschütz M and West K W 1970 *Phys. Rev. Lett.* **24** 383
[4] Trounov V A, Malyshev A L, Chernyshov D Yu, Korsukova M M, Gurin V N, Aslanov L A and Chernyshev V V 1993 *J. Phys.: Condens. Matter* **5** 2479
[5] Mandrus D, Sarrao J L, Lacerda A, Migliori A, Thompson J D and Fisk Z 1994 *Phys. Rev.* B **49** 16809
[6] Ogita N, Nagai S, Okamoto N, Udagawa M, Iga F, Sera M, Akimitsu J and Kunii S 2003 *Phys.*

Rev. B **68** 224305
[7] Hasegawa T, Ogita N and Udagawa M, to be published on this issue.
[8] The ABINIT package is a common project of the Universite Catholique de Louvain, Coring Incorporated, and other contributors [http://www.abinit.org]
[9] Kohgi M, Kuwahara K, Ogita N, Udagawa M and Iga F. 2006 *J. Phys. Soc. Jpn.* **75** 085003
[10] Seto M, Yoda Y, Kikuta S, Zhang X W, and Ando M 2000 *Phys. Rev. Lett.* **61** 461
[11] Mössbauer R L 1958 *Z. Phys.* **151** 124
[12] Tsutsui S, Kobayashi Y, Yoda Y, Seto M, Indoh K and Onodera H 2004 *J. Magn. Magn. Mater.* **272-276** 199
[13] Alekseev P A, Ivanov A S, Dorner B, Schober H, Kikoin K A, Mishchenko A S, Lazukov V N, Konovalova E S, Paderno Yu B, Rumyantsev A Yu and Sadikov I P 1989 *Europhys. Lett.* **10** 457
[14] Tsutsui S, Umemura J, Kobayashi H, Yoda Y, Onodera H, Sugawara H, Kikuchi D, Sato H, Sekine C and Shirotani I 2006 *Physica* B **383** 142
[15] Tsutsui S, Umemura J, Kobayashi H, Tazaki T, Nasu S, Kobayashi Y, Yoda Y, Onodera H, Sugawara H, Matsuda T D, Kikuchi D, Sato H, Sekine C and Shirotani I 2006 *Hyperfine Interact.* **168** 1073
[16] Sturhahn W, Teellner T S, Alp E E, Zhang X, Ando M, Yoda Y, Kikuta S, Seto M, Kimball C W and Dabrowski B 1995 *Phys. Rev. Lett.* **74** 3832
[17] Kobayashi H, Tsutsui S, Baron A Q R and Kunii S 2005 *Physica* B **359-361** 974
[18] Mizumaki M, Tsutsui S and Iga F, to be published on this issue.
[19] Lipkin H J 1995 *Phys. Rev.* B **52** 10073

Journal of Physics: Conference Series **176** (2009) 012034 doi:10.1088/1742-6596/176/1/012034

Temperature dependence of Sm valence in SmB_6 studied by X-ray absorption spectroscopy

Masaichiro Mizumaki[1], Satoshi Tsutsui[1] and Fumitoshi Iga[2]

[1] Japan Synchrotron Radiation Research Institute, SPring-8, Sayo, Hyogo 679-5198, Japan

[2] Institute for Advanced Materials Research, Hiroshima University, Higashi-Hiroshima 739-8530, Japan

E-mail: mizumaki@spring8.or.jp

Abstract. X-ray absorption spectra of a valence-fluctuated Kondo insulator SmB_6 have been studied near the Sm L_3-edge at temperatures below 300 K. The average valence of Sm ions in SmB_6 was deduced as 2.59 at 300 K; it shows a characteristic temperature behaviour with three anomalies at T = 15, 50 and 150 K. These temperatures correspond to the characteristic energies of SmB_6 in macroscopic measurements. The present results imply that the Sm-4f electrons strongly hybridize with the conduction electrons and play an important role in the gap formation.

1. Introduction

A Kondo insulator SmB_6 has attracted much attention because of its anomalous physical properties, such as the pressure- and magnetic field sensitive temperature dependence of electrical resistivity [1]. These properties have been associated with the small, temperature-dependent energy gap at the Fermi level, which originates from the valence fluctuations of Sm ions in SmB_6 [2]. The average valence of Sm ions is ~2.6, as estimated by X-ray absorption spectroscopy (XAS), photoemission spectroscopy, and Mössbauer measurements [3]. This value clearly indicates that the hybridization between 4f electrons and the conduction band plays an important role in the formation of the insulating gap. Band structure calculation indicated that the electronic configuration of Sm^{2+} has to be taken into account for other electronic orbitals [4 - 6]; that configuration was proposed as $4f^5 5d^1$ [7]. However, the electronic configuration of Sm site and the origin of the insulating gap are still under debate. Negative thermal expansion and temperature variation of the lattice constant suggest significant temperature dependence of the Sm site properties. XAS and lattice constant measurements revealed that the average valence changes gradually below 150 K [8]. Observation of the negative thermal expansion between ~10 and 80 K also indicated a valence change at lower temperatures [9]. However, detailed temperature dependence of the Sm valence has not been documented yet. In this paper, we report the electronic structure and the temperature dependence of the average valence of SmB_6 by XAS at the

Sm L_3-edge. We discuss the correlation between the temperature dependence of the average Sm valence and the previously reported anomalous properties, such as the negative thermal expansion.

2. Experimental procedure

XAS spectra around the Sm L_3-edge of SmB_6 were measured at beam lines BL01B1 and BL39XU of SPring-8 in Japan. The beam line consists of a double-crystal monochromator equipped with a Si 111 crystal and a Si mirror for eliminating the higher harmonics [10]. The spectra were recorded in the transmission mode, using ionization chambers, at various temperatures between 2 and 300 K. The energy resolution was ~1 eV around the Sm L_3-edge. The energy was calibrated with the XAS spectrum of Sm_2O_3 at room temperature. The method of deducing the average valence was reported in [11].

3. Results and discussion

Figure 1 shows the temperature dependence of XAS spectra at the Sm L_3-edge of SmB_6. The spectra consist of two peaks characteristic of a mixed valence system. The low-energy one at $E = 6713$ eV corresponds to the divalent component and the high-energy peak at $E = 6720$ eV to the trivalent component. Previous Mössbauer measurements have demonstrated that the Sm valence state is an intermediate valence state with a single-line spectrum [3]. This disagreement with our results reflects different time scales of XAS and Mössbauer measurements: 10^{-12} s for XAS and 10^{-8} s for ^{149}Sm Mössbauer spectroscopy. Therefore, this inconsistency indicates that the Sm valence fluctuates at a 10^{-8} s scale.

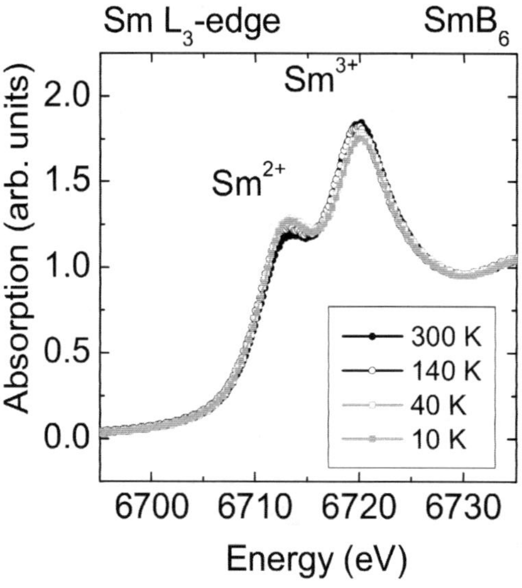

Figure 1. Temperature dependence of XAS spectra in SmB_6 around Sm L_3-edge.

The divalent component increases with decreasing temperature. The temperature dependence of the XAS spectra indicates a stabilization of the divalent state at low temperature. This temperature dependence does not contradict the previously suggested non-magnetic ground state [3,12,13].

Figure 2 shows the temperature dependence of the average valence of Sm ions in SmB_6 which shows three characteristic temperatures. The average valence of Sm ion is 2.59 ± 0.01 at 300 K and is constant between 300 K and 120 K (= T_1). Then it gradually decreases upon cooling down to 50 K (= T_2) and stabilizes at 2.50 ± 0.01 between 50 and 15 K (= T_3). At lower temperatures, the Sm valence increases to 2.52 ± 0.01 at 2 K. These results generally agree with the previous measurements. An anomaly at T_1 is consistent with the previous neutron diffraction results, because the Sm^{2+} ionic radius is larger than the Sm^{3+} one [11]. The T_2 and T_3 anomalies are consistent with the temperature range of

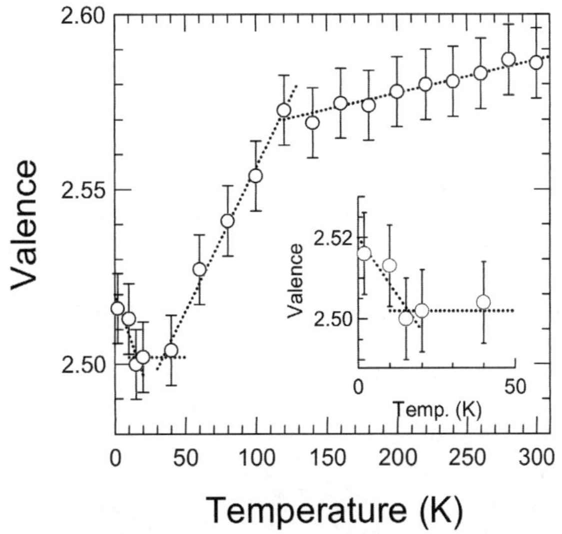

Figure 2. Temperature dependence of the average valence of Sm ions in SmB_6. The dotted line is a guide for the eyes. The inset zooms into the low-temperature region.

the negative thermal expansions [9], as well as with temperature dependence of the longitudinal elastic constant C_{11} [14].

The three characteristic temperatures of figure 2 are in reasonable agreement with the characteristic energies in SmB_6: T_1 corresponds to the energy gap Δ = 100 K (or 130K) estimated from electrical resistivity (or photoemission spectroscopy) [2,7]. This correspondence implies that the Sm valence state correlates with formation of the gap Δ. The magnetic susceptibility shows a peak at $T = T_2$ corresponding to another gap [15]. This temperature also agrees with the activation energy E_d estimated from electrical resistivity measurements [7]. Moreover, the temperature dependence of the phonon energy, which was estimated from nuclear resonant inelastic scattering (NRIS), showed a minimum at 120 K (= T_1) and a maximum at 50 K (= T_2) [16]. The negative slope between T_1 and T_2 apparently correlates with the Sm ionic radius. Since the average valence reflects the ionic radius of Sm, our result suggests that the Sm phonon energy correlates with the guest free space at the Sm site. At $T = T_3$, the thermal expansion coefficient shows a small peak, which might relate to the change in the Sm valence [9]. Below $T = T_3$, the increase in the average valence of Sm ions is consistent with the positive thermal expansion. To conclude, we find that the anomalous temperature dependence of the lattice constants and thermal expansion does correlate with the average valence of Sm ions.

4. Summary

XAS measurements around the Sm L_3-edge were performed at various temperatures, on a valence-fluctuated Kondo insulator SmB_6. The temperature dependence of the Sm valence state, particularly below the temperature T_2 = 50 K, has not been reported previously. This dependence agrees with the temperature dependences of various parameters, including the thermal expansion, elastic constant, insulator gap and hybridization gap. Our results suggest the importance of the Sm valence in understanding the physical properties of SmB_6.

Acknowledgments

The authors appreciate assistance of Dr. N. Kawamura and Dr. T. Uruga in the experiments. The present work was carried out under the approval of JASRI (Proposal Nos. 2007B1351 and 2008A1972). This work was supported by a Grant-in-Aid from Scientific Research Priority Area "Skutterudite" of the Ministry of Education, Science and Technology (No.15072206).

References

[1] Nickerson J C, White R M, Lee K N, Bachmann R, Geballe T H, and Hull Jr. G W 1971 *Phys. Rev.* B **3** 2030.
[2] Souma S, Kumigashira H, Ito T, Takahashi T and Kunii S 2002 *Physica* B **312-313** 329
[3] Cohen R L, Eibschutz and West K W 1970 *Phys. Rev. Lett.***23** 383
[4] Yanase A and Harima H 1992 *Prog. Theor. Phys. Suppl.* **108** 19
[5] Antonov V N, Harmon B N and Yaresko A N 2002 *Phys. Rev.* B **66** 165208
[6] Antonov V N, Harmon B N and Yaresko A N 2002 *Phys. Rev.* B **66** 165209
[7] Derr J, Knebel G, Braithwaite D, Salce B, Flouquet J, Flachbart K, Gabani S and Shitsevalova N 2008 *Phys. Rev.* B **77** 193107
[8] Trounov V A, Malyshev A L, Chernyshov D Yu, Korsukova M M, Gurin V N, Aslanov L A, and Chernysev V V 1993 *J. Phys. Condens. Matter* **5** 2479
[9] Mandrus D, Sarrao J L, Lacerda A, Migliori A, Thompson J D and Fisk Z 2002 *Phys. Rev.* B **49** 16809.
[10] Uruga T, Tanida H, Yoneda Y, Takeshita K, Emura S, Takahashi M, Harada M, Nishihata Y, Kubozono Y, Tanaka T, Yamamoto T, Maeda H, Kamishima O, Takabayashi Y, Nakata Y, Kimura H, Goto S and Ishikawa T 1999 *J. Synchr. Radiat.* **6** 143
[11] Mizumaki M, Tsutsui S, Tanida H, Uruga T, Kikuchi D, Sugawara H and Sato H 2007 *J. Phys. Soc. Jpn.* **76** 053706
[12] Tarascon J M, Ishikawa Y, Chevalier B, Etourneau B, Hagenmuller P and Kasaya M 1980 *J. Physique* **41** 1141
[13] Trounov V A, Malyshev A L, Chernyshov D Yu, Korsukova M M, Gurin V N, Aslanov L A and Cherysev V V 1993 *J. Phys.: Condens. Matter* **5** 2479
[14] Nakamura S, Goto T, Kasaya M and Kunii S 1991 *J. Phys. Soc. Jpn.* **60** 4311
[15] Caldwell T 2004 "*Nuclear magnetic resonance studies of field effects on single crystal SmB_6*" PhD thesis, Florida State Unversity, p 60
[16] Tsutsui S *et al* 2009, to be published in this conference proceedings

Effect of pressure and magnetic field on the electrical resistivity of TbB_6

Takeshi Sakai[1], Gendo Oomi[2] and Satoru Kunii[3]

[1]Department of General Education, Ariake National College of Technology, Omuta, Fukuoka 836-8585, Japan

[2]Department of Physics, Kyushu University, Ropponmatsu, Fukuoka 810-8560, Japan

[3]Faculty of Engineering, Tohoku-Gakuin University, Sendai, Miyagi 980-8511, Japan

E-mail: sakai@ariake-nct.ac.jp

Abstract. Electrical resistivity of a single crystal of TbB_6 was studied under hydrostatic pressures up to 2.1 GPa and magnetic fields up to 9 T. The Néel temperature, T_N, decreases linearly with increasing pressure: $\partial \ln T_N/\partial P$ = -3.14×10^{-2} GPa^{-1} at zero external field. This pressure dependence of T_N weakens as external fields increase. At ambient pressure, the magnetoresistance at 4.2 K is positive up to 4.8 T and becomes negative above 4.8 T. The positive magnetoresistance observed at ambient pressure is suppressed by applying pressure, which enhances the negative magnetoresistance. These results are interpreted in terms of the reduction of the scattering of conduction electrons, due to disordered magnetic moment being suppressed by delocalization of 4f electrons at high pressure, and the magnetic field variation of the large transition probability between the ground state and the excited levels.

1. Introduction

The rare-earth hexaborides, RB_6, with the cubic CaB_6-type structure have attracted much interest because of the variety of magnetic properties and other cooperative phenomena originating from magnetic and quadrupolar orderings. The interplay among these different types of interaction in RB_6 can give rise to unexpected magnetoelastic properties and complex magnetic structures [1, 2, 3]. In the RB_6 series, heavy rare-earth hexaborides (TbB_6, DyB_6 and HoB_6), which have been less investigated because of their incongruent melting property, exhibit the anomalous softening of the transverse C_{44} mode of the elastic constant at low temperatures. For DyB_6 and HoB_6, noticeable softening is possibly induced by the quadrupolar orderings. Note that a crossover from the first-order transition, accompanied by a structural phase transition into a second-order-type transition in the temperature-dependent thermal expansion, was detected for DyB_6 by the application of pressure [4], which was ascribed to possible short-range order of the quadrupolar moment. In TbB_6, the observed anomalous softening was ascribed to the Van Vleck contribution in the strain susceptibility for the Γ_3 ground state [2]. The anomaly around 19 K, observed in electrical resistivity, magnetic susceptibility and lattice constants, was due to the antiferromagnetic ordering. Therefore, the temperature dependence of electrical resistivity, $\rho(T)$, of TbB_6 at ambient pressure shows an abrupt increase at 19.5 K, in a

Journal of Physics: Conference Series **176** (2009) 012035 doi:10.1088/1742-6596/176/1/012035

striking contrast to $\rho(T)$ of DyB_6 which has the Γ_8 ground state [5]. In this work, we have studied the pressure and the magnetic field dependence of the electrical resistivity in order to understand whether or not there is a crossover of the electronic and magnetic states in TbB_6.

2. Experimental details

A single crystal of TbB_6 was prepared by the crucible-free vertical floating zone method [6, 7]. Electrical resistivity under high pressure was measured using a standard DC four-probe method in the temperature range from 4.2 K to room temperature and in magnetic fields up to 9 T applied parallel to the [110]-direction. Hydrostatic pressures up to 2.1 GPa were generated using a piston cylinder device, and a 1:1 mixture of Fluorinert, FC 70 and 77 as a pressure transmitting medium. The pressure was changed only at room temperature to minimize an internal strain in the specimen and the load was controlled within ±1% throughout the measurement. The details of the high pressure system have been reported elsewhere [8].

3. Results and discussion

Figure 1 shows the temperature dependence of the electrical resistivity, $\rho(T)$, of TbB_6 at various pressures up to 2.1 GPa. At ambient pressure (= 0 GPa), $\rho(T)$ decreases smoothly as temperature decreases to about 20 K. The onset of a resistivity anomaly, which is caused by the antiferromagnetic (AF) transition, is observed as an abrupt increase in the $\rho(T)$ curve at 19.5 K (= T_N). The data obtained at ambient pressure are in good agreement with those reported in the literature [5]. Upon application of pressure, as shown in figure 1, $\rho(T)$ decreases in the whole paramagnetic temperature range. The anomalies of ρ at T_N are clearly observed in the $\rho(T)$ curves under all external pressures. At room temperature, ρ decreases linearly with increasing pressure (see the inset of figure 1) at a rate $\partial \ln\rho/\partial P$ = □5.7×10^{-2} GPa^{-1}, which is as same as that for DyB_6 [9].

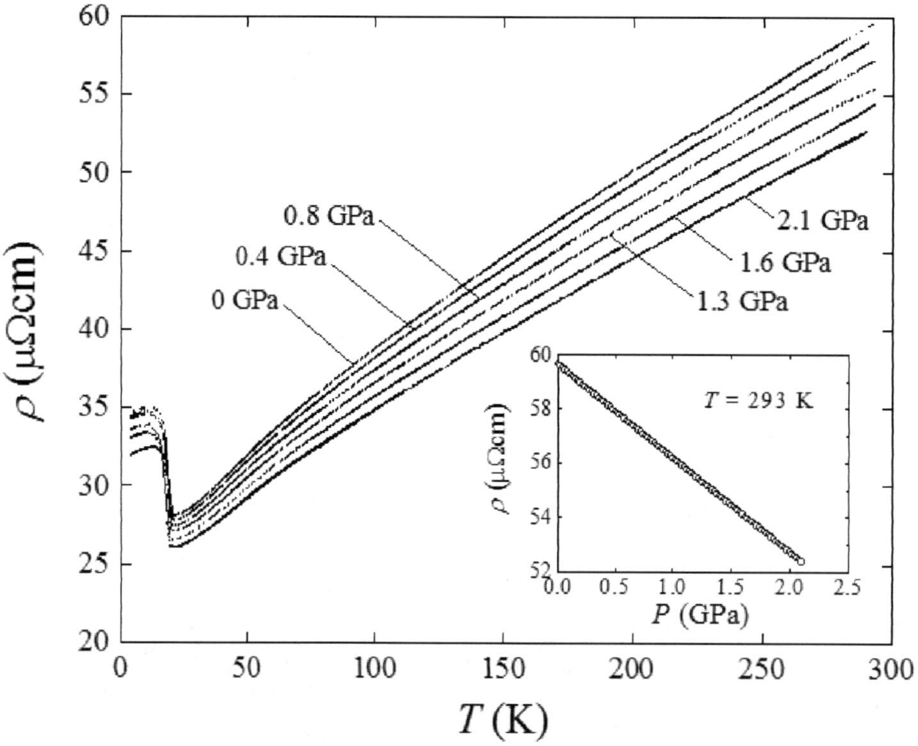

Figure 1. Temperature dependence of electrical resistivity in TbB_6 at various pressures. Inset shows pressure dependence of resistivity at room temperature.

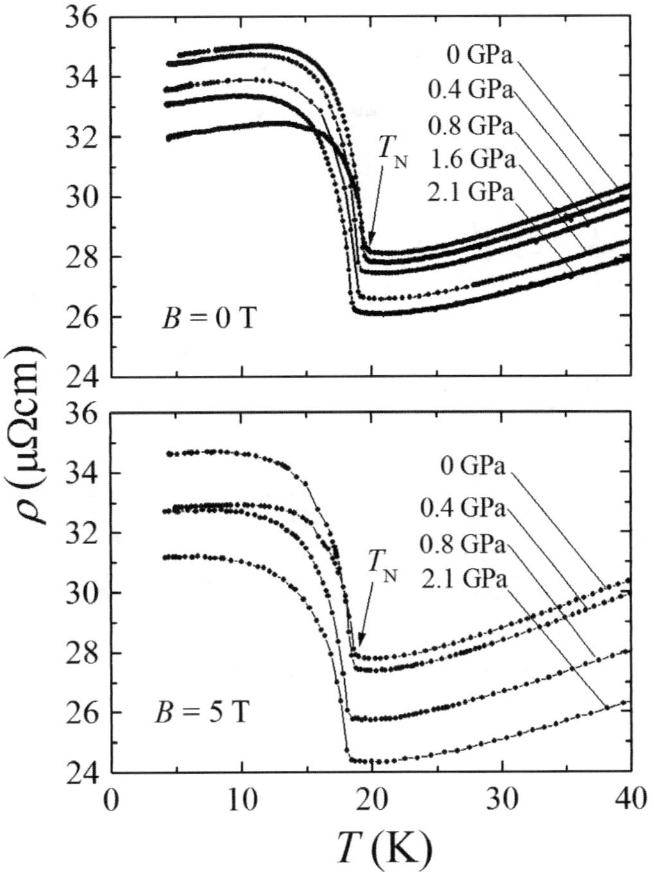

Figure 2. Temperature dependence of electrical resistivity in a zero magnetic field and in 5 T at temperatures below 40 K under various pressures.

In figure 2, the temperature dependence of ρ below 40 K was observed without an external magnetic field and in a field of 5 T. In the paramagnetic temperature range, the decrease of ρ at 2.1 GPa is ~7 % in zero field and increases to ~13 % in 5 T. Another effect is that the abrupt increase in $\rho(T)$ below T_N is enhanced at 0.4 GPa. Such marked pressure effects on the $\rho(T)$ curve (in the former case amplified by an external field) may be attributed to a drastic change in the electronic states of TbB_6. The origin of this resistivity anomaly at T_N has not been clarified yet. It may be caused by the superzone effects in TbB_6 which depends not only on the width of the energy gaps but also on their positions in the reciprocal space [10].

Figure 3 shows T_N in various magnetic fields as a function of pressure, where T_N was determined from the inflection point of $\rho(T)$ (see in figure 2). The value of T_N decreases linearly with increasing pressure at a rate $\partial T_N/\partial P \approx$ □0.51 $\mathrm{KGPa^{-1}}$ under ambient pressure, as deduced by a least square fit shown by a solid line in figure 3. This value is comparable with □0.13 K $\mathrm{GPa^{-1}}$ estimated from the Ehrenfest relation, $dT_N/dP = T_N V_m (\Delta\alpha_V/\Delta C)$, where V_m is molar volume, $\Delta\alpha_V$ and ΔC are the change in thermal expansion coefficient and specific heat at T_N, respectively. This pressure dependence of T_N with negative derivative is in sharp contrast to the previous result for DyB_6, in which T_N increases with pressure [9]. To see this more clearly, the value of dT_N/dP is shown as a function of pressure in figure 4. Application of an external field reduces the dT_N/dP value in accordance with a linear field dependence up to 5 T, and it is almost 0 (*i.e.*, pressure independent) at 7 T. A noticeable deviation from the linear field dependence of dT_N/dP is found between 5 T and 7 T indicating a significant change in the electronic state of TbB_6 in this field range.

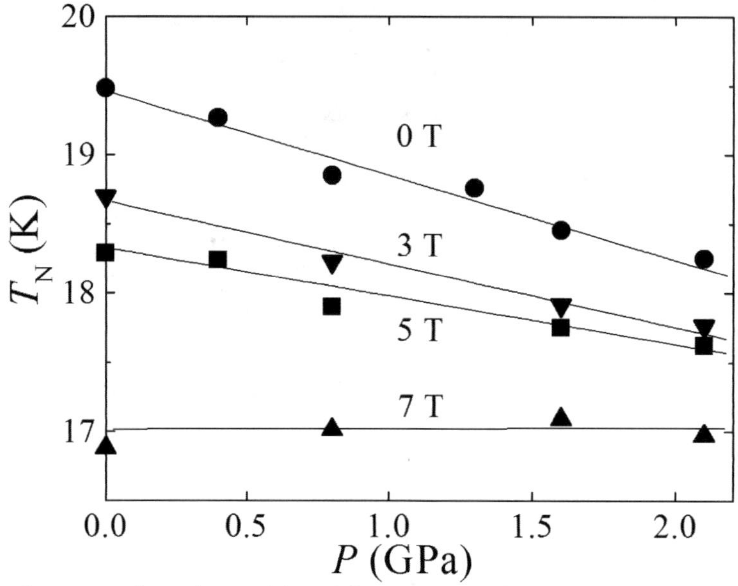

Figure 3. Pressure dependence of the Néel temperature of TbB_6 in zero and high fields.

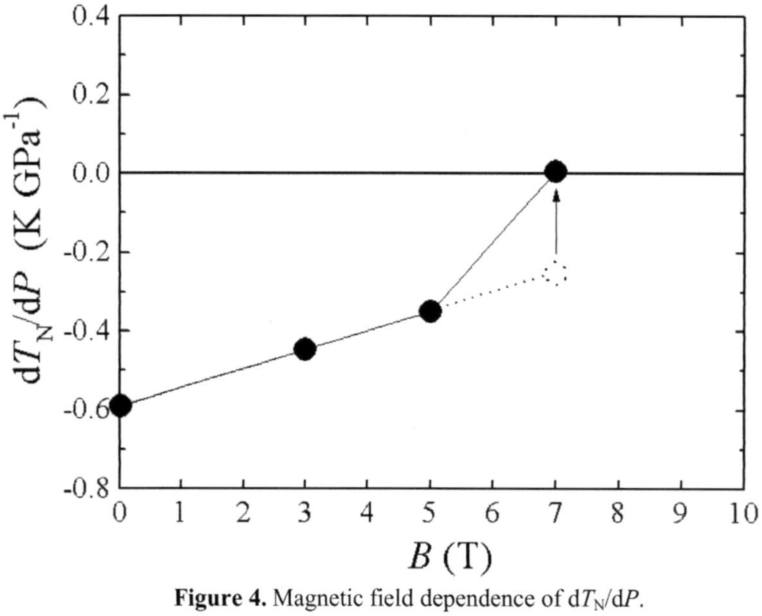

Figure 4. Magnetic field dependence of dT_N/dP.

In order to confirm the fact mentioned above, we investigated the transverse magnetoresistance (MR) $\Delta\rho/\rho_0 \equiv [\rho(B)-\rho_0]/\rho_0$ at 4.2 K as a function of the magnetic field B up to 9 T applied along [110] direction, as shown in figure 5. At ambient pressure, $\Delta\rho/\rho_0$ is positive up to ~4.8 T, and changes its sign into negative above 5 T. The critical field B_C where the sign changes (shown by arrows in figure) is shifted to a lower field by applying pressure. The pressure dependence of B_C is displayed as the inset of figure 5.

Interesting temperature dependence of electrical resistivity under pressure and magnetic field was observed in the TbB_6 system: (i) large decrease of $\rho(T)$ under pressure in the whole paramagnetic temperature range, which is further enhanced by magnetic field; (ii) negative pressure derivative of T_N; (iii) abrupt increase of ρ at T_N, which is enhanced by pressure of 0.4 GPa;

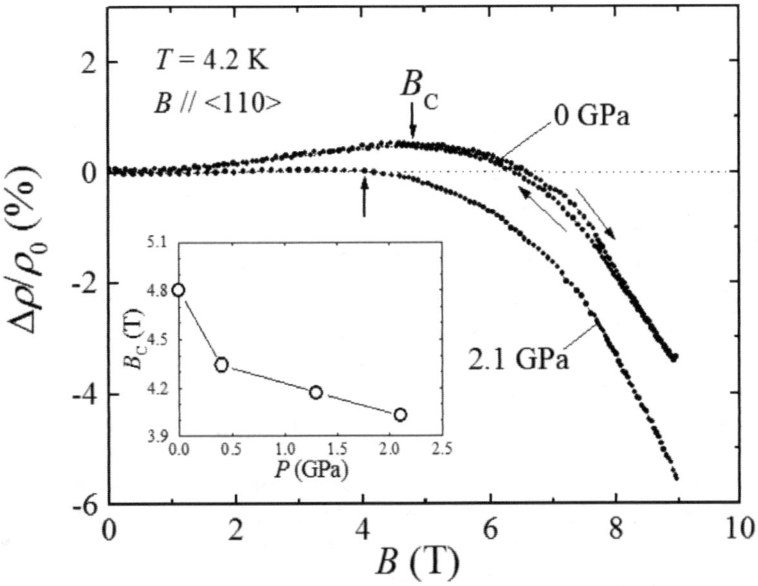

Figure 5. The transverse magnetoresistance at 4.2 K as a function of magnetic fields at different pressures.

(iv) sign reversal of the magnetoresistance. The results (i) and (ii) can be interpreted by considering the effect of pressure on the resistivity based on the Matthiessen's rule. The total resistivity above T_N can be represented by a sum of three contributions

$$\rho(T) = \rho_i + \rho_{ph}(T) + \rho_{spd}(T). \tag{1}$$

Here ρ_i represents temperature-independent residual resistivity, $\rho_{ph}(T)$ is due to interactions of conduction electrons with thermally excited phonons, and $\rho_{spd}(T)$ arises from spin disorder (SPD) scattering processes. Usually, $\rho_{ph}(T)$ is best accounted for in terms of Bloch Grüneisen formula yielding a continuous increase with increasing temperature. The pronounced anomalies controlled by either pressure or temperature arise only from changes of the magnetic contribution to $\rho(T)$ because the exchange interaction between the 4*f* and the conduction electrons leads to spin-dependent scattering affecting the total resistivity. From this point of view, the origin of the marked decrease in $\rho(T)$, induced under pressure by an external field in the paramagnetic temperature range, is mainly due to the reduction of the spin-dependent scattering term $\rho_{spd}(T)$ which is attributable to the delocalization of 4*f* electrons of Tb.

For the result (iii), there is a possible mechanism which could be responsible for the rapid enhancement of ρ at T_N for well localized moment systems, such as heavy rare earth metals. In this mechanism, the antiferromagnetic ordering introduces additional Brillouin zones because the periodicity of the magnetic structure is different from that of the crystal structure [10]. If these zone boundaries (so called superzone boundaries, SZB) are located very close to the Fermi surface then changes are induced in the conduction electron system. In other words, the effective mass of conduction electron, m^*, varies from the paramagnetic to antiferromagnetic state due to an additional distortion of the Fermi surface by either pressure or magnetic field. If m^* increases due to the presence of the SZB then the ρ-value should rise. For TbB_6, this contribution to ρ at T_N would be very marked. According to Elliott and Wedgwood [11], the resistivity tensor is

$$\rho_{ii} = \rho_{ii}^0/(1-\delta_{ii}), \tag{2}$$

where ρ_{ii}^0 is the unperturbed value. If the SZB (a) cuts and (b) touches the Fermi surface then, to order Δ/E_f,

$$\text{(a)}\ \delta_{zz} = 3\pi l\Delta/4k_f E_f, \tag{3}$$

(b) $\delta_{zz} = 1-(l / k_f)^3 + (3\pi l\Delta/8k_f E_f)$. (4)

Here Δ is the energy gap at the SZB, E_f is the Fermi energy, l and k_f are the magnitude of the wave vectors at the SZB and at the Fermi surface, respectively. The effect is largest when the superzone cuts the surface. Equations (3) and (4) qualitatively account for the fact that $\Delta\rho$ is enhanced significantly only at 0.4 GPa and is almost constant under pressure above 0.4 GPa, where $\Delta\rho$ is the resistivity change due to the antiferromagnetic ordering of TbB_6 at T_N. If the boundary with largest l only touches, but not cuts, the Fermi surface at pressures of up to 0.4 GPa, its position will depend on pressure, which will significantly enhance $\Delta\rho$. However, if the boundary cuts the Fermi surface above 0.4 GPa, the pressure effect of $\Delta\rho$ is not expected any longer.

The last point to discuss is the result (iv). The ground state of 4*f* CEF levels for TbB_6 is expected to be a nonmagnetic Γ_3 doublet, and the first excited state of magnetic Γ_5 triplet locates above the ground state from the elastic constant measurement [2]. Figure 5 reveals positive magnetoresistance in low fields, indicating the nonmagnetic ground state. The negative magnetoresistance in higher fields is attributable to the field variation of the large transition probability between the ground state and the excited Γ_5 levels. As the applied field increases, the population of higher lying levels increases, and the Γ_5 level contribution to magnetoresistance gradually becomes dominant above 5T. However, additional measurements, in particular concerning the specific heat and the elastic constant under strong magnetic field and high pressure, are needed to clarify the origin of the negative magnetoresistance of TbB_6.

4. Conclusion

In this paper, we have measured electrical resistivity of TbB_6 under high pressures and strong magnetic fields. The results suggest the antiferromagnetically ordered states in TbB_6 become unstable under applied pressure. This effect could be due to the delocalization of 4*f* electrons in Tb atoms, and it contrasts the case of DyB_6 where T_N and B_C increase as increasing pressure. Large negative magnetoresistance observed in strong magnetic fields is attributable to the field variation of the large transition probability between the ground state and the excited Γ_5 levels in TbB_6, which is enhanced by external pressure.

References

[1] Sera M, Kobayashi S, Hiroi M, Kobayashi N and Kunii S 1996 *Phys. Rev. B* **54** R5207
[2] Nakamura S, Goto T and Kunii S 1995 *J. Phys. Soc. Jpn.* **64** 3941
[3] Gorshunov B, Sluchanko N, Volkov A, Dressel M, Knebel G, Loidl A and Kunii S 1999 *Phys. Rev.* B **59** 1808
[4] Sakai T, Oomi G and Kunii S 1999 *Physica* B **259-261** 904
[5] Segawa K, Tomita A, Iwashita K, Kasaya M, Suzuki T and Kunii S 1992 *J. Magn. Magn. Mater.* **104-107** 1233
[6] Kunii S 1988 *J. Phys. Soc. Jpn.* **57** 361
[7] Takahashi K and Kunii S 1997 *J. Solid State Chem.* **133** 198
[8] Oomi G, Kagayama T 1997 *Physica* B **239** 191
[9] Sakai T, Oomi G and Kunii S, *Physica* B **230-232** 741
[10] Taylor K N R and Darby M I 1972 *Physics of Rare Earth Solids* (Chapman & Hall, London) p 374
[11] Elliott R J and Wedgwood F A 1963 *Proc. Phys. Soc.* **81** 846

Novel physical properties of rare earth higher borides

Takao Mori
International Center for Materials Nanoarchitectonics (MANA), National Institute for Materials Science (NIMS), 1-1 Namiki, Tsukuba 305-0044, Japan

mori.takao@nims.go.jp

Abstract. Boron forms various compounds with metal atoms occupying voids in the boron framework. We are particularly interested in the rare earth borides, of which many new and varied compounds have been discovered within the span of 10 years, and where the shell of f electrons supplies additional interesting properties. The following recently emerging topics in these borides will be covered: 1. Unexpectedly strong magnetic coupling discovered in dilute f-electron insulators. 2. Attractive high temperature thermoelectric properties. 3. Discovery of intrinsic building defects.

1. Introduction

The traditional focus on boron-rich cluster compounds has tended to be on their striking structural and mechanical properties. These compounds are attractive as materials for their typical high hardness and great stability in extreme environments such as high temperatures, which originate from the particular strong covalent bonding of boron. Boron forms various compounds with metal atoms occupying voids in the boron framework. We are particularly interested in the rare earth borides, of which many new and varied compounds have been discovered within the span of 10 years [1], and where the shell of f electrons supplies additional properties. As a new development, novel functions in the form of exciting electronic and magnetic physical properties have been found in the newly discovered rare earth boron compounds [1].

Rare earth B_{12} icosahedra-containing compounds are magnetically dilute semiconducting/insulating f-electron compounds, but it has been revealed that they display a wide variety of magnetism with surprisingly strong magnetic coupling [1-4]. A new aspect discovered for the homologous series of 2 dimensional (2D) spin glasses will be reported in this paper.

We have also been investigating the potential of these novel borides as possible thermoelectric materials, since the boron cluster compounds generally exhibit low thermal conductivity [5-7]. Analysis on the origin/mechanism of this intrinsic low thermal conductivity has recently been done on some rare earth borides, comparing effects of "crystal complexity" (the "amorphous concept"), possible "rattling" phenomena, and disorder. It was concluded that disorder was playing a dominant role [8]. In regard to the recently discovered rare earth borides, in a striking new development, the anomalous and attractive n-type behavior was observed in a homologous series of rare earth boron carbonitrides, $REB_{17}CN$, $REB_{22}C_2N$, and $REB_{28.5}C_4$ without any need for extreme doping [9,10]. These compounds have similarities in crystal structure to boron carbide which is an excellent p-type material [11] and these rare earth boron carbonitrides may embody the long awaited n-type counterpart.

Morphology of the samples of these compounds was investigated and the effects will be presented and discussed.

As the third topic, intrinsic building defects were discovered to be generated for some layered borides like $TmAlB_4$ and found to have a significant effect on the physical properties, resulting in, i.e. multiple transitions at low temperatures. This phenomenon will be reviewed.

2. Experimental details

Details on the synthesis of the borides dealt with in this paper have been published elsewhere (for a review see ref. [1] and references therein), but the basic approach can be described as the following.

First of all powders of REB_n are prepared with the borothermal reduction method at 2000 K in BN crucibles:

$RE_2O_3 + (2n+3)B \rightarrow 2REB_n + 3BO$

Binary rare earth borides like REB_{50} can be made directly with this one step. For ternary borides, in the next step the desired amounts of boron, carbon, and hexagonal BN, in the case of the rare earth boron carbonitrides, $REB_{17}CN$, $REB_{22}C_2N$, and $REB_{28.5}C_4$, and silicon in the case of $REB_{44}Si_2$, are added and fired again at a reaction temperature of around 1900 to 2100 K. Crystals of rare earth borosilicides, $REB_{44}Si_2$, can also be grown by the floating zone (FZ) method. Samples are characterized by X-ray diffraction (XRD) measurements and chemical analysis. Magnetic measurements were made using a superconducting quantum interference device (SQUID) magnetometer. ac susceptibility measurements were made after reducing the residual field below 1.0 Oe by resetting the magnet. Both the in-phase component χ' and out-of-phase component χ'' were obtained. Resistivity and thermoelectric power were measured with an ULVAC ZEM-2 by using the four probe method. Thermal conductivity was measured by a laser flash method with an ULVAC TC-7000. Scanning electron microscope (SEM) pictures were taken with a JEOL SM-67F 10 kV.

3. Properties of rare earth higher borides

3.1. Magnetism of B_{12} icosahedral rare earth borides

The magnetism of rare earth borides like tetraborides REB_4, hexaborides REB_6, and dodecaborides REB_{12} has attracted a lot of attention over the years [12], with recent interesting discoveries still being made such as the frustrated Shastry-Sutherland-type magnetic spin system revealed in RB_4 (R = Dy, Er) [13,14] and the complex magnetic structure discovered in the dodecaborides [15]. These compounds are all good metals in the case of trivalent rare earth elements, and their magnetic coupling has basically been described by the Ruderman-Kittel-Kasuya-Yoshida (RKKY) mechanism with possible secondary effects from dipole-dipole interaction and frustration.

As an emerging novel phenomenon, it has been found that borides, which contain the B_{12} icosahedra as a structural building block, can exhibit unexpectedly strong magnetic interactions, although they are relatively magnetically dilute, f-electron insulators [1]. These compounds will be discussed in the next subsections.

3.1.1 RB_{50}-type compounds

The TbB_{50} compound was discovered to exhibit the first magnetic transition in the higher borides [2]. TbB_{50} has a transition temperature of $T_N = 17$ K, which is surprisingly high considering that these are magnetically dilute insulating f-electron compounds. Investigation of the rare earth series indicates that the magnetic coupling mechanism is different from that of conventional f-electron systems [1]. It has been proposed that the B_{12} icosahedron is playing a role to mediate the interaction [1-4]. REB_{50} and isostructural $REB_{44}Si_2$ compounds were also discovered to have similar transitions.

The invariance of the critical magnetic field with regards to non-magnetic dilution, indicated that the transition is dimer-like with magnetic ions forming pairs [4]. Electron spin resonance (ESR) results support one dimensionality of the system which indicates that magnetic dimer pairs are formed along a bond alternating chain parallel to the *c*-axis [16].

To summarize, RB_{50}-type compounds are found to be one dimensional dimer-like magnetic systems with surprisingly strong magnetic interaction for a dilute, localized f-electron insulator.

3.1.2 $REB_{17}CN$, $REB_{22}C_2N$, and $REB_{28.5}C_4$

$REB_{22}C_2N$ was found to exhibit irreversible magnetic behavior [17]. For example, as a characteristic physical property, the relaxation of isothermal remanent magnetization I_m following the stretched exponential dependence was observed (figure 1):

$$I_m \propto \exp[-C(\omega t)^{-(1-n)}/(1-n)]$$

(e.g., $HoB_{22}C_2N$, C = 0.033, 1-n = 0.10). Furthermore, the existence of wait time effects was also observed in the thermal remanent magnetization σ_{TRM} as can be seen in figure 2 which is another strong indication of a spin glass system.

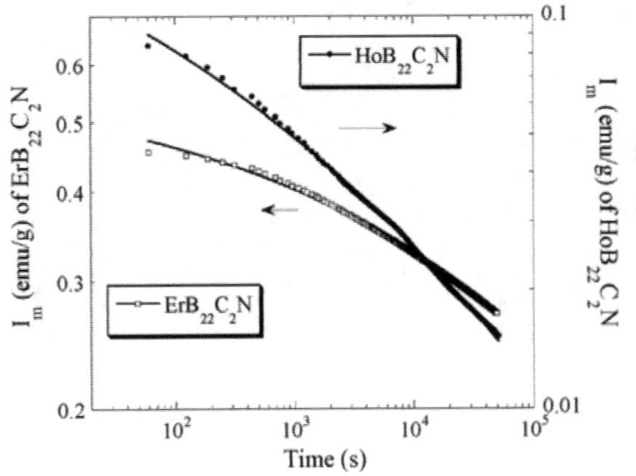

Figure 1. Isothermal remanent magnetization of $REB_{22}C_2N$.

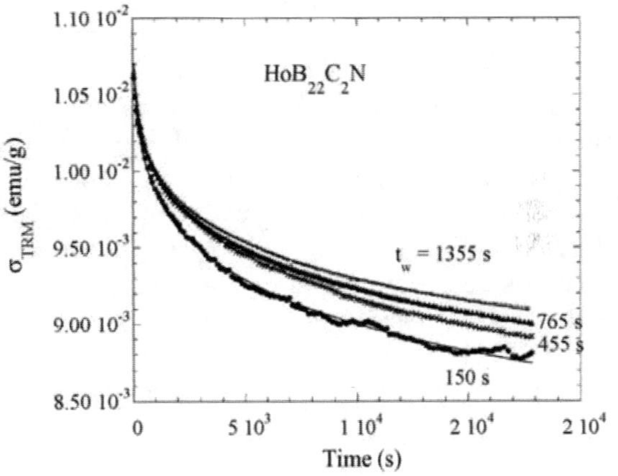

Figure 2. Thermal remanent magnetization of $HoB_{22}C_2N$ for different wait times.

Further investigations established $REB_{22}C_2N$ as the first spin glass system observed for a non-doped boride [3]. The uniqueness of this system exhibiting spin glass behavior without doping among all the borides was laid down to strong **frustration** in the system originating from the triangular configuration of the rare earth atoms.

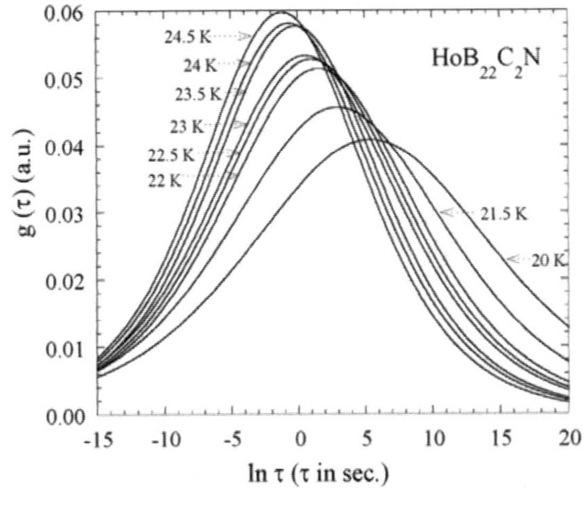

Figure 3. Relaxation time distribution of $HoB_{22}C_2N$.

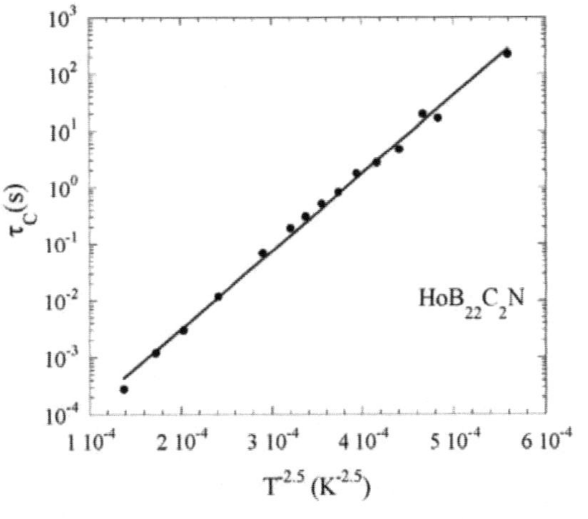

Figure 4. Temperature dependence of τ_C.

Strong frequency dependence of the ac susceptibility was found, but could not be analyzed satisfactorily by the dynamical scaling theory of a three dimensional spin glass. A more detailed investigation of the relaxation behavior was carried out by a Cole-Cole analysis on the frequency dependence of the in-phase component χ' and out-of-phase component χ'' of the ac susceptibility. A relaxation time distribution was determined as shown in figure 3. The temperature dependence of the median relaxation time τ_C could be described well in terms of a generalized Arrhenius law (figure 4);

$$\ln(\tau_C/\tau_0) \propto T^{-(1+\phi\nu)}, \quad \tau_0 = 5.3\text{x}10^{-6}\ \text{s}, \quad \mathbf{1+\phi\nu = 2.5}$$

The relaxation exponent $1+\phi\nu$ takes a value close to 2 and together with the temperature dependence observed for the width parameter of the relaxation time distributions it was demonstrated that this was not a super paramagnetic system (which also exhibits large frequency dependence) but a **new 2 dimensional spin glass** system [3]. Actually, it can be considered to be a dilute triangular lattice magnetic system. It was indicated that interaction within the triangular layers was dominant, despite it being the second nearest separation with significantly longer distances. This matches perfectly our theory of the B_{12} icosahedra mediating the magnetic interaction.

Analogous magnetic behavior was observed for the homologous series of $REB_{17}CN$, $REB_{22}C_2N$, and $REB_{28.5}C_4$ and it was indicated that the magnetism is governed in the two dimensional metal layers [18]. However, in this paper, we consider this in more detail. The structure of the homologous series is plotted in figure 5. The compounds have a layered structure along the c-axis with B_{12} icosahedra and C-B-C chain layers residing in between B_6 octahedral and rare earth atomic layers. The B_{12} icosahedra and C-B-C chain layers increase successfully along the series of $REB_{17}CN$, $REB_{22}C_2N$, and $REB_{28.5}C_4$. In the limit of the boron icosahedra layers going to infinity, the compound is actually analogous to boron carbide.

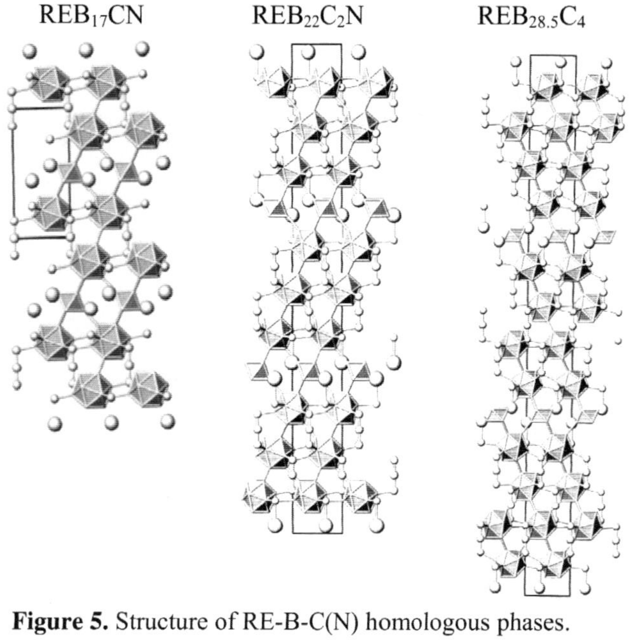

Figure 5. Structure of RE-B-C(N) homologous phases.

One difference observed between the compounds in regard to the amount of boron icosahedra and C-B-C chain layers was that the cusp temperature of the zero field cooled susceptibility T*, decreased monotonically as the number of boron icosahedra and C-B-C chain layers were increased. E.g., T* = 28.8 K, 22.5 K, and 19.3 K for $HoB_{15.5}CN$, $HoB_{22}C_2N$, and $HoB_{28.5}C_4T$, respectively. This was attributed in ref. 18 as possibly due to the change in spacing of the rare earth atoms. However, the absolute differences in the lattice constant *a(b)* are only about 0.5% and in any imaginable functions of magnetic interaction, this would be too small to explain the large differences in T* via interaction only in *ab* plane. Therefore, it is indicated that some correlation or effect exists beyond the rare earth layers. As noted above, spin glass behavior was thought to be caused by disorder from partial occupancies of the rare earth sites and frustration of magnetic interaction. The origin of disorder in these structures can also be surmised to come from their layered stacking nature. The more stacking layers the polytype member of homologous series has, the more frequent are mistakes in layer sequences and disorder in the structure. This means that the disorder becomes stronger as the boron icosahedra and C-B-C chain layers increase along the series, and this serves to depress T* more than would be expected from just a simple change in the magnetic interaction in the rare earth layer. In this way the observed anomalous dependency can be explained.

3.1.3 $GdB_{18}Si_5$

A three-dimensional long-range order antiferromagnetic transition was discovered to occur in $GdB_{18}Si_5$. This is interesting because it is the only long range ordering observed for the higher borides and occurs despite the relatively low partial occupancy of the gadolinium sites [19]. The details of the magnetic ordering are under investigation at present.

An interesting dependence at low magnetic fields (applied along the *c*-axis) was observed when the field was varied in-plane, with a reorientation of the spin, a spin flip, appearing to occur at fields below 300 G. The origin of this behavior of course can be attributed to the round spin of the gadolinium ions, but the actual observation of a spin flip at such low magnetic fields is unusual and interesting [20].

3.2. Thermoelectric properties of novel higher borides.

3.2.1 $REB_{17}CN$, $REB_{22}C_2N$, and $REB_{28.5}C_4$

As noted in the introduction, the n-type RE-B-C(N) compounds are generating excitement because it may be developed as the long awaited high temperature thermoelectric counterpart to p-type boron carbide [9,10]. After the attractive thermoelectric properties of boron carbide were discovered, various rigorous efforts were made over the years however it was not possible to obtain similar n-type materials, unless through doping, for example, β-boron by extremely large amounts of transition metals like vanadium [21,22]. Boron carbide doped with nickel has also been reported to show n-type but only at a limited temperature range, and not at high temperatures [23]. Asides from the boron icosahedra cluster compounds, we note that divalent alkaline-earth hexaborides like CaB6 have been found to exhibit n-type behavior as first shown by Paderno *et al* [24,25] and Etourneau *et al* [26] (recently other groups have also reported and further investigated this behavior [27]). However, they are a different class of compounds compared to the boron icosahedra compounds, since they are basically dirty metals/small gap semiconductors with relatively large thermal conductivity. The n-type behavior is not unexpected, especially for inadvertently doped samples.

As noted above about the crystal structure of the RE-B-C(N) series of compounds, in the limit of the boron icosahedra layers going to infinity, the compound is actually analogous to boron carbide. This similarity of the structural blocks is also why we consider them to be compatible n-type counterparts to boron carbide which has been established to be an excellent p-type high temperature thermoelectric material. Recently, the origin of the anomalous n-type behavior is also becoming clearer [28].

These compounds will not melt stably and therefore, large crystals cannot be grown so sintered powders are compacted in various ways for measurements. Conventional hot press methods and cold pressing and annealing methods yield samples with densities of only around 50% of the theoretical value. Spark plasma sintering (SPS) has been found to easily increase the density to ~70% [10]. The morphologies of a typical hot pressed and SPS-prepared RE-B-C(N) sample are shown in figure 6.

The SPS sample shows an increase of the grain size compared to hot pressed sample. A particularly interesting feature is that the grains in the SPS sample appear very thin and plate-like. Considering the anisotropic structure of the compound (figure 5) we speculate that the growth is along the in-plane directions of the layered structure (the *ab* plane).

With such densification by SPS there is large improvement in electrical conductivity with resistivities decreasing by close to 2 orders of magnitude (figure 7). Accompanying the decrease in resistivity, the absolute values of the Seebeck coefficients also show small decreases, but the effect is much smaller than the resistivity improvement. In fact, the power factor at 1000 K increases by close to 2 orders.

The thermal conductivities of two SPS RE-B-C(N) samples (both approximately 70% of theoretical density) are compared with conventionally hot pressed samples as shown in figure 7. There is a sizable increase of the measured thermal conductivity compared to the hot pressed samples, however, this

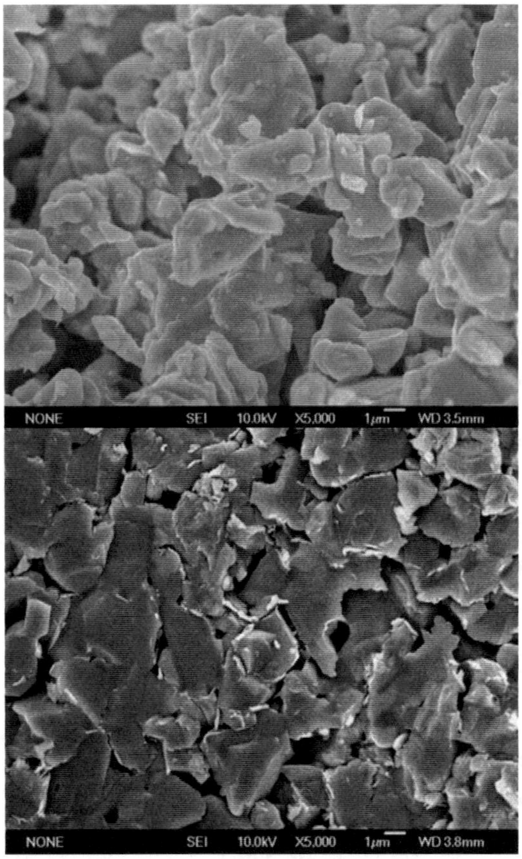

Figure 6. SEM pictures of hot pressed and SPS $HoB_{17}CN$.

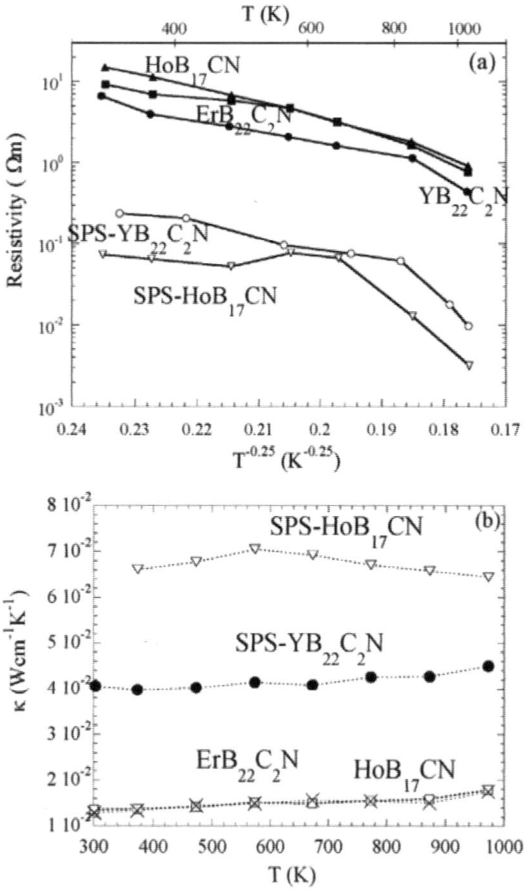

Figure 7. Resistivity and thermal conductivity of hot pressed and SPS RE-B-C(N) samples.

cannot be attributed simply to the effect of simple densification, because the two SPS samples have similar densities (in regard to their theoretical densities).

SPS-$HoB_{17}CN$ actually exhibits almost 1.5 times the thermal conductivity of SPS-$YB_{22}C_2N$. From this result, the following can be surmised. First of all, it should be assumed that there is a very large anisotropy in the thermal conductivity of the rare earth boron carbonitrides. This is reasonable because of the crystal structure as noted before, and a hypothesis can be made that the thermal conductivity is low in the *c*-axis direction through the boron cluster layers, while being relatively high in-plane within the rare earth layers. As the grain size increases through SPS along this relatively high thermal conductivity direction and the sample becomes less "uniform" (more anisotropic), it becomes easier for significant differences in the thermal conductivity to manifest for different samples.

In such anisotropic systems, control of the nano/microstructure of the samples can be a powerful tool to enhance or tune the thermoelectric properties. Further investigations should continue into the processing of this promising system which is hoped to be a viable n-type counterpart to boron carbide, which is a well established high temperature p-type material.

A striking result was obtained when small amounts of metal borides were doped into samples [10]. The densities of the doped and nondoped samples are all similar, being ~50% of theoretical density, but a significant decrease up to 2 orders in the resistivity, is obtained through the doping. The Seebeck coefficient and thermal conductivities do not show a large change, so this doping method is revealed to be an even more effective way to increase the figure of merit compared to simple densification. Since the percent of doping is small, it is unlikely that a percolation effect is the origin for the large reduction in resistivity, where the metal boride particles provide connecting channels of high electrical conductivity.

The investigation of the morphology of the samples (figure 8) reveals the origin of the behavior. The metal doped samples obviously have larger grain size. Therefore, it is indicated that the presence of metal borides appear to promote the growth of larger grains in the RE-B-C(N) samples. We note that compared to typical compounds in which in some cases it is preferred to inhibit grain growth so as

Figure 8. SEM pictures of $YB_{22}C_2N$ (left) and $ErB_{22}C_2N$:ErB_4 (right).

to depress thermal conductivity, in the case of the boron cluster compounds, they are compounds which typically have intrinsic low thermal conductivity. Therefore, the grain growth is not necessary a detrimental effect when the overall effect is taken into account.

As noted above, control of the morphology of anisotropic compounds is a very important aspect to control their properties. The small addition of metal borides has been shown to change the morphology and this effect will be investigated in detail further.

3.3. Building defects in layered borides

The rare-earth metal aluminoboride system $R$$AlB_4$ has been attracting increasing attention with recent discoveries. Multiple magnetic transitions were reported to occur in $TmAlB_4$ at low temperatures below an antiferromagnetic transition temperature T_N [29] while crystals with two different structure types were reportedly obtained from the same flux for $YbAlB_4$ and $LuAlB_4$ [30]. $TmAlB_4$ takes a well-known $YCrB_4$-type structure (space group *Pbam*) [31], and we have found that X-ray diffraction measurements point to an intrinsic tiling variation existing in the crystals [32] due to the presence of fragments of the closely related $ThMoB_4$-type structure (space group *Cmmm*) [33]. Both structures have similarities to the AlB_2-type structure, with a planar boron network build of pentagonal and heptagonal rings having differently sized rare-earth and aluminum atoms. The difference between the $YCrB_4$-type (α-type) and $ThMoB_4$-type (β-type) structures is in the orientation of the pairs of condensed pentagonal rings. The previous work indicates that the existence of an intrinsic intergrown nanostructure influences physical properties and is the origin of multiple magnetic anomalies which are observed in $TmAlB_4$. This is striking because there are more than a hundred compounds reported with this type of structure, and it is possible that such indicated phenomenon of the intrinsic tiling variation is actually ubiquitous.

4. Conclusions

Interesting and striking physical properties have been discovered for novel rare earth higher borides. Topics covered in this paper deal with the magnetic behavior and unexpectedly strong coupling discovered for rare earth borides containing the B_{12} icosahedra. It is newly pointed out that in the homologous series of two dimensional spin glasses, $REB_{17}CN$, $REB_{22}C_2N$, and $REB_{28.5}C_4$, contrary to what was assumed previously, it is indicated that some correlation or effect exists beyond the rare earth and octahedral layer. Stronger disorder in the polytypes with more boron icosahedra layer stacking can cause a more rapid depression of T* than normally expected.

The potential of novel rare earth borides as mid to high temperature thermoelectric materials is also a topic of interest. Morphology is indicated to play a large role in affecting the thermoelectric properties. The increase in the power factor of rare earth borocarbides doped with metal borides which was previously observed is indicated to originate from an increase in the grain growth.

Finally, the phenomena of building defects observed in layered borides and their effect on physical properties were briefly reviewed.

Acknowledgments

Dr. T. Nishimura is thanked greatly for his help with some of the experiments.

References

[1] Mori T 2008 *Handbook on the Physics and Chemistry of Rare Earths* vol 38, ed K A Gschneidner Jr, J C Bunzli and V Pecharsky V (Amsterdam: Elsevier) p 105
[2] Mori T and Tanaka T 1999 *J. Phys. Soc. Jpn.* **68** 2033
[3] Mori T and Mamiya H 2003 *Phys. Rev.* B **68** 214422
[4] Mori T 2004 *J. Appl. Phys.* **95** 7204
[5] Slack G A, Oliver D W and Horn F H 1971 *Phys. Rev.* B **4** 1714
[6] Cahill D G, Fischer H E, Watson S K, Pohl R O and Slack G A 1989 *Phys. Rev.* B **40** 3254
[7] Wood C, Emin D and Gray PE 1985 *Phys. Rev.* B **31** 6811
[8] Mori T, Martin J and Nolas G 2007 *J. Appl. Phys.* **102** 073510
[9] Mori T and Nishimura T 2006 *J. Solid State Chem.* **179** 2908
[10] Mori T, Nishimura T, Yamaura K and Takayama-Muromachi E 2007 *J. Appl. Phys.* **101** 093714
[11] Wood C and Emin D 1984 *Phys. Rev.* B **29** 4582
[12] e.g. Gignoux D and Schmitt D 1997 *Handbook of Magnetic Materials* vol 10, ed K H J Buschow (Amsterdam: Elsevier) p 239
[13] Watanuki R, Sato G, Suzuki K, Ishihara M, Yanagisawa T, Nemoto Y and Goto T 2005 *J. Phys. Soc. Jpn.* **74** 2169
[14] Michimura S, Shigekawa A, Iga F, Sera M, Takabatake T, Ohoyama K and Okabe Y 2006 *Physica* B **378-380** 596
[15] Siemensmeyer K, Flachbart K, Gabani S, Mat'as S, Paderno Y, and Shitsevalova N 2006 *J. Solid State Chem.* **179** 2748
[16] Mori T 2006 *J. Appl. Phys.* **99** 08J309
[17] Mori T and Leithe-Jasper A 2002 *Phys. Rev.* B **66** 214419
[18] Mori T, Zhang F and Leithe-Jasper A 2004 *J. Solid State Chem.* **177** 444

[19] Mori T and Zhang F 2002 *J. Phys.: Condens. Matter* **14** 11831
[20] Mori T 2005 *Polyhedron* **24** 2803
[21] Werheit H, deGroot K, Malkemper W and Lundstrom T 1981 *J. Less Common Met.* **82** 163
[22] Slack G A, Rosolowski J H, Hejna C, Garbauskas M and Kasper J S 1987 *Proc. 9th Int. Symp. Boron, Borides and Rel. Comp., University of Duisburg, Germany*, ed H Werheit, University of Duisburg: Duisburg, p 132
[23] Liu C H 2001 *Materials Letters* **49** 308
[24] Paderno Yu B 1969 *Poroshk. Metallurgija* **11** 70
[25] Paderno Yu B 1970 *Electron. Technol.* **3** 175
[26] Etourneau J 1970 *J. Solid State Chem.* **2** 332
[27] e.g., Takeda M, Terui M, Takahashi N and Ueda N 2006 *J. Solid State Chem.* **179** 2823
[28] Mori T, Nishimura T, Budnyk S, Schnelle W, Burkhardt U and Grin Yu to be published
[29] Mori T, Okada S and Kudou K 2005 *J. Appl. Phys.* **97** 10A910
[30] Macaluso R T, Nakatsuji S, Kuga K, Thomas E L, Machida Y, Maeno Y, Fisk Z and Chan J Y 2007 *Chem. Mater.* **19** 1918
[31] Kuz'ma Y 1970 *Sov. Phys. Crystallogr.* **15** 312
[32] Mori T, Borrmann H, Okada S, Kudou K, Leithe-Jasper A, Burkhardt U and Grin Yu 2007 *Phys. Rev.* B **76** 064404
[33] Rogl P and Nowotny H 1974 *Monatsh. Chem.* **105** 1082

Reliable measurement of Seebeck coefficient in semiconductors

Helmut Werheit, Udo Kuhlmann, Bernhard Herstell and Walter Winkelbauer

Experimental Physics, University Duisburg-Essen, D-47048 Duisburg, Germany

helmut.werheit@koeln.de and helmut.werheit@uni-duisburg-essen.de

Abstract. Reliable measurement of the Seebeck coefficient requires painstaking precautions. Unfortunately, these requirements are not met even in commercially available equipment, and therefore the obtained data often contain systematic errors. We describe equipment for reliable Seebeck measurements and present results on YB_{66} and $B_{4.3}C$ boron carbide as an example.

1. Introduction

The temperature gradient δT in a solid implies a gradient in the density of mobile carriers, and accordingly, an electric potential gradient δV. Their relation yields the Seebeck coefficient $S = \delta V/\delta T$. Preliminary information on this material parameter, in particular on its sign, can be easily obtained using simple probe tips held at different temperatures. In contrast to the Hall effect, the sign of the Seebeck effect unambiguously corresponds to the prevailing type of charge carriers, irrespective of the mechanism of their transport. However, painstaking precautions must be taken if reliable values of the Seebeck coefficient are required, e.g., for quantitative characterization of charge transport. Unfortunately, even some commercially available equipment does not meet these requirements and hence introduces systematic errors [1-3].

The following requirements must be considered in reliable Seebeck measurements:

- Temperature and potential must be taken at the same point inside the sample. Mechanically attached probes cause a temperature jump between the probe and sample.
- The high Seebeck effect of semiconductors is particularly important for elucidating the electronic transport or for technical applications. Pressing metal probes creates Schottky barriers, whereas only ohmic contacts are appropriate for semiconductor measurements.
- Contact potentials occur at any junction between conducting solids. They are inevitable even in soldered probes. Theoretically, sign and value of contact potential depend on the position of the Fermi levels, potentially shifted or bended near the contact. In practice, the conditions are much more complicated, because there are additional, hardly controllable surface effects. Reliable Seebeck measurements require elimination of the contact potentials.
- In general, the Seebeck coefficient S depends on temperature. Accordingly, the measured potential difference δV usually varies nonlinearly with δT. Therefore, $S(T_0)$ must be determined for $S(T)_{T \to T_0}$

2. Problems of Seebeck effect measurements

Usually the Seebeck coefficient S = ΔV/ΔT is determined measuring V(T) at two temperatures. For this case, figure 1 demonstrates the possible influence of the contact potential; linear variation of S(T) presumed. As shown, even the sign of S can be incorrect at unfavorable conditions.

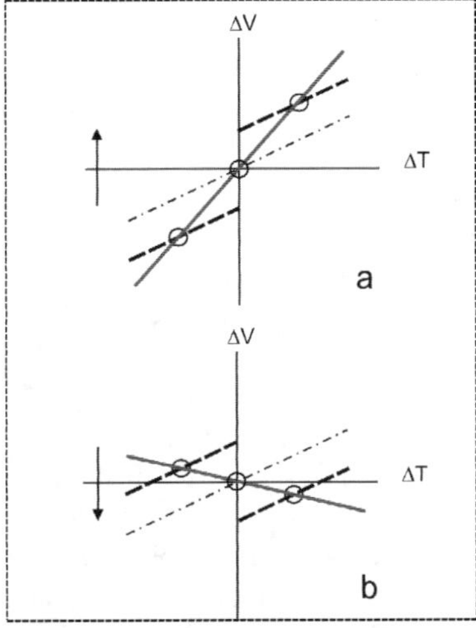

Figure 1. 1. Possible effects of the contact potential ΔU if S = ΔV/ΔT is obtained from measurements at two different temperatures. (a) positive contact potential; (b) negative contact potential. Circles correspond to the measured data; arrows to contact potential ΔU; broken lines to Seebeck effect curves shifted by ΔU; solid red line to false Seebeck effect and dash-dotted line to the true Seebeck measurements unaffected by the contact potential.

In most cases, the Seebeck coefficient depends on temperature. Then, when the measurements are taken at two temperatures only, another problem arises. This is demonstrated in figure 2. Choosing small ΔT, would solve this problem, but then the ΔV becomes too small for reliable measurement.

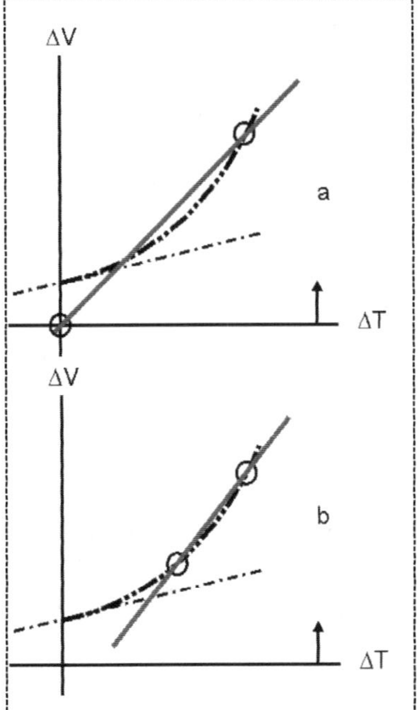

Figure 2. Problems in obtaining temperature dependent Seebeck coefficients by measuring at two different temperatures. (a) $T = T_0$, $T > T_0$; (b) $T_1 > T_0$, $T_2 > T_0$. Circles correspond to the experimental data; arrows to contact potential ΔU; dash-double-dotted black line to real Seebeck effect shifted by ΔU; solid red line to false Seebeck effect and dash-dotted line to the true Seebeck effect at T_0 without contact potential.

Therefore, measurements at two temperatures are not sufficient for determining qualitatively and quantitatively reliable Seebeck coefficients. As demonstrated in figure 3, more or less continuous data $U_{Seebeck}$ (±T) are needed for deriving the accurate asymptote representing $S(T_0) = \lim (\Delta V/\Delta T)$ for

$T \rightarrow T_0$. A reversal of the sign of ΔT is useful.

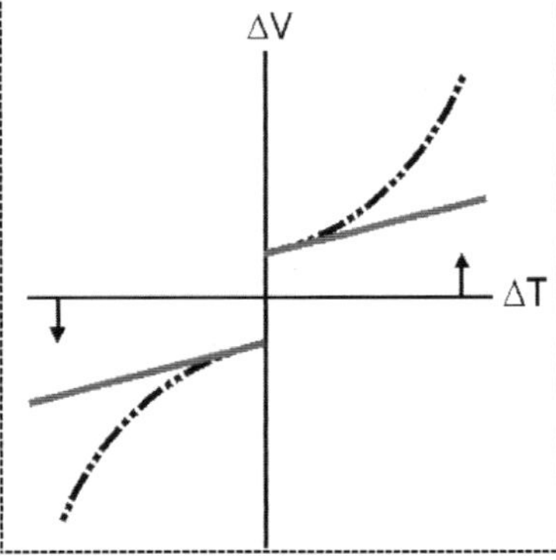

Figure 3. A reliable Seebeck measurement. Black dash-double dotted curve is measured thermoelectric voltage V(T); arrows present contact potential ΔU and solid red line an asymptote to the black curve at T_0.

3. Equipment for reliable Seebeck measurements

The realization of a quasi-continuous curve $\Delta V(\Delta T)$ requires a dense sequence of simultaneously measured data for ΔV and ΔT tapped at exactly the same measuring points within the sample. Accordingly, the same probes must be used for both data.

We report on equipment for measuring Seebeck coefficients of boron-rich solids exemplary for high-ohmic semiconductors. Temperatures and potentials are determined at the very same small probes with ohmic contacts in samples with resistances up to 10^{12} Ω. Thermocouples of suitable metal combinations were used as probes. They were affixed to the sample by capacitor discharge. In this way, ohmic Cu and Pt contacts were realized to boron-rich solids.

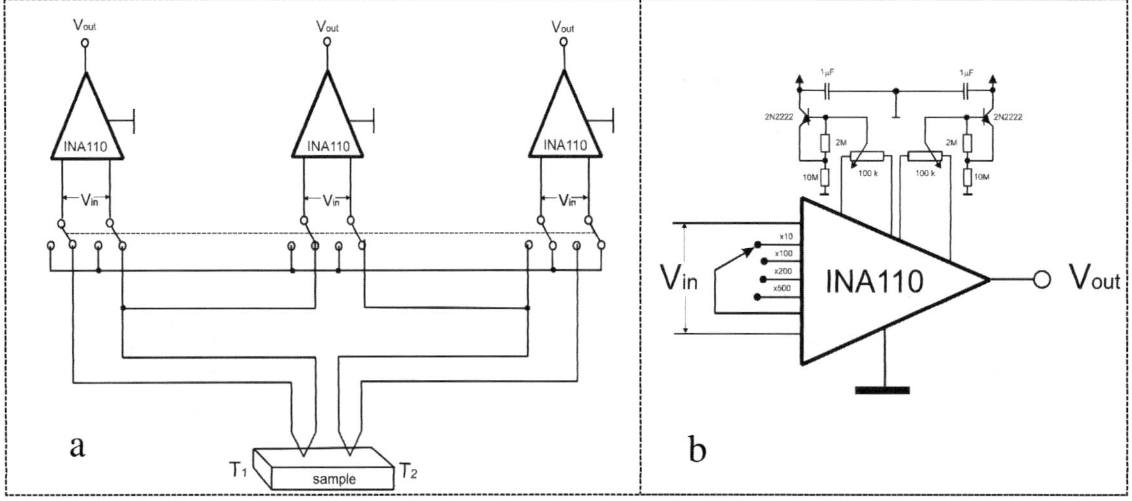

Figure 4. Electronic circuits for measuring Seebeck coefficients. (**a**) complete circuit; T_1, T_2 are temperatures applied to the sample; V_{in}, V_{out} are the input and output voltages, respectively, of the INA 110 amplifier. (**b**) wiring scheme of INA 110 amplifier (Burr-Brown corporation)

The electronic circuits are displayed in figure 4. The electrical leads for measuring temperature via thermocouples and for measuring the thermoelectric voltage must be electrically isolated from one another, in spite of using the same probes. This was realized with FET amplifiers INA 110 (Burr-Brown Corporation) having input resistances of 10^{15} Ohms. Very important was the small drift of this amplifier (2 μV/K only), largely avoiding the compensation of offset voltages after the equipment had reached thermal equilibrium (i.e., after about two hours). The lowest amplifier stage (10×) was sufficient for obtaining reliable data. The simultaneous recording of data was enabled by trigger impulses initiating the measuring devices integrated into a bus system.

The data recorded in a dense sequence were edited in a computer, were the calibration curves of the thermocouples had been approximated by high-order polynoms. The edited data were displayed on the computer screen, similar to figure 3, during several cycles of increasing and decreasing temperatures in the range ±1 to 3 K around the relevant temperature.

The central temperature was controlled in a thermostat, and the heating and cooling was provided by small internal heaters having small heat capacity. All the relevant data were displayed on the computer screen as well. This way the whole procedure was supervised, contact voltages were discerned, and reliable values for the Seebeck coefficient could easily be determined.

Pt/PtRh thermocouples are usually usable below ~2030 K. However, in boron-rich solids, a boron-platinum eutectic is formed at ~1070 K, destroying the thermocouple and modifying the sample. Hence, for measuring the Seebeck coefficient of boron carbide (melting point ~2720 K) up to about 2000 K, a modification of the equipment was required.

For determining the temperatures at the sample, we used the boron carbide itself as thermistor, whose temperature-dependent resistivity was thoroughly calibrated. Tungsten tips were used as contacts. As an example, the Seebeck coefficient of boron carbide was measured up to ~2000 K. This temperature limit was due to decomposition of the hexagonal BN sample holder.

4. Examples of measurement

The high input resistance of the amplifier allows measurements even on highly-resistive samples. As an example, in figure 5 the Seebeck coefficient of YB_{66} is displayed, whose resistivity at low temperatures is about $10^9\ \Omega^{-1}cm^{-1}$[4].

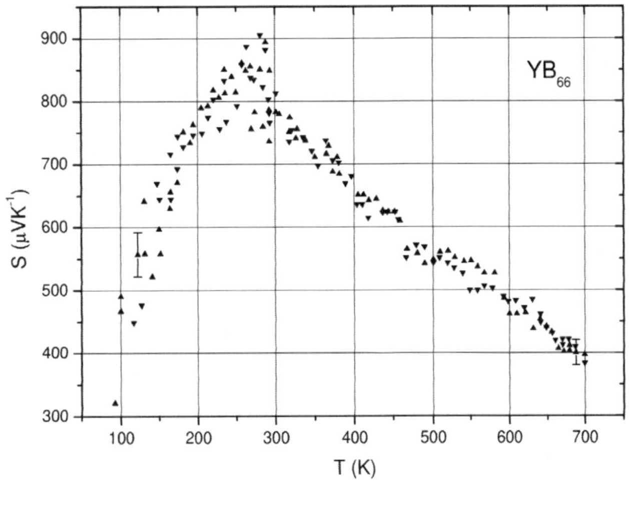

Figure 5. Seebeck coefficient of YB_{66}.; ▲ T increasing; ▼ T decreasing.

Figure 6 presents the results obtained on $B_{4.3}C$ boron carbide by both methods described above (for moderate and high temperatures, respectively). In spite of a somewhat greater error margin, the Seebeck data obtained by the modified high-temperature method satisfactorily fit to the data measured with the method for moderate temperatures. For more details of the Seebeck coefficient of boron carbide see [5,6], and references therein. The measured prismatic samples had ~12 mm length and 4 mm thickness. The probe distance was ~8 mm and the probe size was <1 mm.

Journal of Physics: Conference Series **176** (2009) 012037 doi:10.1088/1742-6596/176/1/012037

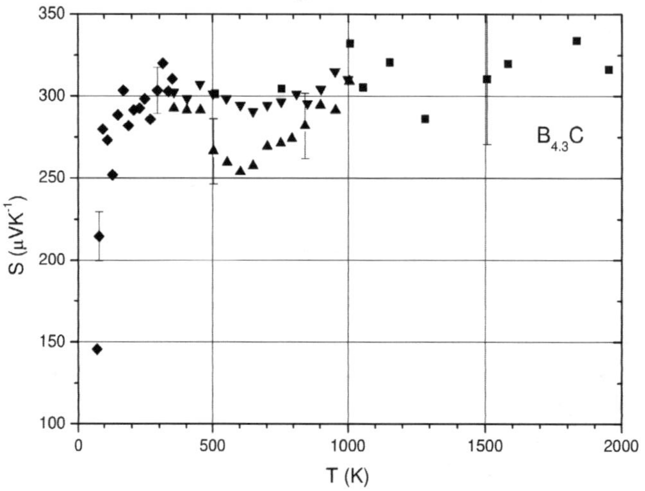

Figure 6. Seebeck coefficient of $B_{4.3}C$ boron carbide. ◆, method as described above, T = 75 - 350 K; ▲, method as described above, T = 350 - 1000 K, T increasing; ▼, method as described above, T = 350 - 1000 K, T decreasing; ■ modified method for high T, T = 450 - 1950 K

5. Summary

We have described a method of reliable measurement of the Seebeck coefficients in semiconductors, including high-resistivity materials. The method avoids systematic errors caused by contact potential, irrespective of its sign and quantity. Electric potential and temperatures are obtained at the very same probes, and therefore their distance does not need to be measured.

References

[1] Boffoué O, Jacquot A, Dauscher A, Lenoir B and Stölzer M 2005 *Rev. Sci. Instr.* **76** 053907

[2] Burkov A T, Heinrich A, Konstantinov P P, Nakama T and Yagasaki K 2001 *Meas. Sci. Technol.* **12** 264

[3] Jacquot A, König J and Böttner H 2006 *Proc. 25th Int. Conf. Thermoelectrics* ed P Rogl (Piscataway, NJ: IEEE, Catalog Nr. 06TH8931) p 184

[4] Werheit, H., Kuhlmann, U and Tanaka, T 1991 *Boron-Rich Solids, Proc. 10th Int. Symp. Boron, Borides Rel. Compds.*; *AIP Conf. Proc. 231* eds D Emin, T L Aselage, A C Switendick, B Morosin and C L Beckel (AIP: New York) p 125

[5] H. Werheit 2000 *Boron Compounds, in Landolt-Börnstein, Numerical Data and functional relationships in Science and Technology Group III* Vol. 41D, ed O Madelung (Springer, Berlin) p 1

[6] H. Werheit 2006 *Proc. 25th Int. Conf. Thermoelectrics* ed P Rogl (Piscataway, NJ: IEEE, Catalog Nr. 06TH8931) p 159

Effect of pressure on the anomalous magnetoresistance and antiferromagnetism of single crystal GdB_4

Gendo Oomi[1], Masashi Ohashi[2] and Beongki Cho[3]

[1] Department of Physics, Kyushu University, 4-2-1 Ropponmatsu, Fukuoka 810-8560, Japan

[2] Faculty of Environment Design, Kanazawa University, Kakuma-machi, Kanazawa 920-1192, Japan

[3] Center for Frontier Materials and Department of Materials Science and Engineering, Gwangju Institute of Science and Technology (GIST), Gwangju 500-712, South Korea

E-mail : oomi@rc.kyushu-u.ac.jp

Abstract. Electrical resistance of single-crystalline GdB_4 has been measured at high pressures and high magnetic fields. The antiferromagnetic ordering temperature T_N increases with pressure at a rate 0.3 K/GPa, but the hidden transition temperature T_a is almost pressure independent. The magnitude of magnetoresistance at 4.2 K (about 6000% at 8.5 T) is not affected by applying pressures below 3 GPa. This fact suggests similar origins of the hidden transition and of the extremely large magnetoresistance.

1. Introduction

Materials exhibiting extremely large magnetoresistance (MR) have been extensively studied because of the wide range of applications [1]. The origins of large MR (so-called giant magnetoresistance, GMR, or tunneling magnetoresistance, TMR) are studied both experimentally and theoretically. In these studies, high pressure is well known as an efficient tool because it significantly affects the transport properties [2]. Recently, we have reported that the GMRs of Fe/Cr magnetic multilayers and the TMR of Co-Al-O granular films are enhanced by pressure [3,4]. In order to understand GMR and TMR phenomena, it is worthwhile to document pressure effects on other large-MR materials.

The tetraborides RB_4 (R: rare earth elements) having tetragonal crystal structure have been investigated mainly because of potential applications, such as n-type thermionic emitter. These materials show magnetic ordering due to the RKKY interaction [5]. Recently, extraordinary large positive MR, as much as several thousand per cent, has been reported in antiferromagnetic GdB_4 (T_N = 41 K) [6], and its origin is yet unclear. The GMR or TMR in many systems, such as magnetic multilayers or granular materials, strongly depends on the spin-dependent scattering of conduction electrons. It is important to test whether or not this is true for GdB_4.

In this work, we measured electrical resistance of single-crystalline GdB_4 under high pressures and high magnetic fields in order to examine the relation between the large MR and the stability of antiferromagnetism at high pressure.

2. Experimental details

Single crystal of GdB_4 was prepared by solution growth method [6]. The residual resistivity ratio (RRR) was as high as 200. Pressures up to 3 GPa were generated using a piston-cylinder device. The pressure was automatically kept constant throughout the measurements. The details of the high-pressure apparatus were reported previously [7]. Electrical resistance was measured with a conventional four-probe method at constant pressure. The magnetoresistance was measured at 4.2 K by applying the magnetic fields up to 9 T in a superconducting magnet.

3. Results and discussion

Figure 1 shows the temperature dependence of electrical resistivity $\rho(T)$ at 3 GPa; ρ decreases with decreasing temperature and shows a cusp-like anomaly around 41 K(= T_N) due to antiferromagnetic ordering. The shape of the $\rho(T)$ curve is not affected much by pressure. In the inset of figure 1, $\rho(T)$ at 0.1 and 3.0 GPa is shown in the narrow temperature range $35 < T < 50$ K. The sudden decrease at T_N is clearly seen, and T_N increases with pressure. In order to deduce the T_N value, we differentiated $\rho(T)$ with respect to T as shown in figure 2. There are two anomalies in the temperature dependence of $d\rho(T)/dT$ at T_N and T_a (~10 K). The anomaly at T_N is sharper than that at T_a, which was also observed in the temperature dependence of the heat capacity, but less clear than that at T_N.

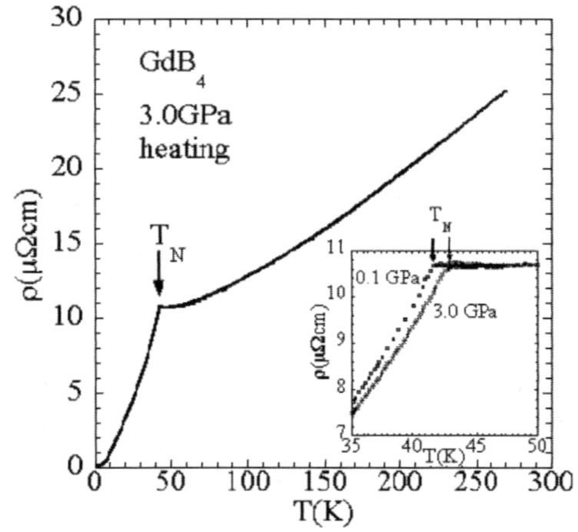

Figure 1. Electrical resistivity of GdB_4 at 3 GPa. Inset zooms into $\rho(T)$ near T_N.

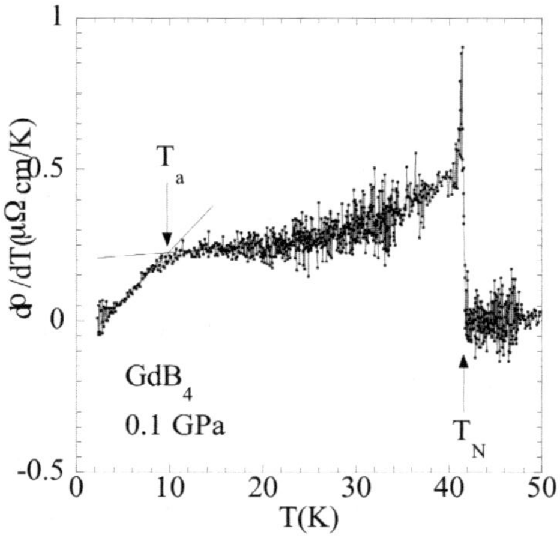

Figure 2. First derivative of $\rho(T)$ as a function of temperature.

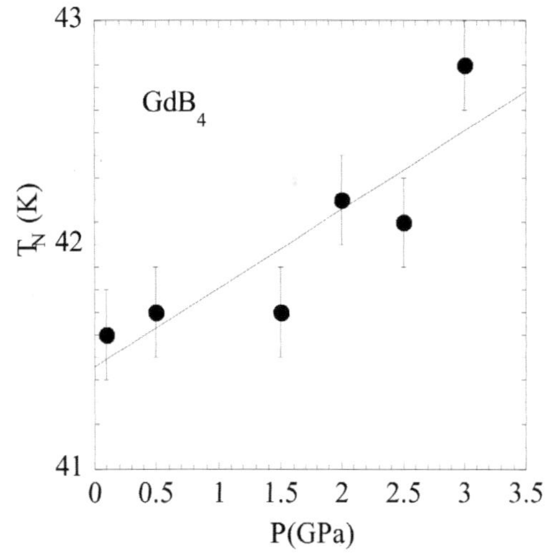

Figure 3. Antiferromagnetic ordering temperature T_N as a function of pressure.

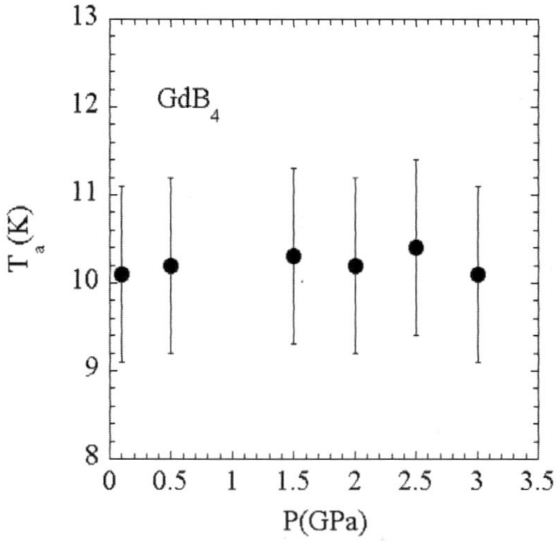

Figure 4. Hidden transition temperature T_a as a function of pressure.

Journal of Physics: Conference Series **176** (2009) 012038 doi:10.1088/1742-6596/176/1/012038

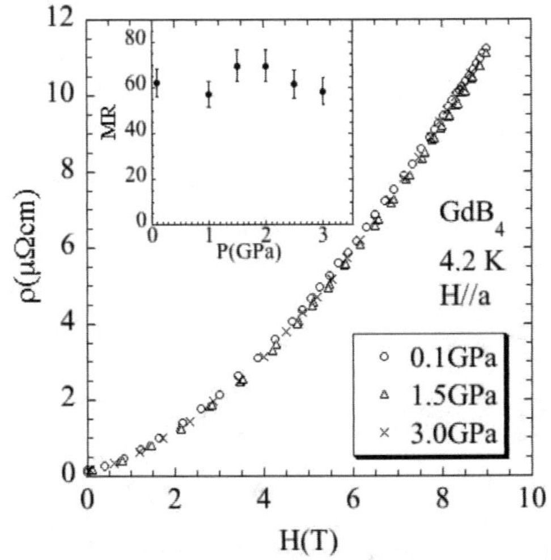

Figure 5. Electrical resistivity at high pressures as a function of *H*. The inset shows MR at 8.5 T.

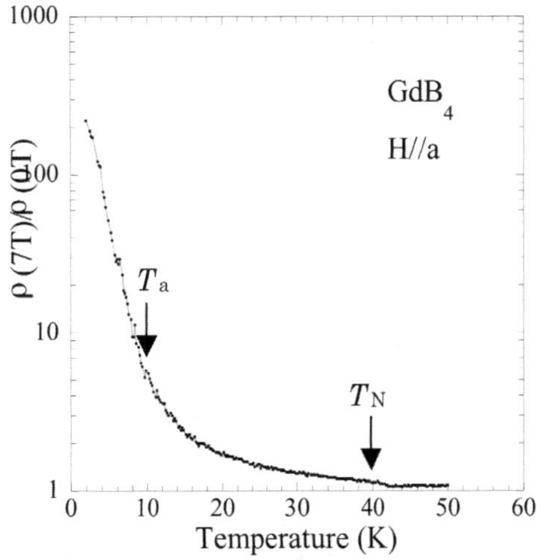

Figure 6.Temperature dependence of MR at 7 T at ambient pressure. T_a and T_N are marked by arrows.

Below temperature T_a, $\rho(T)$ shows T^2 dependence with the coefficient of 47×10^{-3} μΩcmK^{-2}, which is moderately large compared with that of heavy fermions [8]. Figures 3 and 4 show pressure dependences of T_N and T_a. Temperature T_N slightly increases with pressure at a rate $dT_N/dP = 0.3$ K/GPa, which is an order of magnitude smaller than that of DyB_6 [9]. On the contrary, T_a is pressure independent within the experimental error.

Figure 5 shows electrical resistivity at 4.2 K and at high pressures as a function of magnetic fields *H*. Resistivity increases smoothly with increasing *H*, indicating absence of metamagnetic transition below 9 T. This field dependence $\rho(H) \sim H^{1.5}$ is similar to the $\rho(H) \sim H^2$ behavior of normal metals at low fields and to the MR of α-Ce [10]. It is different from those of magnetic multilayers [4] or layered antiferromagnetic compounds such as $CeNiGe_2$ [11], which shows a large negative MR of the order of several tens per cent. Therefore, the extremely large positive MR of GdB_4 below 9 T is not due to the spin-dependent scattering. It has been reported that MR of GdB_4 obeys the so-called Kohler rule [6]. According to this rule, a pure material with low resistivity at $H = 0$ is required for large MR [12]. Such case was reported for α-Ce, in which the MR was enhanced by decreasing the resistivity under applied pressure [13]. In this sense, our result implies that GdB_4 is a typical large-MR material because its RRR value exceeds 200.

MR curve is weakly affected by pressure as shown in figure 5. The magnitude of MR at 4.2 K is $R(8.5T)/R(H = 0) \cong 6000$ % at ambient pressure. The inset of figure 5 reveals that MR is pressure independent below 3 GPa. If the origin of the large MR of GdB_4 is spin-dependent scattering of conduction electrons then the antiferromagnetic spin arrangement should be important. In order to examine this idea, we plot the temperature dependence of MR in figure 6. MR increases rapidly around T_a, but not at T_N, implying that the large MR of GdB_4 is related not only to the purity but also to the "hidden phase transition". The details of this transition at $T=T_a$ are not clear and further studies are highly desired.

4.Conclusions

In this work, we have measured the electrical resistivity of single-crystalline antiferromagnetic tetraboride GdB_4 under high pressures and high magnetic fields. The main results are summarized as follows:

1) Electrical resistivity of GdB_4 is almost pressure-independent below 3 GPa.
2) The antiferromagnetic ordering temperature T_N increases with pressure at a rate 0.3 K/GPa.
3) The hidden transition temperature T_a is pressure independent and is related to the large MR.
4) The magnitude of MR is not affected by applying pressures below 3 GPa.

5.Acknowledgment
This work was supported by a Grant-in-Aid for Scientific Research from the Japanese Ministry of Education, Science and Culture (No.20900131).

References

[1] (Eds) Maekawa S and Shinjo T 2002 *Spin Dependent Transport in Magnetic Nanostructures* (Taylor & Francis, London)
[2] Oomi G, Sakai T, Uwatoko Y, Takanashi K and Fujimori H 1997 *Physica B* **239** 19
[3] Kaji S, Oomi G, Mitani S, Takahashi S, Takanashi S and Maekawa S 2003 *Phys. Rev.* B. **68** 054429
[4] Suenaga K, Higashihara S, Ohashi M, Oomi G, Hedo M, Uwatoko Y, Saito K, Mitani S and Takanashi K 2007 *Phys. Rev. Lett.* **98** 207202
[5] Fisk Z, Maple M B, Johnston D C and Woolf L D 1981 *Solid State Commun.* **39** 1189
[6] Cho B K, Rhyee J-S, Kim J Y, Emilia M and Canfield P C 2005 *J. Appl. Phys.* **97** 10A923
[7] Honda F, Kaji S, Minamitake I, Ohashi M, Oomi G, Eto T and Kagayama T 2002 *J. Phys.: Condens. Matter* **14** 11501
[8] Kagayama T, Oomi G, Takahashi H, Mori N, Ōnuki Y and Komatsubara T 1991 *Phys. Rev.* B **44** 7690
[9] Sakai T, Oomi G and Kunii S 1997 *Physica* B **230-232** 741
[10] Miyagawa H, Oomi G, Ohashi M, Maezawa K and Kagayama T 2006 *J. Alloys Compd.* **408-412** 230
[11] Ohashi M *et.al*, to be submitted.
[12] Meaden G T 1965 *Electrical resistance of metals* (Plenum Press, 1965) p 132
[13] Sakigawa Y, Ohashi M and Oomi G 2007 *J. Phys. Soc. Jpn.* **76** Suppl. A 66

Journal of Physics: Conference Series **176** (2009) 012039 doi:10.1088/1742-6596/176/1/012039

^{11}B NMR study on rare earth ternary borides $RRuB_2$

Yutaka Kishimoto[1], Yu Kawasaki[1], Yukiichi Ideta[1], Sigeki Endou[1], Tetsu Tanaka[1], Masayoshi Tanabe[1], Takashi Ohno[1], Goutam Ghosh[2], A. K. Tyagi[3] and Laxmi C. Gupta[4]

[1]Department of Quantum Materials Science, Institute of Technology, University of Tokushima, 2-1 Minami-josanjima, Tokushima 770-8506, Japan
[2]UGC-DAE Consortium for Scientific Research, Mumbai Centre, BARC, Trombay, Mumbai 400085, India
[3]Bhabha Atomic Research Centre, Mumbai 400085, India
[4]Max Planck Institute for the Physics of Complex Systems, Nöthnitzer Strasse 38, D 01187, Dresden, Germany

yutaka@pm.tokushima-u.ac.jp

Abstract. ^{11}B NMR measurements have been performed on rare earth ternary borides, $YRuB_2$, $LuRuB_2$ and $(Tm_{0.4}Y_{0.6})RuB_2$ to investigate their electronic states. From the measurements of temperature dependence of spin-lattice relaxation rate $1/T_1$, $YRuB_2$ and $LuRuB_2$ are found to be Bardeen-Cooper-Schrieffer (BCS) superconductors with the energy gap $2\Delta(0)$=3.52 k_BT_c and the superconducting transition temperatures T_c=7.6 K ($YRuB_2$) and 10.2 K ($LuRuB_2$). A very broad and temperature dependent spectrum is observed in $(Tm_{0.4}Y_{0.6})RuB_2$. The magnitude of $1/T_1T$ is considerably larger in $(Tm_{0.4}Y_{0.6})RuB_2$ than in $YRuB_2$ and $LuRuB_2$. These facts suggest the contribution of 4*f* electrons to the hyperfine field at ^{11}B nuclei in $(Tm_{0.4}Y_{0.6})RuB_2$.

1. Introduction

The superconductivity and magnetism in the rare earth (R) ternary boride have attracted much interest. Relatively high superconducting transition temperature T_c in rare earth ternary borides $YRuB_2$ (T_c=7.7 K) and $LuRuB_2$ (T_c=10.2 K) with the orthorhombic $LuRuB_2$-type structure has been a matter of interest [1, 2]. These T_c's are comparable to those in the strong coupling superconductor such as C15 Laves phase compound HfV_2 (T_c = 9.2 K) and ZrV_2 (T_c = 8.5 K) [3], indicating strong electron-phonon coupling and high density of states of conduction electrons in $YRuB_2$ and $LuRuB_2$. Among $RRuB_2$, the depression of T_c in $(Lu_{1-x}R_x)RuB_2$ caused by the magnetic rare earth ions and reentrant superconductivity observed in $(Lu_xTm_{1-x})RuB_2$ and $(Lu_xEr_{1-x})RuB_2$ have attracted much attention [1, 4]. $YRuB_2$ with no 4*f* electron and $LuRuB_2$ with a closed 4*f* shell are important as reference materials to a rare earth ternary boride $RRuB_2$. To investigate the electronic states, we have performed ^{11}B NMR measurements and reported that the reference materials $YRuB_2$ and $LuRuB_2$ are BCS superconductors with energy gap $2\Delta(0)$ = 3.52 k_BT_c [5, 6]. In the normal state, the difference in the magnitude of $1/T_1T$ between $YRuB_2$ ($1/T_1T = 6.0\times10^{-3}$ s^{-1} K^{-1}) and $LuRuB_2$ ($1/T_1T = 2.1\times10^{-3}$ s^{-1} K^{-1}) is explained by the difference between $YRuB_2$ and $LuRuB_2$ in the density of states of conduction electrons at the Fermi level.

In this paper, we report ^{11}B NMR measurement on $(Y_{0.6}Tm_{0.4})RuB_2$ with 4*f* electrons and compare the results with those in reference materials, $YRuB_2$ and $LuRuB_2$.

2. Experimental details

Polycrystalline $YRuB_2$, $LuRuB_2$ and $(Tm_{0.4}Y_{0.6})RuB_2$ samples were prepared in an argon arc furnace by melting high-purity elements Y, Tm, Lu, Ru and B. The obtained pellets were turned over several times to ensure good homogeneity. They were further annealed at 1050 °C. The obtained sample had single phase orthorhombic $LuRuB_2$-type structure, as deduced from the powder X-ray diffraction patterns recorded using Cu *Kα* radiation. For NMR measurements, samples were crushed into 200 mesh powder. Transition temperatures T_c=7.6 K for $YRuB_2$ and T_c=10.2 K for $LuRuB_2$ were determined by the AC susceptibility measured with NMR coil at NMR frequency. These values are almost the same as the previously reported values [1,2], indicating good quality of each sample. The superconducting transition was not observed in $(Tm_{0.4}Y_{0.6})RuB_2$ above 4.2 K. In the temperature range 4.2-100 K, the NMR spectrum and spin-lattice relaxation rate $1/T_1$ of ^{11}B nuclei were measured in $YRuB_2$, $LuRuB_2$ and $(Tm_{0.4}Y_{0.6})RuB_2$ with a phase coherent pulsed NMR spectrometer. $1/T_1$ was measured with the saturation-recovery method. The measurements on $YRuB_2$ and $LuRuB_2$ were performed at 8.5 MHz. To improve the poor S/N ratio, ^{11}B NMR in $(Tm_{0.4}Y_{0.6})RuB_2$ was measured at 16 MHz.

3. Results and Discussion

3.1. $YRuB_2$ and $LuRuB_2$

A narrow central peak accompanied by a wide wing, broadened by a distribution in the electric field gradient, is observed in the field swept ^{11}B NMR spectra of $YRuB_2$ and $LuRuB_2$. The peak position and line width does not depend on temperature in the normal state. Knight shift decreases and line width increases at the superconducting transition. The spin-lattice relaxation rate $1/T_1$ were measured at the peak field of the NMR spectrum corresponding to the central (+1/2 ↔ -1/2) transition of ^{11}B (I=3/2) nuclei. A good agreement is found between the recovery data of the nuclear magnetization after the saturation pulse and the theoretical multi-exponential type curve [7].

As shown in figure 1, in the normal state, $1/T_1$ in $YRuB_2$ and $LuRuB_2$ are proportional to the temperature T. The relations, $1/T_1T = 6.0\times10^{-3}\ s^{-1}\ K^{-1}$ ($YRuB_2$) and $1/T_1T = 2.1\times10^{-3}\ s^{-1}\ K^{-1}$ ($LuRuB_2$) are obtained by the least-squares method. This difference in the magnitude of $1/T_1T$ between $YRuB_2$ and $LuRuB_2$ can be explained by the difference between $YRuB_2$ and $LuRuB_2$ in the density of states of conduction electrons at the Fermi level [6]. In the superconducting state, $1/T_1$ shows a small coherence peak just below T_c and agree with the calculated BCS curves with the energy gap $2\Delta(0)=3.52\ k_BT_c$ and energy breadth $\delta/\Delta(T)=0.10$ [8] as indicated by the solid curves.

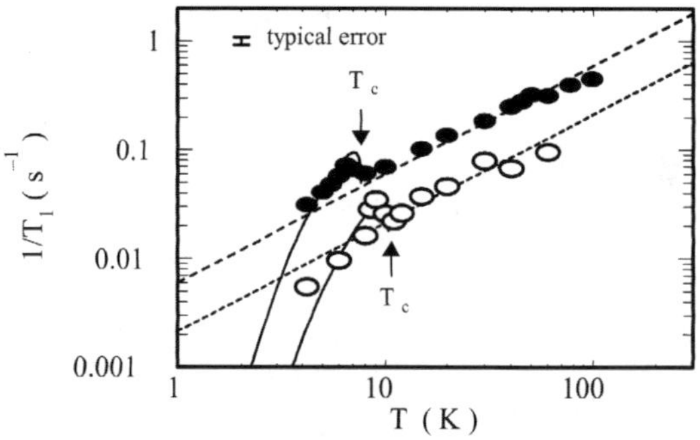

Figure 1. Temperature dependence of $1/T_1$ in $YRuB_2$ (closed circles) and $LuRuB_2$ (open circles). The straight dotted lines show the relations, $1/T_1T = 6.0\times10^{-3}\ s^{-1}\ K^{-1}$ ($YRuB_2$) and $1/T_1T = 2.1\times10^{-3}\ s^{-1}\ K^{-1}$ ($LuRuB_2$). The solid curves below T_c indicate BCS curves with $2\Delta(0)=3.52\ k_BT_c$ and $\delta/\Delta(T)=0.10$.

3.2. $(Tm_{0.4}Y_{0.6})RuB_2$

As shown in figures 2 (a) and (b), with decreasing temperature, the line width increases and Knight shift decreases. The NMR spectra in $(Tm_{0.4}Y_{0.6})RuB_2$ are very broad (~ 1 kOe at 4.2 K and ~ 200 Oe at 100 K) compared with those (~ 20 Oe) in $YRuB_2$ and $LuRuB_2$. In $(Tm_{0.4}Y_{0.6})RuB_2$, 40% of Y^{3+} ions in $YRuB_2$ are randomly substituted by magnetic Tm^{3+} ions with $4f$ electrons. The distribution of magnetic Tm^{3+} ions in a periodic array throughout the lattice might broaden the ^{11}B NMR spectrum.

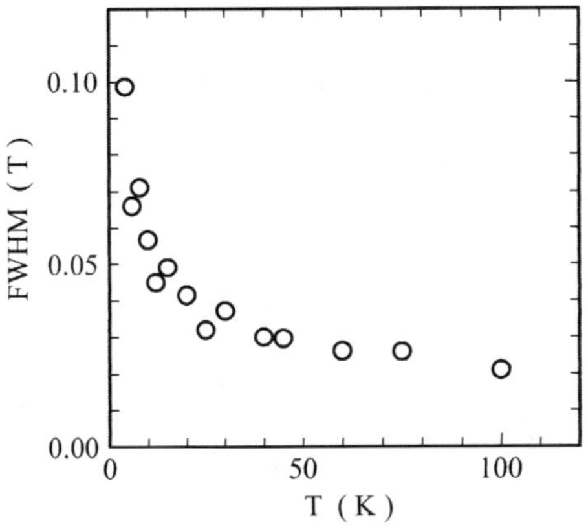

Figure 2 (a). Temperature dependence of line width of ^{11}B spectrum in $(Tm_{0.4}Y_{0.6})RuB_2$.

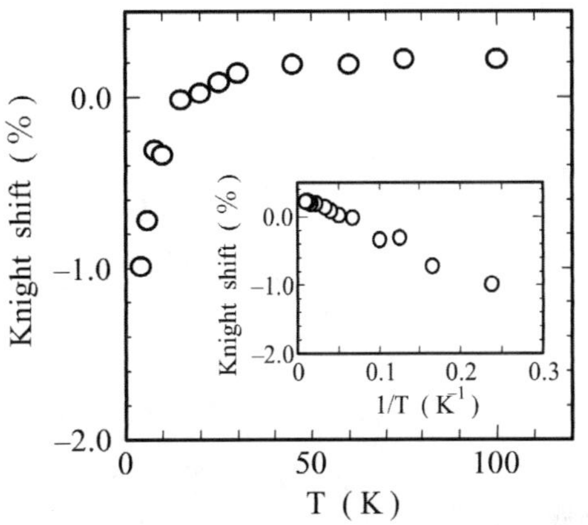

Figure 2 (b). Temperature dependence of Knight shift of ^{11}B in $(Tm_{0.4}Y_{0.6})RuB_2$. Inset shows the Knight shift plotted against $1/T$.

Previously, we reported ^{11}B NMR study in a rare earth ternary boride RRh_3B_2 (R=La, Ce, Nd, Sm, Eu and Gd) [9]. The comparison of results between $(Tm_{0.4}Y_{0.6})RuB_2$ and magnetic RRh_3B_2 will characterize $4f$ electronic states in $(Tm_{0.4}Y_{0.6})RuB_2$. In $GdRh_3B_2$, $NdRh_3B_2$ and $CeRh_3B_2$ the line width ~1 kOe is observed in ^{11}B NMR spectrum at low temperatures near the magnetic ordering temperature. In $(Tm_{0.4}Y_{0.6})RuB_2$, however, magnetic ordering is not observed above 4.2 K and the $4f$ magnetic moment is small compared with that in $GdRh_3B_2$, $NdRh_3B_2$ and $CeRh_3B_2$. Therefore, the line width of ~1 kOe in $(Tm_{0.4}Y_{0.6})RuB_2$ is considered to originate from $4f$ magnetic moment of Tm^{3+} ions and to be enhanced by its spatial distribution in the lattice. In $GdRh_3B_2$, $NdRh_3B_2$ and $CeRh_3B_2$, the line width and Knight shift are proportional to the magnetic susceptibility of $4f$ electrons χ_{4f} and are temperature dependent. In $(Tm_{0.4}Y_{0.6})RuB_2$, therefore, the temperature dependence of line width and Knight shift is associated with χ_{4f} of Tm^{3+} ions. Actually, as shown in the inset of figure 2(b), the Knight shift is proportional to $1/T$, indicating the Curie law type temperature dependence of χ_{4f}. The hyperfine field at ^{11}B in $(Tm_{0.4}Y_{0.6})RuB_2$ might arise from $4f$ electrons by the same mechanism as in RRh_3B_2.

$1/T_1$ in $(Tm_{0.4}Y_{0.6})RuB_2$ has been measured at the peak field of the NMR spectrum. The recovery of nuclear magnetization after the saturation pulse is multi-exponential type, and we can use the same theoretical function for the $1/T_1$ fitting as that in $YRuB_2$ and $LuRuB_2$. As shown in figure 3, $1/T_1$ is proportional to temperature, $1/T_1T = 0.19\ s^{-1}\ K^{-1}$. The magnitude of $1/T_1T$ is much larger than those in $YRuB_2$ and $LuRuB_2$. This also suggests the contribution of $4f$ electrons to the hyperfine field at ^{11}B in $(Tm_{0.4}Y_{0.6})RuB_2$.

Above the magnetic ordering temperature T_m, the relation $1/T_1$=constant is expected in the system where the spin-lattice relaxation is dominated by the fluctuation of well-defined localized moment. Among RRh_3B_2, $4f$ moment in $GdRh_3B_2$ and $NdRh_3B_2$ is well localized at Gd^{3+} and Nd^{3+} ions, respectively. Actually, $1/T_1=1.73\times10^3\ s^{-1}$ and $0.405\times10^3\ s^{-1}$ are observed above T_m in $GdRh_3B_2$ (T_m=90 K) and $NdRh_3B_2$ (T_m=10.5 K), respectively [9]. On the other hand, the relation $1/T_1 = A + BT$ is observed in $CeRh_3B_2$ (T_m=115 K) where the hybridization between Ce $4f$ and conduction electrons is strong compared to $GdRh_3B_2$ and $NdRh_3B_2$. Here, $A = 1.66\ s^{-1}$ is the constant arising from the

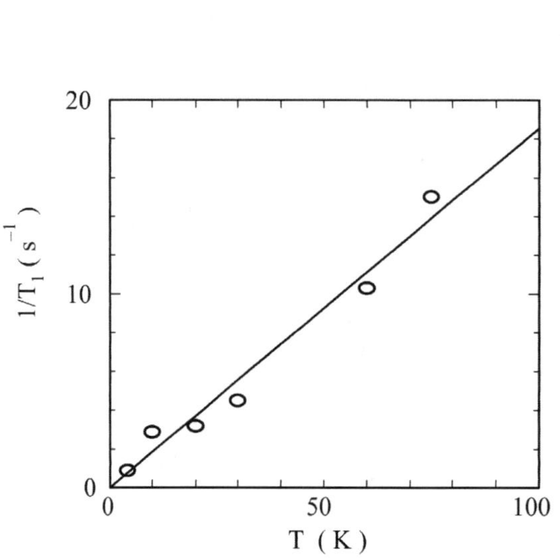

Figure 3. Temperature dependence of $1/T_1$ in $(Tm_{0.4}Y_{0.6})RuB_2$.
The solid line indicates the relation, $1/T_1T$=0.19 s^{-1} K^{-1}.

fluctuation of 4*f* moment of Ce^{3+} ions and B = 1.48 × 10^{-2} s^{-1} $K^{-1}$is a Korringa term due to the conduction electrons. The magnitude of the Korringa term is comparable to that of the localized moment term above T_m in $CeRh_3B_2$. In the same way, an itinerant nature of 4*f* electrons of Tm^{3+} ions in $(Tm_{0.4}Y_{0.6})RuB_2$ might be responsible to the Korringa type relation $1/T_1T$=0.19 s^{-1} K^{-1}. To clarify the contribution of 4*f* electrons to conduction electrons in $(Tm_{0.4}Y_{0.6})RuB_2$, NMR studies of other $RRuB_2$ and rare earth ternary borides are necessary.

4. Conclusion

^{11}B NMR measurements have been performed to study the electronic states in rare earth ternary borides $YRuB_2$, $LuRuB_2$ and $(Tm_{0.4}Y_{0.6})RuB_2$,. Reference materials $YRuB_2$ and $LuRuB_2$ are proved to be simple metals in the normal state and to be BCS superconductors with energy gap $2\Delta(0) = 3.52\ k_BT_c$.

On the other hand, in $(Tm_{0.4}Y_{0.6})RuB_2$ having 4*f* electrons, large and temperature dependent line width, as well as the Knight shift have been observed. These facts are associated with the hyperfine field at ^{11}B site due to 4*f* electrons. The observed $1/T_1T$=0.19 s^{-1} K^{-1} is considerably large compared with those in $YRuB_2$ and $LuRuB_2$, suggesting a considerable contribution of 4*f* electrons to conduction electrons.

References

[1] H. C. Ku and R. N. Shelton, *Solid State Communications* **40** (1981), 237.
[2] W. H. Lee, S. Appl and R. N. Shelton, *J. Low Temp. Phys.* **68** (1987), 147.
[3] Y. Kishimoto, T. Ohno, T. Hihara, K. Sumiyama and K. Suzuki, *Phys. Rev. B* **64** (2001), 024509.
[4] R. N. Shelton and H. E. Horng, *Phys. Rev. B* **33** (1986), 1671.
[5] Y. Kishimoto, Y. Kawasaki, M. Tanabe, T. Tanaka, T. Ohno, G. Ghosh, A. K. Tyagi and L. C. Gupta, *J. Magn. Magn. Mater.* **310** (2007), 581.
[6] Y. Kishimoto, Y. Kawasaki, M. Tanabe, Y. Ideta, T. Ohno, G. Ghosh, A. K. Tyagi and L. C. Gupta, *J. Phys. Chem. Solids* **69** (2008), 3153.
[7] A. Narath, Phys. Rev. **162** (1961), 320.
[8] L. C. Hebel and C. P. Slichter, *Phys. Rev.* **113** (1959), 1504.
[9] Y. Kishimoto, Y. Kawasaki and T. Ohno, *J. Phys. Soc. Jpn.* **73** (2004), 1970.

Competition between ferromagnetic and antiferromagnetic interactions in $Pr_{1-x}Gd_xB_4$

Toshihiko Kobayashi, Tetsuro Ikemoto, Ryuta Watanuki and Kazuya Suzuki*
Department of Advanced Materials Chemistry, Yokohama National University, 79-5 Tokiwadai, Hodogaya-ku, Yokohama 240-8501, Japan

E-mail: ksuzuki@ynu.ac.jp

Abstract. We have prepared single crystals of $Pr_{1-x}R_xB_4$ (R = La, Ce and Gd) and measured the magnetic susceptibility and specific heat in order to elucidate the origin of the ferromagnetic interaction in PrB_4 and the competition of the ferro- and antiferromagnetic interactions in this system. The ferromagnetic transition temperature decreases with x at nearly the same rate in $Pr_{1-x}La_xB_4$ and $Pr_{1-x}Gd_xB_4$, suggesting that the magnetic moments of Gd do not affect the ferromagnetic order. In $Pr_{0.82}Gd_{0.18}B_4$ and $Pr_{0.8}Gd_{0.2}B_4$, ferromagnetic to antiferromagnetic transitions appear with decreasing temperature. The development of antiferromagnetic correlation between Pr and Gd breaks the ferromagnetic order and induces antiferromagnetic correlation among Pr atoms. These results indicate that two types of magnetic interactions with different origin are present in PrB_4.

1. Introduction

Investigation of basic physical properties of rare earth metal tetraboride RB_4 family started more than 30 years ago and continued extensively within the next decade [1-4]. After that, the discovery of the geometrical frustration of quadrupole moments of 4f electrons in DyB_4 [5] triggered the re-examination of physical properties of rare earth metal tetraborides RB_4 in the viewpoint of multipolar ordering [6-8]. Among RB_4 containing magnetic R elements, only PrB_4 shows a ferromagnetic transition with the transition temperature T_c = 23 K [2]. The origin of the strong ferromagnetic interactions appearing in PrB_4 and its relatively high T_c in comparison with that considered from the de Gennes factor have been studied but are still under discussion [9]. In order to investigate these problems, we have prepared single crystals of $Pr_{1-x}R_xB_4$ (R = La, Ce and Gd), where LaB_4 and CeB_4 are Pauli paramagnets and GdB_4 is an antiferromagnet (T_N = 42 K) [2], by a self-flux method and have studied physical properties by means of magnetic susceptibility and specific heat measurements.

2. Experimental

Single crystals of $Pr_{1-x}R_xB_4$ (R = La, Ce and Gd) were prepared by arc melting with excess amounts of rare earth metals to boron. Crystals were grown in a melt after one cycle of heating-cooling process where the excess rare earth metals acted as flux. The crystals were picked up from the melt by dissolving the excess rare earth metals in diluted hydrochloric acid. The size of the single crystals obtained in this method is up to 2 × 2 × 0.5 mm^3. The stoichiometry of the single crystals was examined using the (004)-X-ray diffraction peak by applying the Vegard's law; and it was very close to the starting ratio of Pr : R. The X-ray diffraction peaks are as narrow as those of PrB_4 so that there is no distinct distribution of the chemical composition x in a crystal. The magnetic susceptibility and magnetization curves of $Pr_{1-x}R_xB_4$ were measured using a SQUID magnetometer in the temperature range from 1.8 to 300 K, and the magnetization curves were recorded at 1.8 K in fields up to 5 T. The specific heat measurement was carried out by the heat-relaxation method down to 2.0 K. The magnetic transition temperatures of $Pr_{1-x}R_xB_4$ are defined by the heat capacity (x = 0, 0.15, 0.2, 0.25 and 0.3) or the magnetization (x = 0.18 and 0.22).

3. Results and discussion

Figures 1 and 2 show the temperature dependence of the magnetization divided by magnetic field *M*/*B* for zero-field cooling (ZFC) and field-cooling (FC) processes, as well as the magnetization curves at 1.8 K for $Pr_{1-x}Gd_xB_4$ (x = 0, 0.15, 0.18, 0.2, 0.22, 0.25 and 0.3).

The magnetization of PrB_4 (x = 0) steeply increases with decreasing temperature and saturates below the ferromagnetic transition temperature of T_c = 23.0 K. The easy axis of the magnetization is the *c*-axis as a result of the crystalline electric field splitting of the 4f states of Pr. It should be noted that the magnetizations below T_c after the FC or ZFC processes show nearly the same behaviours and there is no hysteresis in the magnetization curves. Such behaviours characteristic for "soft" magnets are unusual for the ferromagnets with strong single-ion magnetic anisotropy, because significant energy is required to rotate magnetic domains toward the direction of the applied magnetic fields.

The ferromagnetic behaviour is found for $Pr_{0.85}Gd_{0.15}B_4$ (T_c = 19.5 K), $Pr_{0.82}Gd_{0.18}B_4$ (T_c = 19.0 K) and $Pr_{0.8}Gd_{0.2}B_4$ (T_c = 18.2 K). In $Pr_{0.85}Gd_{0.15}B_4$ (x = 0.15), difference of magnetization in the FC and ZFC process appears below 13 K, provably originating from the pinning of magnetic domains by the introduction of Gd. The magnetization curves of $Pr_{0.85}Gd_{0.15}B_4$ exhibit a small hysteresis with the coercive force of 0.05 T. In $Pr_{0.82}Gd_{0.18}B_4$ (x = 0.18) and $Pr_{0.8}Gd_{0.2}B_4$ (x = 0.2), the magnetization shows a ferromagnetic plateau below T_c and then it abruptly decreases to get into an antiferromagnetic state at T_N = 7.5 and 10.7 K, respectively. The antiferromagnetic states below T_N are confirmed by the presence of the metamagnetic transition seen in the magnetization curves at the critical magnetic field of B_c = 0.4 T for x = 0.18 and B_c = 0.5 T for x = 0.2. These ferromagnetic to antiferromagnetic transitions suggest that the evolution of the antiferromagnetic correlation between the magnetic moments of Pr and Gd breaks the ferromagnetic order of Pr and even induces the antiferromagnetic correlation among the magnetic moments of Pr.

The magnetization of $Pr_{0.78}Gd_{0.22}B_4$ (x = 0.22), $Pr_{0.75}Gd_{0.25}B_4$ (x = 0.25) and $Pr_{0.7}Gd_{0.3}B_4$ (x = 0.3)

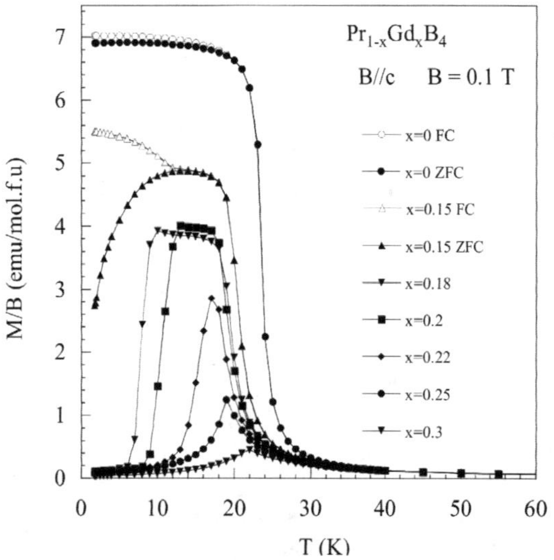

Figure 1. Temperature dependence of the magnetization divided by magnetic field M/B for $Pr_{1-x}Gd_xB_4$.

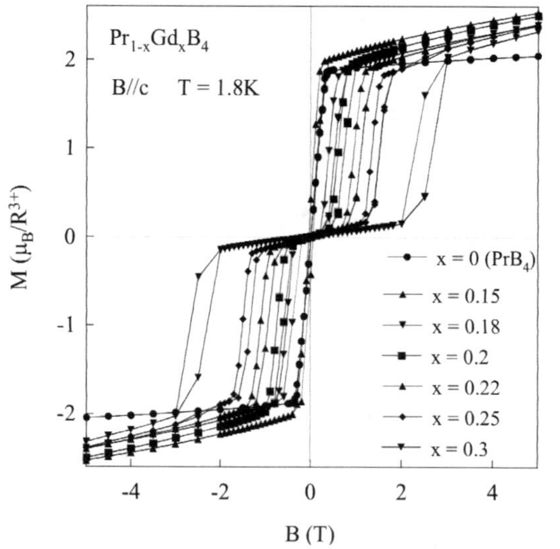

Figure 2. Magnetization curves of $Pr_{1-x}Gd_xB_4$.

shows a sharp cusp at T_N = 17.0, 18.8 and 21.5 K, respectively. The low-temperature phase is antiferromagnetic, which is also assured by the presence of metamagnetic transition at B_c = 1.0, 1.8 and 2.5 T, respectively. The sharp cusp of the magnetization indicates that, with decreasing temperature, the ferromagnetic correlation among the magnetic moments of Pr is developed but, before occurrence of ferromagnetic order, the Pr-Gd antiferromagnetic interactions overcome the Pr-Pr ferromagnetic interactions to go into the antiferromagnetic phase without undergoing the ferromagnetic state.

Above the critical magnetic field of the metamagnetic transition B_c for $Pr_{1-x}Gd_xB_4$ with x > 0.18, the magnetization gradually increases with magnetic field, in contrast to the behaviour of PrB_4. In the antiferromagnetic phase, the magnetic moments of Pr and Gd are both antiferromagnetically arranged. Above B_c, the magnetic moments of Pr are directly flipped to the direction of the external magnetic field because of the Ising-like single-ion anisotropy of Pr 4f moments due to the crystalline electric field. On the other hand, the single-ion anisotropy of Gd is very small since the Gd^{3+} ion does not have orbital angular momentum *L*. In this case, the magnetic moments of Gd are considered to have a spin-flop state above B_c, namely, the magnetic moments of Gd are nearly perpendicular to the direction of the external magnetic field. The increase of the magnetization above B_c can be explained in terms of the contribution of the spin-flopped magnetic moments of Gd

Figure 3 is the temperature dependence of the magnetic specific heat of $Pr_{1-x}Gd_xB_4$ (x = 0, 0.15, 0.2, 0.25 and 0.3). For all the compounds, the main peaks, corresponding to the transition from paramagnetic to ordered phase, are λ-type, indicating that the transitions are of second order. In comparison with PrB_4, the specific heat of $Pr_{0.85}Gd_{0.15}B_4$ has large values below T_c and shows a broad peak at around 5 K. The low-temperature contribution of the specific heat can be attributed to the remaining magnetic entropy of Gd below T_c, namely, the magnetic moments of Gd are not ordered but paramagnetic even though they feel strong internal magnetic field from surrounding ferromagnetic moments of Pr.

The specific heat of $Pr_{0.8}Gd_{0.2}B_4$ shows additional two peaks at 12.5 and 10.7 K, the latter corresponding to the antiferromagnetic transition temperature T_N. The peak shape of the specific heat at 12.5 K suggests that the transition is of first order. From our limited experimental data, further discussion of the intermediate phase is limited. Complicated magnetic phases are expected to exist between the ferromagnetic and antiferromagnetic phases in this system.

Figure 4 shows the summarized phase diagram of $Pr_{1-x}Gd_xB_4$. The variations of the ferromagnetic transition temperature of $Pr_{1-x}La_xB_4$ and $Pr_{1-x}Ce_xB_4$ are also illustrated.

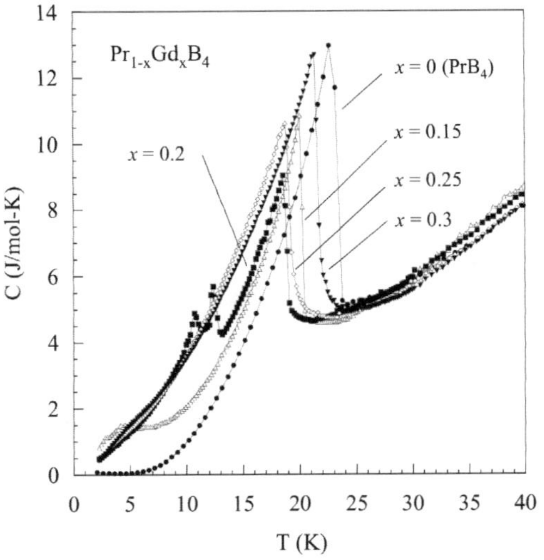

Figure 3. Magnetic specific heat of $Pr_{1-x}Gd_xB_4$.

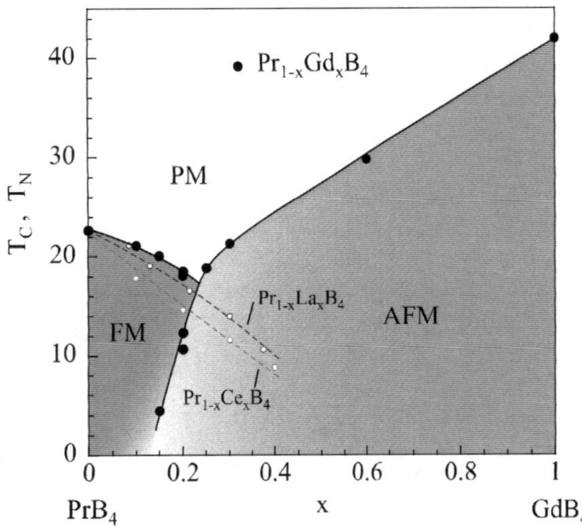

Figure 4. Phase diagram of $Pr_{1-x}Gd_xB_4$.

Considering all the experimental results, we may conclude that there exist two, almost independent types of magnetic interactions in PrB_4. One is the antiferromagnetic interaction, which commonly exists in magnetic RB_4 and is mediated by conduction electron (the RKKY interaction). Another is additional ferromagnetic interaction explicitly appearing only in PrB_4. The transition temperature is determined by the competition of these two types of interactions.

In $Pr_{1-x}R_xB_4$, the elements La, Pr and Gd are present as R^{3+} ions, while Ce as nonmagnetic R^{4+} ions. Therefore, the Pr ions are replaced by nonmagnetic ions both in $Pr_{1-x}La_xB_4$ and $Pr_{1-x}Ce_xB_4$, and the conduction carrier density is changed in $Pr_{1-x}Ce_xB_4$ but not in $Pr_{1-x}La_xB_4$. As seen in figure 4, the ferromagnetic transition temperature drops more rapidly in $Pr_{1-x}Ce_xB_4$ than in $Pr_{1-x}La_xB_4$. Therefore, the increase of the carrier density enhances the antiferromagnetic interaction among the Pr ions and increases the competition between ferro- and antiferromagnetic interactions, resulting in the rapid drop of T_c in $Pr_{1-x}Ce_xB_4$.

In $Pr_{1-x}Gd_xB_4$, the ferromagnetic transition temperature T_c shows nearly the same variation with x as in $Pr_{1-x}La_xB_4$. The result means that the Gd^{3+} ions behave like nonmagnetic La^{3+} ions and the magnetic moments of Gd do not affect the ferromagnetic ordering, namely, the magnetic moments of Gd stay in the paramagnetic state even the magnetic moments of Pr are in the ferromagnetic state.

If ferro- and antiferromagnetic interactions in $Pr_{1-x}Gd_xB_4$ directly compete with each other, it is expected that T_c drops much more rapidly and disappears, and then, antiferromagnetic order appears with increasing x. Alternatively, the ferromagnetic transition disappears and instead the spin-glassy state appears in a specific range of x, as a result of frustration due to the competing interactions. As shown in the phase diagram, even within the ferromagnetic state, the antiferromagnetic correlation between Pr and Gd seems to develop independently of the ferromagnetic correlation. Therefore, the ferromagnetic interaction in PrB_4 is not from the RKKY-type interaction. The direct exchange interaction between the Pr 4f-electrons might be an origin, because it can be enhanced in the case the Pr $4f^2$ energy level is closer to the Fermi energy than in other rare earth compounds. This conclusion originates from the fact that the Ce 4f electron is completely delocalized to result in Ce^{4+}. Similar situation of the Pr 4f state is known in skutterudite compound $PrOs_4Sb_{12}$, where the Pr 4f level is close to the Fermi level. In that case, the interactions between the Pr 4f electron and conduction electron cause the heavy Fermion superconductivity [10].

4. Summary

We have found the transitions from ferromagnetic to antiferromagnetic state in $Pr_{0.82}Gd_{0.18}B_4$ and $Pr_{0.8}Gd_{0.2}B_4$, where the ferromagnetic and antiferromagnetic correlations develop independently. The origin of the ferromagnetic interaction is still unclear but it is completely different from the RKKY-type origin of antiferromagnetic interactions.

References

[1] Fisk Z, Cooper, A S, Schmidt P H and Castellano R N 1972 *Mater. Res. Bull.* **7** 285
[2] Buschow K H J and Creyghton J H N 1972 *J. Chem. Phys.* **57** 3910
[3] Fisk Z and maple M B 1981 *Solid State Commun.* **39** 1189
[4] Etourneau J 1985 *J. Less-common Met.* **110** 267
[5] Watanuki R, Saito G, Suzuki K, Ishihara M, Yanagisawa T, Nemoto Y and Goto T 2005 *J. Phys. Soc. Jpn.* **74** 2169
[6] Okumura D, Matsumura T, Nakao H and Murakami Y 2005 *J. Phys. Soc. Jpn.* **74** 2434
[7] Okumura D, Matsumura T, Nakao H and Murakami Y 2006 *J. Phys. Soc. Jpn.* **75** 198
[8] Yoshii S, Yamamoto T, Hagiwara M Michimura S, Shigekawa A, Iga F, Takabatake T and Kindo K 2008 *Phys. Rev. Lett.* **101** 087202
[9] Wigger G A, Felder E, Monner R, Ott H R, Pham L and Fisk Z 2005 *Phys. Rev.* B **72** 014419
[10] Bauer E D, Frederick N A, Ho P-C, Zapf V S and Maple M B 2002 *Phys. Rev.* B **65** 100506

Ultrahigh temperature ceramics (UHTCs) based on ZrB_2 and HfB_2 systems: powder synthesis, densification and mechanical properties

Guo-Jun Zhang[1], **Wei-Ming Guo**[1], **De-Wei Ni**[1,2] and **Yan-Mei Kan**[1]
[1]State Key Laboratory of High Performance Ceramics and Superfine Microstructures, Shanghai Institute of Ceramics, Shanghai 200050, China
[2]Graduate School of the Chinese Academy of Sciences, Beijing 200049, China

E-mail: gjzhang@mail.sic.ac.cn

Abstract. Phase pure ZrB_2 and HfB_2 powders were synthesized by boro/carbothermal reduction (BCTR) of ZrO_2 and borothermal reduction of HfO_2, respectively. Combining the thermodynamic calculations and XRD results, the reaction process of BCTR of synthesized ZrB_2 powder was shown. The ZrB_2 powder synthesized at 1650 °C showed a better sinterability, due to its smaller particle size and lower oxygen content. Most of the HfB_2 particles obtained at 1600 °C presented nearly column-like morphology with 1-1.5μm size. Using the synthesized HfB_2 powder and commercial α-SiC, hot-pressed HfB_2-20 vol.% SiC composite has high flexural strength of 993 ±90 MPa and fracture toughness of 6.29±0.65 MPa·m$^{1/2}$, which were significantly higher than the reported values in literature.

1. Introduction

Ultrahigh temperature materials (UHTMs) are a family of materials that are chemically and physically stable at high temperatures and in reactive atmospheres. The characteristics of UHTMs allow them to be used in extreme environments including those associated with hypersonic flight, atmospheric re-entry, and rocket propulsion [1]. The UHTMs includes carbon/carbon composites, refractory metals, and refractory ceramic compounds (such as carbides, borides and nitrides). The carbon/carbon composites and tungsten alloy were successfully installed on X-43A vehicle that flew at Mach 7 in March 2004 as part of the Hyper-X program [2]. However, due to the poor oxidation resistance of the carbon/carbon composites and high density of the tungsten alloy, interest in refractory ceramic compounds characterized by high melting points, high electrical and thermal conductivities, high hardness, and good chemical inertness and oxidation resistance, has increased significantly in recent years, especially transition metal diborides with the relatively good oxidation resistance [3-4]. Of the transition metal diborides, zirconium diboride (ZrB_2) and hafnium diboride (HfB_2) are identified as the most promising candidates for high temperature applications such as nose caps, sharp leading edges, vanes and similar objects for use in high velocity flight or on future generations of reentry vehicles [3-5]. It has been shown that addition of SiC to diborides ceramics results in significant improvements in oxidation resistance and mechanical properties compared to diborides alone [6-10]. Hot-pressed HfB_2-20 vol.%SiC ceramics was flown on NASA-funded hypervelocity slender-body flight tests, as small radius nose tips, in 1997 during the SHARP-B1 program and as components of leading-edge strakes in 2000 during the SHARP-B2 program, which successfully showed the non-ablating performance of borides ceramics [11].

Because of the strong covalent bonding and low self-diffusion, high temperatures and external pressures are required to densify diborides [12]. Two ways typically are used to enhance densification. The first way is that alternative processing routes, such as reactive hot pressing [13-15], spark plasma sintering [16-17], and polymer precursors [18-19] have been investigated to improve the purity and densification behavior of diborides. The second way is that various additives have been studied to reduce sintering temperature. For example, the addition of metals (Ni) or ceramics (Si_3N_4, ZrN, Yb_2O_3 and VC) as sintering aids has been proven to enhance the sinterability and densification of ZrB_2-SiC ceramics [20-25]. Grain boundary phases deriving from the sintering aids often drastically deteriorate the elevated temperature strength and other mechanical properties [26-27]. Therefore, study on the preparation of MB_2-based ceramics from fine starting powders with high sinterability and high purity without adding sintering additives is very significant. However, most of the current commercially available powders have coarse particle size and demonstrate poor sinterability. In order to improve sinterability, pulverization is generally used to decrease the particle size. However, impurities are introduced, especially oxygen from the surface oxidation of particles, which significantly deteriorates the densification behavior and ceramic properties [28-29].

In this work, we carried out the synthesis of ZrB_2 powder via boro/carbothermal reduction (BCTR) of ZrO_2 with B_4C and carbon in vacuum, where the characterization of powder morphology and sinterability was investigated. On the other hand, highly active HfB_2 powder was synthesized by borothermal reduction of HfO_2 using amorphous boron. To reflect the performance of synthesized HfB_2 powder compared with commercially available powders, HfB_2-20 vol.%SiC composite was prepared by hot pressing at 2000 °C followed by characterization of microstructures and mechanical properties.

2. Experimental Procedure

The raw materials used in this study were ZrO_2 (D_{50}=0.6 μm, CSG Holding Co., Ltd, Shenzhen, China), HfO_2 (D_{50}=0.25μm, Found Star Science and Technology Co., Ltd, Beijing, China), B_4C (D_{50}=1.5μm, Jingangzuan Boron Carbide Co., Ltd, Mudanjiang, China), carbon (D_{50}=1.5μm), α-SiC (D_{50}=0.45μm, Changle Xinyuan Carborundum Micropowder Co. Ltd, Changle, China), and amorphous boron (average particle size <1μm, Dandong Chemical Engineering Institute Co. Ltd., Dandong, China), with purities of 99.8 %, 96 %, 99 %, 99 %, 98.5 % and 96 %, respectively. The ZrO_2, B_4C and carbon were used as starting powders to synthesize ZrB_2 based on the boro/carbothermal reduction. The HfO_2 and boron were used as starting powders to synthesize HfB_2

powder based on the borothermal reduction. First the starting mixtures were mixed for 24 h in a polythene bottle using ethanol and Si_3N_4 balls, and dried by rotary evaporation. After being dried, the powder mixtures were dry pressed into disks. Then the disks were heated at a rate of 10 °C/min to the desired temperature and held for 60 min in vacuum. The powder synthesis was conducted using a graphite crucible in a resistance-heated graphite element furnace.

To evaluate the ZrB_2 sinterability, the densification process of the synthesized powder was performed by pressureless sintering in the temperature range of 1900-2100 °C for 2 h. Firstly, as-prepared ZrB_2 powder was uniaxially pressed at a pressure of 30 MPa for 60 s. Cold isostatic pressing with an applied pressure of 300 MPa for 120 s was subsequently conducted. Powder compacts were heated to 1650 °C with a heating rate of 15 °C/min in the graphite element furnace under vacuum. Then 60 min isothermal hold at 1650 °C was used for the recovery of the vacuum (~5 Pa). After 60 min at 1650 °C, the furnace was backfilled with argon and heated at 30 °C/min to the desired sintering temperature.

Using the as-synthesized HfB_2 and commercially available α-SiC, HfB_2-20 vol.% SiC composite was hot pressed at 2000 °C under a pressure of 30 MPa. The specimen was heated from room temperature to 1650 °C with a heating rate of 8 °C/min under vacuum. Above 1650 °C, the furnace was backfilled with argon and heated at 15 °C/min to 2000 °C, followed by a 1h dwell, and then cooled to room temperature naturally. The application of pressure was only at the highest temperature.

Phase composition was determined by X-ray diffraction (XRD, D/max 2550 V, Tokyo, Japan). The relative amounts of the phases were quantified using K value method [30]. Morphology of the synthesized powder was characterized by scanning electron microscope (SEM, Hitachi S-570, Tokyo, Japan). Oxygen content was determined by a Nitrogen/Oxygen Determinator (TC600, Leco Corporation, St. Joseph, MI). Particle size distribution was analyzed using a laser particle size analysis (*Zeta Plus*). The hardness and fracture toughness were measured by the indentation method (Wilson-Wolpert Tukon2100B, USA), using a load of 5 kg for 10 s on a polished surface. Flexural strength was measured by a 3-point bending test (test bars 2.5 mm×2 mm×25 mm) with a span of 20mm.

3. Results and discussion

3.1. Boro/carbothermal reduction synthesis of ZrB_2 and pressureless densification

The ZrB_2 powder was synthesized according to boro/carbothermal reduction reaction:

$$2ZrO_2 + B_4C + 3C \rightarrow 2ZrB_2 + 4CO \qquad (1)$$

Figure 1 showed the XRD patterns of the synthesized ZrB_2 powder at different temperatures via BCTR of ZrO_2 with B_4C and carbon based on stoichiometric proportion of the reaction (1). Both ZrB_2 and ZrC phase were present in the synthesized powder. To understand the reason why ZrC phase also remained in the synthesized powder, the reaction process was firstly explored.

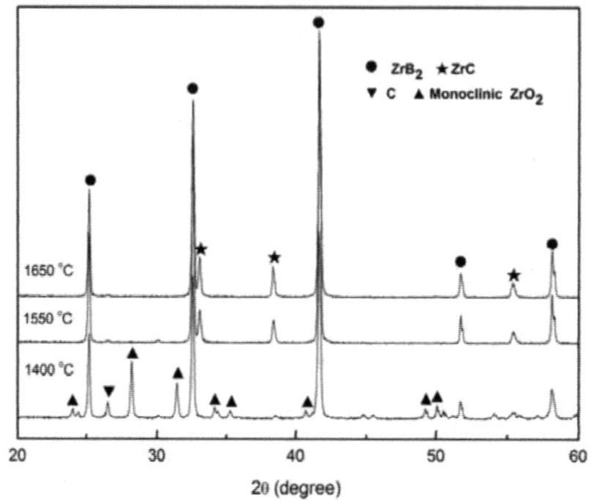

Figure 1. XRD patterns of the synthesized ZrB_2 powder at different temperatures according to stoichiometric proportion of the reaction (1).

Compared to reaction (1), previous studies have shown that ZrO_2 could react with B_4C at lower temperatures [12]:

$$7ZrO_2 + 5B_4C \rightarrow 7ZrB_2 + 3B_2O_3 + 5CO \tag{2}$$

Then the product B_2O_3 of reaction (3) continues to react with ZrO_2 and carbon to form ZrB_2 [31]:

$$ZrO_2 + B_2O_3 + 5C \rightarrow ZrB_2 + 5CO \tag{3}$$

Favorability of reaction is determined based on the sign of change of the Gibbs free energy. When it is negative, the reaction becomes favourable. In standard state ($pCO=1.013\times10^5$ Pa), reactions (1)-(3) become favorable above 1424 °C, 1218 °C and 1509 °C, respectively (figure 2(a)). In the present study, the synthesis of ZrB_2 powder was conducted in vacuum. The CO product was constantly removed by the pumping, and the furnace pressure remained below 11 Pa during the whole reduction process. Assuming a CO partial pressure of 11 Pa, reactions (1)-(3) are thermodynamically favorable above 894 °C, 784 °C and 938 °C, respectively (figure 2(b)). High vacuum markedly decreased the starting temperature of the above reactions. Observing the reaction (3), the intermediate B_2O_3 product played an important role in the synthesis of ZrB_2. In fact, B_2O_3 has an unusually low melting point (450 °C) and a high vapor pressure. The vapor pressure of B_2O_3 at 1527 °C reached 344 Pa, leading to its rapid vaporization [32]. According to reaction (2), the deficiency of B_2O_3 meant the excess of ZrO_2 and carbon, which would form ZrC phase based on the reaction:

$$ZrO_2 + 3C \rightarrow ZrC + 2CO \tag{4}$$

Therefore, the presence of ZrC phase in the synthesized powder was mainly attributed to the vaporization of intermediate B_2O_3 product. Though minimum favorable temperature of reaction (4) was 1065 °C under 11 Pa pressure, its Gibbs' free energy value did not show significantly more negative, compared to reactions (1)-(3) (figure 2(b)). So the ZrC phase appeared at relative high temperatures.

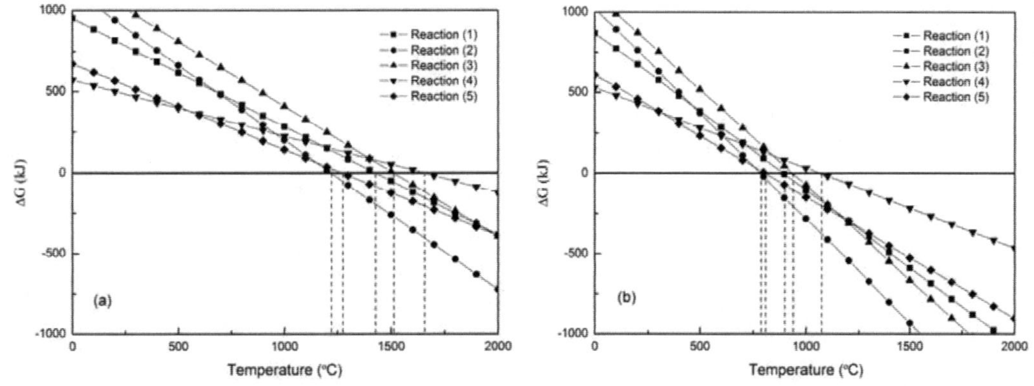

Figure 2. The change in Gibbs' free energy of reaction as a function of temperature for reactions (1)-(5) under the standard state (a) and 11 Pa pressure (b).

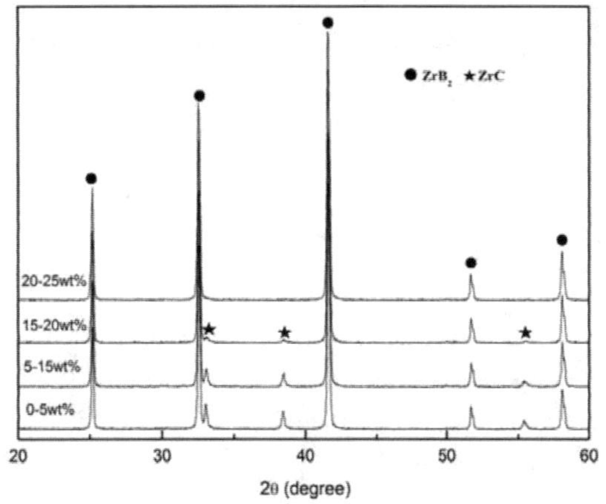

Figure 3. XRD patterns of the synthesized ZrB_2 powder at 1650°C with different amounts of B_4C in excess.

In order to eliminate ZrC phase, the B_2O_3 loss could be compensated by a substantial excess of B_4C. At the same time, due to the presence of element C in B_4C, the amount of raw carbon should be decreased. Figure 3 showed the XRD patterns of the synthesized powder by adjusting the starting amounts of B_4C and carbon. With the increase of B_4C, ZrC phase gradually decreased. When B_4C was in a 20-25 wt% excess, the ZrC phase disappeared and only ZrB_2 phase existed in powder.

XRD patterns of the synthesized powder with starting B_4C in excess of 20-25 wt% at different temperatures for 1 h were shown in figure 4. With the increase of temperature, a peculiar phenomenon occurred in XRD patterns. The ZrC phase was detected at 1550 °C, but disappeared at 1650 °C. The formation of ZrC should be based on the reaction (4). Due to the presence of enough boron, the ZrC phase could be consumed by the reaction with B_2O_3 and carbon:

$$ZrC + B_2O_3 + 2C \rightarrow ZrB_2 + 3CO \quad (5)$$

Thermodynamic calculation showed the minimum favorable temperature of reaction (5) is 806 °C under 11 Pa pressure (figure 2(b)). Finally, ZrC phase disappeared at 1650 °C, leaving pure ZrB_2 powder.

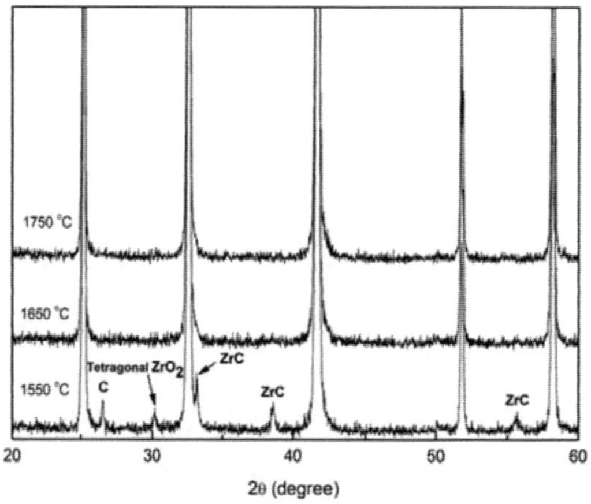

Figure 4. XRD patterns of the synthesized ZrB_2 powder at different temperatures with B_4C in 20-25 wt% excess.

From figure 4, pure ZrB_2 powder was obtained above 1650 °C. The ZrB_2 powder synthesized at 1650 °C and 1750 °C was referred to as Z6 and Z7, respectively. Figure 5 showed the SEM images of Z6 and Z7 powder. SEM observation did not demonstrate remarkable difference between the Z6 and Z7 powder. However, it can be found that there was a tendency for the particle size and particle morphology with temperature increase, that is, most of the ZrB_2 particles obtained at 1650 °C presented nearly spherical-like with smaller size. When temperature reached 1750 °C, the ZrB_2 powder showed nearly column-like with increasing size. The oxygen content of Z6 and Z7 powder had no obvious difference and was about 0.7 wt%.

Table 1 lists the relative densities of the ZrB_2 ceramics by pressureless sintering in the temperature range of 1900-2100 °C for 2 h using Z6 and Z7 powder. For convenient comparison, the reported results for the pressureless sintered ZrB_2 ceramics from pure commercial ZrB_2 powder were also listed in table 1 [29,33]. The decrease in particle size would result in a higher driving force for densification during pressureless sintering [29,33-34]. Therefore, the Z6 powder showed better sinterability when compared to the Z7 powder as well as the commercial ZrB_2 powder used by Fahrenholtz et al.. The surface oxygen contamination of ZrB_2 particles, in the form of B_2O_3 and ZrO_2, evidently inhibited the densification [35]. Reducing the oxygen content of ZrB_2 powder could enhance densification. Zhang et al. have shown that the relative densities of ZrB_2 increased from 53.8 % to 63.6 % at 1850 °C by pressureless sintering, with the reduction of oxygen content from 3 % to 2 % [34]. So, compared to fine commercial ZrB_2 powder used by Zhu et al., the Z6 powder showed better sinterability due to its lower oxygen content. Accordingly, due to its smaller particle size and lower oxygen content, the synthesized ZrB_2 powder at 1650 °C in this work showed better sinterability.

Figure 5. SEM morphologies of the ZrB_2 powder synthesized at 1650 °C (a) and 1750 °C (b) for 1 h.

Table 1. Relative densities of the ZrB_2 ceramics pressureless sintered at different temperatures for 2 h.

Sample	Particle size (μm)	Oxygen content (wt%)	Relative density (%)		
			1900 °C	2000 °C	2100 °C
Z6	0.5~1	~0.7	~70	~85	~93
Z7	1~1.5	~0.7	~66	~80	~91
Fahrenholtz et al.[33]	~2.0	~0.9			~78
Zhu et al. [29]	~0.5	~3.0			~70

3.2. Borothermal reduction synthesis of HfB_2 and hot-pressed densification

HfB_2 is very expensive. consequently, the research on HfB_2-based ceramics is relatively less at present compared to ZrB_2. Therefore, it would be very attractive for large-scale synthesis of HfB_2 powders in a convenient and low-cost way. From above boro/carbothermal reduction synthesis of ZrB_2 powder, it is found that some impurities, such as ZrC, may exist in the final products. Therefore, to more easily prepare the pure HfB_2 powder, another potentially more useful method is conducted according to the borothermal reduction reaction using HfO_2 and boron.

Figure 6 showed XRD pattern of the synthesized HfB_2 powder at 1600 °C for 1 h according to stoichiometric proportion by forming B_2O_3 product. In addition to HfB_2 phase, however, the monoclinic HfO_2 was detected, which did not match the theoretical analysis. One rational explanation was that boron sublimated in the form of not only B_2O_3, but also boron suboxide such as B_2O_2 [36-37], which resulted in the lack of boron and the excess of HfO_2 phase. This analogous phenomenon in the synthesis of HfB_2 powder by carbothermal reduction of HfO_2 with B_4C and graphite was also found in our previous study [30]. Based on the data of XRD, HfO_2 mass fraction was estimated as about 9.7% using the K value method, which meant that there's about 10% boron loss. Figure 7 showed the XRD pattern of the synthesized powder at 1600 °C for 1h with starting boron in excess of 10 wt%, where the HfB_2 powder was referred to as HB10. The XRD pattern showed only hexagonal HfB_2 phase and no evidence of impurities was observed. This indicated that the excess of 10 wt% boron could compensate boron loss which caused by the sublimation of boron in the form of boron-rich oxides.

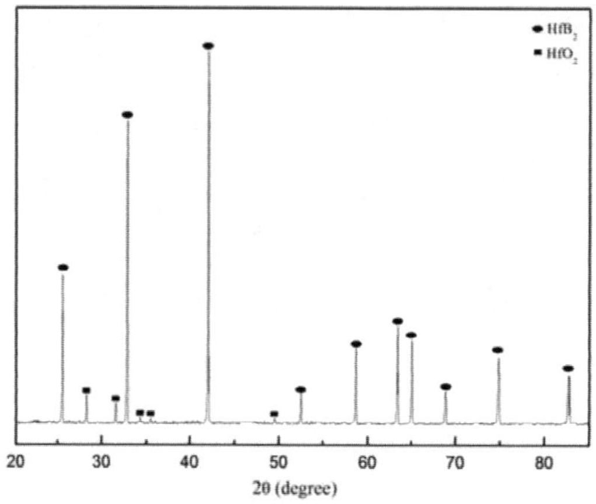

Figure 6. XRD pattern of the synthesized HfB_2 powder at 1600 °C for 1 h according to stoichiometric proportion by forming B_2O_3 product..

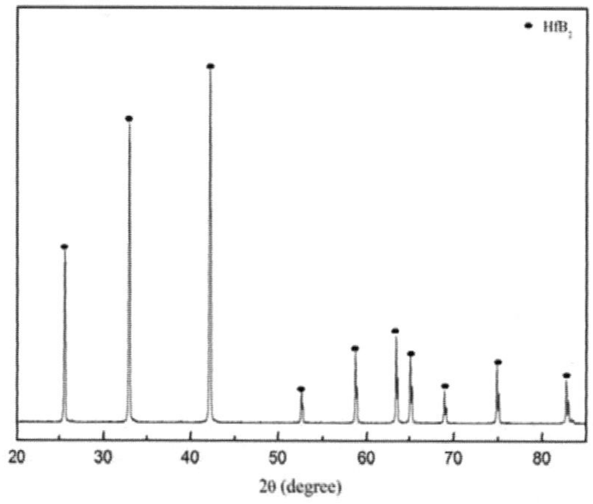

Figure 7. XRD pattern of the synthesized HfB_2 powder at 1600 °C for 1h with starting boron in excess of 10 wt%.

The oxygen content of the HB10 was 0.79 wt%. The SEM micrograph of the HB10 was shown in figure 8. The as-synthesized HfB_2 powder had a particle size of about 1-1.5 μm with an equiaxed shape, and had a very narrow diameter distribution, but soft agglomeration appears. Figure 9 was the results of laser particle size analysis. The laser particle size analysis confirmed that the mean diameter and polydispersity of the as-prepared HfB_2 powder HB10 are 1.37 μm and 0.005, respectively, which was in accordance with the above SEM observation.

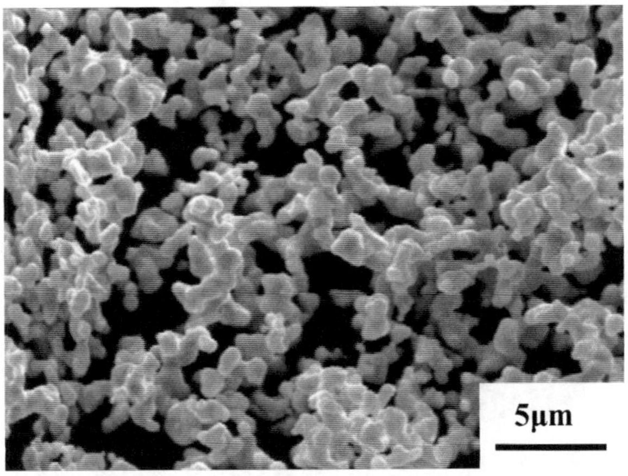

Figure 8. SEM morphology of the as-synthesized HfB_2 powder HB10

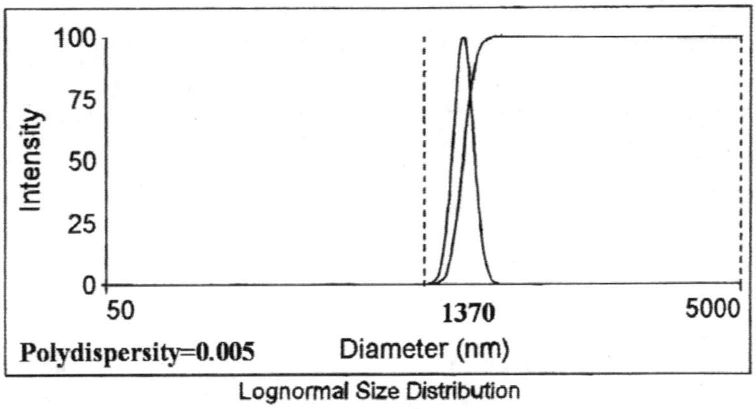

Figure 9. The size distribution of the as-synthesized HfB_2 powder HB10

The small particle size and uniform diameter distribution is believed to be helpful for further ceramic sintering process to obtain materials with uniform microstructure and high performance. To determine the lowest temperature for reaction completion, heat treatment of HB10 was carried out at 1400 °C and 1500 °C with different holding time successively. However, we found the existence of HfO_2 impurity in the as-prepared HfB_2 powder heat treated both at 1400 °C and 1500 °C, which seems disagree with the thermodynamic calculation. In fact, the thermodynamic calculation can just give out a possibility for a reaction at a certain temperature, but we cannot know whether the reaction could complete or not at that temperature from the thermodynamic calculation. Therefore, heat treatment carried out at 1600 °C or over is necessary to ensure reaction complete. However, it is known that grain grows up and sintering activity decreases as heat-treatment temperature rises. So the HfB_2 powder synthesized at 1600 °C was chosen in the present study.

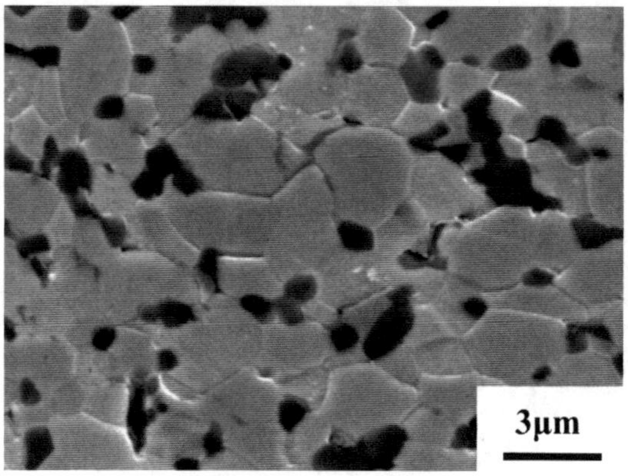

Figure 10. SEM micrographs of polished surfaces of HBS after acid corrosion

Table 2. The starting composition and mechanical properties of the HBS ceramics

Sample	Raw meterials (vol.%)	Vickers' hardness (GPa)	Flexural strength (MPa)	Fracture toughness ($MPam^{1/2}$)
HBS	$80HfB_2$+20SiC	20.2±1.0	993±90	6.29±0.65
65# [38]	$80HfB_2$+20SiC	17±0.7	453±46	4.1±0.2
HS [39]	$80HfB_2$+20SiC	17.3	476±46	4.2±0.6
A-HP [40]	$77HfB_2$+20SiC+3HfN	22.3±0.9	650±55	3.8±0.3

The HfB_2-20 vol.% SiC composite was referred to as HBS. Figure 10 showed the SEM micrograph of the polished surface of materials HBS after acid corrosion. The SiC particulates were mainly equiaxed and homogeneously distributed within the diboride skeleton. The average grain size of HfB_2 was estimated as about 2 μm, which was smaller than that of previous studies [38-39]. The mechanical properties of HBS, including Vickers' hardness, bend strength and fracture toughness, were summarized in table 2. The hardness of HBS was 20.2±1.0 GPa, which was higher than the reported value for HfB_2-20 vol.% SiC [38-39], but slightly lower than the value for HfB_2-20 vol.% SiC-3 vol.% HfN reported by Bellosi et al. (22.3 GPa) [40]. The bending strength of HBS ceramics was 993 ± 90 MPa, which was significantly higher than those reported in literature for HfB_2-20 vol.% SiC materials (450-480 MPa) [38-39]. The fracture toughness for HBS was 6.29±0.65 $MPa{\cdot}m^{1/2}$, which was substantially higher than those reported in literature for HfB_2-20 vol.% SiC-based materials (3.8-4.15 $MPa{\cdot}m^{1/2}$) [38-40].

4. Summary

Phase pure ZrB_2 and HfB_2 powders were synthesized by boro/carbothermal reduction (BCTR) of ZrO_2 and borothermal reduction of HfO_2, respectively. Combining the thermodynamic calculations and XRD results, the reaction process of BCTR of synthesized ZrB_2 powder was shown. Compared to the synthesized powder at 1750 °C as well as the commercially available powder used in literature reported work, the ZrB_2 powder synthesized at 1650 °C showed a better sinterability due to its smaller particle size and lower oxygen content. Most of the HfB_2 particles obtained at 1600 °C presented nearly column-like morphology with 1-1.5μm size. Using the as-synthesized HfB_2 powder and a commercial α-SiC, full HfB_2-20 vol.% SiC composite was prepared by hot pressing. The composite has high flexural strength of 993 ±90 MPa and fracture toughness of 6.29±0.65 MPa·m$^{1/2}$, which is significantly higher than the reported values in literature.

Acknowledgements
The authors thank the financial supports from the Chinese Academy of Sciences under the Program for Recruiting Outingstanding Overseas Chinese (Hundred Talents Program) and the National Natural Science Foundation of China (No. 50632070).

References

[1] Fahrenholtz W G and Hilmas G E 2004 *NSF-AFOSR Joint Workshop on Future Ultra-High Temperature Materials* (NSF, Arlington, VA 22230)
[2] Ohlhorst C W, Glass D E, Bruce W E, Lindell M C, Vaughn W L and Smith R W 2005 *Development of X-43A Mach 10 Leading Edges* (56th International Astronautical Congress, Fukuoka, Japan) IAC-05-D2.5.06.
[3] Gasch M J, Ellerby D T and Johnson S M 2005 *Ultra high temperature ceramic composites Handbook of Ceramic Composites* ed N P Bansal (New York: Springer) p 197
[4] Spring A, Guo W M, Zhang G J, Wang P L and Krstic V D 2008 *J. Am. Ceram. Soc.* **91** 2763
[5] Opeka M M, Talmy I G, Wuchina E J, Zaykoski J A and Causey S J 1999 *J. Eur. Ceram. Soc.* **19** 2405
[6] Opila E, Levine S and Lorincz J 2004 *J. Mater. Sci.* **39** 5969
[7] Guo W M, Zhang G J, Kan Y M and Wang P L 2008 *J. Alloys. Compd.* doi:10.1016/j.jallcom.2008.04.006
[8] Guo W M, Zhou X J, Zhang G J, Kan Y M, Li Y G and Wang P L 2008 *J. Alloys. Compd.* doi:10.1016/j.jallcom.2008.02.108
[9] Chamberlain A L, Fahrenholz W G, Hilmas G E and Ellerby D T 2004 *J. Am. Ceram. Soc.* **87** 1170
[10] Hwang S S, Vasiliev A L and Padture N P 2007 *Mater. Sci. Eng. A* **464** 216
[11] Gasch M, Johnson S and Marschall J 2008 *J. Am. Ceram. Soc.* **91** 1423
[12] Fahrenholtz W G and Hilmas G E 2007 *J. Am. Ceram. Soc.* **90** 1347
[13] Zhang G J, Deng Z Y, Kondo N, Yang J and Ohji T 2000 *J. Am. Ceram. Soc.* **83** 2330
[14] Wu W W, Zhang G J, Kan Y M and Wang P L 2006 *J. Am. Ceram. Soc.* **89** 2967
[15] Wu W W, Zhang G J, Kan Y M and Wang P L 2008 *J. Am. Ceram. Soc.* **91** 2501
[16] Wu W W, Zhang G J, Kan Y M, Wang P L, Vanmeensel K, Vleugels J and der Biest O V 2007 *Scripta Mater.* **57** 317
[17] Medri V, Monteverde F, Balbo A and Bellosi A 2005 *Adv. Eng. Mater.* **7** 159
[18] Zhou X J, Zhang G J, Li Y G, Kan Y M and Wang P L 2007 *Mater. Lett.* **61** 960
[19] W. M. Guo, Zhou X J, Zhang G J, Kan Y M, Li Y G and Wang P L 2008 *Mater. Lett.* **62** 3724

[20] Meléndez-Martínez J J, Domínguez-Rodríguez A, Monteverde F, Melandri C and de Portu G 2002 *J. Eur. Ceram. Soc.* **22** 2543

[21] Monteverde F, Bellosi A 2002 *Scripta Mater.* **46** 223

[22] Monteverde F, Guicciardi S and Bellosi A 2003 *Mater. Sci. Eng. A* **346** 310

[23] Monteverde F, Bellosi A 2005 *Solid State Sci.* **7** 622

[24] Zou J, Zhang G J, Kan Y M and Wang P L 2008 *Scripta Mater.* **59** 309

[25] Guo W M, Zhang G J, Zou J, Kan Y M and Wang P L 2008 *Adv. Eng. Mater.* **10** 759

[26] Wuchina E, Opeka M, Causey S, Buesking K, Spain J, Cull A, Routbort J and Guitierrez-Mora F 2004 *J. Mater. Sci.* **39** 5939

[27] Bellosi A, Monteverde F, Fabriche D D and Melandri C 2000 *J. Mater Proc. Manuf. Sci.* **9** 156

[28] Baik S and Becher P F 1987 *J. Am. Ceram. Soc.* **70** 527

[29] Zhu S M, Fahrenholtz W G, Hilmas G E and Zhang S C 2007 *Mater. Sci. Eng. A* **459** 167

[30] Ni D W, Zhang G J, Kan Y M and Wang P L 2008 *J. Am. Ceram. Soc.* **91** 2709

[31] Karasev A I 1973 *Powder Metall. Met. Ceram.* **12** 926

[32] Fahrenholtz W G 2005 *J. Am. Ceram. Soc.* **88** 3509

[33] Fahrenholtz W G, Hilmas G E, Zhang S C and Zhu S M 2008 *J. Am. Ceram. Soc.* **91** 1398

[34] Zhang S C, Hilmas G E and Fahrenholtz W G 2006 *J. Am. Ceram. Soc.* **89** 1544

[35] Zhu S M, Fahrenholtz W G, Hilmas G E and Zhang S C 2007 *J. Am. Ceram. Soc.* **90** 3660

[36] Tanaka T, Okada S, Ishizawa Y 1994 *J. Alloys Compd.* **205** 281

[37] Tanaka T, Okada S, Yu Y, Ishizawa Y 1997 *J. Solid State Chem.* **133** 122

[38] Gasch M, Ellerby D, Irby E, Beckman S, Gusman M and Johnson S 2004 *J. Mater. Sci.* **39** 5925

[39] Weng L, Zhang X H, Han J C, Han W B, Hong C Q 2008 *J. Alloys. Compd.* doi:10.1016/j.jallcom.2008.05.093

[40] Bellosi A, Monteverde F and Sciti D 2006 *Int. J. Appl. Ceram. Technol.* **3** 32

Journal of Physics: Conference Series **176** (2009) 012042 doi:10.1088/1742-6596/176/1/012042

Thermoelectric properties of B-doped $SrTiO_3$ singe crystal

Jun Okamoto[1], Gen Shimizu[1], Shugo Kubo[1], Yasuji Yamada[1*], Hiroyuki Kitagawa[1], Akiyuki Matsushita[2], Yuh Yamada[3] and Fumihiro Ishikawa[4]

[1] Department of Materials Science, Shimane University, Matsue, Shimane 690-8504, Japan
[2] National Institute for Materials Science, Tsukuba, Ibaraki 305-0047, Japan
[3] Department of Physics, Niigata University, Niigata, Niigata 950-2181, Japan
[4] Graduate School of Science and Technology, Niigata University, Niigata, Niigata 950-2181, Japan

*E-mail: yamadaya@riko.shimane-u.ac.jp

Abstract. Effect of boron on the electric and thermoelectric properties of $SrTiO_3$ has been studied. Boron-doped $SrTiO_3$ was prepared by vacuum annealing of $SrTiO_3$ single crystals in presence of boron vapors. The crystals show low resistivity ~0.1 Ω·cm and high Seebeck coefficient of several hundreds μV/K at a room temperature. It was estimated that almost every boron atom incorporated in $SrTiO_3$ crystal provided a charge carrier. Temperature dependence of Seebeck coefficient and figure of merit *ZT* were measured as a function of carrier concentration.

1. Introduction

$SrTiO_3$ (STO) has a simple lattice structure and highly degenerate electronic orbitals. This results in the narrow conduction band and therefore, in relatively large electron mass and a large power factor of 2-3 × 10^{-3} W/m·K^2, comparable to that of the practical thermoelectric material BiTe [1]. However, because of the high thermal conductivity of ~10 W/m·K, the dimensionless figure of merit *ZT* of $SrTiO_3$ is only 0.1.

Ion substitution can increase *ZT* through reduction of the lattice thermal conductivity and of electrical resistivity. Indeed, yttrium substitution for Sr provides carrier doping and reduces thermal conductivity, thereby enhancing *ZT* [2]. Recently, multilayered structures of STO, containing 24 layers of a 4-unit-cell Nb-doped and non-doped STO, were produced by pulsed laser deposition [3]. They exhibited a high value of $ZT = 2.4$, which is attributed to the quantum confinement of electron wave function in a layered structure. Therefore, the electric, thermal and thermoelectric properties of STO and its structure have drawn much attention.

Recently, we found an effect of boron on electric and thermoelectric properties of STO crystal heat-treated with boron vapors: the resistivity was significantly reduced from insulating to conducting, and very high Seebeck coefficient was observed. This boron-related change of STO properties has not been recognized. In this paper, we present resistivity, carrier density and Seebeck coefficient of the

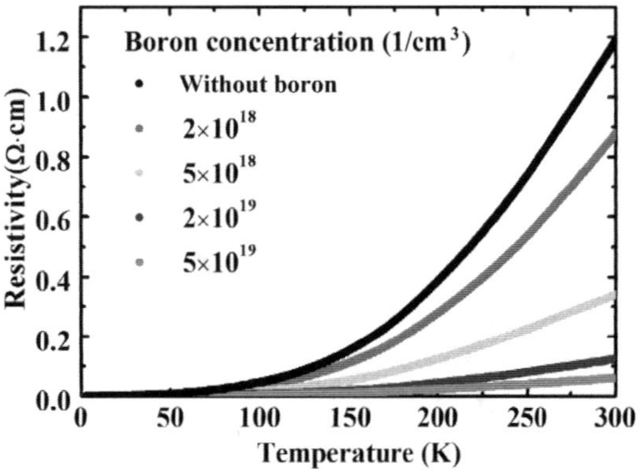

Figure 1. Resistivity of the boron-doped and undoped STO single crystals as a function of temperature for various boron concentrations.

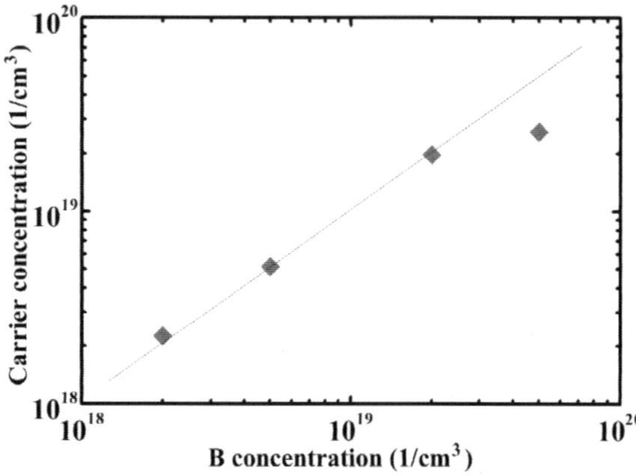

Figure 2. Carrier concentration at 300 K determined by the Hall measurement as a function of boron concentration.

boron-doped STO. In addition, the origin of the conductivity, in terms of boron doping or oxygen vacancies, is discussed by comparing STO samples with and without the boron doping.

2. Experimental procedure

We used single crystals of STO with dimensions $1 \times 1 \times 0.035$ cm^3. The largest surfaces were parallel to the crystallographic (100) plane and were optically polished. The crystals were heated at 900 °C in a vacuum chamber, wherein boron was evaporated by electron beam heating. During evaporation, the vacuum was kept at 10^{-4} Pa. Evaporation rate was monitored by a film thickness monitor placed near the STO plate and held constant at 1.16 mol/s·cm^2. Doping amount of boron was controlled by changing the evaporation time. After the evaporation, the plate was kept at the same temperature for 600 s to diffuse the boron atoms from the surface into the bulk. To characterize the STO samples, we measured Seebeck coefficient, electrical resistivity by four-probe method, as well as carrier concentration and Hall mobility by Hall effect in the Van der Pauw configuration.

3. Results

After the heat treatment with boron vapors, no deposition of boron was detected on the surface by X-ray diffraction. A very weak B_2O_3 peak, however, appeared on the plate exposed for long period. Resistivity was lower for the samples annealed with boron than without boron, and it decreased with increasing evaporation time. These facts indicate that boron deposited on the plate has incorporated into STO and introduced a substantial amount of carriers to the STO crystal. Secondary ion mass spectroscopy indicated inhomogeneous boron distribution in the samples. Nevertheless, we define the boron concentration as an averaged value, i.e., assuming that the boron homogeneously distributed in the crystal. This assumption is made for convenience of presenting the results and also for understanding the relationship between the boron doping and carrier density.

Figure 1 shows the resistivity of the STO plate as a function of temperature for various boron concentrations. In the figure, we also plotted resistivity of the undoped sample, i.e., treated at the same conditions without boron vapors. The boron-doped samples had lower resistivity than the undoped sample. Their resistivity showed metallic temperature dependence and decreased as boron concentration increased. The Hall measurement indicated that carriers were electrons for all samples. Carrier concentration of the doped samples at 300 K is presented in figure 2 as a function of boron concentration. It is proportional to boron concentration except for the highest concentration. The sample with highest concentration contained B_2O_3 on the surface. Carrier concentration of the undoped sample was of the order of 10^{16} cm^{-3}, about two orders of magnitude lower than that of the doped samples. Hall mobility of the doped samples was typically less than 10 cm^2/V·s, but it was ~100

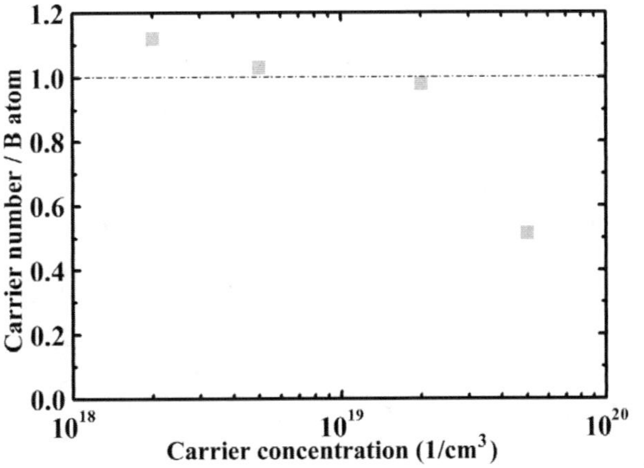

Figure 3. Carrier concentration dependence of carrier number per boron atom.

$cm^2/V{\cdot}s$ in the undoped sample. Figure 3 displays carrier number per boron atom as a function of the carrier concentration.

Figure 4 shows Seebeck coefficient as a function of carrier concentration. Seebeck coefficient was as high as 700 μV/K and decreased as carrier concentration increased. Dimensionless figure of merit, *ZT* was evaluated using the obtained Seebeck coefficient, resistivity and the reported thermal conductivity value of 11.2 W/m·K [3]. The thermal conductivity is reported as almost constant below carrier concentration of 10^{21} cm^{-3}. Therefore, we assume constant thermal conductivity in the evaluation. Figure 5 shows dimensionless figure of merit *ZT* including the previously reported values [1, 3, 4]. Temperature dependence of Seebeck coefficient is depicted in figure 6.

4. Discussion

Although the boron-doped and undoped samples show comparable resistivity in figure 1, carrier concentration of the boron-doped samples was 30-300 times higher. This indicates that the origin of the lower resistivity of the boron-doped samples is not merely the oxygen deficiency, created by annealing at high temperature in vacuum. In addition, the almost proportional dependence of carrier concentration on boron concentration shown in figure 2 implies that boron atoms incorporated in the STO crystal and acted as carrier dopants. Figure 3 indicates that one boron atom provided almost one

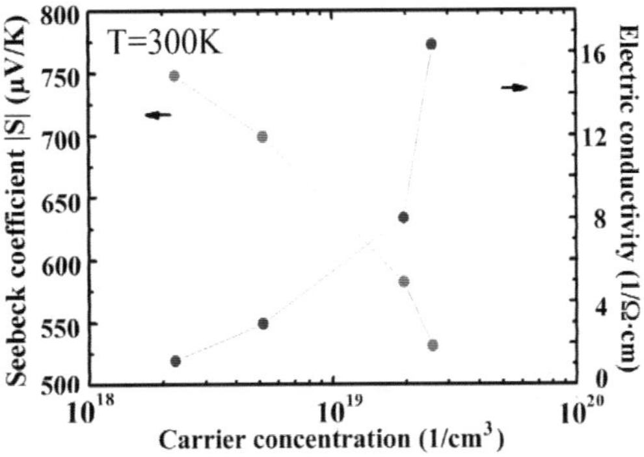

Figure 4. Seebeck coefficient as a function of carrier concentration.

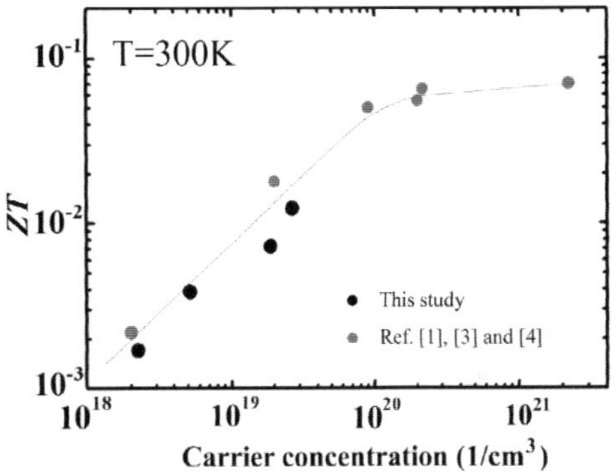

Figure 5. Dimensionless figure of merit *ZT* as a function of carrier concentration.

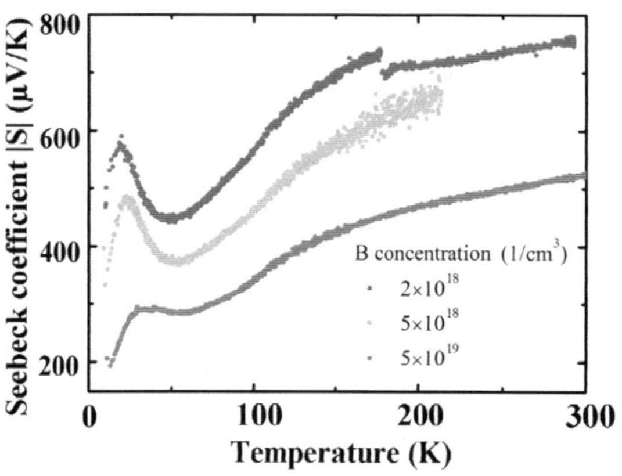

Figure 6 Temperature dependence of Seebeck coefficient.

electron, as long as B_2O_3 was not formed on the sample surface. Therefore, we conclude that high-temperature annealing with boron vapors incorporates boron atoms and introduces carriers into STO crystal and that the oxygen vacancies contribute less to the doping than boron atoms.

The Seebeck coefficient and dimensionless figure of merit measured on the boron-doped STO crystals are comparable to reported values on STO. This indicates that heat treatment of STO with boron vapors is a valid process of STO doping, and that STO is a potential thermoelectric material.

5. Conclusions

STO single crystals annealed in a vacuum at 900 °C with boron vapors exhibit low resistivity ~0.1 Ω·cm, lower than that of the STO annealed without boron. The resistivity decreased as boron evaporation time increased, and the estimated carrier density in boron-exposed samples was about two orders of magnitude lower than that of the sample without boron. Therefore, the measured electrical properties originated predominantly from boron interaction with STO. The carrier concentration calculated from the Hall measurement is almost proportional to the assumed boron concentration in the STO crystal. It implies that one boron atom provided one carrier in the STO. Consequently, these results indicate that boron was introduced into the STO crystal and acted as a carrier dopant.

Seebeck coefficient measured from the boron doped STO crystal was more than several hundreds μV/K. Dimensionless figure of merit also showed a similar behavior as in a typical carrier-doped STO. Quite a few studies have been conducted on the interaction between boron and STO. Therefore, further investigation on boron-doped STO is necessary for revealing the carrier doping mechanism and the potential for thermoelectric applications.

References

[1] Okuda T, Nakanisi K, Miyasaka S and Tokura Y 2001 *Phys. Rev.* B **63** 113104
[2] Obara H, Yamamoto A, Lee C, Kobayashi K, Matsumoto A and Funahashi R 2004 *Jpn. J. Appl. Phys.* **43** L540
[3] Ohata H, Kim S-W, Mune Y, Mizoguchi T, Nomura K, Ohta S, Nomura T, Nakanishi Y, Ikuhara Y, Hirano M, Hosono H and Koumot K 2007 *Nature Mater.* **6** 129
[4] Frederikse H P R, Thurber W R and Hosler W R 1964 *Phys. Rev.* **134** A442

Journal of Physics: Conference Series **176** (2009) 012043 doi:10.1088/1742-6596/176/1/012043

Preparation of ultrafine boride powders by metallothermic reduction method

Katsuhiro Nishiyama*[1], Takanobu Nakamur[2], Shigenori Utsumi[1], Hideki Sakai[2], and Masahiko Abe[2]

[1]Department of Mechanics and Systems, Faculty of System Engineering, Tokyo University of Science, Suwa, 5000-1 Toyohira, Chino, Nagano, 391-0292, Japan
[2]Department of Pure and Applied Chemistry, Faculty of Science and Technology, Tokyo University of Science, 2641 Yamazaki, Noda, Chiba, 278-8510, Japan

*E-mail: nsym@rs.suwa.tus.ac.jp

Abstract. Ultrafine TiB_2, ZrB_2, and ReB_2 powders were prepared by the metallothermic reduction method using Mg. Obtained TiB_2, ZrB_2, and ReB_2 powders were characterized by X-ray diffraction, scanning electron microscopy, transmission electron microscopy, dynamic light scattering, and scratch test. The morphology of the powders was a laminar hexagonal single crystal. The mean particle sizes of TiB_2 and ZrB_2 increased with increasing reaction temperature, and that of ReB_2 was 0.11 μm at 1093 K. The obtained ReB_2 powder can scratch the surface of polycrystalline synthetic diamonds (H_K = 7000 kg/mm^2). The obtained powders will be useful for developing of new superhard alloys and pigments.

1. Introduction

Boron combines with various elements forming a large group of more than 200 compounds. Metal borides are mostly covalent compounds with metallic appearance, high electric and thermal conductivities, high melting point and high hardness [1]. In particular, TiB_2 and ZrB_2 which combine with group-IV metals have high hardness and heat resistance. Recently, the synthesis of ReB_2 which has hardness comparable to that of diamond, has been reported [2]. However, high temperature was necessary for the synthesis of metal diboride powders by usual methods [3]. At present, commercially available boride powders have a large particle size of several tens μm. Even after mechanical crushing treatment, such as ball milling or vibration milling, the smallest average particle size is 1-2 μm. Impurities coming from a vessel in the crushing treatment are also a serious problem.

The preparation methods of metal boride powders are roughly classified into the following five groups [1,4-8]: (i) direct reaction, (ii) borothermic reduction, (iii) carbothermic reduction, (iv) borocarbide reduction, and (v) metallothermic reduction [8, 9]. Commercially available metal diboride powders are mainly prepared by the methods (ii)-(iv). High temperature is necessary for these methods between 2073 and 3273 K [1,6], and hence, the formation of coarse particles is unavoidable. Therefore, establishing a preparation method for ultrafine metal diboride powders is necessary for the developments of a new superhard alloy and a pigment having these characteristics of metal borides.

This paper reports the syntheses of ultrafine TiB_2, ZrB_2, and ReB_2 particles by the metallothermic reduction method using Mg.

2. Thermodynamic consideration

Thermodynamic consideration of TiB_2 and ZrB_2 formations were carried out. Since no thermodynamic data on ReB_2 could be found in the literature, we estimated the free energies of TiB_2 and ZrB_2 formation by metallothermic reduction. The chemical equations of metallothermic reduction are listed as eqs. (1)-(10) when Mg, Si, Zn, Al, and Ca are considered as a metal reduction reagent. The equations of direct reaction, borothermic reduction, carbothermic reduction and borocarbide reduction are shown in figure 1.

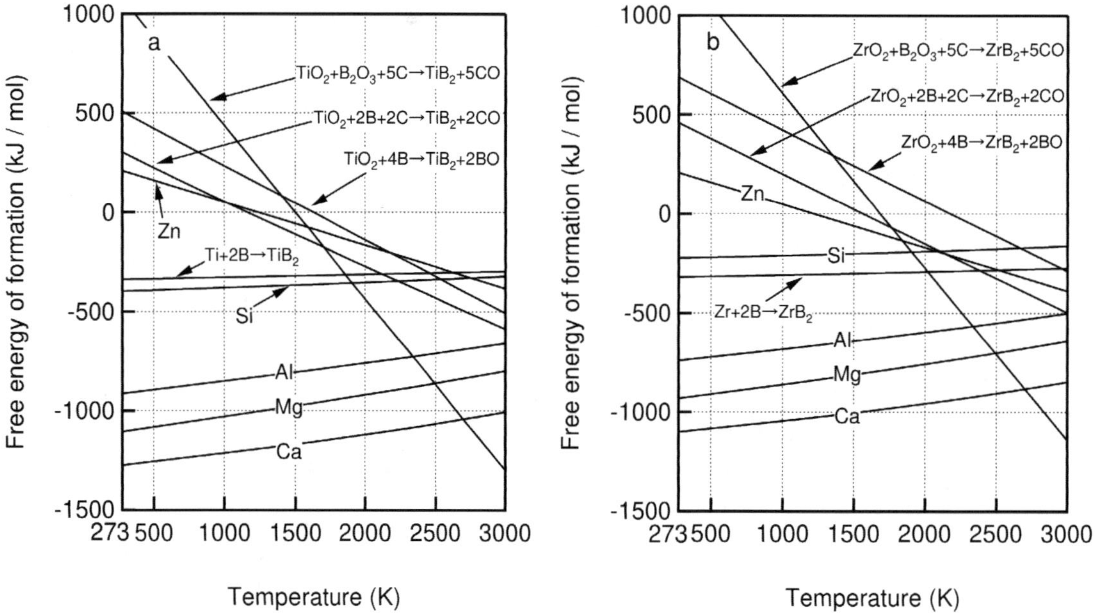

Figure 1. Temperature dependences of the free energy of formation of TiB_2 (a) and ZrB_2 (b) by the usual methods and metallothermic reduction.

$TiO_2 + B_2O_3 + 5Mg \rightarrow TiB_2 + 5MgO$

$$\Delta G^0_{Mg} = 1.72\times10^{-1}T - 1.04\times10^{-2}T\ln T + 7.95\times10^{-6}T^2 + 3.45\times10^{2}T^{-1} - 1.14\times10^{3} \quad (1)$$

$TiO_2 + B_2O_3 + 2.5Si \rightarrow TiB_2 + 2.5SiO_2$

$$\Delta G^0_{Si} = 9.34\times10^{-2}T - 9.70\times10^{-3}T\ln T + 4.69\times10^{-6}T^2 + 9.65\times10^{2}T^{-1} - 4.12\times10^{2} \quad (2)$$

$TiO_2 + B_2O_3 + 5Zn \rightarrow TiB_2 + 5ZnO$

$$\Delta G^0_{Zn} = 1.42T - 1.99\times10^{-1}T\ln T - 2.34\times10^{-6}T^2 + 2.10\times10^{3}T^{-1} + 4.72\times10 \quad (3)$$

$TiO_2 + B_2O_3 + 10/3Al \rightarrow TiB_2 + 5/3Al_2O_3$

$$\Delta G^0_{Al} = 1.59\times10^{-1}T - 1.07\times10^{-2}T\ln T + 7.04\times10^{-6}T^2 + 5.84\times10^{2}T^{-1} - 9.44\times10^{2} \quad (4)$$

$TiO_2 + B_2O_3 + 5Ca \rightarrow TiB_2 + 5CaO$

$$\Delta G^0_{Ca} = 2.17\times10^{-1}T - 2.05\times10^{-2}T\ln T + 1.54\times10^{-7}T^2 + 3.67\times10^{2}T^{-1} - 1.30\times10^{3} \quad (5)$$

$ZrO_2 + B_2O_3 + 5Mg \rightarrow ZrB_2 + 5MgO$

$$\Delta G^0_{Mg} = 1.74\times10^{-1}T - 1.19\times10^{-2}T\ln T + 9.43\times10^{-6}T^2 + 3.43\times10^{2}T^{-1} - 9.63\times10^{2} \quad (6)$$

$ZrO_2 + B_2O_3 + 2.5Si \rightarrow ZrB_2 + 2.5SiO_2$

$$\Delta G^0_{Si} = 9.61\times10^{-2}T - 1.12\times10^{-2}T\ln T + 6.17\times10^{-6}T^2 + 9.63\times10^2T^{-1} - 2.73\times10^2 \quad (7)$$

$ZrO_2 + B_2O_3 + 5Zn \rightarrow ZrB_2 + 5ZnO$

$$\Delta G^0_{Zn} = 1.42T - 2.00\times10^{-1}T\ln T + 8.62\times10^{-7}T^2 + 2.10\times10^3T^{-1} - 2.22\times10^2 \quad (8)$$

$ZrO_2 + B_2O_3 + 10/3Al \rightarrow ZrB_2 + 5/3Al_2O_3$

$$\Delta G^0_{Al} = 1.61\times10^{-1}T - 1.22\times10^{-2}T\ln T + 8.52\times10^{-6}T^2 + 5.83\times10^2T^{-1} - 7.69\times10^2 \quad (9)$$

$ZrO_2 + B_2O_3 + 5Ca \rightarrow TiB_2 + 5CaO$

$$\Delta G^0_{Ca} = 2.19\times10^{-1}T - 2.20\times10^{-2}T\ln T + 1.69\times10^{-7}T^2 + 3.66\times10^2T^{-1} - 1.13\times10^3 \quad (10)$$

Figure 1 shows the free energies of TiB_2 and ZrB_2 formations as a function of temperature. It is obvious that Mg, Al, and Ca are suitable for metallothermic reduction, because the free energies using Mg, Al, and Ca are smaller than free energies by usual methods at relatively low temperatures. When Al and Ca are used, it is difficult to remove the by-product of Al_2O_3 by acid treatment and to deal with Ca in the atmosphere. Therefore, Mg was finally selected as a metal reduction reagent because MgO formed as a slug can be easily removed by acids.

3. Experimental details

3.1. Starting reagents

Powdered TiO_2 (1-2 μm; Soekawa Chemical Co.), ZrO_2 (2 μm; Soekawa Chemical Co.), ReO_2 (Soekawa Chemical Co.), B_2O_3 (40-60 mesh; Soekawa Chemical Co.), and Mg (<200 mesh; Yamaishimetal Co.) were used as starting reagents. The 30 wt% of MgO (200 nm; Soekawa Chemical Co.) was added as a negative catalyst [9].

3.2. Metallothermic reduction and purification procedure

Figure 2 shows a flow chart of the preparation process of aimed products.

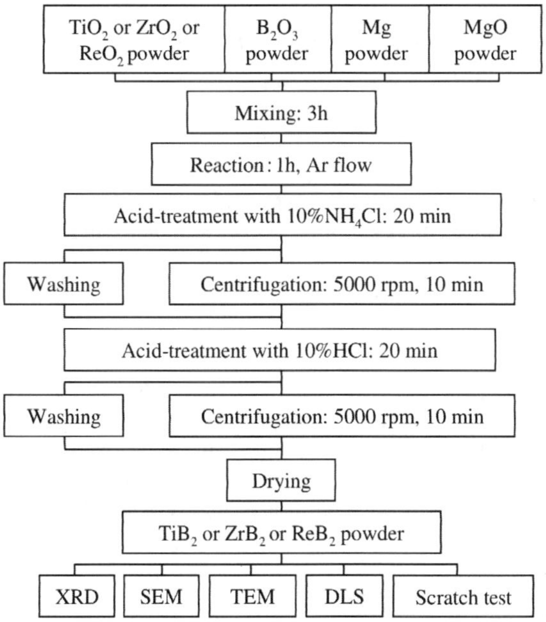

Figure 2. Preparation process flow chart of the metal diboride products.

The starting powders were weighed in the molar ratio of MO_2: B_2O_3: Mg = 1: 1: 5, where M is Ti, Zr, or Re. The powders were mixed by an Al_2O_3 ball mill at 120 rpm for 3 h, and then placed into a graphite boat with an internal volume of $30 \times 30 \times 400$ mm^3. Figure 3 shows a stainless steel tubular furnace for metallothermic reduction. The graphite boat with the mixed powder was put at one end of the tubular furnace, and then the central part of the furnace was heated to the reaction temperature. Metallothermic reduction was carried out for 1 h in an Ar flow (1 L/min) after moving the boat to the heating zone. After the reduction, the boat was cooled to room temperature at the other end of the furnace. The aimed products were obtained by the acid treatment with boiling 10% of NH_4Cl and boiling 10% of HCl for 20 min and centrifugation at 5000 rpm, followed by washing with distilled water 5 times and drying.

3.3. Characterization of the products
The obtained powders were identified by X-ray diffraction (XRD: Rigaku, MiniFlex II). The morphology of the powders was observed with a scanning electron microscope (SEM: JEOL, JSM-5910LV) and a transmission electron microscope (TEM: HITACHI, H-7650, accelerating voltage 120 kV). Particle size distributions were measured by dynamic light scattering (DLS: NICONMP, 380ZLS). Hardness test was performed on the obtained ReB_2 powder using synthetic polycrystalline diamonds.

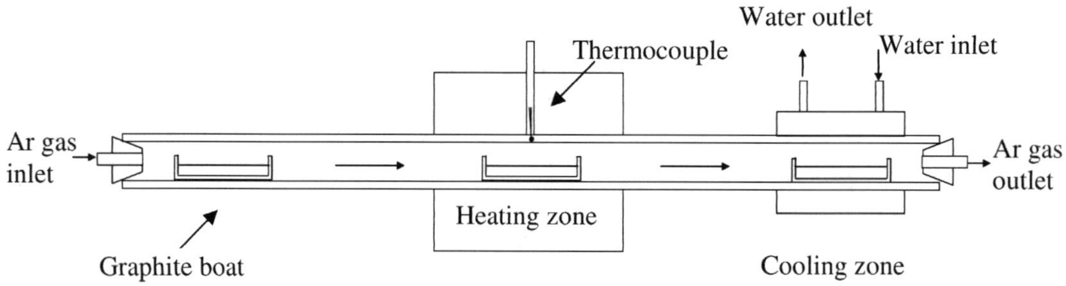

Figure 3. Schematic illustration of the tubular furnace.

4. Results and discussion

4.1. XRD patterns of the products
Figure 4 shows the XRD patterns of the products obtained from TiO_2-containing powders at 923, 973, and 1073 K before and after purification. Diffraction peaks were observed due to TiB_2, MgO, and other compounds such as TiO_2, Ti_2O_3, and $Mg_3B_2O_6$; TiO_2 and $Mg_3B_2O_6$ were formed as by-products by metallothermic reduction. Because boron was used for the formation of $Mg_3B_2O_6$, unreacted TiO_2 remained. Even after purification, the diffraction peaks of Ti_2O_3 and Ti_2O_3, which are resistant to acids, were observed.

Figure 5 shows the XRD patterns of the products obtained from ZrO_2-containing powders at 923, 973, and 1073 K before and after purification. Diffraction peaks due to ZrO_2 and $Mg_3B_2O_6$ were also observed in addition to those of ZrB_2 and Mg. The by-product ZrO_2 insoluble to acid remained after purification, however, the amount of ZrO_2 apparently decreased with increasing reaction temperature.

Figure 6 shows the XRD patterns of the products obtained from ReO_2-containing powders at 1073 K before and after purification. Diffraction peaks of the aimed products ReB_2 were prominent. However, many diffraction peaks were also present from by-products such as Re and Re_7B_3, which remained after purification. More detailed investigations are necessary on the mixing ratio of starting reagents, reaction temperature, and purification process.

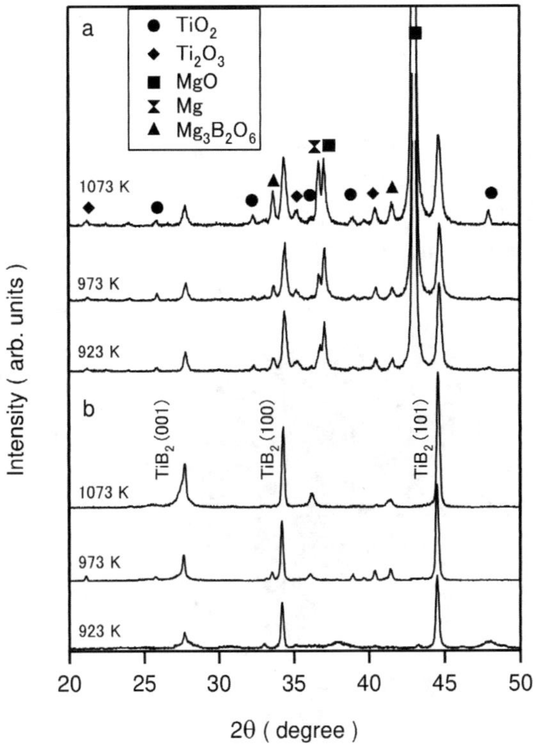

Figure 4. XRD patterns of the products obtained from TiO_2-containing powder (a) before and (b) after purification.

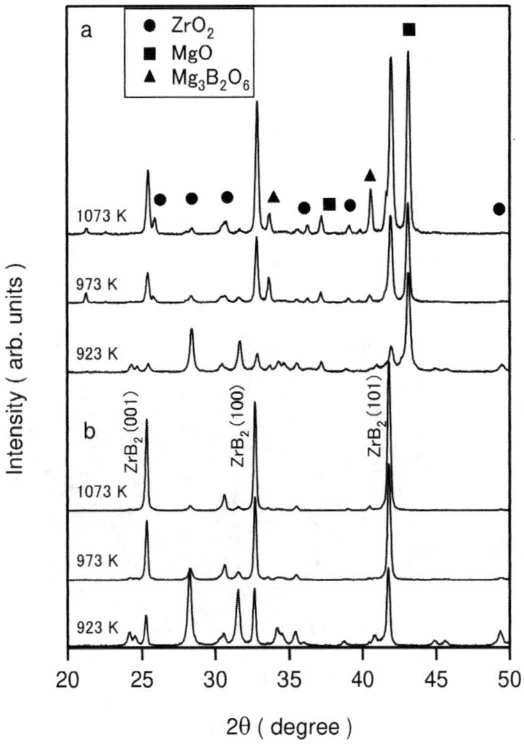

Figure 5. XRD patterns of the products obtained from ZrO_2-containing powder (a) before and (b) after purification.

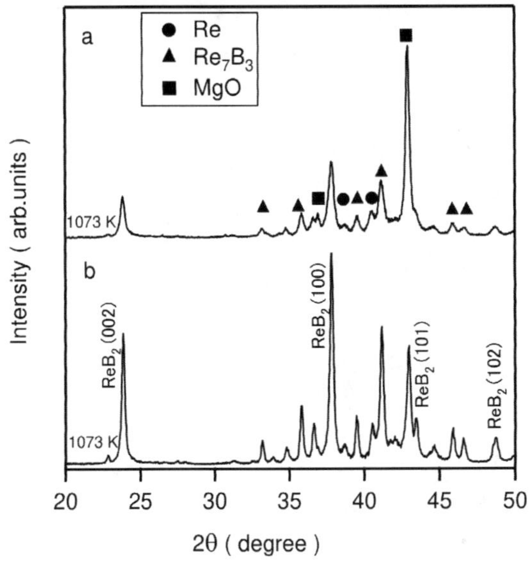

Figure 6. XRD patterns of the products obtained from ReO_2-containing powder (a) before and (b) after purification.

4.2. Morphology and particle size of products

Figures 7 and 8 show SEM and TEM images of purified products prepared at 1073 K. These boride particles are laminar hexagonal single crystals and have the crystal habit characteristic of the hexagonal system. The particle size was 0.1-0.4 μm indicated by the SEM and TEM images. Furthermore, the crystals was very thin (hundreds nm) as revealed by their transparency in the TEM images.

Figure 9 shows the particle size distributions of TiB_2, ZrB_2, and ReB_2. The particle sizes of TiB_2 and ZrB_2 increase with increasing reaction temperature because of the coalescence of reacted particles. The mean particle size of ReB_2 powder prepared at 1073 K was 0.11 μm. It is concluded that submicron powders of TiB_2, ZrB_2, and ReB_2 powders were obtained.

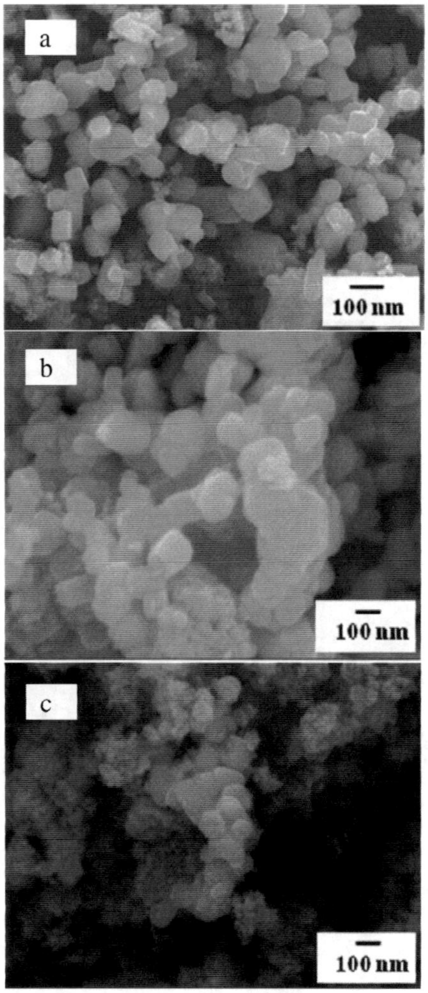

Figure 7. SEM images of (a) TiB_2, (b) ZrB_2, and (c) ReB_2 obtained after purification.

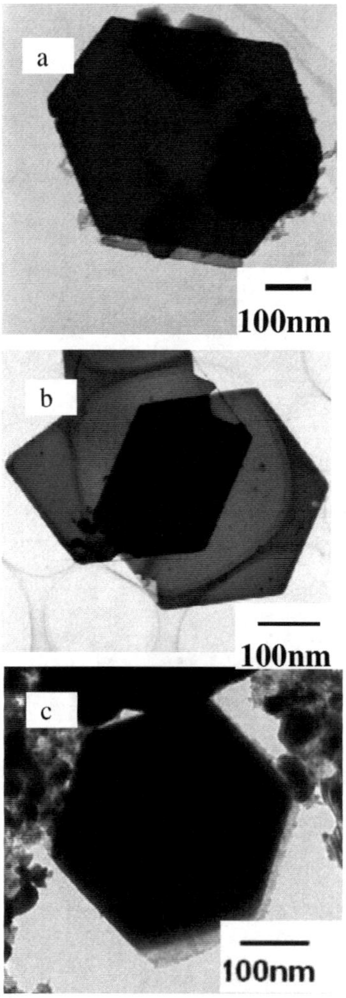

Figure 8. TEM images of (a) TiB_2, (b) ZrB_2, and (c) ReB_2 obtained after purification.

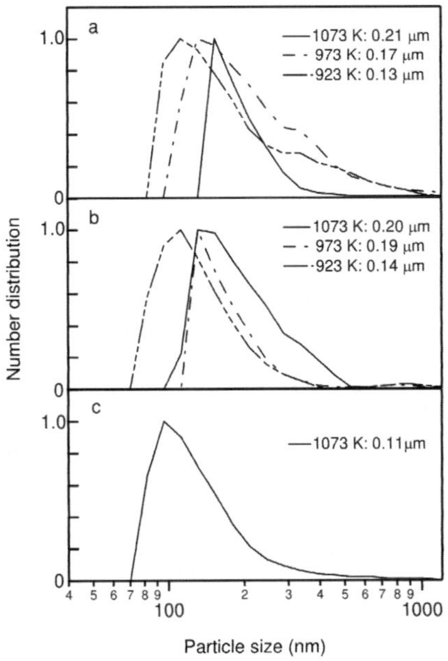

Figure 9. Particle size distributions of (a) TiB_2, (b) ZrB_2, and (c) ReB_2 powders.

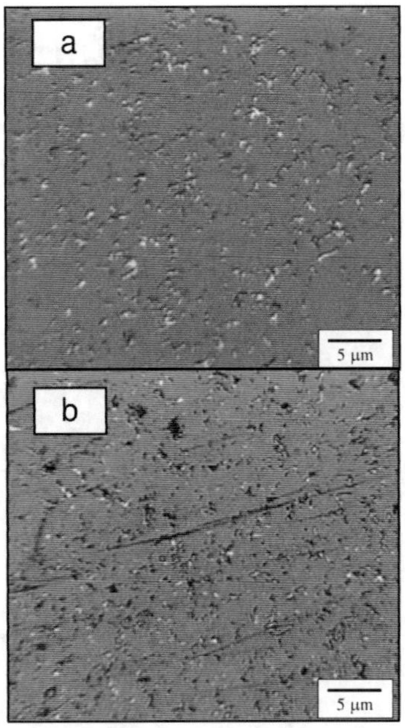

Figure 10. Surface of synthetic diamonds (a) before and (b) after scratch test.

4.3. Scratch test on ReB_2

Scratch test was carried out using high pressure sintered synthetic diamonds which have the Knoop hardness of 7000 kg/mm^2. In this test, the obtained ReB_2 powder was sandwiched between two synthetic diamonds, and then ReB_2 powder rubbed by them. Figure 10 shows the surface of polycrystalline synthetic diamonds before and after the test. We can see the scratches on the surface of the diamond after the test revealing that the hardness of ReB_2 is comparable to that of synthetic diamond.

5. Conclusion

Ultrafine TiB_2, ZrB_2, and ReB_2 powders were prepared by the metallothermic reduction method using Mg. Laminar hexagonal single crystals were obtained of TiB_2, ZrB_2, and ReB_2 with submicron sizes and thickness of several nanometers. ReB_2 powder could scratch the surface of a synthetic diamond, indicating that the hardness of ReB_2 is higher than or equal to a synthetic diamond which has Knoop hardness of 7000 kg/mm^2.

It has been reported that laminar TiB_2 particles dispersed in resin improved the hardness of the resin film surface due to their hardness and particle shape which leads to increase in the face-to-face contact area between particles [11]. Thus, boride powders obtained in this study should be promising materials as a pigment of paint coating as well as a superhard alloy for mobile phone and automobile frames, and a converter duct wall.

References

[1] Atouda T 1973 *Titanium Diboride, Titanium-Zriconium* **21** 145
[2] Chung H-Y 2007 *Science* **316** 436
[3] Peshev P 1968 *J. Less-Common Metals* **14** 23
[4] Reddev D D 1994 *J. Alloy. Compd.* **206** 34

[5] Reddev D D 1996 *J. Alloy. Compd.* **244** 48
[6] Welham N J 2000 *J. Am. Ceram. Soc.* **83** 1290
[7] Hwang Y 2002 *Mater. Lett.* **54** 1
[8] Sundaram V 1997 *Mater. Lett.* **12** 2657
[9] Nishiyama K 1990 *J. Jpn. Soc. Powder Powder Metall.* **37** 500
[10] Smithells C J 1983 *Metals Reference Book* (6th ed.) (Butterworths)
[11] Nishiyama K 2003 *Surf. Coat. Int. B: Coat. Trans.* **86** 169

Effect of Mo_2C content on the properties of TiC/TiB_2 base cermets

Ken-ichi Takagi, Ken Osada, Wataru Koike and Takuya Fujima

Department of Mechanical Engineering, Faculty of Engineering, Musashi Institute of Technology, 1-28-1 Tamazutsumi, Setagaya-ku, Tokyo, 158-8557, Japan

E-mail: ktakagi@sc.musashi-tech.ac.jp

Abstract. The effects of Mo_2C content on the microstructure and mechanical properties of TiC/TiB_2 base cermets were studied using the model cermets with the compositions of TiC/TiB_2-(11-17)Mo_2C-24Ni (mass%). TiC and TiB_2 ratio is set to molar ratio of 59:41 that is near quasi-eutectic composition. As a result, both transverse rupture strength and hardness of the cermets showed maxima for the cermet containing 13% Mo_2C. The cermet achieved remarkable microstructural refinement and still maintained characteristic core-rim structure of the TiC base cermets. TiC/TiB_2 cermets, in addition to TiCN base cermets, are a good alternative material to cemented carbides.

1. Introduction

During the last 80 years, WC-Co cemented carbides have been used in various applications such as wear resistant and cutting tool materials. As shown in figure 1 [1], People's Republic of China holds about 86% of world tungsten mine production. The dependence in the tungsten resources on People's Republic of China has increased recently. Paucity and uneven distribution of such resources as tungsten, cobalt, etc., have become a crucial issue around the world. Therefore, development of alternative materials containing less or no tungsten than cemented carbides has become an important research topic in the field of wear-resistant hard materials.

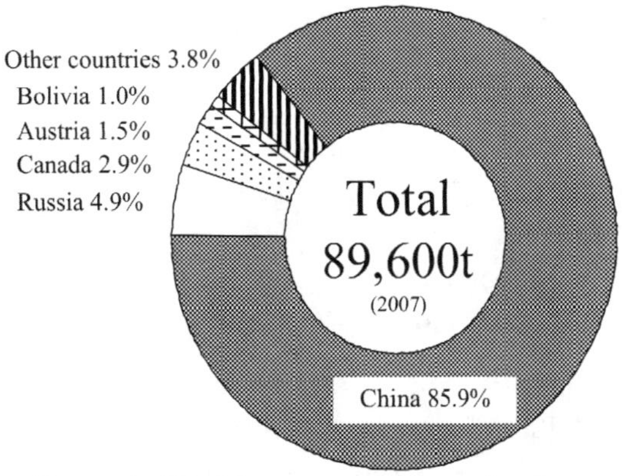

Figure 1. World tungsten mine production.

TiC base cermets were first studied as heat-resistant alloys in the 1950ies. Since then, the addition of Mo_2C (Mo) and TiN (N) to the cermets has remarkably improved their mechanical properties and microstructure. Hence, TiC cermets were developed as cutting tool materials. Today, TiC base cermets are the most promising alternative material to cemented carbides [2, 3]. However, difficulties in the large-scale sintering and in the machinability by a diamond wheel hinder many potential applications.

Borides are more stable than nitrides at high temperatures. Duschanek *et al* [4] summarized the thermodynamic calculation and experimental results on B-C-Ti systems. No ternary compounds exist in this system. Mutual solid solubilities among the binary borides and carbides are very small. Cola *et al* [5] and Singh *et al* [6] studied TiC/TiB_2-Ni cermets, which showed relatively good mechanical properties after addition of Mo_2C and Ni/Mn binders. However, the phase formation and relationship between the mechanical properties and microstructure of the TiC cermets upon addition of boride are still unclear.

Preliminary investigation suggested that TiC/TiB_2-19Mo_2C-24Ni (mass%) cermets had not only fine microstructure, but also some unknown large grains which might degrade the fracture strength. In this investigation, we studied the effects of Mo_2C content (< 19%) on the microstructure and mechanical properties of the TiC/TiB_2 base cermets.

2. Experimental procedure

Table 1 shows the specimen compositions used in this investigation. TiC and TiB_2 ratio is set to molar ratio of 59:41 which is near quasi-eutectic composition [4].

Table 1. Composition of TiB_2 Added TiC base Cermets.

Symbol	Composition (mass%)
17Mo_2C	TiC/TiB_2-17Mo_2C-24Ni
15Mo_2C	TiC/TiB_2-15Mo_2C-24Ni
13Mo_2C	TiC/TiB_2-13Mo_2C-24Ni
11Mo_2C	TiC/TiB_2-11Mo_2C-24Ni

The powder mixtures were ball-milled in acetone to an average particle size of about 1 μm. Raw material powders were TiC (Japan New Metals, particle size 1.5-2.5 μm), TiB_2 (Japan New Metals, average particle size 3.38 μm), Mo_2C (Japan New Metals, average particle size 1.57 μm) and carbonyl nickel (INCO type 123, average particle size 3.70 μm). The powder mixtures were dried and pressed into green compacts, which were sintered in vacuum for 1 h at 1673 and 1723 K. Mechanical properties were evaluated through the sintered density, transverse rupture strength (TRS) and Rockwell A scale hardness of the sintered compacts. The TRS test was conducted on 4.0 × 8.0 × 25 mm^3 test specimens in three-point loading mode with a 20 mm span. X-ray diffraction (XRD) analysis and scanning electron microscopy (SEM) were used for microstructural analysis.

3. Results and discussion

3.1. Sintered density

Figure 2 shows the density of the sintered cermets as a function of Mo_2C content. The density increases with increasing Mo_2C content and is temperature independent.

3.2. Transverse rupture strength

Transverse rupture strength (TRS) of the cermets is shown in figure 3. TRS shows a maximum of 0.83 GPa for the cermet containing 13% Mo_2C and sintered at 1673 K, and then decreases with increasing

Mo_2C content at both sintering temperatures. The cermets sintered at 1673 K have higher TRS than those sintered at 1723 K for all Mo_2C concentrations.

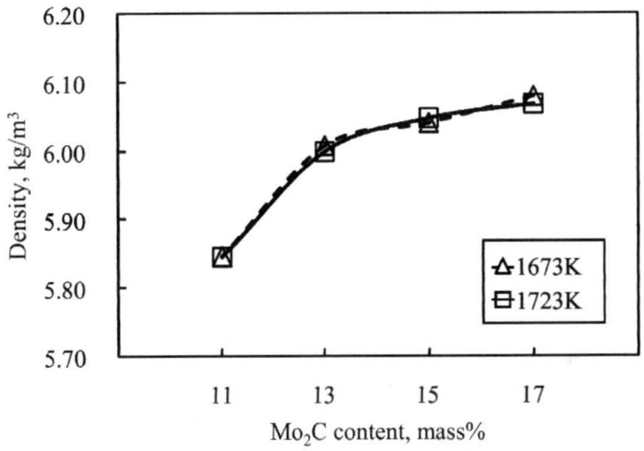

Figure 2. Density of TiC/ TiB_2 base cermets Sintered at 1673 and 1723K as a function of Mo_2C content.

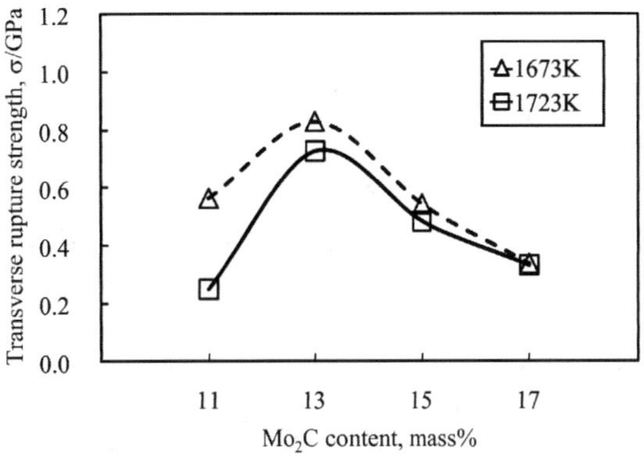

Figure 3. Transverse rupture strength of TiC/TiB_2 base cermets sintered at 1673 and 1723K as a function of Mo_2C content.

3.3. Hardness

Figure 4 shows that Rockwell A scale hardness of the cermets has a maximum of 92.6 R_A at 13% Mo_2C and then decreases with increasing Mo_2C content at both sintering temperatures. The cermets sintered at 1673 K have higher hardness than those sintered at 1723 K for all Mo_2C concentrations, same as TRS. The cermet containing 13% Mo_2C has better combination of TRS and hardness than other cermets.

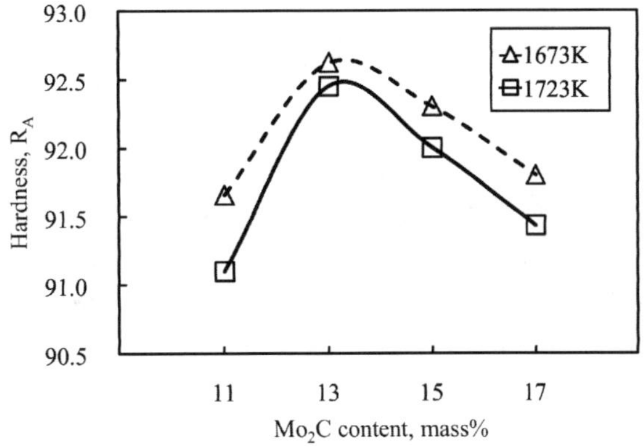

Figure 4. Hardness of TiC/TiB_2 base cermets as a function of Mo_2C content

3.4. Phase formation

CuKα XRD analysis was performed to identify the phases formed during sintering of the cermets (see figure 5). All the cermets mainly consist of TiC, TiB_2, Mo_2C and Ni. Small amounts of unknown phases were also detected in these cermets. No ternary compounds or other kinds of carbides and borides, except for TiC and TiB_2, are detected in accordance with Duschanek *et al* [4].

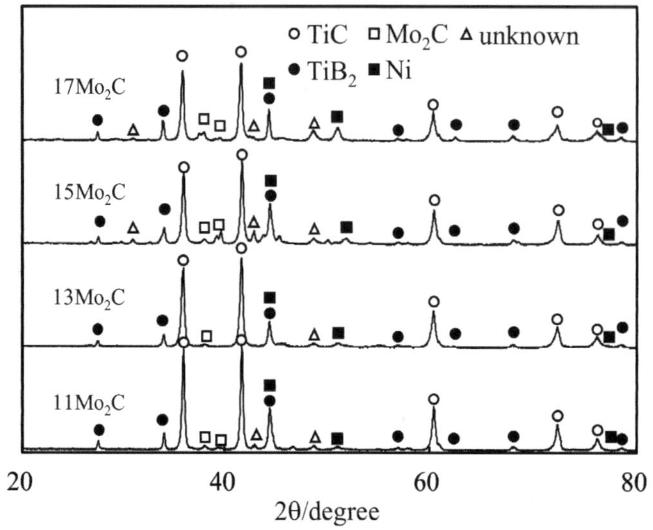

Figure 5. X-ray diffraction results of TiC/TiB_2 base cermets Sintered at 1673K

3.5. Microstructure

Microstructure of the cermets was studied by SEM. Figure 6 shows back scattered electron images of the cermets. Boride-added cermets have finer microstructure than the cermet with no boron shown in figure 7; they still maintain core-rim structure characteristic of TiC base cermets. The coarse white grains are observed in cermets with 15 and 17 Mo_2C. The white grains might be Mo-Ni intermetallic compound, as suggested by the electron probe microanalysis (EPMA), and their white color indicates presence of heavy atoms such as molybdenum. Unknown peaks observed in XRD patterns are thought to result from the white phase. Some micro-pores were also observed at the fractured surfaces. Degradation of the mechanical properties of the cermets is attributed to these large white grains and micro-pores. Further enhancement of mechanical properties needs the reduction of micro-pores by the improvement of wettability of the binder phase.

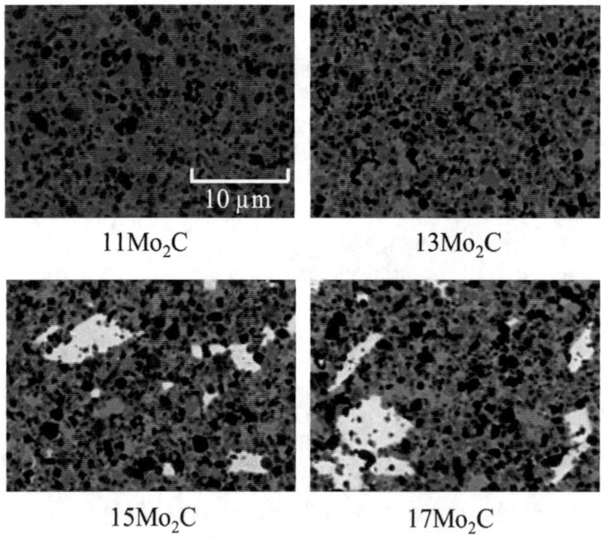

Figure 6. Back scattered electron images of TiC/TiB_2 base cermets Sintered at 1673K

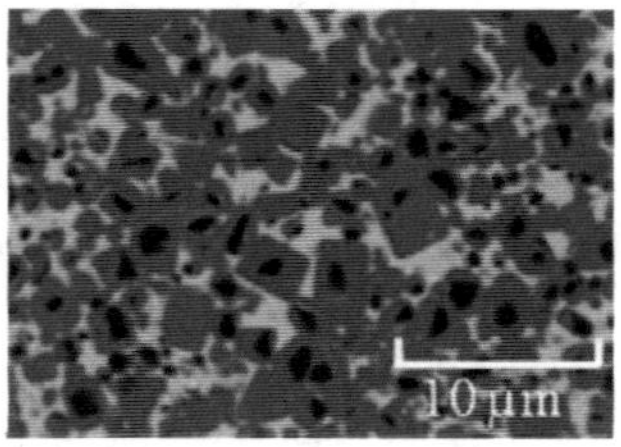

Figure 7. Back scattered electron images of TiC-$19Mo_2C$-24Ni cermet.

Journal of Physics: Conference Series **176** (2009) 012044 doi:10.1088/1742-6596/176/1/012044

4. Conclusion

The effects of Mo_2C content on the microstructure and mechanical properties of the 59TiC/41TiB_2 (molar ratio) base cermets were studied using the model cermets with composition of TiC/TiB_2- (11-17)Mo_2C-24Ni (mass%).

Both transverse rupture strength and hardness showed maxima for the cermet containing 13% Mo_2C. This cermet had fine microstructure without large intermetallic grains, which degrade the mechanical properties.

TiC cermet containing TiB_2 achieved remarkable microstructural refinement and still maintained characteristic core-rim structure of the TiC base cermets.

TiC/TiB_2 cermets are good alternative material to cemented carbides, in addition to TiCN base cermets.

References

[1] http://minerals.usgs.gov/minerals/pubs/mcs/2008/mcs2008.pdf
[2] Lyubimov V D, Élinson D S and Shveikin G P 1992 *Soviet Powder Metall. and Met. Ceram.*, **30** 957
[3] Pastor H 1999 *Adv. Sci. Technol. Sinter.* Stojanović et al (New York: Kluwer Academic/Plenum Publishers) p 461
[4] Duschanek H, Rogl P and Lukas H L 1995 *J. Phase Equilibria* **16** 46
[5] Cola P L, Vallauri D, Maizza G, Amato I and Shcherbakov V A 2001 *Int. J. Self-Propag. High-Temp. Synth.*, **10** 451
[6] Singh M, Rai K N and Upadhyaya G S 2001*Mater. Chem. Phys.* **67** 226

Properties of boride-added powder metallurgy magnesium alloys

Atsushi Tanaka, Syota Yoshimura, Takuya Fujima and Ken-ichi Takagi

Department of Mechanical Engineering, Tokyo City University
1-28-1 Tamazutsumi, Setagaya, Tokyo 158-8557, Japan

E-mail: ktakagi@tcu.ac.jp

Abstract. Magnesium alloys with metallic borides, magnesium diboride (MgB_2) or aluminum diboride (AlB_2), were investigated regarding their mechanical properties, transverse rupture strength (TRS) and micro Vickers hardness (HV). The alloys were made from pure Mg, Al and B powders by mechanical alloying and hot pressing to have boride content of between 2.0 and 20 vol%. The alloy with AlB_2 exhibited an obvious improvement of HV around a boride content of 6 vol% though the other alloy, with MgB_2, did not. TRS showed moderate maxima around the same boride content region for the both alloys. X-ray diffraction measurements indicated an intermetallic compound, $Mg_{17}Al_{12}$, formed in the alloy with AlB_2, which was consistent with its higher hardness.

1. Introduction

Magnesium alloys have attracted much interest as structural materials because of their lower density and higher specific strength even than widely used aluminum alloys.[1] These characteristics satisfy the criteria required from structural materials for various vehicles like car, airplane and so on. Their weight saving will reduce the energy consumption as well as carbon dioxide emission.[2] In spite of these advantages, the use of wrought magnesium alloys is very limited because of their insufficient strength and low workability.

Powder metallurgy is expected to overcome these problems. This method is known to strengthen materials by its microstructure, fine particle dispersion, with high workability.[3] Powder metallurgy enables to form complicated and accurate sintered bodies by its forming method. Dispersion of hard boride particle or microcrystals by powder metallurgy is expected to improve the mechanical properties of Mg alloy with high workability. Boride-dispersed Mg alloys, however, have not been adequately studied so far. [4, 5, 6]

In this work, we focus on MgB_2 and AlB_2 as the particle to disperse in Mg alloy. They both have a hexagonal crystal structure, low density (MgB_2: 2.6 g/cm^3, AlB_2: 3.2 g/cm^3), high melting point and high hardness.[7, 8] We investigated the boride-content dependence of mechanical properties for the Mg alloys and discussed in their microstructure.

2. Experimental procedure

Samples were prepared by powder metallurgy: raw material powders were mixed, mechanically alloyed and then sintered by hot pressing. The Mg alloy with MgB_2 was made from Mg and B

Table 1. Chemical composition of sample alloys (vol%).

Magnesium	100	98	96	94	92	85	80
Boride in Alloy-1	0.0	2.0	4.0	6.0	8.0	15	20
Boride in Alloy-2	0.0	2.0	4.0	6.0	8.0	15	20

powders. The other alloy with AlB_2 was from Mg, Al and B powders.

Raw material powders were pure Mg (99.9 mass%, 180 μm, Kojundo Chemical Laboratory Co., Ltd.), Al (99.9 mass%, 10 μm, Kojundo Chemical Laboratory Co., Ltd.) and B (96.8 mass%, 0.8 μm, H. C. Starck Co., Ltd.). All the powders were used without further purification.

Table 1 shows the boride content for both Alloy-1 (Mg-B alloy) and Alloy-2 (Mg-Al-B) that was contolled between 2 vol% and 20 vol%. Here all the added boron is assumed to form MgB_2 in Alloy-1 and AlB_2 in Alloy-2. We mixed the raw material powders with 3 mass% of zinc stearate as lubricant agent, then put the mixture into a stainless steel milling pot with stainless steel balls of 8-mm diameter. The volume ratio between the powder and balls was 1/20. The mixed powder was mechanically alloyed by a planetary ball mill (Itou Seisakujo Co., Ltd.) that rotated the pot in argon atmosphere at 200 rpm for 20 hours. The alloyed powder was then investigated by X-ray diffraction.

Each mechanically alloyed powder was hot-pressed under a pressure of 98 MPa, in Ar atmosphere, at 600 °C for 30 minutes. Then the sintered bodies were machined to 5.0 x 10 x 30 mm blocks for X-ray diffraction (XRD) measurement and mechanical properties investigation.

Transverse rupture strength (TRS) and micro Vickers hardness (HV_{100}) were measured on the sintered bodies. The former was performed with the three-point bending method, and the latter was carried out with a diamond indenter of square-based pyramid form loading 0.98 N for 10 seconds. HV_{100} for each alloy was obtained by averaging 30 measured values on three specimens, 10 points for each. Microstructure and composition analyses were performed on the alloys after the hot pressing by a scanning electron microscope (S-4100, Hitachi Co. Ltd.) and an X-ray diffractometer (RAD-rR, Rigaku Co. Ltd.).

3. Results and discussion

XRD patterns of the alloys with 6 vol% boride, as representative samples, before and after the mechanical alloying and of sintered bodies are shown in figure 1. The mechanically alloyed powders exhibited broadened diffraction peaks of raw materials and newly reacted borides. The peaks were sharpened after the sintering, indicating further alloying and reduction of strain caused by mechanical alloying. An intermetallic compound, $Mg_{17}Al_{12}$, was formed in the Alloy-2.

Figure 2 shows transverse rupture strength (TRS) of Alloy-1 and Alloy-2. The datum for pure Mg made by same process is shown in comparison. Both Alloy-1 and Alloy-2 generally have high TRS compared to the pure Mg. As shown in the figure, boride improves TRS more for the Alloy-1 than for the other. The boride content dependence of TRS exhibits a moderate maximum around the content of 6 vol%.

Figure 3 shows the micro Vickers hardness (HV_{100}) of Alloy-1 and Alloy-2. The two alloys exhibited a quite different behavior compared to the TRS results. Though the hardness of Alloy-1 was totally independent on the boride content, that of Alloy-2 was clearly improved as the boride content increased. The boride addition obviously increased the hardness of Alloy-2 in high boride content region over 6 vol%, which finally reached 250 HV that was quite high for

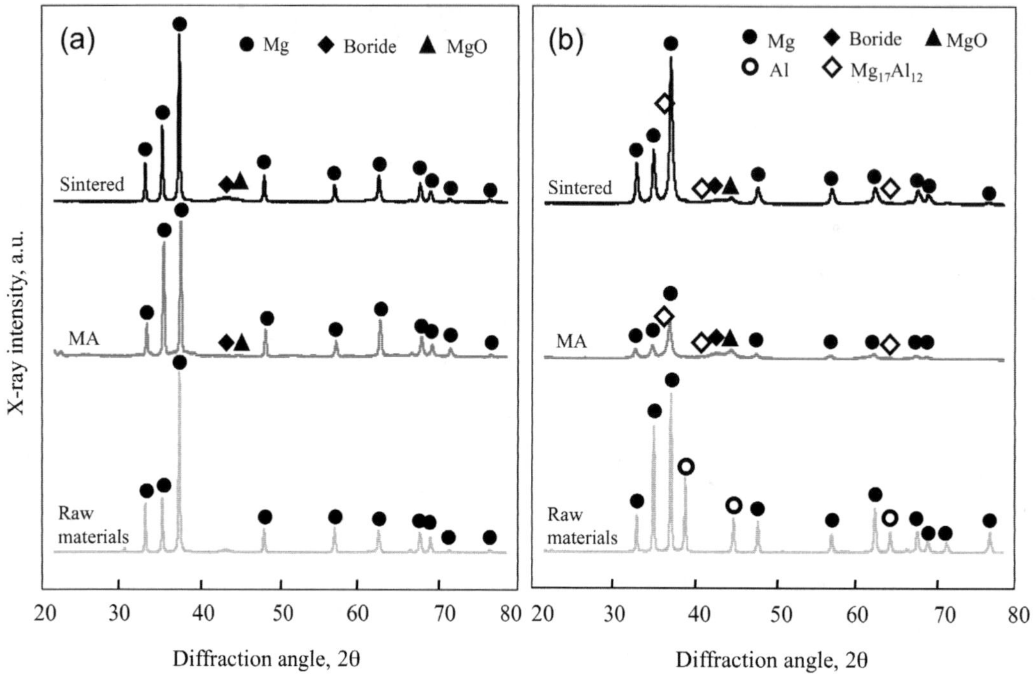

Figure 1. XRD patterns of (a) Alloy-1 and (b) Alloy-2 with 6% of boride: before and after mechanical alloying and after sintering.

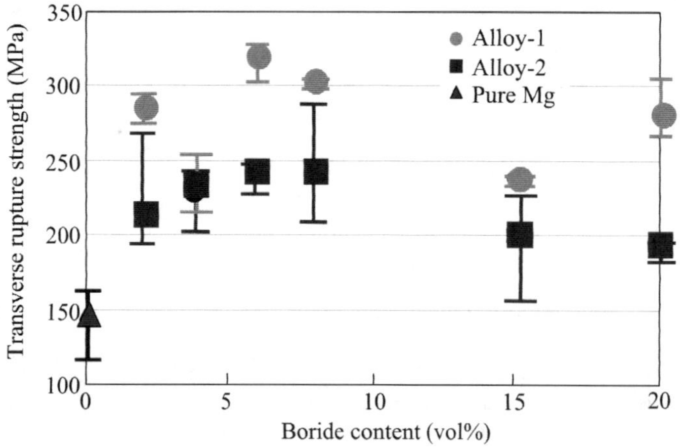

Figure 2. Boride content dependence of TRS for Alloy-1, Alloy-2 and pure Mg.

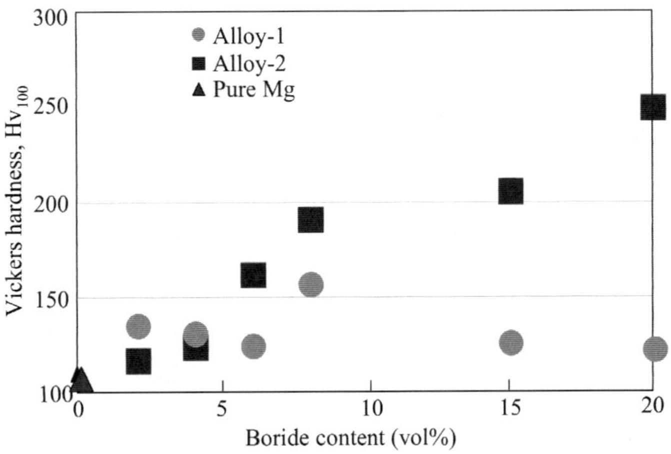

Figure 3. Boride content dependence of micro Vickers hardness for Alloy-1, Alloy-2 and pure Mg.

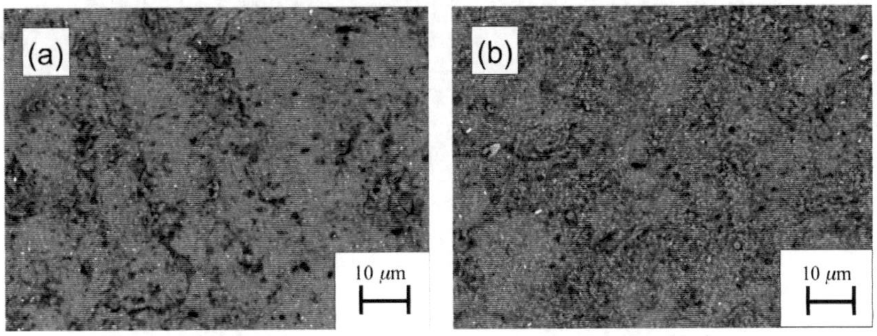

Figure 4. SEM micrograph of (a) Alloy-1 and (b) Alloy-2, both containing 6 % boride.

a Mg alloy.

Figure 4 shows SEM micrographs of (a) Alloy-1 and (b) Alloy-2 containing 6 vol% boride. Alloy-1 consisted of two phases: a light gray phase and a dark one. Energy-dispersive X-ray spectroscopy (EDX) revealed that the light gray phase consisted of Mg, and the dark phase of both Mg and B that was indicated to be boride by XRD in figure 1. This figure also shows that the boride was formed mostly around magnesium grain boundary, which could have a role of the grain growth inhibition by its pinning effect.

Alloy-2 in figure 4(b) exhibited finer microstructure than Alloy-1. Here light gray Mg phase was separated into smaller grains than in Alloy-1 by dark gray grain boundary. The grain boundary consisted of Mg, B and Al according to the EDX result. Since Al was indicated to form $Mg_{17}Al_{12}$ by XRD, the particles in the boundary were borides and the intermetallic compound that is a very hard material which can improve hardness.[9] The hard but brittle $Mg_{17}Al_{12}$ was supposedly formed and finely milled during the mechanical alloying process; it then brought out the pinning effect in shorter range than in Alloy-1.

The results showed that the boride addition improves TRS and HV of the magnesium alloys. The remarkable improvement in the hardness of the Alloy-2 is mainly attributed to the formation of $Mg_{17}Al_{12}$ observed in figure 4. XRD peaks corresponding to the $Mg_{17}Al_{12}$ become obvious

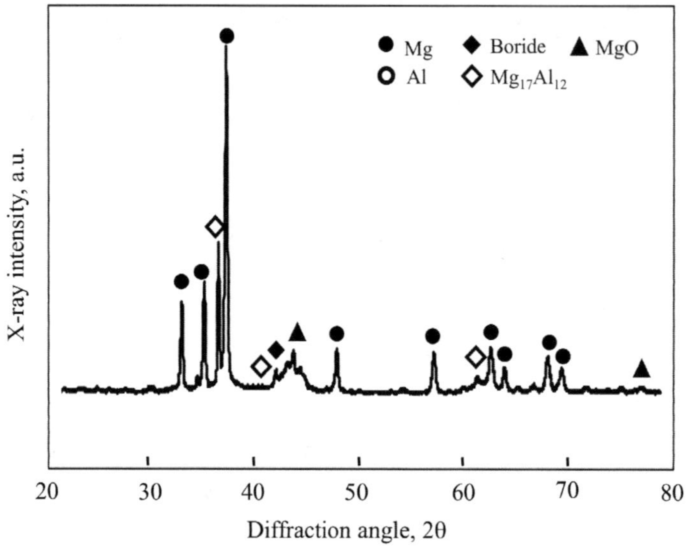

Figure 5. XRD pattern of Alloy-2 containing 20 % boride.

in high boride content region as shown in figure 5. Aluminum did not fully react with boron but with magnesium to form the intermetallic, which resulted in TRS reduction and hardness improvement in high boride content region. The maxima of TRS for both alloys and steep improvement of hardness of Alloy-2 occur around the boride content of 6 vol%. There can be some interaction between the microstructure and the intermetallic compound formation which directly defines the mechanical properties, which is to be studied in future work.

4. Conclusion

Magnesium alloys with dispersed MgB_2 (Alloy-1) and MgB_2 and AlB_2 (Alloy-2) dispersed were investigated for their hardness (HV_{100}) and TRS. The alloys were prepared by powder metallurgical technique: mechanical alloying and hot pressing.

The boride addition surely improved both TRS and HV_{100} of the alloys. The alloys with 6 vol% exhibited the highest TRS. Though HV_{100} of Alloy-1 was almost independent of the boride content for more than 2 vol% boride, the hardness of Alloy-2 increased for boride content over 6 vol%. The formation of hard $Mg_{17}Al_{12}$ and the finer microstructure is supposed to improve the hardness of Alloy-2. Formation of $Mg_{17}Al_{12}$ lowers TRS of Alloy-2 compared to Alloy-1.

The mechanism of the coincidence of TRS maximization on both alloys and steep hardness improvement of Alloy-2 is still unclear. Further analysis about microstructure and particle dispersion is expected to clarify this mechanical property improvement.

Acknowledgments

We also express our gratitude to Ms. Shindo and Mr. Hamamura for their great help on instrumental analysis.

References

[1] Kojima Y 2000 *Handbook of Advanced Magnesium Technology* (Tokyo: The Japan Magnesium Association) p 55
[2] Matsuoka S 2007 *J. Jpn. Soc. Technol. Plasticity* **48** 355

[3] German R M 1994 *Powder Metallurgy Science* (Princeton, NJ, USA: Metal Powder Industries Federation) p 88
[4] Nishiyama K, Ogawa T, Takagi K, Omori M and Umekawa S 1994 *J. Jpn. Soc. Powder Powder Metall.* **41** 156
[5] Tatsuzaki Y, Momozawa N, Abe M, Takagi K and Nishiyama K 1998 *J. Jpn. Soc. Powder Powder Metall.* **45** 430
[6] Kamado S 2007 *J. Jpn. Soc. Technol. Plasticity* **48** 358
[7] Pearson W B 1967 *A handbook of Lattice Spacings and Structures of Metals and Alloys* (Oxford: Pergamson Press) p 403
[8] Thompson R 1970 *Progress in Boron Chemistry* (Oxford: Pergamon Press) vol.2 p 173
[9] Samsonov G V 1965 *High-Temperature Compounds of Rare Earth Metals with Nonmetals* (New York: Consultants Bureau) p 1

Journal of Physics: Conference Series **176** (2009) 012046 doi:10.1088/1742-6596/176/1/012046

Effect of double addition of V and Cr on the properties of Mo_2NiB_2 ternary boride-based cermets

Yuusuke Shiota, Yuuta Miyajima, Takuya Fujima and Ken-ichi Takagi

Department of Mechanical Engineering, Tokyo City University
1-28-1 Tamazutsumi, Setagaya-ku, Tokyo 158-8557, Japan

E-mail: ktakagi@tcu.ac.jp

Abstract. The effect of double addition of V and Cr on the mechanical properties and microstructure of Mo_2NiB_2 base cermet was investigated. Total additional amount of V and Cr was fixed to 12.5 mass% and the fraction between the two additives was varied. Transverse rupture strength (TRS) and Rockwell A-scale hardness (HRA) were measured on the cermets and discussed together with their microstructure obtained by X-ray diffraction (XRD) and backscattered electron images (COMP). Addition of 2.5-mass% Cr and 10-mass% V showed the highest mechanical properties. Microstructural analysis revealed that brittle orthoronbic-M_5B_3 phase was formed in high V fraction. The corrosion resistance of the cermets against hydrochloric acid was superior to that of JIS SUS 304. The resistance against nitric acid decreased with decreasing Cr content and was lower than that of SUS 304.

1. Introduction

Tungsten carbide based cermets are the most widely used wear resistant hard materials in the world. In spite of its importance for various industries, tungsten ore production is quite uneven across countries, that is, 86% is in China[1]. Reduction of scarce and unevenly distributed resources such as tungsten, indium, cobalt and dysprosium is a recent crucial issue all over the world. Now wear resistant hard material with less or without tungsten is a hot research topic.

Ternary boride-based cermet is a promising candidate as a substitute material for cemented carbides due to its excellent mechanical properties[2]. Transition metal borides show high hardness, high melting point and electrical conductivity, which make the borides suitable for a new wear resistant material.

Although the brittleness of the metal borides requires their practical application to be in a form of cermet materials in which the borides are used as hard phase, high chemical reactivity of the borides forms a new brittle phase between the hard phase and metallic binder phase. Because of these characteristics, it was difficult to develop cermets with a hard boride phase. A recently developed technique, Reaction Boronizing Sintering, has realized the new type of boride base cermet[3]. Cermets with Mo_2FeB_2 and Mo_2NiB_2 for its hard phase, ternary boride-based cermets, have been developed by this methodology and are now commercially available.

The Mo_2NiB_2 cermet shows lower mechanical properties and higher corrosion resistance by using Ni for base metal in comparison with Mo_2FeB_2 cermet. Mo_2NiB_2 takes the orthorhombic-M_3B_2 type crystal structure (here, M stands for metallic element), which causes its grain growth in the plate-like shape during sintering and brittleness for the bulk material.

Table 1. Chemical composition of sample cermets. Nickel balances the total fraction for all the cermets.

Cermet	B	Mo	Cr	V	Ni
A	6.0	55.9	7.5	5.0	bal.
B	6.0	55.9	5.0	7.5	bal.
C	6.0	55.9	2.5	10.0	bal.
D	6.0	55.9	0.0	12.5	bal.

(mass%)

It is known that the addition of some amount of chromium (Cr) or vanadium (V) into Mo_2NiB_2 changes its crystalline structure to tertragonal-M_3B_2 type, which leads to microstructural refinement and improved mechanical properties of Mo_2NiB_2 cermet[2, 4]. The V-added cermet exhibits finer boride grain within it and better mechanical properties. The Cr-added cermet, on the other hand, realizes better corrosion resistance by forming an insulating passivation layer on its surface. Addition of both V and Cr into Mo_2NiB_2 cermets has a prospect of coincident realization of good mechanical properties and corrosion resistance.

We, therefore, investigated the mechanical properties, corrosion resistance and microstructure of the V-Cr double added Mo_2NiB_2 cermet in this study. The amount of added V and Cr was fixed to 12.5 mass% in total and the their fraction dependence was investigated since the both single addition of V or Cr showed maximal mechanical properties at 12.5 mass% addition in previous studies[4] and they were indicated to replace the same site in the boride crystal: Ni site.

2. Experimental procedure

Table 1 shows the chemical composition of the cermets we prepared. Content of Cr was varied from 0.0 up to 7.5 mass% with trading off that of V to keep their total content as 12.5 mass%.

All the samples were made from following powders: carbonyl nickel (99.85 %), pure Mo (99.95 %), MoB (Mo: 89.52 %, B: 10.26 %), VB_2 (V: 68.93 %, B: 29.84 %) and CrB (Cr: 82.99 %, B: 16.54 %). We also added 0.35 mass% of graphite to reduce oxide within the samples by chemical reaction during the sintering. The mixed powders were ball-milled in acetone to reduce the average particle size to about $1\mu m$.

The powder mixtures were dried in vacuum at 50 °C for 12 hours then pressed under a pressure of 98 N/mm^2 to form green compacts. They were sintered in vacuum for 30 min. at 1473 and 1523K.

Rockwell A-scale hardness (HRA) and transverse rupture strength (TRS) measurements were performed on the sintered compacts. We used test specimens with dimension of 4.0 x 8.0 x 25.0 mm for the TRS bending in three point loading mode with 20 mm span. X-ray diffraction (XRD) and backscattered electron images (COMP) were used for the study of the microstructure of the cermets. We used RAD-rR (Rigaku Co. Ltd.) with Cu-Kα for XRD, and JXA-8200 (JEOL) for COMP images.

The corrosion resistance was investigated for the specimens sintered at 1523K by measuring their corrosion weight loss after soaking in the 10 wt% hydrochloric acid (HCl) and nitric acid (HNO_3) at 50 °C for 20 hours. The specimens used for this test had dimensions of 4.0 x 8.0 x 9.0 mm. A stainless steel, JIS SUS 304 (AISI 304 equivalent), was used for comparison.

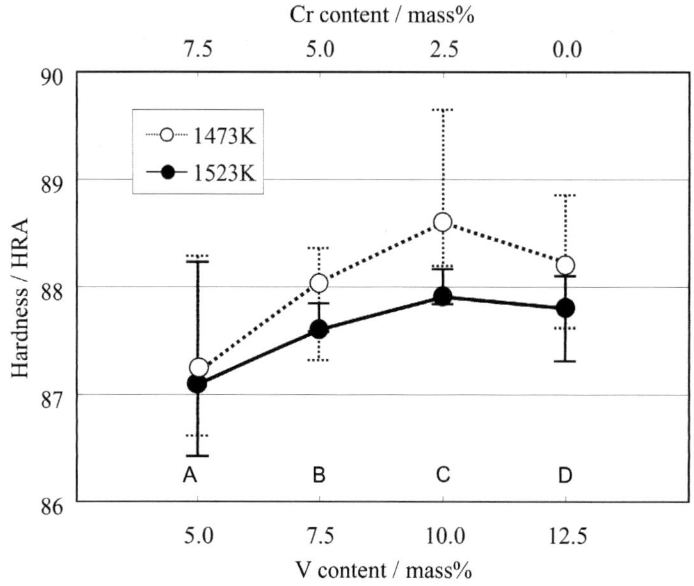

Figure 1. Added V/Cr ratio dependence of hardness. Open and filled circles correspond to alloys sintered at 1473 K and 1523 K, respectively.

3. Results and discussion

As shown in figure 1, HRA exhibits high hardness across all the V and Cr content taking a maximum at 10 mass% of V and 2.5 mass% of Cr. The similar tendency is observed for TRS in figure 2. The cermet sintered at higher temperature (1523 K) shows lower hardness but comparable TRS in comparison with the cermet sintered at lower temperature (1473 K). This indicates overgrowth of the boride grains that surely reduces the hardness but has less impact on TRS.

Figure 3 shows XRD patterns for the cermets sintered at 1523 K. The figure indicates that the main constituents of the cermets are tetragonal-M_3B_2 hard boride phase and Ni-based binder. All the ternary borides have changed their structure from orthorhombic into tetragonal, which improves mechanical properties. In all the cermets except for alloy-A (V: 5.0 %, Cr: 7.5 %), small amount of brittle tetragonal-M_5B_3 type ternary boride is also formed.

COMP images (figure 4) fully support the XRD results. The figure shows gray boride phase dispersed in dark Ni base binder. Tetragonal-M_3B_2 type ternary boride formation is indicated in these micrographs because grains are quasi-spherical, other than plate-like shaped orthorhombic-M_3B_2 crystals. The micrographs also show tetragonal-M_5B_3 phase as a lightest part between the grains[5].

XRD and COMP results, however, do not provide any clear evidence for the forming of some weak structure that can cause the reduction of hardness and TRS in high V region. The reduction of mechanical properties in high V region is not due to the formation of brittle tetragonal-M_5B_3 phase. There could be certain change or tendency in their sintering behavior like the contiguity of the particles. It is, therefore, necessary to investigate them further in detail.

These results suggest a trade-off relation of two dependencies on V amount added into the Mo_2NiB_2 cermet. That is, strengthening of crystalline structure of borides into tetragonal and formation of some brittle or fragile phase are concurrent. This trade-off can cause the maxima in the mechanical properties.

Figure 5 shows the corrosion weight loss in 10 wt% (a) HCl and (b) HNO_3 at 50 °C for 20

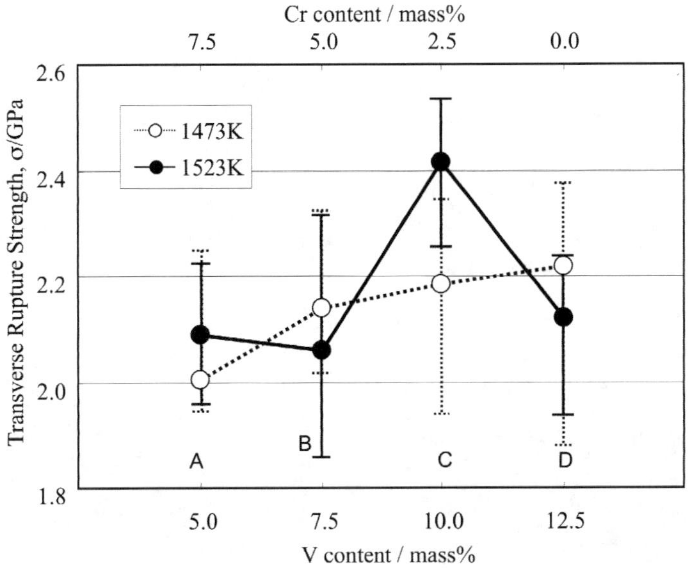

Figure 2. Added V/Cr ratio dependence of TRS. Open and filled circles correspond to alloys sintered at 1473 K and 1523 K, respectively.

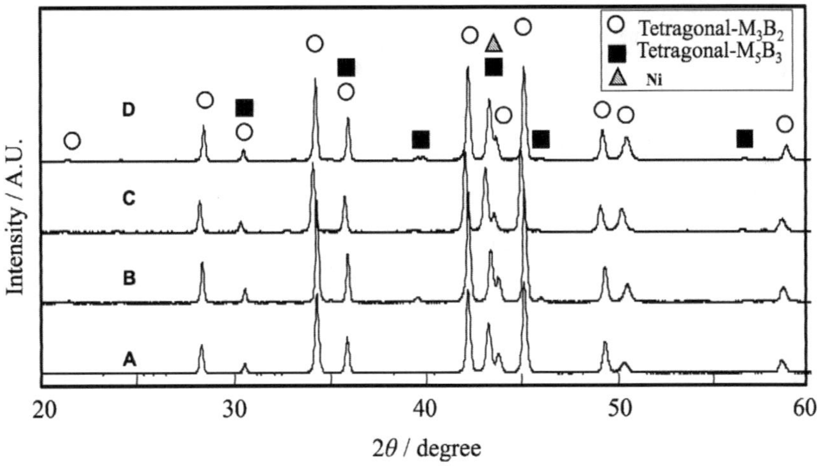

Figure 3. XRD patterns for the cermets sintered at 1523 K.

hours. All the cermets exhibited obviously higher corrosion resistance than SUS 304 in HCl. Their corrosion weight losses in the acid were about a fiftieth of that for SUS 304. Corrosion in the HNO_3, by contraries, obviously depends on the V-additive amount as shown in figure 5(b). That is, higher V-additive amount caused more corrosion weight loss. SUS 304 did not lose its weight in the HNO_3.

The corrosion weight loss of the cermets did not depend on the V and Cr content in HCl. Nickel is well known to have good corrosion resistance against nonoxidizing acids like HCl and to remarkably improve the resistance by Mo addition. It would appear that Ni and Mo played main role for the high corrosion resistance against HCl, which also resulted in the independence

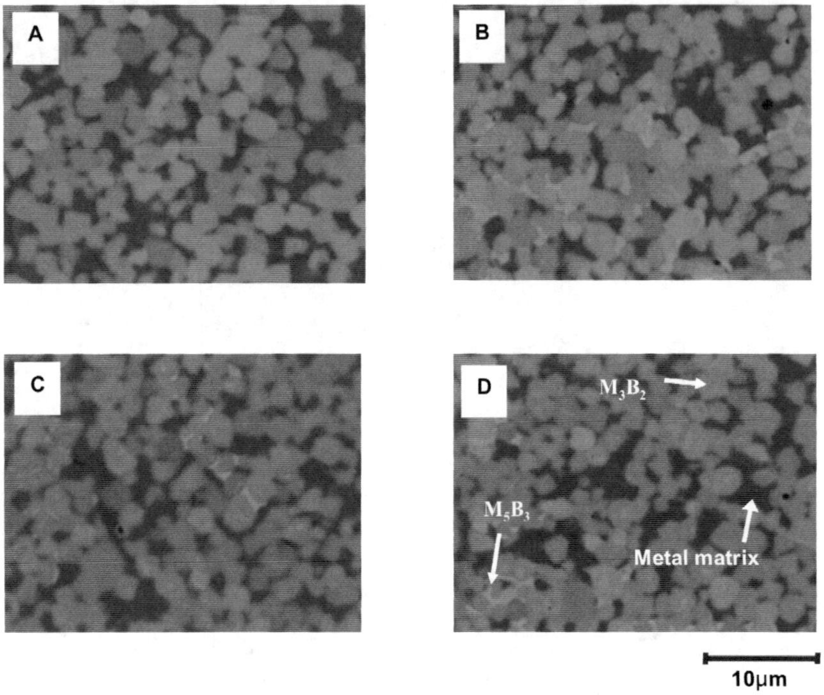

Figure 4. Microstructure of cermets sintered at 1523 K observed by back scattered electron image (COMP).

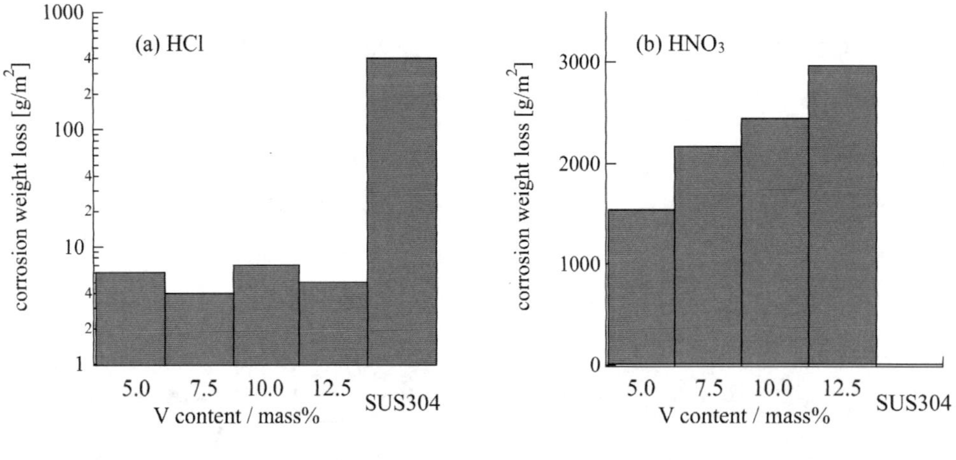

Figure 5. Corrosion weight loss against (a) hydrochloric acid in logarithmic scale and (b) nitric acid in linear scale. All the corrosion tests were carried out in 10 wt% acids at 50 °C for 20 hours.

of the corrosion weight loss from the V and Cr additive amount.

The cermets were much more inferior to SUS 304 in the HNO_3. Ni does not have high corrosion resistance against oxidizing acids like HNO_3. Though Cr-passivation layer has high corrosion resistance against the acids, the Cr content was insufficient for stable passivation layer formation in this work. The corrosion resistance, therefore, simply depended on the Cr content.

4. Conclusion

Mo_2NiB_2 cermet with 12.5 mass% of V and Cr addition was investigated for its mechanical properties and microstructure. TRS and hardness showed maxima at an addition of 10 mass% of V and 2.5 mass% of Cr. Though XRD revealed the formation of hard tetragonal-M_3B_2 that improved mechanical properties, no obvious mechanism of its reduction in high V region was found in XRD and COMP results. Further analysis on microstructure and sintering behavior will illustrate composition dependence of mechanical properties. The cermet with V and Cr addition had both high mechanical properties and high corrosion resistance against a nonoxidizing acid. As the corrosion resistance against an oxidizing acid simply depended on the Cr content, adequate balance between mechanical properties and corrosion resistance is to be selected for practical use.

Acknowledgments

We express our deep thanks to Ms. Shindo and Mr. Hamamura for their great help on microstructural analysis.

References

[1] U.S.Geological Survey 2008 *Mineral Commodity Summaries* (http://minerals.usgs.gov/minerals/pubs/mcs/) p 182
[2] Takagi K 2006 *J. Solid State Chem.* **179** 2809
[3] Takagi K 1994 *Jpn. J. App. Phys. Series* **10** 200
[4] Takagi K, Yamasaki and Komai M 1997 *J. Solid State Chem.* **133** 243
[5] Takagi K and Yamasaki Y 2000 *J. Solid State Chem.* **154** 263

AUTHOR INDEX

AUTHOR INDEX